Fingerprinting Techniques in Food Authentication and Traceability

Food Analysis & Properties

Leo M.L. Nollet
University College Ghent, Belgium

Flow Injection Analysis of Food Additives
Edited by Claudia Ruiz-Capillas and Leo M.L. Nollet

Marine Microorganisms: Extraction and Analysis of Bioactive Compounds
Edited by Leo M.L. Nollet

Multiresidue Methods for the Analysis of Pesticide Residues in Food
Edited by Horacio Heinzen, Leo M.L. Nollet, and Amadeo R. Fernandez-Alba

Spectroscopic Methods in Food Analysis
Edited by Adriana S. Franca and Leo M.L. Nollet

Phenolic Compounds in Food: Characterization and Analysis
Edited by Leo M.L. Nollet and Janet Alejandra Gutierrez-Uribe

Testing and Analysis of GMO-containing Foods and Feed
Edited by Salah E. O. Mahgoub and Leo M.L. Nollet

Fingerprinting Techniques in Food Authenticity and Traceability
Edited by K. S. Siddiqi and Leo M.L. Nollet

For more information, please visit the Series Page: https://www.crcpress.com/Food-Analysis-Properties/book-series/CRCFOODANPRO

Fingerprinting Techniques in Food Authentication and Traceability

Edited by

K. S. Siddiqi and Leo M.L. Nollet

CRC Press
Taylor & Francis Group
Boca Raton London New York

CRC Press is an imprint of the
Taylor & Francis Group, an **informa** business

CRC Press
Taylor & Francis Group
6000 Broken Sound Parkway NW, Suite 300
Boca Raton, FL 33487-2742

Printed on acid-free paper

International Standard Book Number-13: 978-1-138-19767-1 (Hardback)

Visit the Taylor & Francis Web site at
http://www.taylorandfrancis.com

and the CRC Press Web site at
http://www.crcpress.com

Contents

Series Preface

There will always be a need for analyzing methods of food compounds and their properties. Current trends in analyzing methods include automation, increasing the speed of analyses, and miniaturization. The unit of detection has evolved over the years from micrograms to picograms.

A classical pathway of analysis is sampling, sample preparation, cleanup, derivatization, separation, and detection. At every step, researchers are working and developing new methodologies. A large number of papers are published every year on all facets of analysis. So, there is a need for books that gather information on one kind of analysis technique or on analysis methods for a specific group of food components.

The scope of the CRC Series on Food Analysis & Properties aims to present a range of books edited by distinguished scientists and researchers who have significant experience in scientific pursuits and critical analysis. This series is designed to provide state-of-the-art coverage on the following topics:

1. Recent analysis techniques for a range of food components.
2. Developments and evolution in analysis techniques related to food.
3. Recent trends in analysis techniques for specific food components and/or a group of related food components.
4. The understanding of physical, chemical, and functional properties of food.

The book *Fingerprinting Techniques in Food Authentication and Traceability* is the seventh volume in this series.

I am happy to be a series editor of such books for the following reasons:

- I am able to pass on my experience in editing high-quality books related to food.
- I get to know colleagues from all over the world more personally.
- I continue to learn about interesting developments in food analysis.

A lot of work is involved in the preparation of a book. I have been assisted and supported by a number of people, all of whom I would like to thank. I would especially like to thank the team at CRC Press/Taylor & Francis, with a special word of thanks to Steve Zollo, Senior Editor.

Many, many thanks to all the editors and authors of this volume and future volumes. I very much appreciate all their effort, time, and willingness to do a great job.

I dedicate this series to:

- My wife, for her patience with me (and all the time I spent on my computer).
- All patients suffering from prostate cancer; knowing what this means, I am hoping they will have some relief.

Leo M.L. Nollet (Retired)
University College Ghent
Ghent, Belgium

Preface

Food authentication is the process that verifies if a food is in compliance with its label description. This may include, among others, the origin, production method, or processing technologies. The declaration of specific quality attributes in high-value products is of particular interest since these products are often targets of fraudulent labeling. Proof of provenance is an important topic for food safety, food quality, and consumer protection, as well as the compliance with national legislation, international standards, and guidelines. Due to the globalization of food markets and the resulting increase in variability and availability of food products from other countries, consumers are increasingly interested in knowing the geographical origin along with the assumed quality of the products they eat and drink. The quality assurance and the methods used to authenticate foodstuffs are of great interest both from commercial and legal points of view.

Modern instrumentation and advances in basic sciences and in information and communication technologies provide the means for precise measurement and elucidation of origins of food. Since the beginning of the 20th century, organizations that set standards for and control the origins of ingredients and the production processes have appeared all over the world, e.g., the French "Institut National des Appellations d'Origine (INAO)", Italy's "Denominazione di Origine Controllata", Spain's "Denominación de Origen", South Africa's "Wine of Origin", or the United States' "American Viticultural Areas". The production of consumer goods according to these standardized procedures normally results in better products and is rewarded with higher prices at the point of sale. Unfortunately, these financial benefits attract the production of counterfeit food and illegal food trades.

A definition of traceability may be: "The ability to access any or all information relating to that which is under consideration, throughout its entire life cycle, by means of recorded identifications".

Historically, food scares have been with human beings for many years. In the modern livestock production sector, long distance animal transport is increasing. This in turn has not only increased the potential for infections and the spread of diseases related to livestock but also exposed the sector to bioterrorism attacks. These challenges have triggered the importance of animal identification and certification processes. Another risk for food is contamination with radioactive materials.

In addition to the public health risk, food crises lead to economic crises due to direct and indirect (damage to reputation and brand name) costs of product recalls. The indirect cost dominates the recall cost, as the loss of market value and reputation could lead to a total bankruptcy of the brand name. Therefore, traceability is an important component of contemporary supply chains in the production industry in general and in the food sector in particular, as the food sector is sensitive from the human and animal health point of view.

This book details on the current analytical techniques that are being used in respect to food authenticity and traceability. Analytical approaches are mass spectrometry techniques, spectroscopic techniques, separation techniques, and other techniques. The principles of the techniques together with their advantages and drawbacks, and their reported applications are discussed.

In this book, the reader will find four sections.

Section 1 (seven chapters) deals with recent techniques for the analysis of food items concerning authenticity and traceability: *High Resolution Mass Spectrometry, Infrared Spectroscopic Techniques, NMR spectroscopy, Capillary Electrophoresis, Flow injection analysis—Tandem Mass Spectrometry*, and *Ambient Ionization MS*. In Chapter 7, *DNA-based methodologies* are discussed.

In Section 2, introductions to food authentication and food traceability based on fingerprinting techniques are discussed.

Analysis techniques result in a wealth of data. To get meaningful data, multivariate statistical techniques are needed. Such techniques are addressed (Section 3) in four chapters: experimental design, data preprocessing, classification and modeling methods, and validation.

In the last section (seven chapters), numerous applications of fingerprinting techniques are summed up; applications in honey, meat, fruits and vegetables, chocolate, rice, and marine animals.

The editors have the great pleasure to thank all contributors for their excellent contributions. We are aware of all the effort and time spent by them to make this book.

K. S. Siddiqi
Leo M.L. Nollet

A single conversation across the table with a wise man is better than ten years mere study of books.

Henry Wadsworth Longfellow

About the Editors

Professor K. S. Siddiqi graduated as a master of science in 1969 and a doctor of philosophy in 1973 from the Aligarh Muslim University, Aligarh, India. His research interests are bioinorganic chemistry, organometallic chemistry, organoborate chemistry, and nanochemistry. He is the author of more than 100 papers.

He is a life fellow of the Indian Council of Chemists, India; member of the Bulletin of Chemical Society of France; member of the Indian Science Congress Association, India; president of the India nominee to Mizoram University; and has been awarded by the Roche Chemical Company, USA.

Leo M.L. Nollet earned an MS (1973) and PhD (1978) in biology from the Katholieke Universiteit Leuven, Belgium. He is an editor and associate editor of numerous books. He edited for M. Dekker, New York—now CRC Press of Taylor & Francis Publishing Group—the first, second, and third editions of *Food Analysis by HPLC* and *Handbook of Food Analysis*. The last edition is a two-volume book. Dr. Nollet also edited the *Handbook of Water Analysis* (first, second, and third editions) and *Chromatographic Analysis of the Environment*, third and fourth editions (CRC Press). With F. Toldrá, he coedited two books published in 2006, 2007, and 2017: *Advanced Technologies for Meat Processing* (CRC Press) and *Advances in Food Diagnostics* (Blackwell Publishing—now Wiley). With M. Poschl, he coedited the book *Radionuclide Concentrations in Foods and the Environment*, also published in 2006 (CRC Press). Dr. Nollet has also coedited with Y. H. Hui and other colleagues on several books: *Handbook of Food Product Manufacturing* (Wiley, 2007), *Handbook of Food Science, Technology, and Engineering* (CRC Press, 2005), *Food Biochemistry and Food Processing* (first and second editions; Blackwell Publishing—now Wiley—2006 and 2012), and the *Handbook of Fruits and Vegetable Flavors* (Wiley, 2010). In addition, he edited the *Handbook of Meat, Poultry, and Seafood Quality*, first and second editions (Blackwell Publishing—now Wiley—2007 and 2012). From 2008 to 2011, he published five volumes on animal product-related books with F. Toldrá: *Handbook of Muscle Foods Analysis*, *Handbook of Processed Meats and Poultry Analysis*, *Handbook of Seafood and Seafood Products Analysis*, *Handbook of*

Dairy Foods Analysis, and *Handbook of Analysis of Edible Animal By-Products*. Also, in 2011, with F. Toldrá, he coedited two volumes for CRC Press: *Safety Analysis of Foods of Animal Origin* and *Sensory Analysis of Foods of Animal Origin*. In 2012, they published the *Handbook of Analysis of Active Compounds in Functional Foods*. In a coedition with Hamir Rathore, *Handbook of Pesticides: Methods of Pesticides Residues Analysis* was marketed in 2009; *Pesticides: Evaluation of Environmental Pollution* in 2012; *Biopesticides Handbook* in 2015; and *Green Pesticides Handbook: Essential Oils for Pest Control* in 2017. Other finished book projects include *Food Allergens: Analysis, Instrumentation, and Methods* (with A. van Hengel; CRC Press, 2011) and *Analysis of Endocrine Compounds in Food* (Wiley-Blackwell, 2011). Dr. Nollet's recent projects include *Proteomics in Foods* with F. Toldrá (Springer, 2013) and *Transformation Products of Emerging Contaminants in the Environment: Analysis, Processes, Occurrence, Effects, and Risks* with D. Lambropoulou (Wiley, 2014). In the series Food Analysis & Properties, he edited (with C. Ruiz-Capillas) *Flow Injection Analysis of Food Additives* (CRC Press, 2015) and *Marine Microorganisms: Extraction and Analysis of Bioactive Compounds* (CRC Press, 2016). With A.S. Franca, he coedited *Spectroscopic Methods in Food Analysis* (CRC Press, 2017), and with Horacio Heinzen and Amadeo R. Fernandez-Alba he coedited *Multiresidue Methods for the Analysis of Pesticide Residues in Food* (CRC Press, 2017).

List of Contributors

Javed Ahamad
Department of Pharmacy
Faculty of Public Health and Medical Sciences
Mettu University
Mettu, Ethiopia

Javed Ahmad
Department of Pharmaceutics
College of Pharmacy
Najran University
Najran, Kingdom of Saudi Arabia

Mohammad Shabib Akhtar
Department of Clinical Pharmacy
College of Pharmacy
Najran University
Najran, Kingdom of Saudi Arabia

Mohammad Sarwar Alam
Department of Chemistry
School of Chemical and Life Sciences
Jamia Hamdard
New Delhi, India

Saima Amin
Department of Pharmaceutics
School of Pharmaceutical Education and Research
Jamia Hamdard
New Delhi, India

Ammeduzzafar
Department of Pharmaceutics
College of Pharmacy
Al-Jouf University
Aljouf, Kingdom of Saudi Arabia

Silvana M. Azcarate
Universidad Nacional La Pampa (UNLPam)
Instituto de Ciencias de la Tierra y Ambientales de La Pampa (INCITAP) and Consejo Nacional de Investigaciones Científicas y Técnicas (CONICET)
Santa Rosa, Argentina

Showkat Ahmad Bhawani
Department of Chemistry
Faculty of Resource Science and Technology
UNIMAS
Sarawak, Malaysia

Syed Nasir Abbas Bukhari
Department of Pharmaceutical Chemistry
College of Pharmacy
Al-Jouf University
Al-Jawf, Kingdom of Saudi Arabia

José M. Camiña
Universidad Nacional La Pampa (UNLPam)
Instituto de Ciencias de la Tierra y Ambientales de La Pampa (INCITAP) and Consejo Nacional de Investigaciones Científicas y Técnicas (CONICET)
Santa Rosa, Argentina

Khalid Umar Fakhri
Genome Biology Lab
Department of Biosciences
Jamia Millia Islamia
New Delhi, India

Raúl Andrés Gil
Instituto de Química de San Luis (CCT-San Luis)
Consejo Nacional de Investigaciones Científicas y Técnicas (CONCIET)/ Facultad de Química, Bioquímica y Farmacia
Departamento de Química
Área de Química Analítica
Universidad Nacional de San Luis (UNSL)
San Luis, Argentina

Sadaf Jamal Gilani
Department of Pharmaceutical Chemistry
College of Pharmacy
Al-Jouf University
Aljouf, Kingdom of Saudi Arabia

Héctor C. Goicoechea
CONICET-Universidad Nacional del Litoral
Ciudad Universitaria
Santa Fe, Argentina

Ana M. Herrero
Department of Products
Instituto de Ciencia y Tecnología de Alimentos y Nutrición (ICTAN-CSIC)
Ciudad Universitaria
Madrid, Spain

Ahmad Husaini
Department of Molecular Biology
Faculty of Resource Science and Technology
UNIMAS
Sarawak, Malaysia

Syed Sarim Imam
Department of Pharmaceutics
Glocal School of Pharmacy
Glocal University
Saharanpur, Uttar Pradesh, India

Francisco Jiménez-Colmenero
Department of Products
Instituto de Ciencia y Tecnología de Alimentos y Nutrición (ICTAN-CSIC)
Ciudad Universitaria
Madrid, Spain

Nausheen Khan
Department of Pharmaceutics
School of Pharmaceutical Education and Research
Jamia Hamdard
New Delhi, India

Dimitra A. Lambropoulou
Department of Chemistry
Aristotle University of Thessaloniki
Thessaloniki, Greece

Showkat R. Mir
Phyto-pharmaceutical Laboratory
School of Pharaceutical Education and Research
Jamia Hamdard
New Delhi, India

Abdul Moheman
Department of Chemistry
Gandhi Faiz-e-Aam College (Affiliated to M. J. P. Rohilkhand University, Bareilly)
Shahjahanpur, India

Nehal Mohsin
Department of Clinical Pharmacy
College of Pharmacy
Najran University
Kingdom of Saudi Arabia

Basil K. Munjanja
Department of Chemistry
Faculty of Natural and Agricultural Sciences
University of Pretoria
Pretoria, South Africa

Marta Muñoz-Colmenero
University of the Basque Country UPV/EHU
Bilbao, Spain

Leo M.L. Nollet
Retired – University College Ghent,
Gent, Belgium

Anna Ofrydopoulou
Department of Chemistry
Aristotle University of Thessaloniki
Thessaloniki, Greece

Roberto Antonio Olsina
Instituto de Química de San Luis (CCT-San Luis), Consejo Nacional de Investigaciones Científicas y Técnicas (CONCIET) / Facultad de Química, Bioquímica y Farmacia
Departamento de Química
Área de Química Analítica
Universidad Nacional de San Luis (UNSL)
San Luis, Argentina

Semih Otles
Department of Food Engineering
Faculty of Engineering
Ege University
Bornova, Izmir, Turkey

Vasfiye Hazal Ozyurt
Food Engineering Branch
Graduate School of Natural and Applied Sciences
Ege University
Bornova, Izmir, Turkey
Near East University
North Nicosia, Northern Cyprus

Claudia Ruiz-Capillas
Department of Products
Instituto de Ciencia y Tecnología de Alimentos y Nutrición (ICTAN-CSIC)
Ciudad Universitaria
Madrid, Spain

Mohd. Moshahid Alam Rizvi
Genome Biology Lab
Department of Biosciences
Jamia Millia Islamia
New Delhi, India

Md. Rizwanullah
Department of Pharmaceutics
School of Pharmaceutical Education and Research
Jamia Hamdard
New Delhi, India

Marianela Savio
Universidad Nacional La Pampa (UNLPam)
Instituto de Ciencias de la Tierra y Ambientales de La Pampa (INCITAP), and Consejo Nacional de Investigaciones Científicas y Técnicas (CONICET)
Santa Rosa, Argentina

Naiyer Shahzad
Department of Pharmacology and Toxicology
College of Medicine
Umm Al Qura University
Makkah, Kingdom of Saudi Arabia

Apostolos Spyros
NMR Laboratory
Department of Chemistry
University of Crete
Voutes, Heraklion Crete, Greece

Abu Tariq
Chemical Sciences Programme
School of Distance Education
Universiti Sains Malaysia
Penang, Malaysia

SECTION 1

Techniques

CHAPTER 1

High Resolution Mass Spectrometry in Food Analysis

Dimitra A. Lambropoulou and Anna Ofrydopoulou

Aristotle University of Thessaloniki

CONTENTS

1.1 INTRODUCTION

The quality and safety of food products are of a major concern for consumers, governments, and producers throughout the world; thus the new requirements and regulations are stricter than ever before (Hird et al., 2014). The protection of human health requires a systematic monitoring of chemical contaminants found in food products, including toxins, agrochemicals, environmental and industrial contaminants (Knolhoff & Croley, 2016), etc. Trace analysis of these contaminants in food matrices is a challenging task due to some non-trivial difficulties like the high number of analytes to be monitored, the complexity of the matrices, and the multiplicity of interfering compounds (Masiá et al., 2016).

Approaches to measure food contaminants at trace levels have changed significantly over time, moving away from the use of gas chromatography (GC) with selective detectors to the sensitivity and selectivity offered by mass spectrometry (MS). Hence, today, chemical residue analysis is dominated by chromatography coupled to MS. Among the MS techniques, GC-MS continues to be widely used in the determination of volatile, moderate to nonpolar compounds (e.g., persistent organic pollutants, organochlorine pesticides, PCBs, dioxins, etc.). However, based on the polar and ionic characters of many of the food contaminants—including modern, new-generation pesticides as well as the majority of veterinary drugs and toxins, such as mycotoxins—liquid chromatography mass spectrometry (LC-MS) is the technique of choice. Tandem MS (MS/MS) is the most frequently used analytical approach to determine contaminants since it can provide an increased selectivity that helps further distinguish target compounds from potential matrix interferences. Typically, triple quadrupole (QqQ) analyzers are widely used for this scope, run under selected reaction monitoring (SRM), also called multiple reaction monitoring (MRM) mode. Due to their distinct advantages, such as excellent sensitivity and selectivity, these instruments are indispensable tools in official food-testing laboratories—for the routine analysis of a list of known contaminants (e.g., pesticides)—where regulatory control compliance is the main goal of quantification. In spite of the developed methods, the use of the above-described approach has certain limitations, such as the reliance on the availability of reference standards, a limit to the number of contaminants that can be screened in one run, and an inability to screen for unknown contaminants and perform post-acquisition re-interrogation of data other than for those analytes that are preselected into the method (Hernández et al., 2012; Botitsi et al., 2011). Therefore, pretarget screening methods are often insufficient and thus there is an ever-increasing demand for qualitative methods that would be able to screen for a large number of compounds that do not meet the scope of the laboratories. This research field is growing rapidly, and the high resolution mass spectrometry (HRMS) by using time-of-flight (TOF) and Orbitrap mass analyzers and, more recently, by coupling with quadrupole ion analyzer (Q-TOF and Q-Orbitrap) has offered new challenges in food contaminant analysis, providing high sensitivity, nontarget approaches, post-target analysis, and unlimited (theoretically) determination of compounds (De Boevre et al., 2016; Agüera et al., 2017). Current HRMS instruments enable rapid detection and reliable identification of a range of contaminants for even very complex matrices, thanks to the operation carried out in full scan acquisition mode with high mass resolving power (>30,000 FWHM) and high mass accuracy. Its capability to resolve isobaric interferences from the analyte is an excellent tool for obtaining unambiguous qualitative and quantitative results with an adequate sensitivity. Consequently, these instruments are very powerful tools not only for quantitative analysis but also for nontarget screening because of their capability of performing retrospective data evaluation and tentative identification of analytes, such as metabolites and transformation products (TPs), for which, in most cases, reference standards are not available (Wang et al., 2012; Mol et al., 2011; Luz Gómez-Pérez et al., 2015; Martínez-Domínguez et al., 2015, 2016).

Bearing in mind that LC coupled to HRMS is the most flexible and effective technique used for the determination of different groups of contaminants in food matrices by modern food contaminant-testing laboratories, the aim of this chapter is to contribute to this knowledge by summarizing the novel advances in the recently developed MS methods of food contaminant analysis. From a practical point of view, only some recent examples have been chosen to illustrate the advantages and the needs of using HRMS as compared to other techniques.

1.2 HIGH RESOLUTION MASS ANALYZERS

The application of HRMS technologies has substantially enhanced the capabilities for multiresidue analysis of contaminants in food matrices, with the increasing availability of new-generation HRMS instruments such as TOF and Orbitrap mass spectrometers. These instruments can provide not only high-quality data, such as molecular weight and elemental composition, but also the molecular structure of a compound (Ramachandra, 2017) after combining the sensitive full-spectrum mass data (Hernández et al., 2014). The benefit of using HRMS is the full scan acquisition mode (Kaufmann, 2012; Kaufmann et al., 2010; Krauss et al., 2010) with high sensitivity and selectivity, combined with a high resolving power (>100,000 FWHM) and an accurate mass measurement (1–5 ppm) (Righetti et al., 2016).

For now, the current trend is the use of hybrid apparatus with the quadrupole in front (Maurer & Meyer, 2016), such as quadrupole-TOF (Q-TOF) and linear trap quadrupole Orbitrap (LTQ-Orbitrap), offering additional information about compound confirmation and structure elucidation (Hernández et al., 2012).

TOF is a temporally dispersive mass analyzer that separates ions of different masses using the differences in transit time through a drift (a very low pressure tube) by accelerating a group of ions toward the detector. Each ion that exits the source receives an identical high-voltage pulse; those that are having lower masses and are similarly charged have greater velocities, reaching the detector first and getting measured; this happens for each and every one of them (Hird et al., 2014). In TOF analyzers, the time-of-flight of the ion depends on the square root of its m/z and its resolution depends on the capacity of the instrument to eliminate the initial spread of the kinetic energy of the injected ions (La Barbera et al., 2017). The TOF instruments have many advantages, such as full spectral acquisition with better sensitivity than scanning instruments (Hird et al., 2014), an unlimited m/z range, and high-speed acquisition capabilities (Bristow, 2006), thus providing selectivity and sensitivity with a resolving power that reaches 40,000 and an accuracy that is lower than 1 ppm (Ow et al., 2010).

Although there were plenty of examples for stand-alone early TOF instruments, more recently the TOF analyzers are being combined with hybrid instruments (e.g., quadrupole–quadrupole (Qq)-TOF) in order to maximize the mass-resolving power (Xian et al., 2012) and perform MS/MS experiments with mass accuracies for fragment ions (Hernández et al., 2011) higher than 3 ppm (Gosetti et al., 2016). Therefore, the isotopic distribution observed in the spectra makes it feasible to reliably identify the compounds found in the samples (Nácher-Mestre et al., 2013).

The Orbitrap was introduced in the market by Thermo Fischer Scientific Corporation in 2005, and the analyzer consists of a small electrostatic device into which ion packets are injected at high energies to orbit around a central, spindle-shaped electrode. It is regarded as an alternative to Qq-TOF for the identification of unknown compounds and metabolites and for large-scale screening (Denisov et al., 2012; Perry et al., 2008; Perez-Fernandez et al., 2017). Significant technological advancements in the new generation of hybrid instruments, such as Q-Orbitraps, have obviously introduced new possibilities in quantitative analysis, targeted screening, and identification of "unknowns", allowing performance of the specific ion fragmentation, bringing in an additional dimension into the possibilities of unknown compound identification (MS/MS spectra, i.e., spectra of fragment ions by HRMS). The last evolution of Orbitrap technology produced the Q-Exactive, which is equipped with a quadrupole analyzer, and the Orbitrap Fusion, a three-hybrid instrument in which a dual-stage LTQ has been added after the collision cell to offer unmatched flexibility (La Barbera et al., 2017). The Q-Exactive Plus offers resolving power up to 280,000 full width at half maximum (FWHM), with mass accuracy <3 ppm, as well as polarity switching between positive and negative modes as one full cycle in <1 s with acquisition speed up to 12 Hz (Althakafy et al., 2017). This mass analyzer has been considered a cutting-edge technology in terms of high resolution and accuracy (Antignac et al., 2009; Gosetti et al., 2016).

Overall, the Orbitrap/MS has several advantages over QqQ/MS or ion trap mass analyzer (IT/MS) for the screening and identification of different food contaminants: high sensitivity, mass resolution, and mass accuracy for both precursor (MS) and product ions (MS/MS). And, for this reason, its applications have been steadily increased in food analysis in the last decade.

Finally, with regard to classical HRMS instrumentation, such as Fourier transform ion cyclotron resonance (FT-ICR), it should be noticed that although it presents an unsurpassed high accuracy (<1 ppm) and resolving power >750,000 FWHM (La Barbera et al., 2017; Gosetti et al. 2016), it has few applications in food analysis due to the main drawbacks of large size, slow scan speed, complex operation, and high cost.

1.3 TARGET, SUSPECT, AND NONTARGET ANALYSIS

Three different analytical approaches—(i) quantitative target analysis with reference standards, (ii) suspect screening without reference standards, and (iii) nontarget screening of unknowns—have been usually followed for the identification of organic contaminants using HRMS instruments (Oetjen et al., 2017). Target screening methods can be used for the identification of the target analyte by matching the retention time, mass spectra, and MS/MS of the authentic standard with the unknown mass spectra features of the sample (Oetjen et al., 2017). Compared to low resolution MS instruments (LRMS), HR mass spectrometers operating in full scan mode can determine over 500 compounds simultaneously, making the preselection of compound-specific ion currents unnecessary. However, previous studies (Krauss et al., 2010) claimed some limitations for HRMS instruments (e.g., TOF, Q-TOF) in target analysis, such as limited sensitivity and linear dynamic range as compared to the triple quadrupole analyzers, by using the MRM mode. However, in recent years, these deficiencies have been overcome with new Q-TOF and Orbitrap mass analyzers so that equal or better detection limits can be achieved. For example, Q-Orbitrap showed very good peak area, reproducibility, linearity, and satisfying limits of detection for the quantitative determination of pesticide residues in various complex food matrices, such as tomato, pepper, orange, and green tea, by performing full scan MS and simultaneous MS-MS/MS mode. The concentrations detected in real samples by Q-Orbitrap were in agreement with those obtained by the triple quadrupole, thus confirming its good quantitative performance (Rajski et al., 2014; Gomez-Ramos et al., 2015). Similarly, in a study by Zomer and Mol (2015), the Q-Orbitrap combined the full scan and fragmentation approaches by utilizing variable data independent acquisition (vDIA) for the generation of fragment ions, showing an acceptable reproducibility and linearity that covered a range typical for real samples in two food matrices (fruits and vegetables). Quantitative validation of the methodology using a mixture of 184 pesticides demonstrated that this approach was suitable for ca. 93% of the assayed pesticide/matrix/concentration combinations studied in compliance with the guidelines adopted by EU. Other studies also showed good performances of HRMS instruments in quantitative analysis by using TOF mass spectrometers (Hird et al., 2014; Masiá et al., 2016).

Suspect screening or "biased" nontarget approach is an intermediate situation between target and nontarget analysis (Hernández et al., 2014) that requires the construction of a list of suspect chemicals that may be present in a sample, though acquisition of authentic standards for those chemicals is not required. Then, data about the suspected list of compounds are included, such as molecular formula, isotope patterns, exact masses of important adducts, predicted retention times, and exact masses of predicted MS/MS fragments (Oetjen, et al., 2017). In general, suspected screening using a huge database with a large number of compounds or already well-known metabolites/TPs is a commonly employed approach. In case of positive findings, further confirmatory steps based solely on structure-derived information can be performed. Regarding the criteria in screening methods, for the acceptable rates of false positive or negative findings, the United States Department of Agriculture (USDA) Food Safety and Inspection Service (FSIS) requires ≤5% and ≤10%, respectively (Lehotay et al., 2015). Nowadays, a number of studies are being published by using HRMS in suspect screening approach for the determination of pesticides, mycotoxins, fungal metabolites, veterinary drug (Lehner et al. 2011; Wang et al. 2012; Gomez-Perez et al. 2012; Jia et al. 2014; Dzuman et al. 2015), and plant toxins (Mol et al. 2011).

In contrast to suspect screening, the nontarget screening begins without any a priori information on the compounds to be detected (Zedda & Zwiener, 2012). In this case, the undefined mass spectral features measured in HR-MS/MS acquisition represent the presence of some unknown or unexpected chemical species and, thus, novel data analyses can potentially lead to the identification of that chemical (Oetjen et al., 2017). The data is evaluated using statistical methods to examine the most relevant features by comparing the blanks with the different samples. The elemental composition is calculated from the relevant features and the most probable molecular formulae are evaluated by matching the isotope patterns. For identification, the molecular formulae are searched in MS/MS databases or libraries (such as PubChem, ChemSpider, Metlin (a metabolite database), HMDB, etc.) (Gomez et al., 2010). By searching for the accurate mass in databases, the possible candidates for potentially risky compounds can be retrieved. In case, when MS data are not included in chemical databases, *in silico* fragmentation approach has to be used and then the fragments have to be matched against the measured MS fragments (Hill et al., 2008; Pelander et al., 2009; Wolf et al., 2010). This results in a number of proposed compound structures (Zedda & Zwiener, 2012).

The main advantage of nontarget analysis over MS/MS is that there are no limitations in the number of compounds that can be measured in one run (Mol et al., 2016). However, nontarget analysis (Zedda & Zwiener, 2012) demands expertise as it is a rather complicated and time-consuming procedure and entails great complexity when applied to food matrices, beginning without any a priori information on the compounds to be detected (Hug et al., 2014; Ibáñez et al., 2005; Krauss et al., 2010). In addition, data analysis is often another shortcoming that can be improved via the advancement of data processing tools to provide sufficient data quality through chromatographic separations and adequate ion abundance. Consequently, optimization and development are required for the high-throughput applications in food analysis using nontargeted LC/HR-MS (Knolhoff & Croley, 2016).

HRMS target and nontarget analyses have two main fragmentation strategies. The first mode, which is the older and more commonly applied method, is the data-dependent acquisition (DDA or information-dependent acquisition) (Kwok et al., 2017), while the second one, which has been developed recently, is the data independent acquisition (DIA) (Picó et al., 2015; Ruan & Jiang, 2017). In DDA, a survey scan is first performed by the mass spectrometer, from which the precursor ions are selected, isolated, and sequenced by product ion scanning. Under these conditions, the MS/MS spectra are highly selective and interferences are reduced. Data acquisition is carried out in full scan mode, and the instrument switches to MS/MS mode only when any of the parent ions from the target list are detected at the correct retention time. This approach has proved advantageous and is recommended in target analysis (Gomez-Ramos et al., 2015; Agüera et al. 2017).

DIA approach, on the other hand, also known as "all ion mode" MS^{All} (employed in the Orbitrap instruments) or MS^{E} (developed by Waters), has been applied to produce useful MS/MS fragmentations in chemical residue analysis (Pérez-Ortega et al., 2016).

In this case, all the precursor ions are fragmented without preselection by acquiring the mass spectra both at low and high collision energies in an unbiased and parallel manner. Although this operation mode lacks the specificity of DDA, it has the benefit of acquiring all the information all the time without time. Thus, this acquisition mode that offers a high MS/MS acquisition hit rate has proven very attractive for both qualitative and retrospective analysis (Kinyua et al., 2015).

In spite of its distinct advantages, the above acquisition mode has some drawbacks, which are mainly related to the limited dynamic range as well as to the scarce selectivity and quality of DIA MS/MS spectra. For this reason, the use of a new strategy is being increasingly proposed, taking advantage of the capabilities of modern HRMS analyzers. It is known as vDIA or SWATH (sequential windowed acquisition of all theoretical fragment ion mass spectra). In this approach, all precursor ions within a user defined m/z window are deterministically fragmented. Taking into consideration its advantages such as the high sensitivity and selectivity as well as the short overall analysis time, this robust method seems to be a powerful tool for both targeted and nontargeted screening (Zomer & Mol, 2015; Agüera et al. 2017).

1.4 APPLICATION OF LC-HRMS TO THE ANALYSIS OF DIFFERENT CONTAMINANTS IN FOOD SAMPLES

1.4.1 Target Screening

In recent years, the application potential of HRMS has been expanded because of its distinct advantages such as high resolution, accurate mass, and high full scan sensitivity and selectivity, and a wide range of target screening methods using LC-HRMS following generic extraction and cleanup methods with up to hundreds of analytes in the target list, has been developed. Some recent interesting examples are described below (Table 1.1).

Dzuman et al. (2015) described the development, validation, and application of a screening method for the detection and identification of 323 pesticide residues, 55 mycotoxins, and 11 plant toxins represented by pyrrolizidine alkaloids in wheat, leek, and tea matrices by using an optimized QuEChERS (quick, easy, cheap, effective, rugged, and safe) approach coupled to Q-Exactive TM high resolution tandem mass spectrometer. A spectral HRMS/MS library has been established, providing an entire spectrum of fragment ions for each compound, which allows unbiased identification and confirmation of target analytes. Following the SANCO/12571/2013 protocol for validation of screening methods, the limits of quantification (LOQs) were, in general, below 10 mg kg^{-1}, whereas the recoveries were in the acceptable range (70%–120%) for most of the target analytes, except for some highly polar compounds, such as mycotoxin deoxynivalenol-3-glucoside (<45% in all matrices). Apart from spiked samples, the trueness of the method was verified using several certified reference materials (CRMs) and proficiency test samples.

A simple and rapid multiresidue analytical method was developed by Zhang et al. (2016) for the screening and quantification of 90 veterinary drugs in total, belonging to more than 14 families in royal jelly, by using QuEChERS extraction method coupled to ultra performance liquid chromatography (UPLC)-Q-TOF-MS. A homemade database was utilized for the confirmation and identification of the target compounds including the retention time, accurate masses, elemental composition, isotopic pattern data of the target ions, and the characteristic in-source fragment ions. In the screening method, the "all ions" MS/MS mode in a single run was used to acquire data about the precursor and

TABLE 1.1 Recent Applications in LC-HRMS Target Analysis of Food Contaminants

Contaminants	Food Matrix	Extraction Technique	Stationary Phase	Mobile Phase	MS Analyzer (Acquisition Mode)	Recoveries	LOQ (μg kg^{-1})	References
Pesticides (323), mycotoxins (55), plant toxins (11)	Wheat, leek, and tea	QuEChERS	Mycotoxins: Kinetex C_{18} (100 mm×2.1 mm, 2.6 μm) Pesticides: Kinetex C_{18} (50 mm×3 mm, 1.7 μm)	Eluent A: 0.1% (v/v) formic acid in water Eluent B: methanol Flow: 0.3 mL min^{-1}	Q-Exactive Orbitrap-HRMS (full scan)	70–140%	0.7–26.4	Dzuman et al. 2015
Veterinary drugs (90)	Royal jelly	QuEChERS	ACQUITY® UPLC BEH C_{18} (100 mm×2.1 mm, 1.7 μm)	Eluent A: 0.1% (v/v) formic acid in water Eluent B: acetonitrile Flow: -	Q-TOF HRMS (full scan; all ions MS/MS)	70–120%	0.21–20	Zhang et al., 2016
Organic contaminants (182) (pesticides, veterinary drug, and personal care products)	Fish fillet	QuEChERS	Zorbax Eclipse Plus C_{18} (50×2.1 mm; 1.8 μm)	Eluent A: water:methanol 98:2 (v/v), 0.1% (v/v) of formic acid and 5 mmoL L^{-1} of ammonium formate. Eluent B: methanol, 0.1% (v/v) of formic acid and 5 mmoL L^{-1} of ammonium formate. Flow: 0.3 mL min^{-1}	Q-TOF HRMS (full scan; all ions MS/MS)	70–120 (65% of the compounds)	<7.5	Munaretto et al., 2016

(*Continued*)

TABLE 1.1 (*Continued*) Recent Applications in LC-HRMS Target Analysis of Food Contaminants

Contaminants	Food Matrix	Extraction Technique	Stationary Phase	Mobile Phase	MS Analyzer (Acquisition Mode)	Recoveries	LOQ (μg kg^{-1})	References
Pesticides (426), veterinary drugs and pharmaceuticals (117), food-packaging contaminants (42), perfluoroalkyl substances (10), mycotoxins (21), nitrosamines (9), and sweeteners	Different baby food	QuEChERS	Zorbax Rapid RRHD Eclipse-Plus C_{18}) (50 mm×2.1 mm; 1.8 μm)	Eluent A: 0.1% (v/v) formic acid in water Eluent B: 0.1% (v/v) formic acid in acetonitrile Flow: 0.5 mL min^{-1}	Q-TOF HRMS (full scan; full scan: CID)	—	/<10	Pérez-Ortega et al., 2016
Pesticides (134), Mycotoxins (11)	Paprika blend powders	QuEChERS	Mycotoxins: Kinetex C_{18} (100 mm×2.1 mm; 2.6 mm) Pesticides: Kinetex C_{18} (50 mm×3 mm, 1.7 μm)	Mycotoxins Eluent A: 0.1% (v/v) formic acid in water Eluent B: methanol Flow: 0.3 mL min^{-1} Pesticides Eluent A: 5 mM ammonium formate in 0.1% formic acidEluent B: acetonitrile Flow: 0.4 mL min^{-1}	Q-Exactive Orbitrap-HRMS (full scan)	75–120	3–9	Reinholds et al., 2016

(*Continued*)

TABLE 1.1 (*Continued*) Recent Applications in LC-HRMS Target Analysis of Food Contaminants

Contaminants	Food Matrix	Extraction Technique	Stationary Phase	Mobile Phase	MS Analyzer (Acquisition Mode)	Recoveries	LOQ (μg kg^{-1})	References
Sulfonamides (12) and acetylated metabolites (5)	47 different baby foods	QuEChERS, ASE	Hypersil Gold aQ (100 mm×2.1 mm; 1.9 μm)	Eluent A: 0.5 mM oxalic acid and 1 mM ammonium formate in water Eluent B: 0.2 mM oxalic acid in methanol Flow: 0.5 mL min^{-1}	Exactive Orbitrap-HRMS (full scan)	60.9–85.9 75.5–96.6	0.10–0.55	Konak et al., 2017
Pesticides (199)	Spices (black pepper, cardamom, chili, coriander, cumin, and turmeric)	QuEChERS, Clean up: SPE (HLB)	Ascentis Express C_{18} (100×2.1 mm; 2.7 μm)	Eluent A: methanol:water (10:90), 5 mMol/L ammonium formate and 0.2% formic acid Eluent B: methanol:water (90:10), 5 mMol/L ammonium formate and 0.2% formic acid Flow: 0.3 mL min^{-1}	Q-Exactive Orbitrap-HRMS (full scan; vDIA)	70–120% for 170 compounds	<20	Goon et al., 2018

CID, Collision induced dissociation; RRHD, Rapid resolution high definition.

product ions of the target analytes simultaneously. The method showed very good peak shape, linearity, and reproducibility.

Munaretto et al. (2016) evaluated the analytical capabilities of a screening method for the identification of organic contaminants in a fish fillet by using two different scan modes (two different scan methods-full scan and "all ions" MS/MS) for data acquisition by LC-Q-TOF/MS. In general, full scan acquisition was found to be more reliable (84%) for target screening when compared to "all ions" MS/MS, giving better peak shape and high identification scores. However, the latter mode showed better information providing capabilities for the structural identification of compounds because of the fragmentation ions obtained on using different collision energies. Both scan acquisition modes showed good sensitivity with a screening detection limit (SDL) below 10 μg kg^{-1}. The validation parameters evaluated in a fish fillet showed good results for a wide scope. LOQ values were <7.5 μg kg^{-1}, while recoveries were in accordance with the European guides (70%–120%), such as SANTE (2015), for most of the studied compounds (65%).

Pérez-Ortega et al. (2016) showed the potential of a ultra-high performance liquid chromatography (UHPLC)-TOF mass spectrometer for the accurate mass screening of 625 multiclass food contaminants (pesticides (426), veterinary drugs and pharmaceuticals (117), food-packaging contaminants (42), perfluoroalkyl substances (10), mycotoxins (21), nitrosamines (9), and sweeteners) in baby food samples (vegetables, meat). Different acquisition modes (full scan or full scan combined with collision induced dissociation (CID) with no precursor ion isolation) were investigated along with their limitations, such as sensitivity or automated data processing. The authors conclude that for this type of large-scale screening method, the "all ion" mode CID MS/MS is probably the best suited acquisition mode, offering reliable detection and identification of the target compounds within a run. However, the low sensitivity obtained, especially for highly polar compounds, and the matrix effects can be considered as the major limitations of this approach.

Reinholds et al. (2016) investigated the efficiency and performance characteristics (i.e., linearity, recoveries, repeatability) of a screening method based on the Orbitrap-HRMS technique in order to evaluate the effectiveness in the screening of mycotoxins (11) and pesticide (134) compounds in samples of paprika blend powders. The efficiency and detection sensitivity of the used UHPLC-Orbitrap-HRMS technique were compared to the results obtained using a QqQ-MS/MS mass spectrometric detector. For both methods, analytical characteristics were in accordance with the EU legislations. The Orbitrap-HRMS technique proved to be more sensitive as compared to the QqQ-MS/MS method for the quantitative determination of several pesticide contaminants in paprika. However, in the case of mycotoxin, comparable results were obtained. Furthermore, the Orbitrap-HRMS detection demonstrates better accuracy by giving better recoveries for Fumonisin B1 and, especially, for Fusarenon X. The application of the proposed method in real samples by performing the full scan mode confirmed the occurrence of mycotoxins (FB1, OTA, STERIG), insecticides, and fungicides, thus indicating that it is a very effective and attractive technique for the evaluation and control of mycotoxins and pesticide residues, in line with the relevant EU guidelines.

Recently, a highly sensitive and reliable multiresidue method was developed and validated for the determination of sulfonamides (12) and their acetylated metabolites (5) in baby food by using UHPLC-Orbitrap-MS (Konak et al., 2017). The compounds were extracted by using the ASE and the QuEChERS methods. Both methods showed good performance characteristics; however, the optimized ASE method showed higher recovery rates for all the analytes and better repeatability and reproducibility values. The high resolving power and the accurate mass measurements of Orbitrap mass analyzer provided high selectivity and sensitivity, with LOQs 0.10–0.55 μg kg^{-1}.

1.4.2 Nontarget Screening

Nontarget screening methods intend to pick up unexpected contaminants, such as metabolites/TPs, present in food matrices. Nevertheless, its application in food control where the detected compounds are already well established is not so frequent. Nontargeted strategies developed in other research areas (i.e., metabolomics etc.) can be used as a model for food analysis. However, food analysis has different challenges than metabolomics or environmental analysis because of the inherent complexity of food matrices (Diaz et al., 2014; Jandric et al., 2014; Knolhoff & Croley, 2016; Castro-Puyana et al., 2017) and, therefore, further advancements are required in this research field.

Up to now, most of the nontarget approaches have been focused on the identification and authentication of food products and on additive screening, whereas the search for chemical contamination is rather limited (Fu et al., 2017). For example, a nontargeted method has been recently reported by Jandric et al. (2014) to assess the authenticity of fruit juice by using multivariate data analysis methods based on UPLC-Q-TOF-MS. According to this report, metabolic fingerprint analysis has proved to be an effective approach in differentiating between authentic fruit juices and their adulterated mixtures, down to 1% adulteration level. The authors suggested that the proposed UPLC-MS metabolomics approach has potential as a screening tool for the detection of food fraud by adulteration and could thus represent a new strategy in food forensics. Similarly, a metabolomic fingerprinting approach based on UHPLC-Q-TOF-MS has been used for biomarker identification for the correct authentication of Valencia (Spain) oranges by Diaz et al. (2016). The model helped in finding useful marker/s to differentiate between Valencia and foreign classes. Citrusin D was found as the optimal marker in two different years' samples, suggesting that this compound can be used to rapidly differentiate the Valencia oranges from the southern ones by employing well-known techniques for target analysis, like LC-MS/MS. In addition, other potential markers were also found as the representatives of sample differentiation.

Based on LC-IT/TOF, a new analytical strategy was developed by Guo et al. (2015) for the large-scale screening and qualitative identification of illegal adulterants in dietary supplements. The authors demonstrated that the unique combination of dual-polarity detection, retention time, isotopic profile, and accurate MS^n spectra enables a more comprehensive and precise confirmation by matching the automated library against the user-created database and, therefore, it can be used as an efficient strategy for large-scale nontargeted screening of complex food matrices.

Regarding the detection of chemical contaminants in food matrices, an interesting example is the study by Cotton et al., (2014) who showed that untargeted metabolomic-like analyses enable the detection of 12 unknown chlorinated xenobiotics in honey samples, one of which is identified as 2,6-dichlorobenzamide, a metabolite of dichlorobenzyl, a pesticide banned in France since 2010. Finally, by using multivariate statistical analyses in honey samples, six discriminating metabolites were characterized, thanks to the MS/MS experiments highlighting the efficiency of the proposed approach.

Rubert et al., (2017) evaluated the potential of UHPLC-Q-TOF as a predictable tool for an early detection of mycotoxins in wheat samples. For this purpose, a novel untargeted metabolomics strategy was jointly assessed with three standardized approaches

(Fusarium disease severity, PCR assays for *Fusarium* spp. identification, and mycotoxin quantification). Advanced chemometric tools including principal component analysis (PCA) were used for wheat sample clustering and metabolic pathway elucidation. The workflow developed by the authors proved to be a useful analytical approach for distinguishing low and strong infection levels, and it could serve as an effective tool for the early detection of mycotoxins and for Fusarium disease prevention.

Finally, in a more recent study, a novel screening and quantitation method is reported for nontarget multiresidue analysis of pesticides using ultra-HPLC-quadrupole-Orbitrap-MS in different spice matrices (black pepper, cardamom, chili, coriander, cumin, and turmeric) (Goon et al., 2018). Two MS modes were performed including the sequential full scan (resolution=70,000) and vDIA with nine consecutive fragmentation events (resolution=17,500). The optimized sample preparation protocol including the QuEChERS method followed by an SPE-based cleanup step through hydrophilic-lipophilic-balance (HLB) cartridges improved the efficiency of the HRMS analysis by minimizing the false negatives. The validation of the method for 199 pesticides showed good sensitivity (LOQ < 10 ng/g for most of the target pesticides) and acceptable analytical characteristics with recoveries within 70%–120% and precision- relative standard deviations (RSDs) < 20%. The results of Orbitrap-MS method were compared with the results from a triple quadrupole, confirming an equivalent quantitative performance by both the techniques.

1.5 CONCLUSIONS

Many advances in terms of instrumentation characterize the LC-HRMS analyses, including separation, ionization techniques, and mass analyzers. All these significant advances coupled to novel software developments have increased the scope of HRMS in food analysis and converted it as a true alternative to conventional triple quadrupole MS instruments. Modern TOF and Orbitrap-based instruments have been much improved, providing a higher resolution, an accurate mass, and a higher full scan sensitivity and selectivity than their older counterparts. In addition, quantification capabilities have also been improved with enhanced linear ranges and reproducibility, making HRMS very attractive and effective for both target and nontarget screening of different groups of food contaminants. Despite its well-known benefits, considering the high number of compounds that can be simultaneously analyzed, additional research is needed to simplify the quantification step. In addition, further technological development is required to advance software tools to minimize the data treatment and increase the data processing speed in quantitation processes, providing fast and efficient methodologies.

With regard to food safety and monitoring programs, a key benefit of LC-HRMS is the possibility of retrospective analysis of full scan data, which allows laboratories to assess virtually all the compounds present in a sample and therefore to search for "new" or "unexpected" contaminants without any need to rerun samples. Thus, despite the higher investment costs as compared with LC-QqQ instruments, the expansion of LC-HRMS instruments from research into routine analysis is anticipated in the near future.

ACKNOWLEDGMENTS

Ms. A. Ofrydopoulou would like to thank the General Secretariat for Research and Technology (GSRT) and the Hellenic Foundation for Research and Innovation (HFRI) for providing her scholarship through the action "1st Proclamation of Scholarships from ELIDEK for PhD Candidates" - Scholarship Code: 429.

REFERENCES

Agüera, A., Martínez-Piernas, A. B., & Campos-Mañas, M. C. (2017). Analytical strategies used in HRMS. In Applications in High Resolution Mass Spectrometry: Food Safety and Pesticide Residue Analysis (pp. 59–82). doi:10.1016/B978-0-12-809464-8.00003-8.

Althakafy, J. T., Kulsing, C., Grace, M. R., & Marriott, P. J. (2017). Liquid chromatography - quadrupole Orbitrap mass spectrometry method for selected pharmaceuticals in water samples. Journal of Chromatogry A, 1515, 164–171.

Antignac, J. P., Courant, F., & Le Bizec, B. (2009). 7-Advances in chromatography coupled to mass spectrometry-related techniques for analysis of endocrine disruptors in food A2- Shaw, Ian. In Endocrine-Disrupting Chemicals in Food (pp. 149–182). Woodhead Publishing, Cambridge UK.

Botitsi, H. V., Garbis, S. D., Economou, A., & Tsipi, D. F. (2011). Current mass spectrometry strategies for the analysis of pesticides and their metabolites in food and water matrices. Mass Spectrometry Reviews, 30(5), 907–939.

Bristow, A. W. (2006). Accurate mass measurement for the determination of elemental formula--a tutorial. Mass Spectrometry Reviews, 25(1), 99–111.

Castro-Puyana, M., Pérez-Míguez, R., Montero, L., & Herrero, M. (2017). Application of mass spectrometry-based metabolomics approaches for food safety, quality and traceability. TrAC - Trends in Analytical Chemistry, 93, 102–118.

Cotton J., Leroux, F., Broudin, S., Marie, M., Corman, B., Tabet, J. C., Ducruix, C., & Junot, C. (2014). High-resolution mass spectrometry associated with data mining tools for the detection of pollutants and chemical characterization of honey samples, Journal of Agriculture and Food Chemistry, 62, 11335–11345.

De Boevre, M., Ediage, E. N., Van Pouckea, C., & De Saeger, S. (2016). Chapter 4: Untargeted analysis of modified mycotoxins using high-resolution mass spectrometry. doi:10.1039/9781782622574-00050.

Denisov, E., Damoc, E., Lange, O., & Makarov, A. (2012). Orbitrap mass spectrometry with resolving powers above 1,000,000. International Journal of Mass Spectrometry, 325(Supplement C), 80–85.

Diaz R., Pozo, O. J., Sancho, J. V., & Hernandez, F. (2014). Metabolomic approaches for orange origin discrimination by ultra-high performance liquid chromatography coupled to quadrupole time-of-flight mass spectrometry. Food Chemistry, 157, 84–93.

Díaz, R., Gallart-Ayala, H., Sancho, J. V., Nuñez, O., Zamora, T., Martins, C. P.B., Hernández, F., Hernández-Cassou, S., Saurina, J., & Checa, A. (2016). Told through the wine: A liquid chromatography–mass spectrometry interplatform comparison reveals the influence of the global approach on the final annotated metabolites in non-targeted metabolomics. Journal of Chromatography A, 1433, 90–97.

Dzuman, Z., Zachariasova, M., Veprikova, Z., Godula, M., & Hajslova, J. (2015). Multi-analyte high performance liquid chromatography coupled to high resolution tandem mass spectrometry method for control of pesticide residues, mycotoxins, and pyrrolizidine alkaloids. Analytica Chimica Acta, 863(1), 29–40.

Fu, Y., Zhao, C., Lu, X., & Xu, G. (2017). Nontargeted screening of chemical contaminants and illegal additives in food based on liquid chromatography–high resolution mass spectrometry. TrAC - Trends in Analytical Chemistry, 96, 89–98.

Gomez, M. J., Gomez-Ramos, M. M., Malato, O., Mezcua, M., & Fernandez-Alba, A. R. (2010). Rapid automated screening, identifFication and quantification of organic micro-contaminants and their main transformation products in wastewater and river waters using liquid chromatography-quadrupole-time-of-flight mass spectrometry with an accurate-mass database. Journal of Chromatogry A, 1217(45), 7038–7054.

Gomez-Perez, M. L., Plaza-Bolanos, P., Romero-Gonzalez, R., Martinez-Vidal, J. L., & Garrido-Frenich, A. (2012). Comprehensive qualitative and quantitative determination of pesticides and veterinary drugs in honey using liquid chromatography-Orbitrap high resolution mass spectrometry. Journal of Chromatography A, 1248, 130–138.

Gomez-Ramos, M.M., et al. (2015). Liquid chromatography Orbitrap mass spectrometry with simultaneous full scan and tandem MS/MS for highly selective pesticide residue analysis. Analytical and Bioanalytical Chemistry, 407(21), 6317–6326.

Goon, A., Khan, Z., Oulkar, D., Shinde, R., Gaikwad, S., & Banerjee, K. (2018). A simultaneous screening and quantitative method for the multiresidue analysis of pesticides in spices using ultra-high performance liquid chromatography-high resolution (orbitrap) mass spectrometry. Journal of Chromatography A, 1532, 105–111.

Gosetti, F., Mazzucco, E., Gennaro, M. C., & Marengo, E. (2016). Contaminants in water: Non-target UHPLC/MS analysis. Environmental Chemistry Letters, 14(1), 51–65.

Guo B., Wang, M., Liu, Y., Zhou, J., Dai, H., Huang, Z., Shen, L., Zhang, Q., Chen, B. (2015). Wide-scope screening of illegal adulterants in dietary and herbal supplements via rapid polarity-switching and multistage accurate mass confirmation using an LC-it/TOF hybrid instrument. Journal of Agriculture and Food Chemistry, 63, 6954–6967.

Hernández, F., Ibáñez, M., Bade, R., Bijlsma, L., & Sancho, J. V. (2014). Investigation of pharmaceuticals and illicit drugs in waters by liquid chromatography-high-resolution mass spectrometry. TrAC Trends in Analytical Chemistry, 63(Supplement C), 140–157.

Hernández, F., Portolés, T., Pitarch, E., & López, F. J. (2011). Gas chromatography coupled to high-resolution time-of-flight mass spectrometry to analyze trace-level organic compounds in the environment, food safety and toxicology. TrAC Trends in Analytical Chemistry, 30(2), 388–400.

Hernández, F., Sancho, J. V., Ibáñez, M., Abad, E., Portolés, T., & Mattioli, L. (2012). Current use of high-resolution mass spectrometry in the environmental sciences. Analytical and Bioanalytical Chemistry, 403(5), 1251–1264.

Hill, D. W., Kertesz, T. M., Fontaine, D., Friedman, R., & Grant, D. F. (2008). Mass spectral metabonomics beyond elemental formula: Chemical database querying by matching experimental with computational fragmentation spectra. Analytial Chemistry, 80(14), 5574–5582.

Hird, S. J., Lau, B. P. Y., Schuhmacher, R., & Krska, R. (2014). Liquid chromatography-mass spectrometry for the determination of chemical contaminants in food. TrAC Trends in Analytical Chemistry, 59(Supplement C), 59–72.

Hug, C., Ulrich, N., Schulze, T., Brack, W., & Krauss, M. (2014). Identification of novel micropollutants in wastewater by a combination of suspect and nontarget screening. Environmental Pollution, 184(Supplement C), 25–32.

Ibáñez, M., Sancho, J. V., Pozo, Ó. J., Niessen, W., & Hernández, F. (2005). Use of quadrupole time-of-flight mass spectrometry in the elucidation of unknown compounds present in environmental water. Rapid Communications in Mass Spectrometry, 19(2), 169–178.

Jandric Z., Roberts, D., Rathor, M. N., Abrahim, A., Islam, M., & Cannavan, A. (2014). Assessment of fruit juice authenticity using UPLC-QToF MS: A metabolomics approach. Food Chemistry, 148, 7–17.

Jia, W., Chu, X. G., Ling, Y., Huang, J. R., & Chang, J. (2014). High-throughput screening of pesticide and veterinary drug residues in baby food by liquid chromatography coupled to quadrupole Orbitrap mass spectrometry. Journal of Chromatography A, 1347, 122–128

Kaufmann, A. (2012). The current role of high-resolution mass spectrometry in food analysis. Analytical and Bioanalytical Chemistry, 403(5), 1233–1249.

Kaufmann, A., Butcher, P., Maden, K., Walker, S., & Widmer, M. (2010). Comprehensive comparison of liquid chromatography selectivity as provided by two types of liquid chromatography detectors (high resolution mass spectrometry and tandem mass spectrometry): "where is the crossover point?". Analytica Chimica Acta, 673(1), 60–72.

Kinyua, J., Covaci, A., Maho, W., McCall, A.-K., Neels, H., & van Nuijs, A. L. N. (2015). Sewage-based epidemiology in monitoring the use of new psychoactive substances: Validation and application of an analytical method using LC-MS/MS Drug. Testing and Analysis, 7(9), 812–818.

Knolhoff, A. M., & Croley, T. R. (2016). Non-targeted screening approaches for contaminants and adulterants in food using liquid chromatography hyphenated to high resolution mass spectrometry. Journal of Chromatogry A, 1428, 86–96.

Konak, Ü. İ., Certel, M., Şık, B., & Tongur, T. (2017). Development of an analysis method for determination of sulfonamides and their five acetylated metabolites in baby foods by ultra-high performance liquid chromatography coupled to high-resolution mass spectrometry (Orbitrap-MS). Journal of Chromatography B, 1057, 81–91.

Krauss, M., Singer, H., & Hollender, J. (2010). LC-high resolution MS in environmental analysis: from target screening to the identification of unknowns. Analytical and Bioanalytical Chemistry, 397(3), 943–951.

Kwok, W. H., Choi, T. L. S., Tsoi, Y. Y. K., Leung, G. N. W., & Wan, T. S. M. (2017). Screening of over 100 drugs in horse urine using automated on-line solid-phase extraction coupled to liquid chromatography-high resolution mass spectrometry for doping control. Journal of Chromatogry A, 1490, 89–101.

La Barbera, G., Capriotti, A. L., Cavaliere, C., Montone, C. M., Piovesana, S., Samperi, R., Zenezini Chiozzi, R., & Lagana, A. (2017). Liquid chromatography-high resolution mass spectrometry for the analysis of phytochemicals in vegetal-derived food and beverages. Food Research International, 100(Pt 1), 28–52.

Lehner, S. M., Neumann, N. K. N., Sulyok, M., Lemmens, M., Krska, R., & Schuhmacher, R. (2011). Evaluation of LC-high-resolution FT-Orbitrap MS for the quantification of selected mycotoxins and the simultaneous screening of fungal metabolites in food. Food Additives and Contaminants Part a-Chemistry Analysis Control Exposure & Risk Assessment, 28(10), 1457–1468.

Lehotay, S. J., Sapozhnikova, Y., & Mol, H. G. J. (2015). Current issues involving screening and identification of chemical contaminants in foods by mass spectrometry. Trac-Trends in Analytical Chemistry, 69, 62–75.

Luz Gómez-Pérez, M., Romero-González, R., José Luis Martínez, V., & Garrido Frenich, A. (2015). Analysis of pesticide and veterinary drug residues in baby food by liquid chromatography coupled to orbitrap high resolution mass spectrometry. Talanta, 131, 1–7.

Martínez-Domínguez, G., Romero-González, R., & Garrido Frenich, A. (2016). Multiclass methodology to determine pesticides and mycotoxins in green tea and royal jelly supplements by liquid chromatography coupled to orbitrap high resolution mass spectrometry. Food Chemistry, 197, 907–915.

Martínez-Domínguez, G., Romero-González, R., Arrebola, F. J., & Garrido Frenich, A. (2015). Multi-class determination of pesticides and mycotoxins in isoflavones supplements obtained from soy by liquid chromatography coupled to orbitrap high resolution mass spectrometry. Food Control, 59, 218–224.

Masiá, A., Suarez-Varela, M. M., Llopis-Gonzalez, A., & Picó, Y. (2016). Determination of pesticides and veterinary drug residues in food by liquid chromatography-mass spectrometry: A review. Analytica Chimica Acta, 936, 40–61.

Maurer, H. H., & Meyer, M. R. (2016). High-resolution mass spectrometry in toxicology: Current status and future perspectives. Archives of Toxicology, 90(9), 2161–2172.

Mol, H. G. J., Tienstra, M., & Zomer, P. (2016). Evaluation of gas chromatography – electron ionization – full scan high resolution orbitrap mass spectrometry for pesticide residue analysis. Analytica Chimica Acta, 935, 161–172.

Mol, H. G. J., van Dam, R. C. J., Zomer, P., & Mulder, P. P. J. (2011). Screening of plant toxins in food, feed and botanicals using full-scan high-resolution (orbitrap) mass spectrometry. Food Additives and Contaminants - Part A Chemistry, Analysis, Control, Exposure and Risk Assessment, 28(10), 1405–1423.

Munaretto, J. S., May, M. M., Saibt, N., & Zanella, R. (2016). Liquid chromatography with high resolution mass spectrometry for identification of organic contaminants in fish fillet: Screening and quantification assessment using two scan modes for data acquisition. Journal of Chromatography A, 1456, 205–216.

Nácher-Mestre, J., Ibáñez, M., Serrano, R., Pérez-Sánchez, J., & Hernández, F. (2013). Qualitative screening of undesirable compounds from feeds to fish by liquid chromatography coupled to mass spectrometry. Journal of Agricultural and Food Chemistry, 61(9), 2077–2087.

Oetjen, K., Giddings, C. G. S., McLaughlin, M., Nell, M., Blotevogel, J., Helbling, D. E., Mueller, D., & Higgins, C. P. (2017). Emerging analytical methods for the characterization and quantification of organic contaminants in flowback and produced water. Trends in Environmental Analytical Chemistry, 15(Supplement C), 12–23.

Ow, S. Y., Noirel, J., Salim, M., Evans, C., Watson, R., & Wright, P. C. (2010). Balancing robust quantification and identification for iTRAQ: Application of UHR-ToF MS. Proteomics, 10(11), 2205–2213.

Pelander, A., Tyrkko, E., & Ojanpera, I. (2009). In silico methods for predicting metabolism and mass fragmentation applied to quetiapine in liquid chromatography/time-of-flight mass spectrometry urine drug screening. Rapid Communication Mass Spectrometry, 23(4), 506–514.

Perez-Fernandez, V., Mainero Rocca, L., Tomai, P., Fanali, S., & Gentili, A. (2017). Recent advancements and future trends in environmental analysis: Sample preparation, liquid chromatography and mass spectrometry. Analytica Chimica Acta, 983, 9–41.

Pérez-Ortega, P., Lara-Ortega, F. J., García-Reyes, J. F., Gilbert-López, B., Trojanowicz, M., & Molina-Díaz, A. (2016). A feasibility study of UHPLC-HRMS accurate-mass screening methods for multiclass testing of organic contaminants in food. *Talanta*, 160, 704–712.

Perry, R. H., Cooks, R. G., & Noll, R. J. (2008). Orbitrap mass spectrometry: Instrumentation, ion motion and applications. Mass Spectrometry Reviews, 27(6), 661–699.

Picó, Y., Farré, M., & Barceló, D. (2015). Quantitative profiling of perfluoroalkyl substances by ultrahigh-performance liquid chromatography and hybrid quadrupole time-of-flight mass spectrometry. Analytical and Bioanalytical Chemistry, 407(15), 4247–4259.

Rajski, Ł., Gomez-Ramos, M. M., Fernandez-Alba, A. R. (2014). Large pesticide multi-residue screening method by liquid chromatography-Orbitrap mass spectrometry in full scan mode applied to fruit and vegetables, Journal of Chromatography A, 1360, 119e127.

Ramachandra, B. (2017). Development of impurity profiling methods using modern analytical techniques. Critical Reviews in Analytical Chemistry, 47(1), 24–36.

Reinholds, I., Pugajeva, I., & Bartkevics, V. (2016). A reliable screening of mycotoxins and pesticide residues in paprika using ultra-high performance liquid chromatography coupled to high resolution orbitrap mass spectrometry. Food Control, 60, 683–689.

Righetti, L., Paglia, G., Galaverna, G., & Dall'Asta, C. (2016). Recent advances and future challenges in modified mycotoxin analysis: Why HRMS has become a key instrument in food contaminant research. Toxins (Basel), 8(12), 361.

Ruan, T., & Jiang, G. (2017). Analytical methodology for identification of novel per- and polyfluoroalkyl substances in the environment. TrAC Trends in Analytical Chemistry, 95(Supplement C), 122–131.

Rubert, J., Righetti, L., Stranska-Zachariasova, M., Dzuman, Z., Chrpova, J., Dall'Asta, C., & Hajslova, J. (2017). Untargeted metabolomics based on ultra-high-performance liquid chromatography–high-resolution mass spectrometry merged with chemometrics: A new predictable tool for an early detection of mycotoxins. Food Chemistry, 224, 423–431.

Wang, J., Chow, W., Leung, D., & Chang, J. (2012). Application of ultrahigh-performance liquid chromatography and electrospray ionization quadrupole orbitrap high-resolution mass spectrometry for determination of 166 pesticides in fruits and vegetables. Journal of Agricultural and Food Chemistry, 60(49), 12088–12104.

Wolf, S., Schmidt, S., Muller-Hannemann, M., & Neumann, S. (2010). In silico fragmentation for computer assisted identification of metabolite mass spectra. BMC Bioinformatics, 11, 148.

Xian, F., Hendrickson, C. L., & Marshall, A. G. (2012). High resolution mass spectrometry. Analytical Chemistry, 84(2), 708–719.

Zedda, M., & Zwiener, C. (2012). Is nontarget screening of emerging contaminants by LC-HRMS successful? A plea for compound libraries and computer tools. Analytical and Bioanalytical Chemistry, 403(9), 2493–2502.

Zhang, Y., Liu, X., Li, X., Zhang, J., Cao, Y., Su, M., & Sun, H. (2016). Rapid screening and quantification of multi-class multi-residue veterinary drugs in royal jelly by ultra performance liquid chromatography coupled to quadrupole time-of-flight mass spectrometry. Food Control, 60, 667–676.

Zomer, P., & Mol, H. G. J. (2015). Simultaneous quantitative determination, identification and qualitative screening of pesticides in fruits and vegetables using LC-Q-orbitrap™-MS. Food Additives and Contaminants - Part A Chemistry, Analysis, Control, Exposure and Risk Assessment, 32(10), 1628–1636.

CHAPTER 2

Infrared Spectroscopic Techniques

Basil K. Munjanja
University of Pretoria

CONTENTS

2.1 INTRODUCTION

Recently, there has been a growing interest among consumers worldwide to buy high-quality food mainly for health reasons. And, the food producers are responding by assuring them that the food meets their expected criteria (Cozzolino, 2012). A good example is the issue of traceability among different food brands, which has become important because some unscrupulous dealers may imitate these established brands to make profits. Figure 2.1 shows the percentage of food fraud cases recorded from 1980–2010. Hence, it is imperative that the authenticity of these food products be determined.

For a very long time, the authenticity of food was largely determined using chromatographic techniques and isotopic analyses (Luykx and van Ruth, 2008). Lately, the spectroscopic techniques have emerged as the methods of choice because they are fast, nondestructive, and they consume less solvent (Rodriguez-Saona and Allendorf, 2011). Infrared spectroscopy (IR) measures the vibration of molecules having different functional groups in three regions, namely near-infrared (NIR) (760–2,500 nm), mid-infrared (MIR) (2,500–50,000 nm), and far-infrared (FIR) (50,000–500,000 nm) (Zhang et al., 2011). The effects of the various functional groups in different parts of the spectrum gives a unique "fingerprint" that can be used to confirm the identity of a sample (Cozzolino, 2012). MIR spectra are obtained by recording the absorbance of light at the vibrational and rotational frequencies of the atoms within the molecules (Luykx and van Ruth, 2008). This is in contrast to NIR that records molecular combinations and overtone vibrations (Wang et al., 2016). The advantages and disadvantages of both techniques are well documented (Bunaciu, Aboul-Enein, and Hoang, 2015).

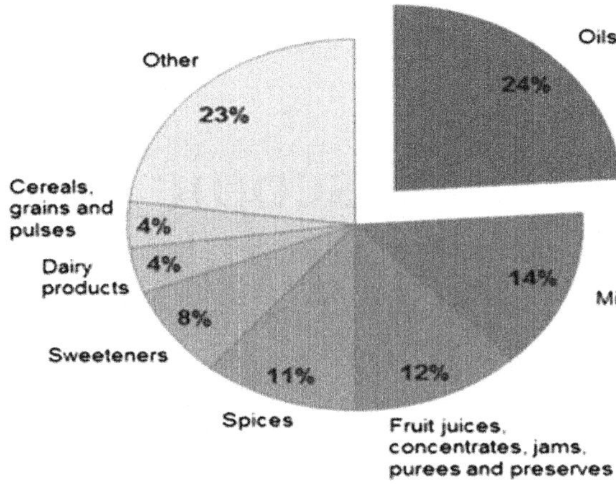

FIGURE 2.1 **(See color insert.)** Percentage of food fraud cases from 1980 until 2010. (Reproduced with permission from Moore et al., Journal of Food Science, 77 (2012), pg 118–126. With permission.)

Most food products are complex and they cannot be differentiated according to their spectral features. Hence, chemometrics is often used for this purpose. The initial step in classification of data is exploratory data analysis (EDA), which is carried out by pattern recognition methods (Gomez-Caravaca, Maggio, and Cerretani, 2016). These can be further classified into supervised and unsupervised methods depending on whether the classes to which the samples belong are known or not (Wang et al., 2016). Examples of supervised techniques include principal component analysis (PCA) and hierarchical cluster analysis (HCA) (Kamal and Karoui, 2015). On the other hand, unsupervised pattern recognition techniques, which are more powerful, rely on previous knowledge of sample information and include linear discriminant analysis (LDA), *k*-nearest neighbors (*k*-NN), partial least squares discriminant analysis (PLS-DA), soft independent modeling class analogy (SIMCA), and artificial neural networks (ANN) (Meza-Márquez, Gallardo-Velázquez, and Osorio-Revilla, 2010; Caredda et al., 2017; Manfredi et al., 2018).

This chapter discusses the applications of MIR and NIR coupled to these different chemometric techniques in the determination of the authenticity of various food items. In each of the subsections, the key aspect discussed is the versatility of the different techniques as shown by the excellent classification rates obtained during the determination of traceability and adulteration. Furthermore, in conclusion, the general limitations and achievements in determining the authenticity of food are also highlighted (Table 2.1).

2.2 CEREALS

Cereal grains such as maize, wheat, rice, and barley are widely consumed worldwide for their high energy values. Most analytical techniques that were used to monitor the quality of cereals suffered from the following drawbacks: slow, destructive, and reliance on empirical relationships (Cozzolino, 2014). For this reason, NIR and MIR techniques coupled to multivariate data analysis (MVA) have been applied to determine the composition of cereals (Hell et al., 2016). Lately, they have also been applied for checking the authenticities and traceabilities of cereals using MVA techniques, such as PCA, PLS-DA,

TABLE 2.1 A Summary of the Different Chemometric Techniques, their Principles, and Applications in the Authenticity of Food

Multivariate Analysis Technique	Description	Application	References
Unsupervised Analyses			
Hierarchical Clustering Analysis (HCA)	Generates tables of distances, which are simplified using algorithms.	Differentiation of commercial juices.	(He, Rodriguez-Saona, and Giusti, 2007)
Supervised Analyses			
Principal Component Analysis (PCA)	Decomposes data matrix with *n* rows (samples) and *p* columns (variables) into the product of scores matrix.	Discrimination of wheat varieties, determination of geographical origin of wheat, authentication of fruit juice.	(Kamal and Karoui, 2015; Miralbes 2008; Vardin et al., 2008; Zhao et al., 2013)
k-nearest neighbors (*k*-NN)	The input consists of the *k* closest training examples in the featured space.	Detection of milk adulteration, detection of honey adulteration.	(Daniel Kelly, Downey, and Fouratier, 2004; Zhang et al., 2014)
Soft Independent Modeling by Class Analogy (SIMCA)	Used to find class boundaries of known classes and to assign new objects to the classes.	Wheat flour classification, detection of wheat adulteration.	(Cocchi et al., 2005; Foca et al., 2009; Kasemsuran, Thanapase, and Kiatsoonthon, 2007)
Partial Least Squares Discriminant Analysis (PLS-DA)	Uses an algorithm that can predict both continuous and discrete variables.	Authentication of lard.	(Chiesa et al., 2016)
Linear Discriminant Analysis (LDA)	Involves searching for linear combination of variables that separates two classes.	Classification of cheese, tracing geographical origin of oils.	(Bevilacqua et al., 2012; Lema-Garcia et al., 2010)
Support Vector Machine	Each data item is plotted in *n*-dimensional space and classification is done by finding the hyper-plane that distinguishes the classes.	Melamine adulteration detection in dairy products, honey adulteration.	(Balabin and Smirnov, 2011; Zhu et al., 2010)

(*Continued*)

TABLE 2.1 (*Continued*) A Summary of the Different Chemometric Techniques, their Principles, and Applications in the Authenticity of Food

Multivariate Analysis Technique	Description	Application	References
Classification and Quantitative Supervised Analyses			
Artificial Neural Networks (ANN)	Comprises artificial neurons that are correlated. The inputs are then converted to output.	Discrimination of season of production and feeding regimen of butters, discrimination of lamb fat quality.	(Gori et al., 2012; Osorio et al., 2009)
Quantitative Supervised Analyses			
Principal Component Regression (PCR)	Analyzes data with a great degree of linearity.	Detection of adulteration in crab meat, adulteration of virgin oil.	(Gayo, Hale, and Blanchard, 2006; Rohman and Che Man, 2010)
Partial Least Squares Regression (PLSR)	Used to find the fundamental relations that explain a significant proportion content between two matrices (X and Y) allowing the overcoming of problems to do with overfitting, multicollinearity, and outliers.	Discrimination of honeys from different botanical origins.	(Kamal and Karoui, 2015; Tahir et al., 2016)

and LDA (Campbell, Sykes, and Glover, 2000; Dowell, 2000; Ziegler et al., 2016; Zhao et al., 2013; Cocchi et al., 2004).

NIRS, in combination with discriminant partial least squares (DPLS), was used to predict the geographical origins of wheat grains and flour produced in different regions of Chile (González-Martín et al., 2014). The method correctly classified 76% of the wheat grain samples and 90%–96% of the flour samples according to their geographical origins. Thus, the results of the study showed that this technique coupled to chemometric tools is very useful for the classification of wheat grain and flour samples according to geographical origins. Similarly, in another study carried out in China, NIR spectroscopy with chemometrics successfully discriminated the wheat varieties from different origins (Zhao et al., 2013). 240 wheat samples from two growing years were analyzed, and the spectra are shown in the Figure 2.2. LDA based on PCA scores and PLS-DA were used to classify samples according to their geographical origins, with significant differences being observed among the wheat varieties from different regions. Moreover, it is worth noting that DPLS was better than LDA for wheat origin determination. In Europe, NIR with chemometrics was used to discriminate 249 wheat varieties from Spain and France, reporting a classification rate of 99.5% for the calibration sample set and 94.5% for the validation set (Miralbes, 2008). It is worth noting that infrared spectroscopic techniques have been successfully utilized to authenticate rice samples with different geographical origins. NIR and multi protocol label switching were used to successfully differentiate South Korean rice samples from foreign ones, although the calibration statistics were not satisfactory (Kim et al., 2003).

The vitreous-ness of durum is a useful indicator of wheat quality in the milling industry. NIR spectroscopy coupled to partial least squares (PLS) was used to successfully classify vitreous and non-vitreous kernels by observing the scattering effects and the starch and protein differences. However, for the difficult kernels, classification rates decreased up to 75% (Dowell, 2000). In a recent study, NIR with PLS-DA was used to discriminate kernels and flour of different wheat species with accuracy values greater than 80% (Ziegler et al., 2016). The potential of NIR in the authentication of corn, according to genotypic classes, was also explored (Campbell, Sykes, and Glover, 2000). To classify two genotypes, PLS was used, whereas PCA was used to classify at least three genotypes. It was observed that the two-class models performed better than the multiclass models.

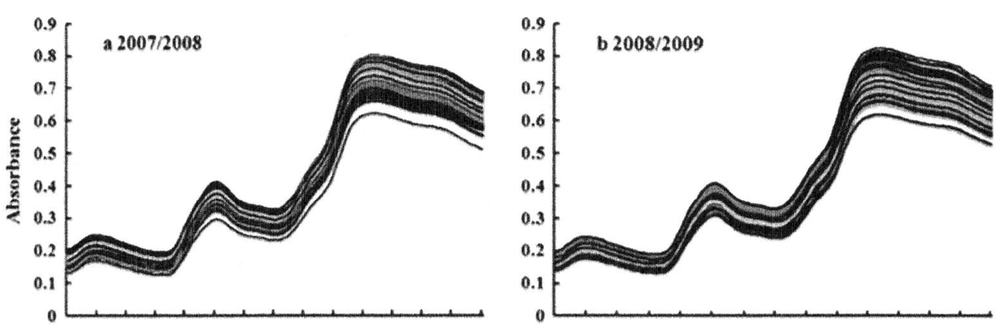

FIGURE 2.2 Raw spectra of wheat kernels from different years and with different geographical origins. (Reproduced with permission from Zhao et al., Food Chemistry, 138 (2013), pg 1902–1907. With permission.)

The quality of flour affects the quality of the products (Moiraghi et al. 2013). For this reason, studies have been carried out to authenticate flour quality in different parts of the world. In Italy, studies have also been carried out to successfully discriminate bread wheat and flour of various baking qualities (Cocchi et al., 2005; Foca et al., 2009). The first study sought to classify wheat flour based on NIR spectra, although the results were not as satisfactory because they were able to discriminate only two out of the five classes that are defined according to the synthetic quality index (ISQ) method.

2.3 MEAT PRODUCTS

There is overwhelming evidence that the geographical origins of meat products need to be determined in order to protect consumers fully (Franke et al., 2005). NIRS coupled to pattern recognition techniques was successfully used to determine the geographical origin of Chinese exported tilapia fillets, obtaining classification rates greater than 75% (Liu et al., 2015). A similar study was carried out to determine the geographical origin of mutton using NIR and SIMCA by preprocessing five-point smooth and multiplicative scatter correction (MSC) in the region of 11995 cm^{-1} and 3999 cm^{-1}. The recognition rates of the validation set were 100%, 83%, 100%, and 100% for the four regions investigated (Ning et al., 2008). In the same way, NIR using chemometrics was used to classify the geographical origins of lamb meat from three pastoral regions and two agricultural regions of China. PCA, DPLS, and LDA were used for data analysis (Sun et al., 2012). DPLS and LDA both gave 100% classification for both agricultural and pastoral regions.

The potential of Fourier transform infrared (FTIR) and chemometrics to discriminate fat samples from the carcasses of lambs from different rearing systems was explored. Fat samples (perirenal and omental) were collected from the carcasses of lambs and the fatty acid compositions of each fat sample were analyzed (Osorio et al., 2009). For the former, perfect classification was obtained using the three data treatment approaches (DPLS, ANN (PCA), ANN support vector machine (SVM)). The ANN models were superior than DPLS, although ANN (SVM) showed greater discriminatory power than ANN (PCA), thus giving better information regarding the functional groups associated with classification. In a related study, NIR with chemometrics was used to authenticate Italian protected designation of origin (PDO) lard with high specificities of 98.2% and 94.4% in the calibration and cross-validation steps, respectively (Chiesa et al., 2016).

MIR spectroscopy was used to detect the adulteration in cooked beef using PCA, partial least squares regression (PLS-R), and LDA (Al-Jowder, Kemsley, and Wilson, 2002). The cross-validated classification success rate was approximately 97%. FTIR-ATR coupled to PLS-DA, and SIMCA was used to detect offal-adulteration in beef burgers (Zhao, Downey, and O'Donnell, 2014). It is worth noting that SIMCA modeling allowed the accurate identification of beef burger samples, whether their thermal history was known or not. However, as the cooking level increased it became more difficult to detect adulteration. Visible NIR (VIS-NIR) was used to successfully quantify and detect adulteration in crab meat using PLS and PCR (Gayo, Hale, and Blanchard, 2006). The best model for both PLS and PCR used the first derivative spectral data gathered from the correlogram, with standard errors of prediction (SEP) of 0.252 and 0.244, respectively. Pork adulteration has also been detected in veal sausages using Fourier transform NIR (FT-NIR) coupled to PCA and SVM (Schmutzler et al., 2015). Laboratory, industrial, and on-site instruments were used, and it was observed that none of them can be regarded as ideal as they all had their own advantages and disadvantages. Thus, the choice of instrument will

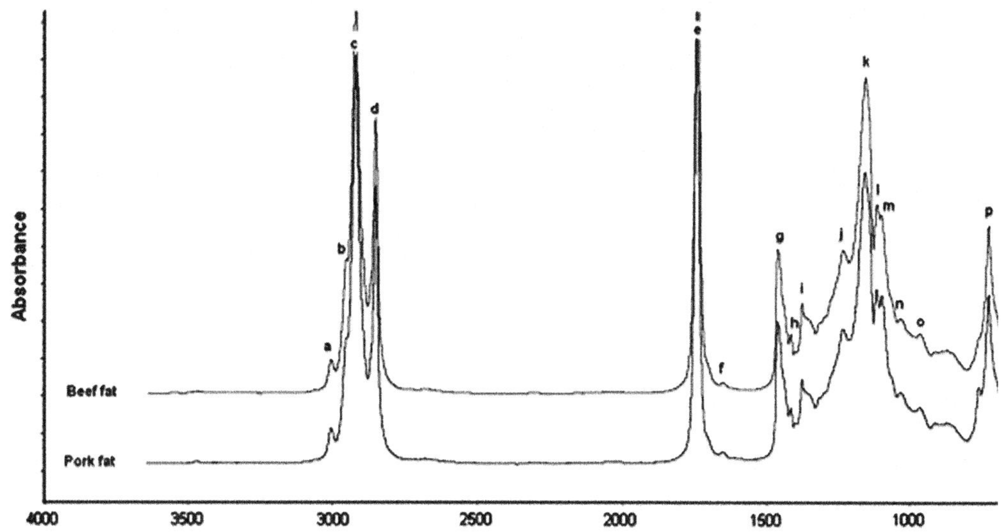

FIGURE 2.3 FTIR spectra of pork fat and beef fat in the mid-infrared region. (Reproduced with permission from Rohman et al., Meat Science, 88 (2011), pg 91–95. With permission.)

depend on the desired use and the required precision, among other factors. Finally, FTIR spectroscopy was used to detect pork adulteration in beef meatballs (Rohman et al., 2011). Although the spectra seemed somewhat similar as shown in Figure 2.3, FTIR spectroscopy could detect differences in peak heights of fats at 1117 cm^{-1} and 1097 cm^{-1} due to C–H bending and C–H deformation vibrations of the fatty acids, respectively. PLS was used to successfully quantify the level of adulteration in the fingerprint region of 1200–1000 cm^{-1}.

2.4 WINE

The authenticity of wine is of paramount importance because it protects consumers from fake products. To maintain the quality of wine, IR spectroscopic techniques have emerged as the method of choice in determining the authenticity of wine because of the significant advantages they possess over traditional fingerprinting techniques (Casale et al., 2010). They are coupled to multivariate data techniques such as PCA, DA, canonical analysis (CA), and cluster analysis (CLA). The studies carried out to date include differentiation based on wine variety, grape variety, and geographical origin (Liu et al., 2008; Tarantilis et al., 2008; Zhang et al., 2010).

The quality and flavor of wine is related to its geographical origin, and it therefore affects the market value. For this reason, it is necessary to classify wine from different geographical origins. Numerous studies have been carried out on this subject using both NIR and MIR. NIR and MIR coupled to chemometric techniques were used to discriminate among Shiraz wines produced in five Australian regions (Riovanto et al., 2011). Spectral data were interpreted using PCA, LDA, and SIMCA. The classification rates using the two IR methods varied according to the classification techniques used. Overall, LDA showed lower classification rates (73%) than SIMCA (95%), and the best classification was obtained using a SIMCA and MIR combination. In a different study, visible-NIR spectroscopy and chemometric techniques (PCA, PLS-DA, and SLDA) were used to classify Riesling wines from four different countries (Liu et al., 2008). It was observed that Riesling wines from Australia were correctly classified with an overall classification rate of 90%. However, lower classification rates were obtained for wine samples from New Zealand and Europe. However, the authors noted that the limitations observed in this study were partly because of the small sample sizes. The same research group applied VIS-NIR spectroscopy coupled to multivariate data techniques to classify the geographical origins of Australian and Spanish Tempranillo wines (Liu et al., 2006). PLS-DA correctly classified 100% and 84.7% of Australian and Spanish wines, respectively, while LDA correctly classified 72% and 85% of Australian and Spanish wines, respectively.

The potential of MIR coupled to chemometrics has been extended to carry out discrimination between the red and white Australian wine varieties by making use of the differences in the wine matrices (Bevin et al., 2008). Spectral data was reduced to a small number of principal components and LDA was used to successfully discriminate among the wine samples, with success rates of over 95% in the validation set. Visible and near infrared spectroscopy was used to discriminate between white Australian wines of different varieties (Chardonnay and Riesling) (Cozzolino, Smyth, and Gishen, 2003). Discrimination was done using PCA, PCR, and DPLS. In addition, accuracy greater than 98% was observed. It is worth noting that IR spectroscopic techniques coupled to chemometrics have been successfully used to distinguish the groups of wines with similar properties, such as white wines, because of the differences in sugars, phenolic compounds, or nonvolatile compounds (Zhang et al., 2010). The first MIR method for monitoring wine authenticity, before and after transport, was developed in 2006 (Bevin et al., 2006). This was done by classifying the various wines using the similarity index (SI). Very high classification rates of 98% were obtained in five of the six winery datasets.

2.5 DAIRY PRODUCTS

The authenticity of dairy products is an important topic for researchers, consumers, and policymakers because they are a major component of the human diet for all age groups (de la Fuente and Juarez, 2005). However, their nutritional benefits can only be realized if their quality is not compromised by economically motivated behavior, such as adulteration with fillers, preservatives, melamine, synthetic milk, and counterfeit dairy products (Abernethy, Rendall, and Holroyd, 2016). NIR and MIR spectroscopic techniques have been successfully used to detect adulteration in various dairy products, such as milk, butter, milk powders, and infant formula (Balabin and Smirnov, 2011; Koca et al., 2010; Botelho et al., 2015). For instance, NIR spectroscopy in the region of 1100–2500 nm was used to detect water and whey adulterations in cow milk samples using DPLS and SIMCA (Kasemsuran, Thanapase, and Kiatsoonthon, 2007). DPLS gave better results than SIMCA with percentage of correct classifications, for the validation set, of 97% and 86.73%, respectively. In a different study to quantify adulterants in powdered milk, a disadvantage of PLS was that it was affected by the spectral differences that led to nonlinearity when all the adulterants were in the same dataset (Borin et al., 2006). For this reason, LS-SVM was used to successfully build robust models. This was confirmed by a later study carried out to determine melamine using MIR and NIR, which showed the nonlinearity of the relationship between the melamine content and MIR/NIR spectra. Thus, nonlinear regression methods, such as Poly-PLS, ANN, SVR, or LS-SVM, were

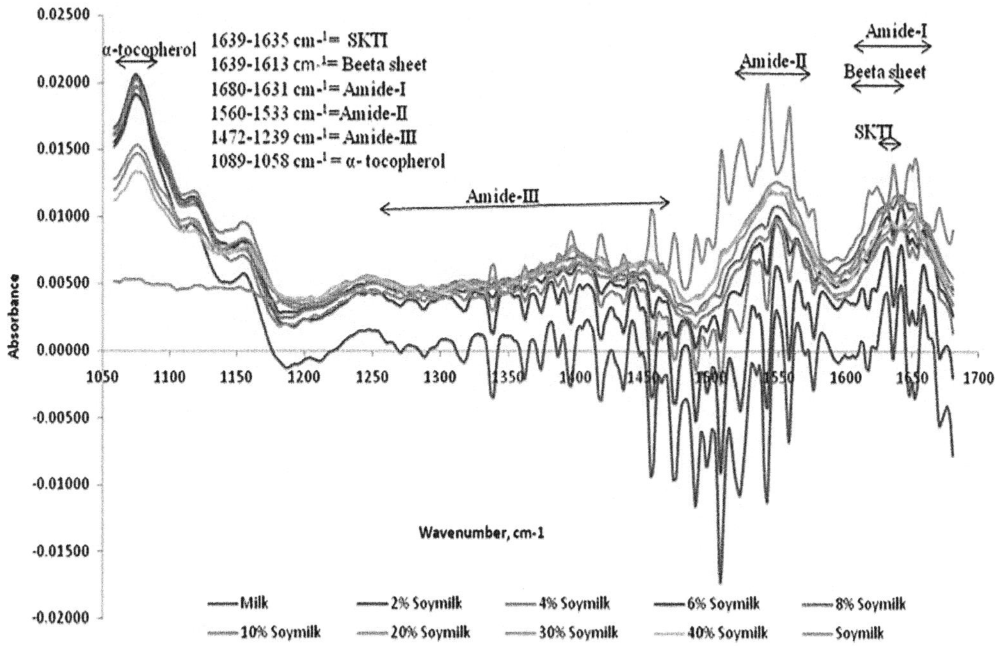

FIGURE 2.4 **(See color insert.)** Typical spectra of milk, soymilk, and milk adulterated with soymilk at different levels in the spectral range 1680–1058 cm^{-1}. (Reproduced with permission form Botelho et al., Food Chemistry, 181 (2015), pg 31–37. With permission.)

used to correctly predict the adulterant content (Balabin and Smirnov, 2011). A recent study carried out to evaluate the performance of pattern recognition techniques, such as IS-SVM and IS k nearest neighbor (IS-*k*-NN), to discriminate adulterated milks based on NIR spectral signals showed that the concentration of the adulterant is an important factor affecting the results (Zhang et al., 2014). Another multivariate supervised classification technique that has been used in conjunction with attenuated total reflectance MIR (ATR-MIR) spectroscopy to detect adulterants in milk is PLS-DA (Botelho et al., 2015). A distinct advantage of this method was that all the adulterants could be detected in the same spectra as shown by Figure 2.4.

FT-MIR coupled to PLS was used to detect whey adulteration in raw milk without any systematic errors (Vieira et al., 2015). In a different study, FT-MIR and FT-NIR spectroscopy were used to detect tetracycline adulteration in milk (Sivakesava and Irudayaraj, 2002b). On comparison of the two methods it was observed that the predictive abilities of the models were better in the latter case as shown by the fewer number of factors and by their standard error of calibration and SEP that were higher than the former. A recent development in the FTIR machine is the use of single-bounce ATR (SB-ATR) sample handling accessory that gives more spectral information because of its very short path length (Jawaid et al., 2013). ATR-FTIR in conjunction with chemometric techniques was used to detect soymilk adulteration in milk. Clear differences were obtained in the absorption of the two milks in the wave number region of 1639–1613 cm^{-1} (Jaiswal et al., 2015). Another variation of the same technique, temperature-controlled ATR-MIR was used to predict butter adulteration with margarine (Koca et al., 2010). A distinctive band around 1741 cm^{-1} allowed clear discrimination between the two.

FTIR spectroscopic techniques were used to discriminate Emmental cheeses according to their geographical origins (Pillonel et al., 2003). LDA applied on PCA scores of NIR spectra gave 100% classification of 20 cheese samples. FTIR spectroscopy followed by LDA analysis of spectral data was used to classify Italian Pecorino cheeses according to their ripening times and manufacturing techniques (Lema-Garcia et al., 2010). An excellent resolution was obtained for both ripening time and manufacturing technique. The application of FTIR was also extended to the discrimination of season of production and the feeding regimen of butters (Gori et al., 2012). Using ANN, the season of production was predicted at 100%. In addition, the ability of the technique to predict feeding regimen was 75%–90% for samples collected in spring, summer, and winter.

In a different study, NIR was successfully used to differentiate cheeses according to the feeding system, regardless of whether or not they contained water (Andueza et al., 2013). A study was carried out to evaluate the potential of NIR spectroscopy to distinguish milk samples according to the feeding system, breed, and altitude of the farm (Valenti et al., 2013). By applying PLS-DA to the spectral datasets, the milk samples could be distinguished according to the feeding system. However, the technique could not distinguish between milk samples from different altitudes and breeds.

2.6 FRUITS AND FRUIT JUICES

FTIR spectroscopy and PCA were used to detect the adulteration of pomegranate juice with grape juice by using differences in their spectra that occurred in the region of 1780–1685 cm^{-1} due to C=O stretching (Vardin et al., 2008). Two principal components were used in the PCA class models to classify adulterated and pure juice samples, with R^2 values of 0.99 being obtained for both class models. In a different study, ATR-FTIR spectroscopy was used to detect adulteration of mango juice with sugar in the region of 4000–650 cm^{-1} using PCA and PLS (Jha and Gunasekaran, 2010). Classification of samples based on adulteration was possible with a significance level of 5%. In addition, the PLS model developed using MSC treated data correctly discriminated between adulterated and unadulterated samples. A distinct advantage of this technique was that it predicted correctly the adulteration at sugar levels as low as 3.6%, which is better than a previous study that reported on adulteration of apple juice in which 100% accuracy was obtained at 10% sugar concentration using NIR and PLS (Leo'n, Kelly, and Downey, 2005). Thus, it is of paramount importance to note that the application of MIR spectroscopy in the determination of apple juice adulteration is highly dependent on the concentrations and ratios of the major chemical components in the apple juice (Daniel Kelly and Downey, 2005). However, in a more recent study, NIR was used to investigate orange juice adulteration in the region of 1100–2498 nm using PCA and FDA to an accuracy of 90% (Twomey, Downey, and McNulty, 1995). The same research group reviewed earlier studies to determine orange juice adulteration (Shilton, Downey, and McNulty, 1998). NIR and principal component-radial basis function neural networks (PC-RBFNN) were used to successfully separate pure bayberry juice from water-adulterated bayberry juice at recognition rates of 97.62% (Xie et al., 2008). However, the technique could not detect the water level in the bayberry juice.

NIR coupled to multivariate data techniques was used to classify apple juices based on apple variety, with classification rates as high as 100% (Reid et al., 2005). Snyder et al. explored the possibility of authenticating concord juice concentration in grape juice blend using FT-IR coupled to SIMCA and partial least squares regression (PLSR). SIMCA discriminated grape juices based on the differences in the IR aromatic bands of their phenolic compounds. In addition, the predictive ability of PLSR was not affected by the differences in the blends. Thus, FTIR coupled to chemometrics can be used to discriminate grape juices based on variety (Snyder et al., 2015). MIR spectroscopy has also been used to classify fruit juices based on ingredients and geographical origins using HCA and SIMCA (He, Rodriguez-Saona, and Giusti, 2007). HCA successfully classified juices according to the differences in their ingredients. In addition, SIMCA successfully classified juices according to their geographical origins with a classification rate of 100%.

2.7 EDIBLE OILS

An important application of IR spectroscopy is in the determination of traceability and adulteration of edible oils, such as virgin olive oil, pumpkin oil, and vegetable oils. Virgin olive oil is one of the most expensive oils whose authenticity needs to be determined always, lest adulteration takes place (Rodriguez-Saona, Giusti, and Shotts, 2016). The major reason why its authenticity is a topic of interest is the high nutritive value derived from its various fatty acids as compared to other vegetable oils (Wang et al., 2016). For this reason, IR techniques have been widely used to detect adulteration in olive oils by making use of the differences in the spectral data at specific wave numbers (Azizian et al., 2015; Downey, McIntyre, and Davies, 2002; Mignani et al., 2011; Ozturk, Yalcin, and Ozdemir, 2010). A study was carried out to detect and quantify sunflower oil adulteration in extra virgin olive oils (EVOOs) using VIS-NIR spectroscopy coupled to HCA, SIMCA, and PLS. SIMCA successfully discriminated between authentic EVOO and sunflower oils at levels as low as 1% (w/w) (Downey, McIntyre, and Davies, 2002). In a more recent study, EVOO adulteration with palm oil was studied using FTIR and DA. Frequencies at the fingerprint region of 1500–1000 cm^{-1} were used for classification and quantification and no misclassifications were obtained (Rohman and Che Man, 2010). In their most recent study, the same research group successfully used the same techniques to determine EVOO adulteration with grapeseed oil, soybean oil, and walnut oil using FTIR in the region of 4000–650 cm^{-1} (Rohman et al., 2017).

False declaration of the geographical origin of EVOOs is fraud. For this reason, IR spectroscopic techniques and chemometrics have been used to authenticate them. FTIR coupled to multivariate techniques was used to distinguish EVOOs from four different European countries (Greece, Italy, Portugal, Spain) (Tapp, Defernez, and Kemsley, 2003). Two approaches were used to analyze the data, namely PLS-LDA and genetic algorithm followed by LDA (GA-LDA). Neither of the techniques could describe the geographical origin of EVOOs in biochemical terms. However, the study showed that FTIR has the ability to potentially discriminate EVOOs according to the country of origin. The possibility of using MIR and NIR to determine the geographical origin of olive oil has been extensively reported in literature (Galtier et al., 2007; Gouvinhas et al., 2015; Bendini et al., 2007). In a recent study, the potential of coupling MIR and NIR was investigated while determining the geographical origin of EVOO from PDO Sabina (Bevilacqua et al., 2012). Classification with PLS-DA gave better prediction with NIR than MIR as shown by Table 2.2, and this was confirmed by SIMCA.

Apart from geographical origin, a recent trend that has been observed is the differentiation of edible oils according to factors such as time of harvesting, genetic variety, and fruity attribute intensity (Downey, McIntyre, and Davies, 2002; Sinelli, Casale, et al., 2010; Sinelli, Cerretani, et al., 2010). ATR-FTIR and LDA were used to classify pumpkin seed oils according to their species and genetic variety (Saucedo-Hernandez et al. 2011).

TABLE 2.2 Application of IR Spectroscopic Techniques in Detecting the Authenticity of Different Food

Type of Food	IR Technique	Chemometric Technique	Objective	Results	References
Wheat flour	NIR	PCA, SIMCA	Classification of different varieties of wheat flour.	Overlap of intermediate quality classes caused misclassification of some data signals.	(Cocchi et al., 2005)
Wheat	NIR	DPLS, PCA	Discrimination of wheat varieties.	Correct classification of 99.5% for calibration set and 94% for validation set.	(Miralbes, 2008)
Wheat	NIR	LDA, PCA, DPLS	Determination of geographical origin.	DPLS was better than LDA in the determination of geographical origin.	(Zhao et al., 2013)
Wheat	NIR	PLS	Differentiation of wheat kernels.	Classification was possible due to scattering effects and protein differences between the kernels.	(Dowell, 2000)
Cheese	FTR-ATR	LDA	Classification according to ripening time and manufacturing technique.	Excellent resolution was obtained for both ripening time and manufacturing technique.	(Lema-Garcia et al., 2010)
Butter	FTIR	ANN, PCA	Discrimination according to season of production and feeding regimen.	100% predictive ability for season of production and >75% for feeding regimen.	(Gori et al., 2012)
Cheese	VIS-NIR	PLS-DA	Authentication of cow feeding regimen.	Technique could authenticate cheese according to cow feeding regimen.	(Andueza et al., 2013)
Cheese	MIR/NIR	PCA/LDA	Determination of geographical origin.	MIR transmission spectra achieved 100% correct classification in LDA.	(Pillonel et al., 2003)
Honey	MIR	PCA	Classification of honey from different floral sources.	Classification accuracy was almost 100%.	(Bertelli et al., 2007)
Honey	NIR	PLS	Confirmation of claimed PDO provenance.	Provenance of honey samples was confirmed with a high degree of accuracy.	(Woodcock, Downey, and Donnell, 2009)

(Continued)

TABLE 2.2 (*Continued*) Application of IR Spectroscopic Techniques in Detecting the Authenticity of Different Food

Type of Food	IR Technique	Chemometric Technique	Objective	Results	References
Honey	ATR-FTIR	PCA, HCA	Differentiation of honey from different botanical origins.	Both PCA and HCA successfully differentiated the honey samples	(Gok et al., 2015)
Honey	NIR	LS-SVM, LDA	Discrimination of honey from different botanical origins	LDA was superior to LS-SVM in both the training and prediction sets.	(Tahir et al., 2016)
Tilapia fillets	NIR	SIMCA	Geographical origin determination.	Classification rates were greater than 80%.	(Liu et al., 2015)
Lamb meat	NIR	D-PLS, LDA	Geographical origin determination.	Very good classification rates of greater than 75% were obtained.	(Sun et al., 2012)
Lamb meat	FTIR	D-PLS, ANN	Differentiation of fat quality according to rearing system.	Perirenal samples were perfectly discriminated. Omental samples had some misclassifications.	(Osorio et al., 2009)
Virgin olive oils	NIR	PLS-DA	Geographical origin determination.	Satisfactory results were obtained.	(Galtier et al., 2007)
Extra virgin olive oils	NIR/MIR	LDA, SIMCA	Discrimination according to cultivar.	High classification rates were obtained.	(Sinelli, Casale, et al., 2010)
Extra virgin olive oil	NIR/MIR	LDA, SIMCA	Classification according to fruit attribute intensity.	Prediction rate varied from 71.6% to 100%.	(Sinelli, Cerretani, et al., 2010)
Extra virgin olive oils	FTIR	PCA, DA	Classification according to cultivar at different maturation stages.	DA allows the classification of oils according to the maturation stage.	(Gouvinhas et al., 2015)
Extra virgin olive oils	NIR/MIR	PLS-DA, SIMCA	Geographical origin determination.	NIR offered better classification than MIR in both the cross-validation and test sets.	(Bevilacqua et al., 2012)

Up to 23 wavelength regions were selected on the spectra and the normalized absorbance peak areas were used as predictors. Excellent resolution was observed for *Curcubita* species and *C. moschata* varieties. FTIR and LDA were also used to predict the genetic variety in EVOO with 88% correct classification and more than 95% assignment probabilities (Concha-Herrera et al., 2009) (Table 2.3).

2.8 HONEY

Honey is a complex substance that contains sugars, proteins, free amino acids, vitamins, and trace compounds, and these can be used to determine the authenticities and traceabilities of honey varieties (Tahir et al., 2016). It is imperative to note that the authenticity of honey guarantees the economic and quality value of the products. For this reason, consumers are very particular about the origin and traceability of honey and a lack of adulteration in it. In this context, powerful analytical techniques such as HPLC and GC have been greatly used to detect adulteration and authenticities of honey products. However, these are expensive and time consuming and, for this reason, IR techniques have evolved as the method of choice. MIR is more suitable than NIR in carrying out these tasks because it is more sensitive and, hence, offers better resolution (Pita-Calvo, Guerra-Rodríguez, and Vázquez, 2017).

The botanical origin of honey depends on its chemical composition (Gok et al., 2015). IR techniques based on various chemometric techniques have been successfully used to authenticate honeys from different botanical origins. Normally, pattern recognition techniques such as PCA, LDA, and CA have been applied to process the data (Ruoff, Luginbuhl, Kunzli, et al., 2006; Ruoff, Luginbuhl, Bogdanov, et al., 2006; Tewari and Irudayaraj, 2005). In China, NIR and chemometric techniques were used to classify honeys from five different botanical origins (Chen et al., 2012). Classification models based on NIR spectra were developed using Mahalanobis distance discriminant analysis (MD-DA) and back propagation artificial neural network (BP-ANN). The former model obtained total correct classification rates of 87.4% and 85.3% for the calibration and validation sets, respectively. By the latter, correct classification rates of 90.9% and 89.3% were obtained for the calibration and validation sets, respectively. In South Africa, NIR coupled to PLS-DA was used to distinguish authentic South African honey from imported honey (Guelpa et al., 2017). The overall classification accuracies obtained were between 93.3% and 99.9% using three different instruments. Mid-infrared diffuse reflectance spectroscopy (DRIFTS) and MVA were used to classify Italian honeys in the region of 4000–600 cm^{-1} (Bertelli et al. 2007). Classification accuracies of approximately 100% were achieved by discriminant and classification tree analyses. The PDO provenance of Corsican honey samples was also confirmed with a high degree of certainty using DPLS analysis of NIR spectra (Woodcock, Downey, and Donnell, 2009).

Infrared spectroscopic techniques have been used to detect that honey adulteration is mainly done with sugars, such as sucrose, glucose, or fructose or with low-cost sweeteners, such as corn syrup, invert sugar syrup, or high fructose syrup (Siddiqui et al., 2017). Despite the numerous advantages that IR techniques possess over other techniques in the detection and quantification of adulteration, one major challenge associated with them is the requirement of advanced chemometric techniques such as discriminant analysis (Wu et al., 2017). Sugar adulteration has been successfully detected using IR techniques coupled to chemometric techniques, such as *k*-NN, PLS, SVM, BP-ANN, and LDA, with very high classification rates that are greater than 90% (Daniel Kelly, Downey,

TABLE 2.3 Application of IR Spectroscopic Techniques in the Detection of Adulteration in Various Food

Adulterant	Food Type	IR Technique	Chemometric Technique	Results	Reference
Water, whey	Cow milk	NIR	DPLS, SIMCA	DPLS correctly classified 100% of adulterants in the validation set. SIMCA performance was better than DPLS at concentration greater than 11%.	(Kasemsuran, Thanapase, and Kiatsoonthon, 2007)
Starch, whey, sucrose	Powdered milk	NIR	PLSR, LS-SVM	LS-SVM models gave better and lower prediction errors than PLSR.	(Borin et al., 2006)
Pseudo-proteins and thickeners	Cow milk	NIR	I-SVM, IS-KNN	Both techniques could discriminate between raw cow milk from adulterated milk.	(Zhang et al., 2014)
Melamine	Liquid milk, infant formula, milk powder	MIR/NIR	ANN, LS-SVM	Both techniques could detect melamine adulteration	(Balabin and Smirnov, 2011)
Margarine	Butter	ATR-FTIR	PLSR	Excellent prediction was obtained.	(Koca et al., 2010)
Sweeteners	Honey	NIR	LS-SVM, SVM, BP-ANN, LDA, kNN	Wavelet transform LS-SVM showed very good detection of honey adulteration with great accuracy.	(Zhu et al., 2010)
Horse meat, fat beef trimmings, textured soy protein	Minced beef	FTIR	SIMCA	100% classification was obtained.	(Meza-Márquez, Gallardo-Velázquez, and Osorio-Revilla, 2010)
Pork	Beef meatballs	FTIR	PLS	Successful quantitation of adulterants was done.	(Rohman et al., 2011)
Beef offals	Beef burger	MIR	PLS-DA, SIMCA	Highest accuracies were observed with SIMCA for both fresh and thawed samples.	(Zhao, Downey, and O'Donnell, 2014)
Surimi-based imitation crab meat	Surimi-based imitation crab meat	VIS/NIR	PLS, PCR	VIS/NIR was used to successfully detect adulteration in crab meat.	(Gayo, Hale, and Blanchard, 2006)
Heart, tripe, kidney, liver	Cooked beef	MIR	PCA, PLS, LDA	Discrimination was possible, but it decreased as cooking level increased.	(Al-Jowder, Kemsley, and Wilson, 2002)
Palm oil	Extra virgin olive oil	FTIR	PLS, PCR, DA	DA was able to classify pure from adulterated samples without any misclassifications.	(Rohman and Che Man, 2010)

and Fouratier, 2004; Rios-Corripio, Roja-Lopez, and Delgado-Macuil, 2012; Sivakesava and Irudayaraj, 2002a). It has however been noted that LS-SVM gave better results than *k*-NN, PLS, or LDA because the former models have better generalization abilities and stronger nonlinear recognition abilities (Zhu et al. 2010).

FTIR-ATR coupled to PLS was used successfully to detect adulteration with corn syrup, high fructose corn syrup, and inverted syrup, with SEP of 1.5–2.1, 2.1–3.0, and 1.4–2.5, respectively (Gallardo-Velázquez et al., 2009). The results are in perfect agreement with an earlier study carried out to investigate the adulteration of honey with corn syrup using FTIR and PLS. The SEP in the validation set was 1.99%–2.22% using PLS on first derivative transformed data (Sivakesava and Irudayaraj, 2001). Near infrared transflectance spectroscopy was also used to detect beet invert syrup and HFCS in artisanal Irish honey when SIMCA and PLS were applied to classify honeys and to quantify adulteration. The best classification model was obtained using raw spectral data. In addition, the preferred model for prediction of percentage adulteration involved PLS1 of multiplicative scatter corrected data and second derivative transformation. Although the sample numbers were small, 100% and 90.9% correct classification were obtained for adulterated and unadulterated honeys, respectively (Daniel Kelly, Petisco, and Downey, 2006b). The same research group applied FTIR and the same chemometric techniques to distinguish between Irish artisanal honey and honey adulterated with various sugar syrups. This combination was unable to unambiguously detect adulteration with high fructose or fully inverted beet syrups possibly because of the similar monosaccharide compositions of honey and these syrups (Daniel Kelly, Petisco, and Downey, 2006a).

2.9 CONCLUSION

Infrared spectroscopic techniques coupled to chemometric techniques play a critical role in the determination of authenticities of various food products. This entails determination of the geographical origin and detection of adulteration. While some studies have been carried out using these techniques for the online monitoring of quality during the processing of products such as wine, more such studies need to be carried out on other foodstuffs. In particular, there is need to distinguish between products based on their storage conditions and transportation. Moreover, in most of the studies, satisfactory results were obtained with small sample sizes; these should also be replicated on a larger scale. To sum up, the advent of IR spectroscopic techniques is a significant breakthrough in the field of food authenticity as classification rates obtained using different chemometric techniques were satisfactory.

REFERENCES

Abernethy, G.A., J.D. Rendall, and S.E. Holroyd. 2016. "Advances in Testing for Adulteration and Authenticity of Dairy Products." In *Advances in Food Authenticity Testing*, edited by G. Downey, 461–90. Amsterdam: Elsevier. doi:10.1016/B978-0-08-100220-9-00017-5.

Al-Jowder, O., E.K. Kemsley, and R.H. Wilson. 2002. "Detection of Adulteration in Cooked Meat Products by Mid-Infrared Spectroscopy." *Journal of Agricultural and Food Chemistry* 50:1325–29. doi:10.1021/jf0108967.

Andueza, D., C. Agabriel, I. Constant, A. Lucas, and B. Martin. 2013. "Using Visible or near Infrared Spectroscopy (NIRS) on Cheese to Authenticate Cow Feeding Regimes." *Food Chemistry* 141:209–14. Elsevier Ltd. doi:10.1016/j.foodchem.2013.02.086.

Azizian, H., M.M. Mossoba, M. Fardin-Kin, A.R. Delmonte, S.R. Karunathilaka, and J.K. Kramer. 2015. "Novel, Rapid Identification, and Quantification of Adulterants in Extra Virgin Olive Oil Using Near-Infrared Spectroscopy and Chemometrics." *Lipids* 50:705–18.

Balabin, R., and S.V. Smirnov. 2011. "Melamine Detection by Mid- and Near-Infrared (MIR / NIR) Spectroscopy: A Quick and Sensitive Method for Dairy Products Analysis Including Liquid Milk, Infant Formula, and Milk Powder." *Talanta* 85:562–68. Elsevier B.V. doi:10.1016/j.talanta.2011.04.026.

Bendini, A., L. Cerretani, F. Di Virgilio, P. Belloni, M. Bonoli-Carbognin, and G. Lercker. 2007. "Preliminary Evaluation of the Application of the FTIR Spectroscopy to Control the Geographic Origin and Quality of Virgin Olive Oils." *Journal of Food Quality* 30:424–37. doi:10.1111/j.1745-4557.2007.00132.x.

Bertelli, D., M. Plessi, A.G. Sabatini, M. Lolli, and F. Grillenzini. 2007. "Classification of Italian Honeys by Mid-Infrared Diffuse Reflectance Spectroscopy (DRIFTS)." *Food Chemistry* 101:1565–70. doi:10.1016/j.foodchem.2006.04.010.

Bevilacqua, M., R. Bucci, A.D. Magri, A.L. Magri, and F. Marini. 2012. "Tracing the Origin of Extra Virgin Olive Oils by Infrared Spectroscopy and Chemometrics: A Case Study." *Analytica Chimica Acta* 717:39–51. doi:10.1016/j.aca.2011.12.035.

Bevin, C.J., R.G. Dambergs, A.J. Fergusson, and D. Cozzolino. 2008. "Varietal Discrimination of Australian Wines by Means of Mid-Infrared Spectroscopy and Multivariate Analysis." *Analytica Chimica Acta* 621:19–23. doi:10.1016/j.aca.2007.10.042.

Bevin, C.J., A.J. Fergusson, W.B. Perry, L.J. Janik, and D. Cozzolino. 2006. "Development of a Rapid 'fingerprinting' system for Wine Authenticity by Mid-Infrared Spectroscopy." *Journal of Agricultural and Food Chemistry* 54:9713–18. doi:10.1021/jf062265o.

Borin, A., M.F. Ferrao, C. Mello, D.A. Maretto, and R.J. Poppi. 2006. "Least-Squares Support Vector Machines and near Infrared Spectroscopy for Quantification of Common Adulterants in Powdered Milk." *Analytica Chimica Acta* 579:25–32. doi:10.1016/j.aca.2006.07.008.

Botelho, B.G., N. Reis, L.S. Oliveira, and M.M. Sena. 2015. "Development and Analytical Validation of a Screening Method for Simultaneous Detection of Five Adulterants in Raw Milk Using Mid-Infrared Spectroscopy and PLS-DA." *Food Chemistry* 181:31–37. doi:10.1016/j.foodchem.2015.02.077.

Bunaciu, A.A., H.Y. Aboul-Enein, and V.D. Hoang. 2015. "Vibrational Spectroscopy Used in Polymorphic Analysis." *TrAC - Trends in Analytical Chemistry* 69:14–22. Elsevier B.V. doi:10.1016/j.trac.2015.02.006.

Campbell, M. R., J. Sykes, and D.V. Glover. 2000. "Classification of Single- and Double-Mutant Corn Endosperm Genotypes by near-Infrared Transmittance Spectroscopy." *Cereal Chemistry* 77:774–78. doi:10.1094/CCHEM.2000.77.6.774.

Caredda, M., M. Addis, I. Ibba, R. Leardi, M.F. Scintu, G. Piredda, and G. Sanna. 2017. "Building of Prediction Models by Using Mid-Infrared Spectroscopy and Fatty Acid Profile to Discriminate the Geographical Origin of Sheep Milk." *LWT - Food Science and Technology* 75:131–36. doi:10.1016/j.lwt.2016.08.053.

Casale, M., P. Oliveri, C. Armanino, S. Lanteri, and M. Forina. 2010. "NIR and UV-Vis Spectroscopy, Artificial Nose and Tongue: Comparison of Four Fingerprinting Techniques for the Characterisation of Italian Red Wines." *Analytica Chimica Acta* 668(2):143–48. Elsevier B.V. doi:10.1016/j.aca.2010.04.021.

Chen, L., J. Wang, Z. Ye, J. Zhao, X. Xue, Y. Van der Hayden, and Q. Sun. 2012. "Classification of Chinese Honeys according to Their Floral Origin by near Infrared Spectroscopy." *Food Chemistry* 135:338–42. doi:10.1016/j.foodchem.2012.02.156.

Chiesa, L., S. Panseri, S. Bonacci, A. Procopio, A. Zecconi, F. Arioli, F.J. Cuevas, and J.M. Moreno-rojas. 2016. "Authentication of Italian PDO Lard Using NIR Spectroscopy, Volatile Profile and Fatty Acid Composition Combined with Chemometrics." *Food Chemistry* 212:296–304. doi:10.1016/j.foodchem.2016.05.180.

Cocchi, M., M. Corbellini, G. Foca, M. Lucisano, M.A. Pagani, L. Tassi, and A. Ulrici. 2005. "Classification of Bread Wheat Flours in Different Quality Categories by a Wavelet-Based Feature Selection / Classification Algorithm on NIR Spectra." *Analytica Chimica Acta* 544:100–107. doi:10.1016/j.aca.2005.02.075.

Cocchi, M., G. Foca, M. Lucisano, A. Marchetti, M. Ambrogina Pagani, L. Tassi, and A. Ulrici. 2004. "Classification of Cereal Flours by Chemometric Analysis of MIR Spectra." *Journal of Agricultural and Food Chemistry* 52:1062–67. doi:10.1021/jf034441o.

Concha-Herrera, V., M.J. Lerma-García, J.M. Herrero-Martínez, and E.F. Simó-Alfonso. 2009. "Prediction of the Genetic Variety of Extra Virgin Olive Oils Produced at La Comunitat Valenciana, Spain, by Fourier Transform Infrared Spectroscopy." *Journal of Agricultural and Food Chemistry* 57:9985–89. doi:10.1021/jf901730p.

Cozzolino, D. 2012. "Recent Trends on the Use of Infrared Spectroscopy to Trace and Authenticate Natural and Agricultural Food Products." *Applied Spectroscopy Reviews* 47:518–30. doi:10.1080/05704928.2012.667858.

Cozzolino, D. 2014. "An Overview of the Use of Infrared Spectroscopy and Chemometrics in Authenticity and Traceability of Cereals." *Food Research International* 60:262–65. doi:10.1016/j.foodres.2013.08.034.

Cozzolino, D., H.E. Smyth, and M. Gishen. 2003. "Feasibility Study on the Use of Visible and near-Infrared Spectroscopy Together with Chemometrics to Discriminate between Commercial White Wines of Different Varietal Origins." *Journal of Agricultural and Food Chemistry* 51:7703–8.

Daniel Kelly, J., and G. Downey. 2005. "Detection of Sugar Adulterants in Apple Juice Using Fourier Transform Infrared Spectroscopy and Chemometrics." *Journal of Agricultural and Food Chemistry*. doi:10.1021/jf048000w.

Daniel Kelly, J.F., G. Downey, and V. Fouratier. 2004. "Initial Study of Honey Adulteration by Sugar Solutions Using Midinfrared (MIR) Spectroscopy and Chemometrics." *Journal of Agricultural and Food Chemistry* 52:33–39. doi:10.1021/jf034985q.

Daniel Kelly, J.F., C. Petisco, and G. Downey. 2006a. "Application of Fourier Transform Midinfrared Spectroscopy to the Discrimination between Irish Artisanal Honey and Such Honey Adulterated with Various Sugar Syrups." *Journal of Agricultural and Food Chemistry* 54:6166–71. doi:10.1021/jf0613785.

———. 2006b. "Potential of near Infrared Transflectance Spectroscopy to Detect Adulteration of Irish Honey by Beet Invert Syrup and High Fructose Corn Syrup." *Journal of Near Infrared Spectroscopy* 14:139–46.

de la Fuente, M.A., and M. Juarez. 2005. "Authenticity Assessment of Dairy Products." *Critical Reviews in Food Science and Nutrition* 45:563–85. doi:10.1080/10408690490478127.

Dowell, F.E. 2000. "Differentiating Vitreous and Nonvitreous Durum Wheat Kernels by Using Near-Infrared Spectroscopy." *Cereal Chemistry* 77:155–58.

Downey, G., P. McIntyre, and A.N. Davies. 2002. "Detecting and Quantifying Sunflower Oil Adulteration in Extra Virgin Olive Oil From the Eastern Mediterranean by Visible and Near-Infrared Spectroscopy." *Journal of Agricultural and Food Chemistry* 50:5520–25. doi:10.1021/jf0257188.

Foca, G., M. Cocchi, M. Li Vigni, R. Caramanico, M. Corbellini, and A. Ulrici. 2009. "Different Feature Selection Strategies in the Wavelet Domain Applied to NIR-Based Quality Classification Models of Bread Wheat Flours." *Chemometrics and Intelligent Laboratory Systems* 99:91–100. Elsevier B.V. doi:10.1016/j.chemolab.2009.07.013.

Franke, B.M., G. Gremaud, R. Hadorn, and M. Kreuzer. 2005. "Geographic Origin of Meat-Elements of an Analytical Approach to Its Authentication." *European Food Research and Technology* 221:493–503. doi:10.1007/s00217-005-1158-8.

Gallardo-Velázquez, T., G. Osorio-Revilla, M. Zuniga-de Loa, and Y. Rivera-Espinoza. 2009. "Application of FTIR-HATR Spectroscopy and Multivariate Analysis to the Quantification of Adulterants in Mexican Honeys." *Food Research International* 42:313–18. doi:10.1016/j.foodres.2008.11.010.

Galtier, O., N. Dupuy, Y. Le Dréau, D. Ollivier, C. Pinatel, J. Kister, and J. Artaud. 2007. "Geographic Origins and Compositions of Virgin Olive Oils Determinated by Chemometric Analysis of NIR Spectra." *Analytica Chimica Acta* 595:136–44. doi:10.1016/j.aca.2007.02.033.

Gayo, J., S.A. Hale, and S.M. Blanchard. 2006. "Quantitative Analysis and Detection of Adulteration in Crab Meat Using Visible and Near-Infrared Spectroscopy." *Journal of Agricultural and Food Chemistry* 54:1130–36. doi:10.1021/jf051636i.

Gok, S., M. Severcan, E. Goormaghtigh, I. Kandemir, and F. Severcan. 2015. "Differentiation of Anatolian Honey Samples from Different Botanical Origins by ATR-FTIR Spectroscopy Using Multivariate Analysis." *Food Chemistry* 170:234–40. doi:10.1016/j.foodchem.2014.08.040.

Gomez-Caravaca, A.M., R.M. Maggio, and L. Cerretani. 2016. "Chemometric Applications to Assess Quality and Critical Parameters of Virgin and Extra-Virgin Olive Oil. A Review." *Analytica Chimica Acta* 913:1–21. doi:10.1016/j.aca.2016.01.025.

González-Martín, M.I., G.W. Moncada, C. González-Pérez, N.Z. San-Matin, F. López-gonzález, I.L. Ortega, and J. Hernández-Hierro. 2014. "Chilean Flour and Wheat Grain : Tracing Their Origin Using near Infrared Spectroscopy and Chemometrics." *Food Chemistry* 145:802–6. doi:10.1016/j.foodchem.2013.08.103.

Gori, A., C. Cevoli, A. Fabbri, M.F. Caboni, and G. Losi. 2012. "A Rapid Method to Discriminate Season of Production and Feeding Regimen of Butters Based on Infrared Spectroscopy and Artificial Neural Networks." *Journal of Food Engineering* 109:525–30. Elsevier Ltd. doi:10.1016/j.jfoodeng.2011.10.029.

Gouvinhas, I., J.M.M.M. De Almeida, T. Carvalho, N. Machado, and A.I.R.N.A. Barros. 2015. "Discrimination and Characterisation of Extra Virgin Olive Oils from Three Cultivars in Different Maturation Stages Using Fourier Transform Infrared Spectroscopy in Tandem with Chemometrics." *Food Chemistry* 174:226–32. Elsevier Ltd. doi:10.1016/j.foodchem.2014.11.037.

Guelpa, A., F. Marini, A. Du Plessis, R. Slabbert, and M. Manley. 2017. "Verification of Authenticity and Fraud Detection in South African Honey Using NIR Spectroscopy." *Food Control* 73:1388–96. doi:10.1016/j.foodcont.2016.11.002.

He, J., L.E. Rodriguez-Saona, and M.M. Giusti. 2007. "Mid-Infrared Spectroscopy for Juice Authentication-Rapid Differentiation of Commercial Juices." *Journal of Agricultural and Food Chemistry* 55:44443–52.

Hell, J., M. Pruckler, L. Danner, U. Henniges, S. Apprich, T. Rosenau, W. Kneifel, and S. Bohmdorfer. 2016. "A Comparison Between Near-Infrared (NIR) and Mid-Infrared (ATR-FTIR) Spectroscopy for the Multivariate Determination of Compositional Properties in Wheat Bran Samples." *Food Control* 60:365–69. doi:10.1016/j.foodcont.2015.08.003.

Jaiswal, P., S.N. Jha, A. Borah, A. Gautam, M.K. Grewal, and G. Jindal. 2015. "Detection and Quantification of Soymilk in Cow – Buffalo Milk Using Attenuated Total Reflectance Fourier Transform Infrared Spectroscopy." *Food Chemistry* 168:41–47. Elsevier Ltd. doi:10.1016/j.foodchem.2014.07.010.

Jawaid, S., F.N. Talpur, S.T.H. Sherazi, S.M. Nizamani, and A.A. Khaskheli. 2013. "Rapid Detection of Melamine Adulteration in Dairy Milk by SB-ATR – Fourier Transform Infrared Spectroscopy." *Food Chemistry* 141:3066–71. Elsevier Ltd. doi:10.1016/j.foodchem.2013.05.106.

Jha, S.N., and S. Gunasekaran. 2010. "Authentication of Sweetness of Mango Juice Using Fourier Transform Infrared-Attenuated Total Reflection Spectroscopy." *Journal of Food Engineering* 101:337–42. doi:10.1016/j.foodeng.2010.07.019.

Kamal, M., and R. Karoui. 2015. "A Review of the Analytical Methods Coupled with Chemometric Tools for the Determination of the Quality and Identity of Dairy Products." *Trends in Food Science & Technology* 46:27–48. doi:10.1016/j.foodchem.2006.05.042.

Kasemsuran, S., W. Thanapase, and A. Kiatsoonthon. 2007. "Feasibility of Near-Infrared Spectroscopy to Detect and to Quantify Adulterants in Cow Milk." *Analytical Sciences* 23:907–10. doi:10.2116/analsci.23.907.

Kim, S.S., M. Rhyu, J.M. Kim, and S. Lee. 2003. "Authentication of Rice Using Near-Infrared Reflectance Spectroscopy." *Cereal Chemistry Journal* 80:346–49. doi:10.1094/CCHEM.2003.80.3.346.

Koca, N., N.A. Kocaoglu-Vurma, W.J. Harper, and L.E. Rodriguez-saona. 2010. "Application of Temperature-Controlled Attenuated Total Reflectance-Mid-Infrared (ATR-MIR) Spectroscopy for Rapid Estimation of Butter Adulteration." *Food Chemistry* 121:778–82. doi:10.1016/j.foodchem.2009.12.083.

Lema-Garcia, M.J., A. Gori, L. Cerretani, E.F. Simo-Alfonso, and M.F. Caboni. 2010. "Classification of Pecorino Cheeses Produced in Italy according to Their Ripening Time and Manufacturing Technique Using Fourier Transform Infrared Spectroscopy." *Journal of Dairy Science* 93:4490–96.

Leo'n, L., J.D. Kelly, and G. Downey. 2005. "Detection of Apple Juice Adulteration Using near-Infrared Transflectance Spectroscopy." *Applied Spectroscopy* 59:593–99.

Liu, L., D. Cozzolino, W.U. Cynkar, M. Gishen, and C.B. Colby. 2006. "Geographic Classification of Spanish and Australian Tempranillo Red Wines by Visible and Near-Infrared Spectroscopy Combined with Multivariate Analysis Geographic Classification of Spanish and Australian Tempranillo Red Wines by Visible and Near-Infrared." *Journal of Agricultural and Food Chemistry* 54:6754–59. doi:10.1021/jf061528b.

Liu, L., D. Cozzolino, W.U. Cynkar, R.G. Dambergs, L.J. Janik, B.K. O'Neil, C.B. Colby, and M. Gishen. 2008. "Preliminary Study on the Application of Visible-near Infrared Spectroscopy and Chemometrics to Classify Riesling Wines from Different Countries." *Food Chemistry* 106:781–86.

Liu, Y., D. Ma, X. Wang, L. Liu, Y. Fan, and J. Cao. 2015. "Prediction of Chemical Composition and Geographical Origin Traceability of Chinese Export Tilapia Fillets Products by near Infrared Reflectance Spectroscopy." *LWT - Food Science and Technology* 60:1214–18. Elsevier Ltd. doi:10.1016/j.lwt.2014.09.009.

Luykx, D.M.A.M., and S.M. van Ruth. 2008. "An Overview of Analytical Methods for Determining the Geographical Origin of Food Products." *Food Chemistry* 107 (March):897–911. doi:10.1016/j.foodchem.2007.09.038.

Manfredi, M., E. Robotti, F. Quasso, E. Mazzucco, G. Calabrese, and E. Marengo. 2018. "Fast Classification of Hazelnut Cultivars Through Portable Infrared Spectroscopy and Chemometrics." *Spectrochimica Acta Part A: Molecular and Biomolecular Spectroscopy* 189:427–35. doi:10.1016/J.SAA.2017.08.050.

Meza-Márquez, O.G., T. Gallardo-Velázquez, and G. Osorio-Revilla. 2010. "Application of Mid-Infrared Spectroscopy with Multivariate Analysis and Soft Independent Modeling of Class Analogies (SIMCA) for the Detection of Adulterants in Minced Beef." *Meat Science* 86:511–19. doi:10.1016/j.meatsci.2010.05.044.

Mignani, A.G., L. Ciaccheri, H. Ottevaere, H. Thienpont, L. Conte, M. Marega, A. Cichelli, C. Attilio, and A. Cimato. 2011. "Visible and Near-Infrared Absorption Spectroscopy by an Integrating Sphere and Optical Fibers for Quantifying and Discriminating the Adulteration of Extra Virgin Olive Oil From Tuscany." *Analytical and Bioanalytical Chemistry* 399:1315–24.

Miralbes, C. 2008. "Discrimination of European Wheat Varieties Using near Infrared Reflectance Spectroscopy." *Food Chemistry* 106:386–89. doi:10.1016/j.foodchem.2007.05.090.

Moiraghi, M., E. de la Hera, G.T. Pérez, and M. Gómez. 2013. "Effect of Wheat Flour Characteristics on Sponge Cake Quality." *Journal of the Science of Food and Agriculture* 93:542–49. doi:10.1002/jsfa.5821.

Ning, Z., Z. Dequan, L. Shurong, and L. Qingpeng. 2008. "Preliminary Study on Origin Traceability of Mutton by Near Infrared Reflectance Spectroscopy Coupled with SIMCA Method." *Transactions of the Chinese Society of Agricultural Engineering* 12. doi:10.3969/j.issn.1002-6819.2008.12.060.

Osorio, M.T., J.M. Zumalacárregui, R. Alaiz-Rodríguez, R. Guzman-Martínez, S.B. Engelsen, and J. Mateo. 2009. "Differentiation of Perirenal and Omental Fat Quality of Suckling Lambs according to the Rearing System from Fourier Transforms Mid-Infrared Spectra Using Partial Least Squares and Artificial Neural Networks Analysis." *Meat Science* 83:140–47. doi:10.1016/j.meatsci.2009.04.013.

Ozturk, B., A. Yalcin, and D. Ozdemir. 2010. "Determination of Olive Oil Adulteration with Vegetable Oils by Near Infrared Spectroscopy Coupled with Multivariate Calibration." *Journal of Near Infrared Spectroscopy* 18:191–201.

Pillonel, L., W. Luginbuhl, D. Picque, E. Schaller, R. Tabacchi, and J.O. Bosset. 2003. "Analytical Methods for the Determination of the Geographic Origin of Emmental Cheese: Mid- and near-Infrared Spectroscopy." *European Food Research and Technology* 216:174–78. doi:10.1007/s00217-002-0628-5.

Pita-Calvo, C., M.E. Guerra-Rodríguez, and M. Vázquez. 2017. "Analytical Methods Used in the Quality Control of Honey." *Journal of Agricultural and Food Chemistry* 65:690–703. doi:10.1021/acs.jafc.6b04776.

Reid, L.M., A. Woodcock, C.P. O'Donnell, J.D. Kelly, and G. Downey. 2005. "Differentiation of Apple Juice Samples on the Basis of Heat Treatment and Variety Using Chemometric Analysis of MIR and NIR Data." *Food Research International* 38:1109–15.

Rios-Corripio, M.A., M. Roja-Lopez, and M. Delgado-Macuil. 2012. "Analysis of Adulteration in Honey with Standard Sugar Solutions and Syrups Using Attenuated Total Reflectance-Fourier Transform Infrared Spectroscopy and Multivariate Methods." *CyTA - Journal of Food* 10:119–22. doi:10.1080/19476337.2011.596576.

Riovanto, R., W.U. Cynkar, P. Berzaghi, and D. Cozzolino. 2011. "Discrimination between Shiraz Wines from Different Australian Regions: The Role of Spectroscopy and Chemometrics." *Journal of Agricultural and Food Chemistry* 59:10356–60. doi:1021/jf202578f.

Rodriguez-Saona, L.E., and M.E. Allendorf. 2011. "Use of FTIR for Rapid Authentication and Detection of Adulteration of Food." *Annual Review of Food Science and Technology* 2:467–83. doi:10.1146/annurev-food-022510-133750.

Rodriguez-Saona, L.E., M.M. Giusti, and M. Shotts. 2016. "Advances in Infrared Spectroscopy for Food Authenticity Testing." In *Advances in Food Authenticity Testing*, edited by G. Downey, 71–116. Amsterdam: Elsevier.

Rohman, A., and Y.B. Che Man. 2010. "Fourier Transform Infrared (FTIR) Spectroscopy for Analysis of Extra Virgin Olive Oil Adulterated with Palm Oil." *Food Research International* 43:886–92. Elsevier Ltd. doi:10.1016/j.foodres.2009.12.006.

Rohman, A., Sismindari, Y. Erwanto, and Y.B. Che Man. 2011. "Analysis of Pork Adulteration in Beef Meatball Using Fourier Transform Infrared (FTIR) Spectroscopy." *Meat Science* 88:91–95. The American Meat Science Association. doi:10.1016/j.meatsci.2010.12.007.

Rohman, A., Y.B. Che Man, A. Ismail, and P. Hashim. 2017. "FTIR Spectroscopy Coupled With Chemometrics of Multivariate Calibration and Discriminant Analysis for Authentication of Extra Virgin Olive Oil." *International Journal of Food Properties* 2912. doi:10.1080/10942912.2017.1336718.

Ruoff, K., W. Luginbuhl, S. Bogdanov, J.O. Bosset, B. Estermann, T. Ziolko, and R. Amado. 2006. "Authentication of the Botanical Origin of Honey by Near-Infrared Spectroscopy." *Journal of Agricultural and Food Chemistry* 54:6867–72. doi:10.1021/jf060770f.

Ruoff, K., W. Luginbuhl, R. Kunzli, M.T. Iglesias, S. Bogdanov, J.O. Bosset, K. Von Der Ohe, and R. Amado. 2006. "Authentication of the Botanical and Geographical Origin of Honey by Mid-Infrared Spectroscopy." *Journal of Agricultural and Food Chemistry* 54:6873–80. doi:10.1021/jf060838r.

Saucedo-Hernandez, Y., M.J. Lerma-García, J.M. Herrero-Martínez, G. Ramis-Ramos, E. Jorge-Rodríguez, and E.F. Simó-Alfonso. 2011. "Classification of Pumpkin Seed Oils According to Their Species and Genetic Variety by Attenuated Total Reflection Fourier-Transform Infrared Spectroscopy." *Journal of Agricultural and Food Chemistry* 59:4125–29. doi:10.1021/jf104278g.

Schmutzler, M., A. Beganovic, G. Böhler, and C.W. Huck. 2015. "Methods for Detection of Pork Adulteration in Veal Product Based on FT-NIR Spectroscopy for Laboratory, Industrial and On-Site Analysis." *Food Control* 57:258–67. doi:10.1016/j.foodcont.2015.04.019.

Shilton, N., G. Downey, and P.B. McNulty. 1998. "Detection of Orange Juice Adulteration by near-Infrared Spectroscopy." *Seminars in Food Analysis* 3:155–61.

Siddiqui, A.J., S.G. Musharraf, M.I. Choudhary, and A. Rahman. 2017. "Application of Analytical Methods in Authentication and Adulteration of Honey." *Food Chemistry* 217:687–98. doi:10.1016/j.foodchem.2016.09.001.

Sinelli, N., M. Casale, V. Di Egidio, P. Oliveri, D. Bassi, D. Tura, and E. Casiraghi. 2010. "Varietal Discrimination of Extra Virgin Olive Oils by near and Mid Infrared Spectroscopy." *Food Research International* 43:2126–31. Elsevier Ltd. doi:10.1016/j.foodres.2010.07.019.

Sinelli, N., L. Cerretani, V. Di Egidio, A. Bendini, and E. Casiraghi. 2010. "Application of Near (NIR) Infrared and Mid (MIR) Infrared Spectroscopy as a Rapid Tool to Classify Extra Virgin Olive Oil on the Basis of Fruity Attribute Intensity." *Food Research International* 43:369–75. Elsevier Ltd. doi:10.1016/j.foodres.2009.10.008.

Sivakesava, S., and J. Irudayaraj. 2001. "Rapid Spectroscopic Technique for Determining Honey Adulteration with Corn Syrup." *Journal of Food Science* 66:787–92.

———. 2002a. "Classification of Simple and Complex Sugar Adulterants in Honey by Mid-Infrared Spectroscopy." *International Journal of Food Science and Technology* 37:351–60.

———. 2002b. "Rapid Determination of Tetracycline in Milk by FT-MIR and FT-NIR Spectroscopy Rapid Determination of Tetracycline in Milk by FT-MIR." *Journal of Dairy Science* 302 (August). doi:10.3168/jds.S0022-0302(02)74099-X.

Snyder, A.B., C.F. Sweeney, L.E. Rodriguez-Saona, and M.M. Giusti. 2015. "Rapid Authentication of Concord Juice Concentration in a Grape Juice Blend Using Fourier-Transform Infrared Spectroscopy and Chemometric Analysis." *Food Chemistry* 147:295–301. Elsevier Ltd. doi:10.1016/j.foodchem.2013.09.140.

Sun, S., B. Guo, Y. Wei, and M. Fan. 2012. "Classification of Geographical Origins and Prediction of ẟ13C and ẟ15N Values of Lamb Meat by Near Infrared Reflectance Spectroscopy." *Food Chemistry* 135:508–14. doi:10.1016/j.foodchem.2012.05.004.

Tahir, H.E., Z. Xiaobo, S. Tinting, S. Jiyong, and A.A. Mariod. 2016. "Near-Infrared (NIR) Spectroscopy for Rapid Measurement of Antioxidant Properties and Discrimination of Sudanese Honeys from Different Botanical Origin." *Food Analytical Methods* 9:2631–41. doi:10.1007/s12161-016-0453-2.

Tapp, H.S., M. Defernez, and E.K. Kemsley. 2003. "FTIR Spectroscopy and Multivariate Analysis Can Distinguish the Geographic Origin of Extra Virgin Olive Oils." *Journal of Agricultural and Food Chemistry* 51(2):6110–15. doi:10.1021/jf030232s.

Tarantilis, P.A., V.E. Troianou, C.S. Pappas, Y.S. Kotseridis, and M.G. Polissiou. 2008. "Differentiation of Greek Red Wines on the Basis of Grape Variety Using Attenuated Total Reflectance Fourier Transform Infrared Spectroscopy." *Food Chemistry* 111:192–96. doi:10.1016/j.foodchem.2008.03.020.

Tewari, J.C., and J. Irudayaraj. 2005. "Floral Classification of Honey Using Mid-Infrared Spectroscopy and Surface Acoustic Wave Based Z-Nose Sensor." *Journal of Agricultural and Food Chemistry* 53:6955–66. doi:10.1021/jf050139z.

Twomey, M., G. Downey, and P.B. McNulty. 1995. "The Potential of NIR Spectroscopy for the Detection of the Adulteration of Orange Juice." *Journal of the Science of Food and Agriculture* 67:77–84.

Valenti, B., B. Martin, D. Andueza, C. Leroux, C. Labonne, F. Lahalle, H. Larroque, et al. 2013. "Infrared Spectroscopic Methods for the Discrimination of Cow's Milk according to the Feeding System, Cow Breed and Altitude of the Dairy Farm." *International Dairy Journal* 32:26–32. Elsevier Ltd. doi:10.1016/j.idairyj.2013.02.014.

Vardin, H., A. Tay, B. Ozen, and L. Mauer. 2008. "Authentication of Pomegranate Juice Concentrate Using FTIR Spectroscopy and Chemometrics." *Food Chemistry* 108:742–48. doi:10.1016/j.foodchem.2007.11.027.

Vieira, S.M., L.M. de Souza, A.S. Franca, L.S. de Oliveira, and W.B. Neto. 2015. "FTMIR-PLS as a Promising Method for Rapid Detection of Adulteration by Waste Whey in Raw Milk." *Dairy Science and Technology*. doi:10.1007/s13594-015-0247-0.

Wang, P., J. Sun, T. Zhang, and W. Liu. 2016. "Vibrational Spectroscopic Approaches for the Quality Evaluation and Authentication of Virgin Olive Oil." *Applied Spectroscopy Reviews* 51(1):763–90. Taylor & Francis. doi:10.1080/05704928.2016.1176034.

Woodcock, T., G. Downey, and C.P.O. Donnell. 2009. "Near Infrared Spectral Fingerprinting for Confirmation of Claimed PDO Provenance of Honey." *Food Chemistry* 114:742–46. doi:10.1016/j.foodchem.2008.10.034.

Wu, L., B. Du, Y. Vander Heyden, L. Chen, L. Zhao, M. Wang, and X. Xue. 2017. "Recent Advancements in Detecting Sugar-Based Adulterants in Honey – A Challenge." *TrAC Trends in Analytical Chemistry* 86 (January):25–38. doi:10.1016/j.trac.2016.10.013.

Xie, L., X. Ye, D. Liu, and Y. Ying. 2008. "Application of Principal Component-Radial Basis Function Neural Networks (PC-RBFNN) for the Detection of Water-Adulterated Bayberry Juice By Near-Infrared Spectroscopy." *Journal of Zhejiang University Science B* 9:982–89.

Zhang, J., X. Zhang, L. Dediu, and C. Victor. 2011. "Review of the Current Application of Fingerprinting Allowing Detection of Food Adulteration and Fraud in China." *Food Control* 22:1126–35. Elsevier Ltd. doi:10.1016/j.foodcont.2011.01.019.

Zhang, L., X. Zhang, L. Ni, Z. Xue, X. Gu, and S. Huang. 2014. "Rapid Identification of Adulterated Cow Milk by Non-Linear Pattern Recognition Methods Based on near Infrared Spectroscopy." *Food Chemistry* 145:342–48. doi:10.1016/j.foodchem.2013.08.064.

Zhang, Y., J. Chen, Y. Lei, Q. Zhou, S. Sun, and I. Noda. 2010. "Discrimination of Different Red Wine by Fourier-Transform Infrared and Two-Dimensional Infrared Correlation Spectroscopy." *Journal of Molecular Structure* 974:144–50. Elsevier B.V. doi:10.1016/j.molstruc.2010.03.021.

Zhao, H., B. Guo, Y. Wei, and B. Zhang. 2013. "Near Infrared Reflectance Spectroscopy for Determination of the Geographical Origin of Wheat." *Food Chemistry* 138:1902–7. doi:10.1016/j.foodchem.2012.11.037.

Zhao, M., G. Downey, and C.P. O'Donnell. 2014. "Detection of Adulteration in Fresh and Frozen Beefburger Products by Beef Offal Using Mid-Infrared ATR Spectroscopy and Multivariate Data Analysis." *Meat Science* 96:1003–11. Elsevier Ltd. doi:10.1016/j.meatsci.2013.10.015.

Zhu, X., S. Li, Y. Shan, Z. Zhang, G. Li, D. Su, and F. Liu. 2010. "Detection of Adulterants such as Sweeteners Materials in Honey Using Near-Infrared Spectroscopy and Chemometrics." *Journal of Food Engineering* 101:92–97. doi:10.1016/j.jfoodeng.2010.06.014.

Ziegler, J.U., M. Leitenberger, C.F.H. Longin, T. Wurschum, R. Carle, and R.M. Schweiggert. 2016. "Near Infrared Reflectance Spectroscopy For the Rapid Discrimination of Kernels and Flours of Different Wheat Species." *Journal of Food Composition and Analysis* 51:30–36. doi:10.1016/j.jfca.2016.06.005.

CHAPTER 3

NMR Spectroscopy

Apostolos Spyros
University of Crete

CONTENTS

3.1 INTRODUCTION

NMR spectroscopy is currently a powerful spectroscopic tool for the qualitative and quantitative analysis and characterization of molecules of every type and size. Although the analytical potential of NMR spectroscopy was apparent from the beginning, it was only in the early 1980s that this potential was realized. The introduction of stable high field superconducting magnets facilitated the use of NMR in the structural characterization of chemical compounds, while developments in computers and electronics made the acquisition and interpretation of NMR spectra a less demanding task and made NMR accessible to chemists, biochemists, and, eventually, food scientists (Spyros and Dais, 2012). Most food products are actually living organisms, or parts thereof, and thus from a chemical point of view, they are simply complex mixtures of organic and inorganic chemical compounds. Although NMR spectroscopy of inorganic compounds is possible, it is the analysis of organic compounds for which NMR spectroscopy has proven to be an invaluable tool. One of the greatest advantages of NMR is its inherent quantitative character (van Duynhoven et al., 2013), that is, the area of suitable NMR signals is directly related to the abundance of chemical compounds in a food matrix, so by the early 1980s it was realized that food scientists could use NMR for the compositional analysis of food. Using NMR spectroscopy (limited only in terms of sensitivity and sample preparation), all organic compounds present in a food sample can be quantified simultaneously, and the NMR spectrum of a compound can be considered its true molecular fingerprint.

TABLE 3.1 NMR Methodologies and Their Main Applications for Food Authentication and Traceability

Technique	Examples of Food Type	Analytical Aspect	Advantages
Liquid state NMR	Liquid food (wine, oils, infusions) or extracts of solid foods	Chemical composition, nutritional information, metabolite fingerprinting	High sensitivity, 2D NMR, quantitative
HR-MAS solid state NMR	Semisolid food (cheese, meat, vegetables)	Chemical composition, metabolite fingerprinting	High sensitivity, 2D NMR, no sample pretreatment
CP-MAS solid state NMR	Solid food with a crystalline or highly immobile phase (seeds, biopolymers)	Structural characterization, chemical composition	No sample pretreatment

The potential of NMR spectroscopy to address quality control procedures, such as the identification of illegal or dangerous chemical compounds (e.g., additives, degradation products, toxic substances, etc.) in food, led to the development of suitable NMR methodologies for food analysis. Food degradation, processing, conservation, and cooking are procedures that alter, in small or big ways, the chemical composition of foods, and thus it was the first area to be exploited by analytical NMR spectroscopy. The success of NMR spectroscopy in tackling these types of food-related themes led to an interest in developing NMR methodologies for addressing more complex food analysis problems, such as authentication and traceability. Food authentication is an even more complex analytical problem, as it is usually associated with identifying minor differences in the chemical profiles of food, associated either with production in a specific geographical region or with a specific production/manufacturing procedure. In this endeavor, powerful analytical methodologies that can provide the full metabolomic fingerprints of food, combined with the ease of application and high analysis speed are needed, since a large number of samples need to be analyzed to obtain the statistical significance for authentication. NMR spectroscopy can provide both the metabolomic profiles and the spectroscopic fingerprints of food and thus has been employed very efficiently in food authentication and traceability studies.

One of the most important attributes of NMR spectroscopy is its versatility. Some NMR spectrometers are available that can be used to analyze samples in the liquid, solid, or even in the gas state, thus matching the incredible variety of existing food matrices and complex foodstuff. Table 3.1 presents an overview of the different magnetic resonance methodologies that can be applied in food analysis, indicating the most suitable food types, general applications, and merits of each technique. The applicability of each NMR technique in food analysis will of course be discussed frequently throughout this chapter in more detail.

3.2 BASIC THEORY

NMR spectroscopy is based on the quantum phenomenon of nuclear spin. Some atomic nuclei, depending on the number of protons and neutrons in the nucleus, possess non-zero values of spin, for example, ^{1}H, ^{13}C, ^{31}P, ^{15}N all have a nuclear spin of ½. Because of spin, these nuclei also have an associated magnetic moment μ and, simply, they can be thought of as small magnets. It should be stressed that this is the classical, as opposed to the more complex but powerful quantum-mechanical, description of the NMR phenomenon (Spyros and Dais, 2012). However, for our purpose, which is limited to a comprehensive view of spin ½ NMR spectroscopy that is suitable for explaining NMR food profiling and fingerprinting to the novice user, this mode of description will suffice.

Away from strong magnetic fields, the nuclear spin (let's assume it is a proton with spin ½), which for our purposes we can imagine as a small magnetic moment vector, is "silent", and it can assume any direction (Figure 3.1A). When an organic sample is positioned under a strong magnetic field B_0 (i.e., inside an NMR magnet), spin ½ nuclei in the sample populate two different energy levels, α and β, as depicted in Figure 3.1B, with level α having a slightly larger spin population. A radio frequency pulse (of frequency ν) creates a momentary secondary field B_1, which pushes the spins from energy level α to energy level β, equating the populations of the two levels (Figure 3.1C). Now we have a spin system out of equilibrium. The excess of spins in β will slowly return to α (Figure 3.1D), via relaxation processes, promptly called T_1 and T_2 relaxation. Relaxation is accompanied by emission of energy with a frequency ν, which is promptly picked up by a suitable receiver, and this constitutes the NMR signal in the time domain (FID-free induction decay). The FID is Fourier transformed into the frequency domain to provide the ^{1}H NMR spectrum, which in our simple example of Figure 3.1 would consist of a single peak at frequency ν, representing a single type of spin in our imaginary sample. However, NMR spectra containing a multitude of signals at different frequencies are obtained from studying real samples that usually contain complex organic compounds possessing many different types of nuclei, and thus spins. The power of NMR spectroscopy stems from the fact that the electronic and chemical bonding environment of nuclei in chemical compounds affects and modifies (a) the resonance frequency ν of their spin, thus producing NMR spectra having many signals of slightly different frequencies (representing slightly different electronic environments) and (b) the multiplicity of the signals produced (singlet, doublet, triplet, etc.) by a phenomenon termed spin-spin scalar coupling (J coupling) that

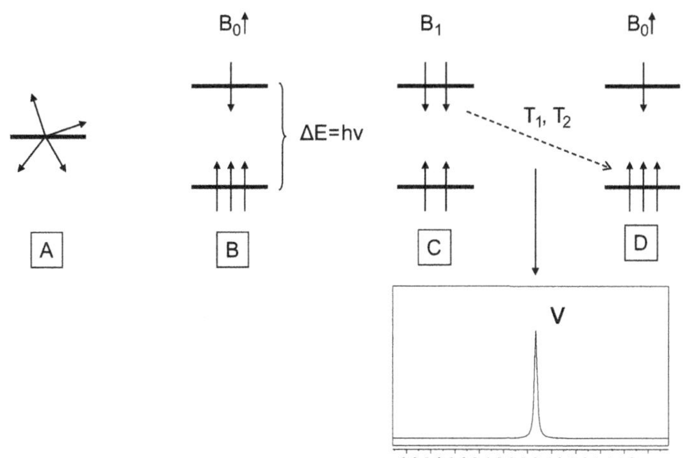

FIGURE 3.1 A simplistic view of the NMR phenomenon for a spin ½ nucleus, such as proton. (A) Sample is outside the magnet. (B) Sample is inside the magnet. (C) Sample is irradiated by an RF pulse, and (D) Spin system returns to equilibrium. Between (C) and (D) the NMR signal is acquired as the FID, which is subsequently Fourier transformed to provide the NMR spectrum.

propagates through chemical bonding. If a spin ½ nucleus has N direct magnetically inequivalent spin neighbors, its multiplicity will be $N+1$. A second coupling interaction exists that depends on the proximity of the spins in space and it is called dipolar coupling. Dipolar coupling is a through space interaction that is mainly relevant in solid state NMR spectroscopy, because in solution it is more or less averaged out by fast diffusional motion. It is however responsible for the nuclear Overhauser effect, which is exploited in solution state NMR for obtaining information about the distance between nuclei in NOE experiments.

The effect of the electronic environment of nuclear spins through spin-spin scalar J coupling on the appearance of NMR spectra can be easily appreciated in the ^1H NMR spectrum of an alcoholic beverage, depicted in Figure 3.2. The main chemical components of the beverage are water and ethanol. We expect one NMR signal for each type of protons in the sample, thus we expect three signals for ethanol (methyl, methylene, and hydroxyl protons) and one signal from water. The ^1H NMR spectrum of the alcoholic beverage shows only three peaks (because water protons and the hydroxyl protons of ethanol have the same frequency) and their assignment is presented in Figure 3.2. The methyl protons (a) of ethanol have two equivalent methylene protons (b) as neighbors ($N=2$) and, thus, they produce a triplet signal ($N+1=3$). Methylene protons (b) have three neighbors ($N=3$), leading to a quadruplet signal. The signal due to the hydroxyl proton (c) is a singlet because of chemical exchange effect with water protons. Note that the triplet signal has a peak ratio of 1:2:1, while for the quadruplet it is 1:2:2:1; the multiplicities are obtained from a Pascal triangle.

The concepts of chemical shift and multiplicity in spectrum interpretation are also valid in solid state NMR spectroscopy, which is used to obtain NMR spectra of solid food. To summarize, the parameters mostly exploited in any form of NMR spectroscopy for obtaining information on the samples studied are the chemical shift δ (measured in ppm), the scalar coupling J (Hz), the dipolar coupling (Hz), and the T_1 and T_2 relaxation times (sec).

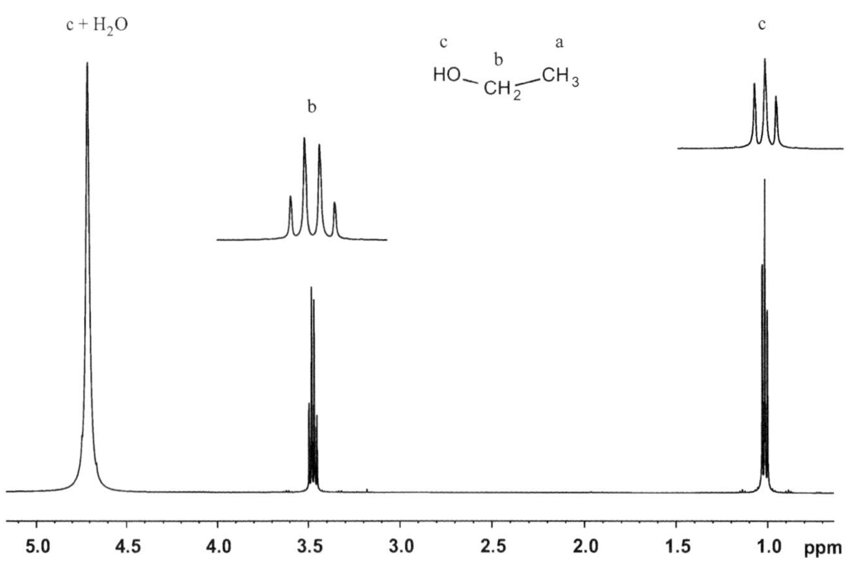

FIGURE 3.2 ^1H NMR spectrum of a typical alcoholic beverage, indicating the assignment and multiplicity of the proton signals of ethanol and water.

TABLE 3.2 Useful 2D NMR Spectroscopy Techniques in Food Analysis

Name	Acronym	Information Obtained
Correlation Spectroscopy	COSY	Proton neighbor connectivity through *J* couplings
J-Spectroscopy	JRES	Determination of *J* couplings
Total Correlation Spectroscopy	TOCSY	Full proton connectivity within a scalar-coupled system.
Nuclear Overhauser Enhancement Spectroscopy	NOESY	Interproton distance information through dipolar coupling
Rotating frame Overhauser Enhancement Spectroscopy	ROESY	Interproton distance information through dipolar coupling
Diffusion Ordered Spectroscopy	DOSY	Spectral editing through diffusion coefficient differences
Heteronuclear Single Quantum Coherence	HSQC	Heteronuclear shift correlation through one-bond X–H coupling.
Heteronuclear Multiple Bond Coherence	HMBC	Heteronuclear shift correlation through two- and three-bond X–H coupling
Multiplicity-edited Heteronuclear Single Quantum Coherence	EHSQC	Heteronuclear shift correlation through one-bond X–H coupling with X nucleus multiplicity editing
Proton-edited Heteronuclear Single Quantum Coherence	HSQC–TOCSY, HSQC–COSY	Heteronuclear shift correlation through one-bond X–H coupling and proton connectivity

A detailed theoretical description of 2D NMR spectroscopy is beyond the scope of this chapter, so only some basic points necessary to comprehend the application of 2D NMR in food analysis will be provided (Spyros and Dais, 2012). The need for a second dimension stems from the fact that in 1D ^{1}H NMR spectra significant signal overlap is observed, especially when complex mixtures, such as food samples, are under study. 2D NMR enhances the selectivity by spreading the available information in a second dimension, that usually is also a chemical shift, resulting in homonuclear ^{1}H–^{1}H 2D experiments (COSY, TOCSY, NOESY) or heteronuclear ^{1}H–^{1}X 2D experiments (HSQC, HMBC), where X is usually ^{13}C, ^{31}P, or ^{15}N. Sometimes the second dimension can involve the *J*-coupling (JRES) or even the diffusion constant of molecules in solution (DOSY). The most useful 2D NMR experiments in food analysis are presented in Table 3.2. For fingerprinting applications, the 2D NMR experiments that are most suitable for obtaining spectroscopic fingerprints of food would be the 2D COSY and JRES experiments, since these are the two least-demanding 2D NMR experiments in terms of experiment duration, but they offer rich molecular information about the proton distribution in a given sample.

3.3 MODES OF APPLICATION

3.3.1 High Resolution Liquid State NMR Spectroscopy

High resolution liquid state NMR spectroscopy is the most commonly used NMR technique for food authentication studies, and the predominant nuclei studied are ^{1}H and ^{13}C (which can be found in all food), followed by ^{31}P, ^{15}N, and ^{23}Na in food containing such nuclei. The wide availability of high resolution NMR spectrometers in universities and research institutes, along with the facile implementation of liquid state NMR, has no doubt played a crucial role. Of equal importance is the fact that some of the most important food are liquids in their natural states, for example, wine, edible oils, beverages, etc., while others such as fat or instant coffee can be easily dissolved in a suitable organic solvent. In these cases, the food can be analyzed by NMR spectroscopy directly, without any sample pretreatment, and the full arsenal of high resolution 1D and 2D NMR experiments can be used for compositional analysis and metabolite identification, which are usually the first steps in food authentication studies (Mannina, Sobolev, and Viel, 2012).

Liquid state high resolution NMR is also widely applied for the metabolite analysis of food extracts. There are several cases where this practice is unavoidable since many food items are complex solid matrices containing different categories of organic compounds (lipids, carbohydrates, proteins, organic acids, etc.) that cannot be dissolved in a common solvent or that contain high molecular weight polymers (polysaccharides, lignin) that are totally insoluble. In this case, the food analyst will have to resort to a sample preparation method—usually liquid-liquid extraction or solid phase extraction—that will allow the isolation of the specific category (or categories) of compounds he or she is interested in and proceed with liquid state NMR analysis. For food authentication studies, the level of molecular details in metabolite profiling needed to observe small differences because of some factors, like geographical origins or pedoclimatic effects, is extremely high. Thus, the application of sample pretreatment and pre-concentration methods is very common in analyzing compounds that are usually found in very low concentrations in food, such as micronutrients, vitamins, or polyphenols.

Alternatively, another popular approach for using NMR spectroscopy in food authentication involves the use of the whole NMR spectrum as the molecular fingerprint of a food, without dealing with the tedious task of assigning spectrum signals to individual compounds, thus transitioning from NMR metabolite profiling to direct NMR spectroscopic fingerprinting. The food NMR fingerprints can then be analyzed with the help of multivariate statistical analysis methods to obtain information regarding differences based on geographical origins, starting material, pedoclimatic conditions, processing, preparation, etc. This approach is advantageous because we can obtain molecular information from those important micro-components that are difficult to identify due to their low concentrations in food, since they contribute small, but observable, modifications to the spectroscopic fingerprints of the food analyzed.

3.3.2 Solid State NMR Spectroscopy

In recent years, solid state NMR has become an attractive technique for studying solid food. The definition of "solid" for NMR is though a bit different compared to the standard meaning of the word and this is causing some confusion, which is not necessarily limited to food scientists only. The term "solid", in NMR jargon, is used to characterize a material whose constituents possess very limited motional mobility at the molecular level. A good example of an NMR solid material would be a crystalline synthetic polymer, or to use a more familiar case, the kernels of certain fruits, which consist mainly of high molecular weight biopolymers, such as cellulose and lignin. Because of the lack of molecular motion, dipolar interactions between the protons (and other nuclei) in a solid are not averaged out (as is conveniently the case in liquids via diffusion and fast motion in solution) leading to conventional NMR spectra containing wide lines, with limited compositional and even spectroscopic information (if any). To circumvent this, solid state NMR spectroscopy requires the application of additional line-narrowing methodologies, namely dipolar decoupling (DD), cross-polarization (CP), and magic angle spinning (MAS), in order to produce the NMR spectra of crystalline solids that provide chemical and compositional information comparable to that of liquids (Bertocchi and Paci, 2008). It should be mentioned that even with the use of the above methodologies ^{1}H NMR spectra of hard crystalline materials are broad and not very useful, so the solid state NMR analysis of food is mostly performed with heteronuclear 1D CP-MAS NMR spectroscopy of the ^{13}C and ^{31}P nuclei. The development of sophisticated 2D NMR methodologies for the analysis of crystalline solids is a very active field of research that has not found any applications so far in food analysis.

Most solid food, for example, meat, cheese, fruits, etc. would be characterized as semisolid in NMR terms. The high water content in such food imparts increased mobility to their low molecular weight chemical constituents, which leads to a partial averaging of dipolar interactions. The residual dipolar coupling can be averaged out by simply rotating the sample at the magic angle at relatively low speeds—of a few KHz—without the need for CP or DD. Thus, semisolid food can be studied by employing a "hybrid" NMR methodology termed as high resolution MAS (HR-MAS) NMR spectroscopy that elegantly combines the elements of liquid state NMR (a deuterated solvent is used for locking) and solid state NMR (MAS) (Valentini et al., 2011). For most semisolid food, the NMR spectra obtained using HR-MAS of the food sample "as is" are very similar, and importantly, have similar resolution to the NMR spectra of solvent extracts from the same food. Furthermore, the high field stability of HR-MAS NMR due to deuterium locking allows the acquisition of all the standard 2D NMR experiments available for liquid state NMR, thus enabling the use of HR-MAS for structural, compositional, and fingerprinting analysis of semisolid food.

3.4 INSTRUMENTATION

In this section, some basic information regarding the different components of instrumentation that can be part of a modern NMR spectrometer will be provided. The type and variety of hardware defines the range of NMR experiments that the spectrometer is capable of accomplishing, and the type of food samples it can analyze. This information is of prime importance, especially in case of planning, for the acquisition of a new NMR spectrometer or for upgrading an existing one.

The heart of any NMR spectrometer is a magnet of a specific field strength, and it should be noted that although field strength is measured in Tesla (SI unit), it is very common in NMR jargon to use the Larmor frequency of proton nuclei positioned inside the magnet (thus, a 14.09 Tesla magnet will power a 600 MHz NMR spectrometer). The magnets used for high resolution liquid state NMR systems, HR-MAS, and CP-MAS solid state systems are almost universally superconducting magnets, in which the field is created by running a strong electric current through superconducting coils that are kept at a temperature of 4 K. Superconducting magnets need a constant supply of cryogens (liquid nitrogen and helium refills) to maintain the required temperature of 4 K in the coils. The recent worldwide shortage of He, leading to price increases, has made the use of He recovery lines an attractive alternative for many instrument NMR facilities, and novel technologies using actively-refrigerated, cryogen-free magnets are being developed, albeit at a much higher cost. Low field NMR spectrometers are based on permanent electromagnets to create a stable (but low intensity) field in the range of 20–80 MHz. Such systems offer spectra of lower resolution and sensitivity, but they have the advantage of affordable prices and very low operational costs, so they are becoming useful for food fingerprinting applications. An important additional component of superconducting magnets is the shim stack, which is a system that optimizes the homogeneity of the magnetic field created by the superconducting coil (B_0) and adjusts it after sample insertion (which obviously introduces small field inhomogeneities). Field optimization is carried out automatically in almost all modern NMR spectrometers.

What really defines the range of NMR experiments that one can perform with a superconducting magnet NMR spectrometer is the probe, which is a removable component that is housed inside the magnet and contains important electronic equipment and hosts the sample during measurements. The probe carries the transmitter and receiver coils that excite the magnetic nuclear spins of atoms inside the sample and then record the NMR spectrum as the spins relax back to equilibrium. There are several types of NMR probes available from the spectrometer manufacturers that are exchangeable in modern magnets, provided that the electronics console can support them. Liquid state probes and solid state HR-MAS and CP-MAS probes are more or less the most frequently used configurations for modern NMR spectrometers, along with specialty probes, such as microprobes (utilizing capillary NMR tubes of smaller volume) and flow probes (for performing hyphenated LC-NMR experiments (Tode, Maoka, and Sugiura, 2009)). It should be noted that the low field liquid state NMR spectrometers are less flexible in their configuration, with the probes usually already built-in inside the spectrometer.

Other useful NMR accessories include autosamplers for handling large numbers of samples in automated NMR spectral acquisition. Needless to say, modern NMR spectrometers are computer-controlled and incorporate significant elements of automation for both setting up and running experiments.

3.5 PRACTICAL ASPECTS

This section will deal with some practical aspects that may be of help when designing and implementing an NMR experiment. In liquid state high resolution NMR, samples are measured in special NMR glass tubes, usually needing at least 700 µL of sample solution for conventional 5 mm probes. The organic solvent used to completely solubilize the sample has to be deuterated (locking at the deuterium frequency is used for magnetic field stabilization purposes), with the most common organic solvents (D_2O, MeOD, $CDCl_3$, DMSO-d_6, acetone-d_6) being readily available commercially. In case of water soluble samples, a 90% H_2O–10% D_2O solvent is often used for economic reasons. Sometimes, the deuterated solvent contains an internal standard to calibrate chemical shifts and/or as a quantification standard (tetramethylsilane, TMS, is used in organic solvents, like 3-(trimethylsilyl)-2,2′, 3,3″-tetradeuteropropionic acid, and TSP-d_4 is used for D_2O). In samples diluted with water, it may be necessary to keep the pH stable by adding a suitable buffer since chemical shifts are sensitive to small pH differences within the samples or sample batches. Running an NMR spectrum is usually pretty straightforward in conventional modern NMR spectrometers, even for the novice user, as most of the experimental setup and the acquisition tasks are semiautomated; at least the common experiments are pre-setup on the spectrometer.

After spectral acquisition follows NMR spectral analysis, a procedure where NMR signals have to be identified and assigned to specific chemical compounds present in the sample. For complex mixtures such as food, this is rarely an easy task, unless one is dealing with samples that have been well characterized in the literature. When unassigned signals are present in the spectrum, it may be necessary to perform further experiments, for example, homonuclear and heteronuclear 2D NMR experiments, in order to elucidate the molecular source of the unknown signals and assign them to hitherto unknown metabolites. Another route is the use of public NMR databases that are available in the internet, such as the Human Metabolome Database (Wishart et al., 2013) (HMDB, www.hmdb.ca) or the Biological Magnetic Resonance Data Bank (Ulrich et al., 2008) (www.bmrb.wisc.edu/metabolomics/). Unknown signals can be matched with the signals of known compounds from the databases and this will aid in the assignment. As an example, Figure 3.3 presents the ^{1}H NMR spectrum of an alcoholic beverage (liquor) along with the standard NMR spectra of some sugar molecules obtained from the BMRB databank. The comparison with the model metabolite spectra easily allows the identification of glucose, fructose, and sucrose as the main sugars present in the soft drink, while the NMR peaks suitable for the quantification of each sugar in the beverage are indicated with arrows.

When the food sample to be analyzed is solid, one has two choices. The first, as already mentioned, is to develop an effective separation and extraction protocol to isolate the chemical compound fraction of interest and to use liquid state NMR on it. A second, and more appealing, choice in recent times is to use solid state NMR spectroscopy directly. When the sample is semisolid (for example, cheese or meat), most of the times

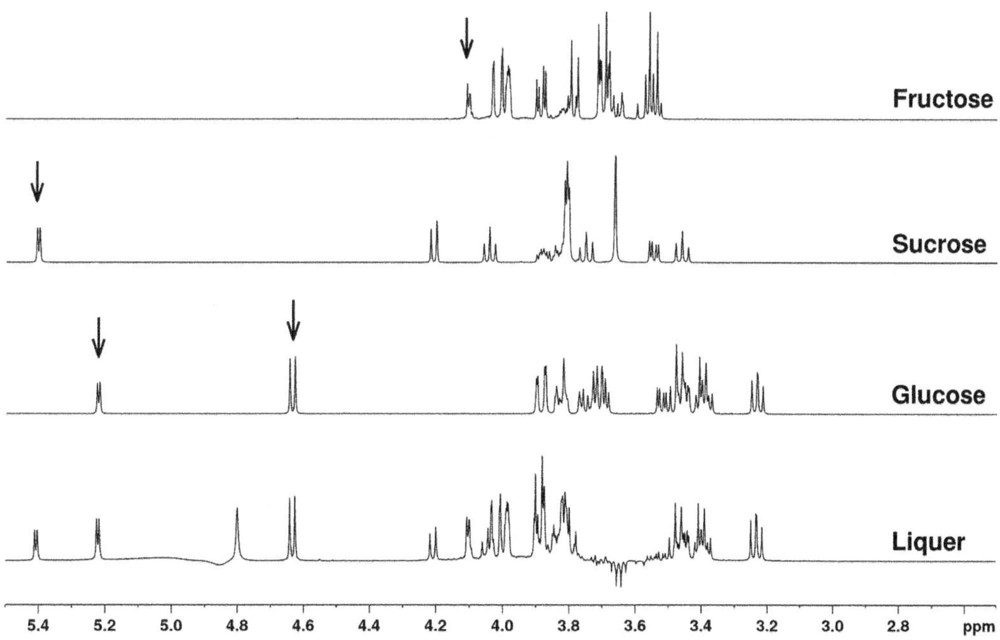

FIGURE 3.3 ^{1}H NMR spectra of an alcoholic beverage (liqueur) and sugars, like sucrose, fructose, and glucose; the latter is downloaded from the Biological Magnetic Resonance Data Bank (Ulrich et al., 2008). Peaks indicated with arrows can be used for quantitative profiling of the sugar molecules in the beverage.

it may be studied by HR-MAS solid state NMR as is and without any pretreatment, other than the addition of a small amount of deuterated solvent (10 μL) in the special sample rotor for locking purposes. The experimental procedure to obtain and analyze the HR-MAS NMR spectra is more or less similar to that in the liquid state NMR. For completely solid samples, like seeds, starch, and lignin, that are insoluble in common solvents one has to resort to CP-MAS solid state NMR spectroscopy (Gidley and Bociek, 1988; Kono et al., 2002). Other than grinding the sample to a fine suspension, usually no other sample treatment is needed. However, solid state NMR profiling usually involves the use of heteronuclear NMR since proton spectra are more or less broad and lack the details needed for metabolite profiling.

3.6 RECENT ADVANCES

The most important advancement concerning NMR spectroscopy in food analysis is, without doubt, the integration of high resolution NMR spectroscopy with chemometrics, which helps in the development of the field of food NMR metabolomics (Mannina, Sobolev, and Viel, 2012). The NMR spectra of complex food contain a very large amount of quantitative chemical composition data for the small- and medium-sized organic molecules (metabolites), thus making the NMR spectra true fingerprints of the food samples at their current states of being. Any factor that affects the food sample (for example, spoilage, adulteration, processing, storage, cooking) will modify its chemical composition and thus will result in a different NMR fingerprint. The same is also true for factors that are related to authentication and traceability and that have affected the food sample in the past (such as type, geographical origin, pedoclimatic conditions, animal or plant specialties, agricultural practices, etc.) and have left a mark in its chemical compositions as can be seen with NMR spectroscopy. Chemometrics is able to harness the rich compositional information provided by NMR spectra by applying multivariate statistical models to analyze the huge collections of NMR spectra, identify the factors that are responsible for their differentiation, and to classify the samples according to these factors. An example of such an analysis is provided in Figure 3.4 where different grades of olive oils are differentiated according to their type with the help of their NMR minor glyceride profiles (Fragaki et al., 2005). The statistical analysis models thus created can be used to authenticate adulterated samples of extra virgin olive oil with olive oils of inferior quality.

The classification depicted in Figure 3.4 was done by quantifying certain metabolites in the NMR spectra and by using these to build the multivariate statistical model, an approach called metabolic profiling. In another approach, which is called metabolic fingerprinting, the whole NMR spectrum can be used for building the statistical model, including the signals of unknown origin that contribute to the variance between samples. Multivariate statistical models for metabolomics are divided in two categories—supervised (PCA: Principal Component Analysis; HCA: Hierarchical Cluster Analysis) or unsupervised (PLS: Partial Least Squares; DA: Discriminant Analysis; NN: Neural Networks)—according to whether the classification system is known to the model before data analysis. Usually unsupervised models are used at first to overview the variance of a dataset, followed by supervised models that offer increased discriminatory power.

Using the metabolomic approach has allowed the increasingly successful application of NMR spectroscopy in dealing with a wide variety of problems in food analysis, including authentication, detection of adulteration, geographical origin, and pedoclimatic factors, such as harvest time, agricultural and farming practices, processing, etc. Several

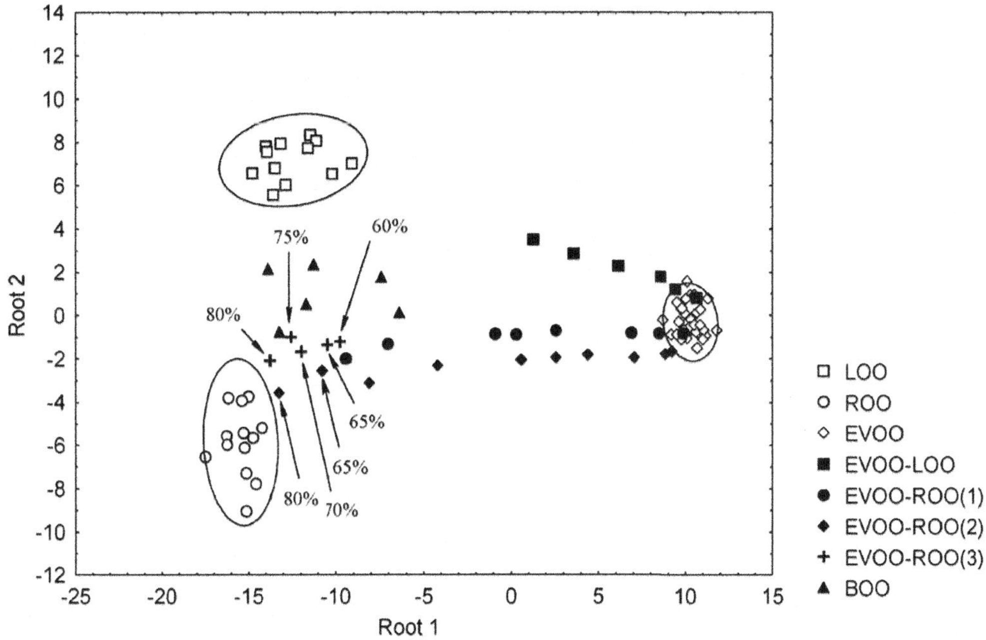

FIGURE 3.4 Discriminant analysis plot for three grades of olive oils—extra virgin olive oil (EVOO), refined olive oil (ROO), and lampante olive oil (LOO) (open symbols) and various adulterated EVOO samples obtained by NMR spectroscopy analysis of their diglyceride profiles. Reprinted with permission from Fragaki et al. (2005). Copyright 2005 American Chemical Society.

detailed presentations in the field of food metabolomics have been published recently and the interested readers can look there for more elaborate details and practical information (Consonni and Cagliani, 2010; Hong, 2011; Mannina, Sobolev, and Viel, 2012; Lamanna, 2013).

Other recent NMR advances that may be benefit the application of the technique in food authentication and analysis come from the hardware's point of view. Modern NMR spectrometers run in a completely automated fashion and also have preprogrammed routines for running advanced 2D NMR experiments with minimal setup, thus facilitating the use of NMR approaches for those who are less familiar with NMR spectroscopy. Furthermore, solid state probes (and especially HR-MAS probes) have become more widely available, thus allowing the direct NMR profiling of semisolid and solid samples, such as fruits, vegetables, meat, and cheese. Finally, the availability of multi-samplers operating in automated mode in the NMR spectrometers is also very beneficial for NMR metabolomics applications, which traditionally depends on a large sample space to derive the high statistical relevance that is needed for building accurate classification models for authentication purposes.

3.7 APPLICATIONS

As already mentioned, application of NMR spectroscopy in its various forms in food analysis has seen an explosion during the last 10–20 years, with more or less routine

TABLE 3.3 Recent Literature Sources for High Resolution NMR Applications in Food Science

Food Type	Reference	Food Type	Reference
General	Spyros and Dais, 2012)	Beverages	(Kidrič, 2008; Valentini et al., 2011)
General	(Alberti, Belton and Gil, 2002)	Alcoholic beverages	(Gil and Duarte, 2006a)
General, instrumentation	(Capozzi and Cremonini, 2008)	Wine	(Hong, 2011)
General	(Mannina, Sobolev, and Viel., 2012)	Beer	(Rodrigues and Gil, 2011)
General	(Lamanna, 2013)	Milk and dairy products	(Belloque and Ramos, 1999)
General	(van Duynhoven et al., 2013)	Fruit juices	(Gil and Duarte, 2006b)
General	(Spyros and Dais, 2009)	Lipids	(Diehl, 2001)
Traditional food products	(Consonni and Cagliani, 2010)	General	(Valentini et al., 2011)
Edible oils	(Hidalgo and Zamora, 2003)	Fruits and vegetables	(Gil and Duarte, 2006c)
Olive oil	(Mannina and Sobolev, 2011)	Olive oil	(Dais and Spyros, 2007)

publications on its analytical applications practically for every type of food. In the following discussion, the focus will be on characteristic applications of NMR spectroscopy in food authentication and traceability that can hopefully serve as representative examples of the field's potential. A list of secondary literature sources that summarize pertinent work in the field can be found in Table 3.3. These references can provide a starting point for food scientists who are interested in evaluating the possibility of using NMR profiling or fingerprinting in a specific analytical problem in food science.

Liquid state high resolution ^1H NMR spectroscopy has found increasing application in the analysis of liquid beverages (Kidrič, 2008), aided by the minimum (if any) sample preparation needed to obtain NMR spectra from liquids, such as wine, coffee (Wei et al., 2010; Wei et al., 2011), beer (Lachenmeier et al., 2005; Almeida et al., 2006), tea (Tarachiwin et al., 2007; Ohno et al., 2011), cocoa (Caligiani et al., 2010), fruit juices (Cuny et al., 2008; Belton et al., 1998), etc. The problem of removing the huge residual signals of water and/or ethanol, which are the main chemical components of such food samples, has been dealt with efficiently using advanced solvent suppression protocols. Figure 3.5 depicts a characteristic ^1H NMR spectrum of wine and the assignment of several low molecular weight components, such as alcohols and organic acids (Godelmann et al., 2013). Quantification of wine metabolites (Košir and Kidrič, 2002) has been used successfully in metabolomic NMR applications dealing with wine authentication problems, such as geographical origin authentication, (Godelmann et al., 2013) cultivar (Anastasiadi et al., 2009), vintage (Ali et al., 2011), wine blending (Imparato et al., 2011), etc.

As already mentioned, another route for the authentication of solid food through NMR fingerprinting is to analyze their solvent extracts. As an example, Figure 3.6 presents the classification of the ^1H NMR profiles of several beef extract samples using

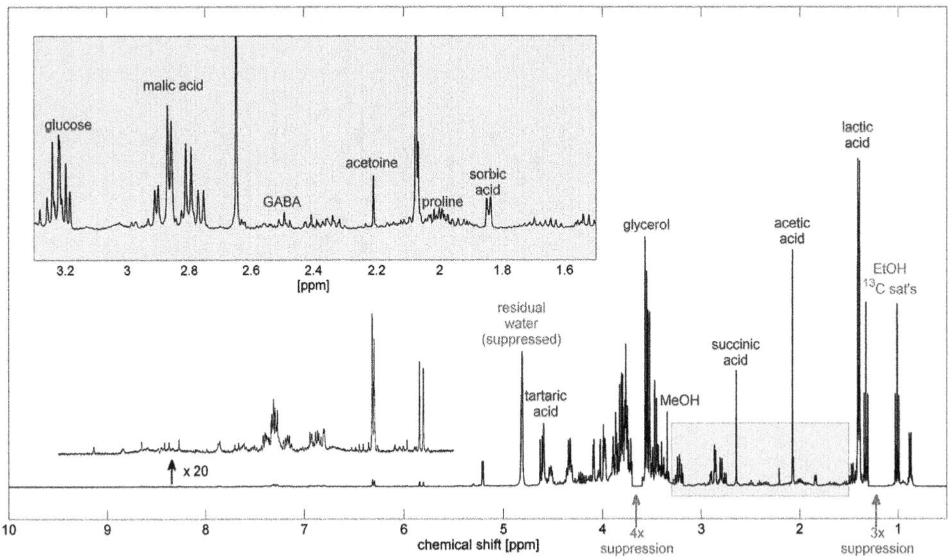

FIGURE 3.5 **(See color insert.)** ^1H NMR spectrum of wine with suppression of the water and ethanol peaks, identifying several low molecular weight metabolites of wine. Reprinted with permission from (Godelmann et al., 2013). Copyright 2013 American Chemical Society.

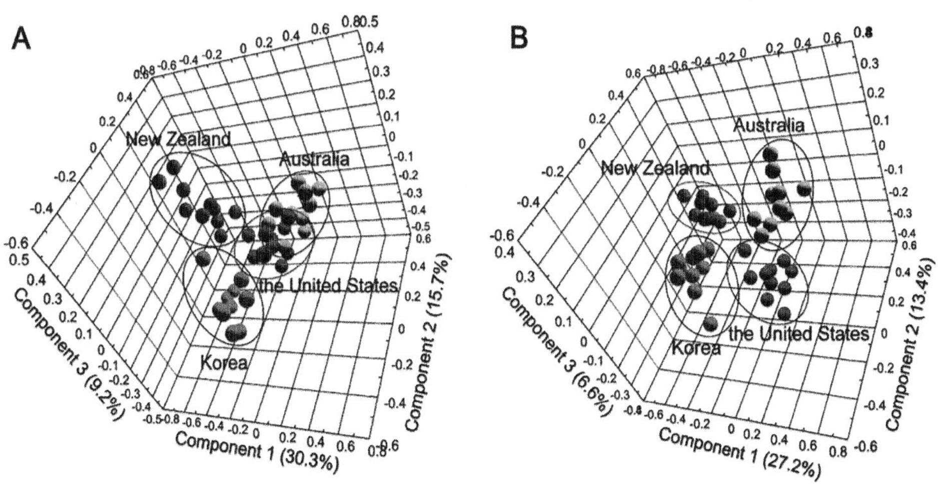

FIGURE 3.6 **(See color insert.)** PCA (A) and OPLS-DA (B) 3D score plots derived from the ^1H NMR spectra of beef sirloin (or chuck) extracts obtained from Australia, South Korea, New Zealand, and the United States. Reprinted with permission from (Jung et al., 2010). Copyright 2010 American Chemical Society.

multivariate statistical analysis tools, leading to the successful authentication of the geographical origins of beef.

HR-MAS NMR spectroscopy has allowed a similar metabolic profiling approach for the authentication of solid food, such as fruits (Vermathen, Marzorati, and Vermathen, 2012; Iglesias et al., 2014), vegetables (Sanchez Perez et al., 2010), meat (Shintu, Caldarelli,

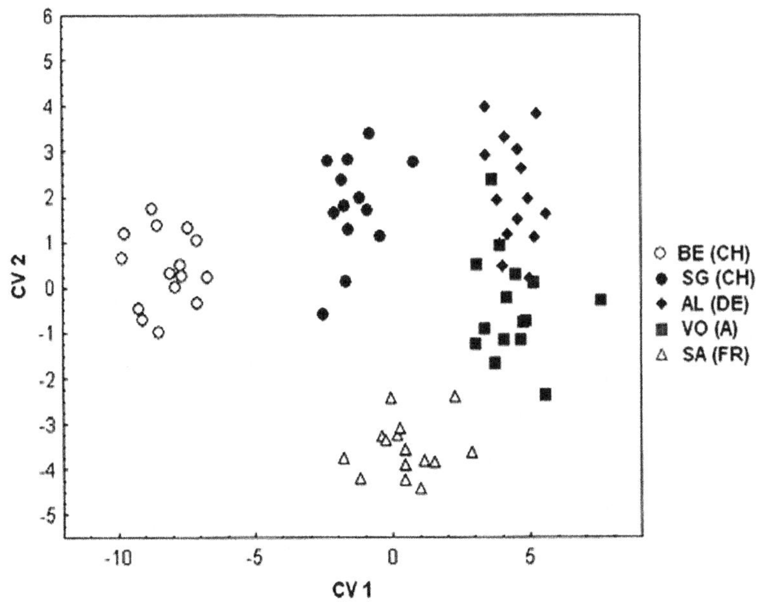

FIGURE 3.7 Canonical analysis plot of Emmental cheeses from Austria (A), France (FR), Germany (DE), and Switzerland (CH) obtained from the analysis of their HR-MAS NMR spectral profiles. Reprinted with permission from Shintu and Caldarelli (2006). Copyright 2006 American Chemical Society.

and Franke, 2007), milk powder (Ishii et al., 2003), and cheese (Shintu and Caldarelli, 2006), without any sample pretreatment. Figure 3.7 depicts the canonical analysis plot of Emmental cheeses from Austria (A), France (FR), Germany (DE) and Switzerland (CH) obtained after the statistical analysis of the HR-MAS NMR spectral fingerprints of the solid cheese samples, demonstrating very good geographical origin authentication.

Another example highlighting the usefulness of solid state NMR spectroscopy in food fingerprinting is a study of the effect of pre-stunning method on meat metabolite profiling. Figure 3.8 depicts the CP-MAS ^{13}C NMR metabolite fingerprints of meat from a CO_2-stunned pig (A) and two pre-slaughter-stressed, electrically-stunned pigs (B) (Bertram et al., 2003). The differences in lactate content observed in these spectra can be correlated to meat quality and pre-slaughter stress. Apart from meat (Bertram et al., 2003; Bertram, Jakobsen, and Andersen, 2004), bread (Mihhalevski et al., 2012) and tea (Martinez-Richa and Joseph-Nathan, 2003; Cai et al., 2016) have also been studied using CP-MAS solid state ^{13}C NMR profiling.

3.8 ANALYTICAL ASPECTS

Although the sensitivity of NMR spectroscopy is generally perceived to be lower than that of other spectroscopic techniques, such as mass spectrometry, fluorescence, IR, and Raman, in the last few years hardware improvements have led to the gradual acceptance of NMR spectroscopy as a quantitative, accurate, precise, and reproducible analytical tool for the analysis of complex mixtures, such as food, plant extracts, and pharmaceutical formulations (van Duynhoven et al., 2013). When NMR spectroscopy is used as a

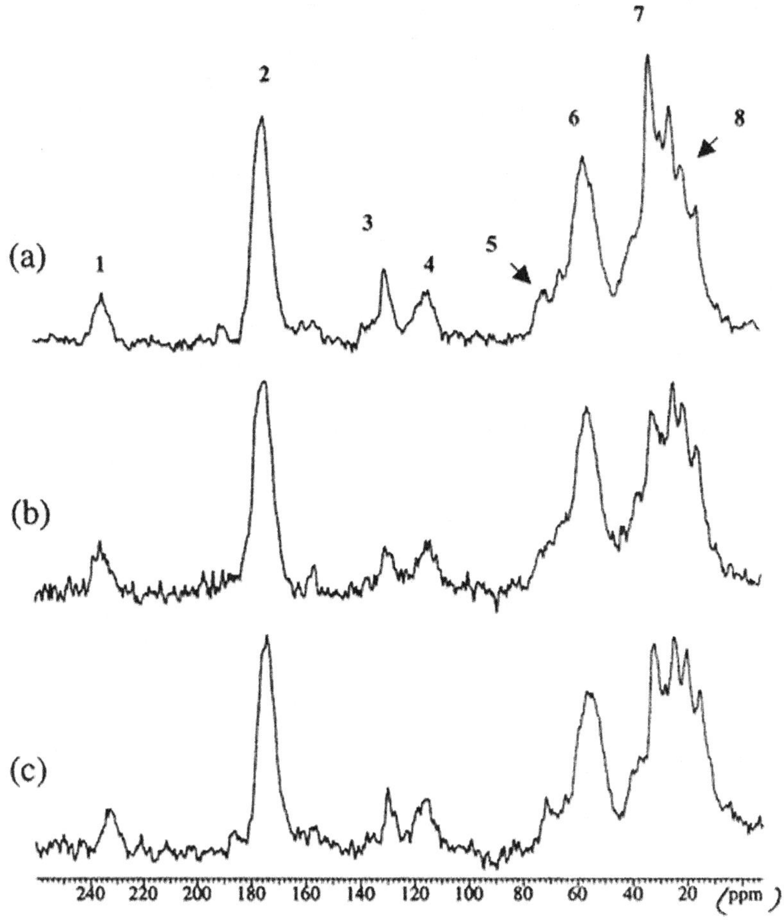

FIGURE 3.8 ^{13}C CP-MAS NMR spectra of meat taken 1 min post-mortem from (A) control, CO_2-stunned pig, (B) pre-slaughter-stressed, electrically-stunned pig, and (C) replicate of (C). Peaks 1 and 4 are assigned to spinning sidebands, whereas peak 2 is assigned to carbonyl carbons. Peak 3 is assigned to unsaturated carbons of fatty acids; peak 5 to C3, C4, C5, and C6 in glycogen, and peak 6 to methyl groups of choline and phosphatidylcholine; peak 7 is assigned to various saturated carbons of methylene chains of the fatty acids; and peak 8 is assigned to the CH_3 group in lactate. Reprinted with permission from (Bertram et al., 2003). Copyright 2003 American Chemical Society.

technique for the quantitative determination of a chemical compound in a food matrix, defining its analytical performance is not as straightforward since it depends on many experimental factors: the magnetic field strength of the spectrometer magnet, the choice of nucleus, the NMR experiment used, the instrument's parameter setup, operator efficiency etc. Proton NMR has by far the highest sensitivity, and is the nucleus of choice for food analysis, provided that it is suitable for the analytical problem at hand. Although ^{13}C NMR has very low sensitivity, it has been used in food analysis occasionally because of the larger ^{13}C NMR spectral width that makes signal overlapping a rarity. ^{31}P NMR is also preferable for the analysis of phosphorus containing food ingredients, such as phospholipids and several other minor food components that can be tagged chemically with phosphorus (Spyros and Dais, 2009).

The sensitivity of the NMR quantification of metabolites in a food matrix can be increased by using a higher magnetic field strength, reducing the volume of the solution (using capillary NMR tubes or microprobes/flow probes), resorting to cryogenically cooled probes (cryoprobes), and by increasing the number of scans done for obtaining a single NMR spectrum (at the expense of increased experimental time, however). For conventional probes, the limit of detection of low molecular weight (<400 g/mol) compounds in typical food samples by ^{1}H NMR spectroscopy is considered to be around 10 μM, (Mannina, Sobolev, and Viel, 2012) meaning that 5–30 μg of an analyte can be detected. In typical quantitative analyses of foods, compounds can be quantified down to levels of quantification (LOQ) of 40 μM, with levels of detection (LOD) as low as 10 μM (Savage et al. 2011; Godelmann et al. 2013). Using capillary tubes and a cryoprobe, an improvement by. a factor of 10 is expected approximately, while with a brand new very high field (750–900 MHz) NMR spectrometer another factor of about 10 is possible, however the latter type of advanced equipment is usually not easily available to food scientists due to its high cost.

Extensive validation studies have shown that NMR has excellent linearity over a wide range of analyte concentrations. For sample concentrations in the mg/mL range, typical residual standard deviations (RSD) of 0.1%–0.4% have been reported using conventional NMR spectrometers (van Duynhoven et al., 2013). However, many metabolites in food exist at even lower concentrations, in the μg/mL range, and reaching such high precision for quantitative NMR of food might be more difficult and might require extensive optimization of the experimental parameters. An extensive discussion of the analytical aspects of quantitative NMR in food analysis has been published recently (van Duynhoven et al., 2013; Lamanna, 2013).

Finally, and as already discussed, one way to obviate the difficulties in quantifying food micro-components by NMR is to skip the quantification part altogether and to use instead the whole NMR spectrum as the spectroscopic fingerprint of a particular food sample. In this case, the presence of micro-components alters the peak intensity and distribution of the NMR spectrum in a measurable way and, thus, small compositional changes can be exploited via NMR spectroscopy in food analysis and authentication purposes.

3.9 FUTURE TRENDS

Although NMR spectroscopy in the last two decades has proven itself to be an excellent tool for the characterization and the compositional analysis of food, there is still a lot of development to come in the following years. It is expected that the field of food metabolomics will continue to develop, aided by (a) hardware developments in the design of NMR probes and electronics that will allow high throughput and cost effective analysis of large collections of food samples; (b) improvements in software that will allow the semiautomatic spectral assignment of NMR spectra to metabolites of interest by integration with metabolite databases. In this respect, further development of the Food Component Database (www.foodb.ca/) and other integrated food metabolite databases will be of prime importance to the food science community.

Furthermore, the NMR analysis and the authentication of solid and semisolid food have only recently been exploited, so a lot of development is expected in this field since the majority of food materials are solid, and food scientists are gaining increasing access to solid state NMR spectrometers. This is especially true for solid state CP-MAS NMR, which is so far under-used in food fingerprinting. The increasing availability of low cost permanent magnet based liquid state NMR spectrometers in research, industry, and state facility labs is expected to increase the interest of the scientific community toward NMR applications in food authentication.

REFERENCES

Alberti, E., P. S. Belton, and A. M. Gil. 2002. Applications of NMR to food science. In *Annual Reports on NMR Spectroscopy*: Academic Press.

Ali, K., F. Maltese, R. Toepfer, Y. H. Choi, and R. Verpoorte. 2011. Metabolic characterization of Palatinate German white wines according to sensory attributes, varieties, and vintages using NMR spectroscopy and multivariate data analyses. *Journal of Biomolecular NMR* 49 (3–4):255–266.

Almeida, C., I. F. Duarte, A. Barros, J. Rodrigues, M. Spraul, and A. M. Gil. 2006. Composition of beer by ^{1}H NMR spectroscopy: Effects of brewing site and date of production. *Journal of Agricultural and Food Chemistry* 54 (3):700–706.

Anastasiadi, M., A. Zira, P. Magiatis, S. A. Haroutounian, A. L. Skaltsounis, and E. Mikros. 2009. ^{1}H NMR-based metabonomics for the classification of Greek wines according to variety, region, and vintage. Comparison with HPLC data. *Journal of Agricultural and Food Chemistry* 57 (23):11067–11074.

Belloque, J., and M. Ramos. 1999. Application of NMR spectroscopy to milk and dairy products. *Trends in Food Science and Technology* 10 (10):313–320.

Belton, P. S., I. J. Colquhoun, E. Katherine Kemsley, I. Delgadillo, P. Roma, M. John Dennis, M. Sharman, E. Holmes, J. K. Nicholson, and M. Spraul. 1998. Application of chemometrics to the ^{1}H NMR spectra of apple juices: Discrimination between apple varieties. *Food Chemistry* 61 (1998):207–213.

Bertocchi, F., and M. Paci. 2008. Applications of high-resolution solid-state NMR spectroscopy in food science. *Journal of Agricultural and Food Chemistry* 56 (20):9317–9327.

Bertram, H. C., H. J. Jakobsen, H. J. Andersen, A. H. Karlsson, and S. B. Engelsen. 2003. Post-mortem changes in porcine M. Longissimus studied by solid-state ^{13}C cross-polarization magic-angle spinning nuclear magnetic resonance spectroscopy. *Journal of Agricultural and Food Chemistry* 51 (7):2064–2069.

Bertram, H. C., H. J. Jakobsen, and H. J. Andersen. 2004. Combined high-field ^{13}C CP MAS NMR and low-field NMR relaxation measurements on post mortem porcine muscles. *Journal of Agricultural and Food Chemistry* 52 (10):3159–3164.

Cai, H., R. Cheng, Y. Jin, S. Ding, and Z. Chen. 2016. Evaluation of oolong teas using ^{1}H and ^{13}C Solid-state NMR, sensory analysis, and multivariate statistics. *Journal of the Chinese Chemical Society* 63 (9):792–799.

Caligiani, A., D. Acquotti, M. Cirlini, and G. Palla. 2010. ^{1}H NMR study of fermented cocoa (Theobroma cacao L.) beans. *Journal of Agricultural and Food Chemistry* 58 (23):12105–12111.

Capozzi, F., and M. A. Cremonini. 2008. Ch. 13 Nuclear magnetic resonance spectroscopy in food analysis. In *Handbook of Food Analysis Instruments*, edited by S. Otles: CRC Press.

Consonni, R., and L. R. Cagliani. 2010. Nuclear magnetic resonance and chemometrics to assess geographical origin and quality of traditional food products. *Advances in food and nutrition research* 59:87–165.

Cuny, M., E. Vigneau, G. Le Gall, I. Colquhoun, M. Lees, and D. N. Rutledge. 2008. Fruit juice authentication by ^{1}H NMR spectroscopy in combination with different chemometrics tools. *Analytical and Bioanalytical Chemistry* 390 (1):419–427.

Dais, P., and A. Spyros. 2007. ^{31}P NMR spectroscopy in the quality control and authentication of extra-virgin olive oil: A review of recent progress. *Magnetic Resonance in Chemistry* 45 (5):367–377.

Diehl, B. W. K. 2001. High resolution NMR spectroscopy. *European Journal of Lipid Science and Technology* 103 (12):830–834.

Fragaki, G., A. Spyros, G. Siragakis, E. Salivaras, and P. Dais. 2005. Detection of extra virgin olive oil adulteration with lampante olive oil and refined olive oil using nuclear magnetic resonance spectroscopy and multivariate statistical analysis. *Journal of Agricultural and Food Chemistry* 53 (8):2810–2816.

Gidley, M. J., and S. M. Bociek. 1988. ^{13}C CP/MAS NMR studies of amylose inclusion complexes, cyclodextrins, and the amorphous phase of starch granules: Relationships between glycosidic linkage conformation and solid-state ^{13}C chemical shifts. *Journal of the American Chemical Society* 110 (12):3820–3829.

Gil, A. M., and I. Duarte. 2006a. High-resolution nuclear magnetic resonance spectroscopy of wine, beer, and spirits. In *Modern Magnetic Resonance*, edited by G. Webb: Springer Netherlands.

Gil, A. M., and I. Duarte. 2006b. High-resolution nuclear magnetic resonance spectroscopy of fruit juices. In *Modern Magnetic Resonance*, edited by G. Webb: Springer Netherlands.

Gil, A. M., and I. Duarte. 2006c. High-resolution magic angle spinning NMR spectroscopy of fruits and vegetables. In *Modern Magnetic Resonance*, edited by G. Webb: Springer Netherlands.

Godelmann, R., F. Fang, E. Humpfer, B. Schütz, M. Bansbach, H. Schäfer, and M. Spraul. 2013. Targeted and nontargeted wine analysis by ^{1}H NMR spectroscopy combined with multivariate statistical analysis. Differentiation of important parameters: Grape variety, geographical origin, year of vintage. *Journal of Agricultural and Food Chemistry* 61 (23):5610–5619.

Hidalgo, F. J., and R. Zamora. 2003. Edible oil analysis by high-resolution nuclear magnetic resonance spectroscopy: Recent advances and future perspectives. *Trends in Food Science & Technology* 14 (12):499–506.

Hong, Y.-S. 2011. NMR-based metabolomics in wine science. *Magnetic Resonance in Chemistry* 49:S13–S21.

Iglesias, M. J., J. G. López, J. F. Collados Luján, F. L. Ortiz, H. B. Pereznieto, F. Toresano, and F. Camacho. 2014. Effect of genetic and phenotypic factors on the composition of commercial marmande type tomatoes studied through HRMAS NMR spectroscopy. *Food Chemistry* 142:1–11.

Imparato, G., E. Di Paolo, A. Braca, and R. Lamanna. 2011. Nuclear magnetic resonance profiling of wine blends. *Journal of Agricultural and Food Chemistry* 59 (9):4429–4434.

Ishii, T., K. Hiramatsu, T. Tanaka, K. Sato, and A. Tsutsumi. 2003. The evaluation of skim milk powder by solid-state ^{31}P magic angle spinning nuclear magnetic resonance spectroscopy (MASNMR). *Milchwissenschaft* 58 (3–4):178–181.

Jung, Y., J. Lee, J. Kwon, K. S. Lee, D. H. Ryu, and G. S. Hwang. 2010. Discrimination of the geographical origin of beef by ^{1}H NMR-based metabolomics. *Journal of Agricultural and Food Chemistry* 58 (19):10458–10466.

Kidrič, J. 2008. Chapter 5 NMR Study of Beverages. In *Annual Reports on NMR Spectroscopy*, edited by A. W. Graham: Academic Press.

Kono, H., S. Yunoki, T. Shikano, M. Fujiwara, T. Erata, and M. Takai. 2002. CP/MAS ^{13}C NMR study of cellulose and cellulose derivatives. 1. Complete assignment of the CP/MAS ^{13}C NMR spectrum of the native cellulose. *Journal of the American Chemical Society* 124 (25):7506–7511.

Košir, I. J., and J. Kidrič. 2002. Use of modern nuclear magnetic resonance spectroscopy in wine analysis: Determination of minor compounds. *Analytica Chimica Acta* 458 (1):77–84.

Lachenmeier, D. W., W. Frank, E. Humpfer, H. Schäfer, S. Keller, M. Mörtter, and M. Spraul. 2005. Quality control of beer using high-resolution nuclear magnetic resonance spectroscopy and multivariate analysis. *European Food Research and Technology* 220 (2):215–221.

Lamanna, R. 2013. Chapter Four - Proton NMR Profiling of Food Samples. In *Annual Reports on NMR Spectroscopy*, edited by A. W. Graham: Academic Press.

Mannina, L., and A. P. Sobolev. 2011. High resolution NMR characterization of olive oils in terms of quality, authenticity and geographical origin. *Magnetic Resonance in Chemistry* 49:S3–S11.

Mannina, L., A. P. Sobolev, and S. Viel. 2012. Liquid state ^{1}H high field NMR in food analysis. *Progress in Nuclear Magnetic Resonance Spectroscopy* 66 (0):1–39.

Martinez-Richa, A., and P. Joseph-Nathan. 2003. Carbon-13 CP-MAS nuclear magnetic resonance studies of teas. *Solid State Nuclear Magnetic Resonance* 23 (3):119–135.

Mihhalevski, A., I. Heinmaa, R. Traksmaa, T. Pehk, A. Mere, and T. Paalme. 2012. Structural changes of starch during baking and staling of rye bread. *Journal of Agricultural and Food Chemistry* 60 (34):8492–8500.

Ohno, A., K. Oka, C. Sakuma, H. Okuda, and K. Fukuhara. 2011. Characterization of tea cultivated at four different altitudes using ^{1}H NMR analysis coupled with multivariate statistics. *Journal of Agricultural and Food Chemistry* 59 (10):5181–5187.

Rodrigues, J. E., and A. M. Gil. 2011. NMR methods for beer characterization and quality control. *Magnetic Resonance in Chemistry* 49:S37–S45.

Sanchez Perez, E. M., M. J. Iglesias, F. L. Ortiz, I. S. Perez, and M. M. Galera. 2010. Study of the suitability of HRMAS NMR for metabolic profiling of tomatoes: Application to tissue differentiation and fruit ripening. *Food Chemistry* 122 (3):877–887.

Savage, A. K., J. P. M. van Duynhoven, G. Tucker, and C. A. Daykin. 2011. Enhanced NMR-based profiling of polyphenols in commercially available grape juices using solid-phase extraction. *Magnetic Resonance in Chemistry* 49:S27–S36.

Shintu, L., and S. Caldarelli. 2006. Toward the determination of the geographical origin of emmental(er) cheese via high resolution MAS NMR: A preliminary investigation. *Journal of Agricultural and Food Chemistry* 54 (12):4148–4154.

Shintu, L., S. Caldarelli, and B. M. Franke. 2007. Pre-selection of potential molecular markers for the geographic origin of dried beef by HR-MAS NMR spectroscopy. *Meat Science* 76 (4):700–707.

Spyros, A., and P. Dais. 2009. ^{31}P NMR spectroscopy in food analysis. *Progress in Nuclear Magnetic Resonance Spectroscopy* 54 (3–4):195–207.

Spyros, A., and P. Dais. 2012. NMR spectroscopy in food analysis, RSC Food Analysis Monographs.

Tarachiwin, L., U. Koichi, A. Kobayashi, and E. Fukusaki. 2007. ^{1}H NMR based metabolic profiling in the evaluation of Japanese green tea quality. *Journal of Agricultural and Food Chemistry* 55 (23):9330–9336.

Tode, C., T. Maoka, and M. Sugiura. 2009. Application of LC-NMR to analysis of carotenoids in foods. *Journal of Separation Science* 32 (21):3659–3663.

Ulrich, E. L., H. Akutsu, J. F. Doreleijers, Y. Harano, Y. E. Ioannidis, J. Lin, M. Livny, S. Mading, D. Maziuk, Z. Miller, E. Nakatani, C. F. Schulte, D. E. Tolmie, R. K. Wenger, H. Yao, and J. L. Markley. 2008. BioMagResBank. *Nucleic Acids Research* 36 (suppl_1):D402–D408.

Valentini, M., M. Ritota, C. Cafiero, S. Cozzolino, L. Leita, and P. Sequi. 2011. The HRMAS–NMR tool in foodstuff characterisation. *Magnetic Resonance in Chemistry* 49:S121–S125.

van Duynhoven, J., E. van Velzen, D. M. Jacobs, and A. Webb Graham. 2013. Chapter Three - Quantification of complex mixtures by NMR. In *Annual Reports on NMR Spectroscopy*: Academic Press.

Vermathen, M., M. Marzorati, and P. Vermathen. 2012. Exploring high-resolution magic angle spinning (HR-MAS) NMR spectroscopy for metabonomic analysis of apples. *Chimia* 66 (10):747–751.

Wei, F., K. Furihata, F. Hu, T. Miyakawa, and M. Tanokura. 2010. Complex mixture analysis of organic compounds in green coffee bean extract by two-dimensional NMR spectroscopy. *Magnetic Resonance in Chemistry* 48 (11):857–865.

Wei, F., K. Furihata, F. Hu, T. Miyakawa, and M. Tanokura. 2011. Two-dimensional ^{1}H–^{13}C nuclear magnetic resonance (NMR)-based comprehensive analysis of roasted coffee bean extract. *Journal of Agricultural and Food Chemistry* 59 (17):9065–9073.

Wishart, D. S., T. Jewison, A. C. Guo, M. Wilson, C. Knox, Y. Liu, Y. Djoumbou, R. Mandal, F. Aziat, E. Dong, S. Bouatra, I. Sinelnikov, D. Arndt, J. Xia, P. Liu, F. Yallou, T. Bjorndahl, R. Perez-Pineiro, R. Eisner, F. Allen, V. Neveu, R. Greiner, and A. Scalbert. 2013. HMDB 3.0—The human metabolome database in 2013. *Nucleic Acids Research* 41 (Database issue):D801–D807.

CHAPTER 4

Capillary Electrophoresis

Semih Otles
Ege University

Vasfiye Hazal Ozyurt
Ege University and Near East University

CONTENTS

4.1 INTRODUCTION

Electrophoresis is one of the most widely utilized techniques for the separation and analysis of ionic substances and an invaluable analytical tool for modern biochemical research. Electrophoresis usually utilizes some form of solid support to prevent convectional distortion of the analyte bands and is found in the paper form (high voltage separation of amino acids and other small organic molecules) and as polyacrylamide or agarose gels.

Capillary electrophoresis (CE) offers high throughput, high resolution, automatic operation, and on-line detection with automatic data acquisition (Sun et al., 2016). Although electrophoresis has the ability to resolve the components of complex systems, the traditional forms of electrophoresis have several disadvantages, such as the speed of separation, poor reproducibility, analyte-dependent differences in staining, etc. (Landers, 2008).

Therefore, this section discusses the fundamental theory, equations, and definitions necessary to comprehend the concepts involved in CE.

4.2 THEORY

The separation mechanisms of electrophoresis are described in this part. These mechanisms help the researcher understand the methodology, analyze the experimental data, identify the relevant separation regimes, and/or design the optimization strategies.

CE is a "finish line" technique and it is useful for defining several performance parameters. It is relatively fast and highly informative and is easily automated.

In a capillary filled with electrolyte, when the voltage is applied, the solution flows along the capillary. Solute ions are pumped to the detector along the capillary due to ionization of the acidic silanol groups inside the capillary (Altria, 1995a). The inner wall of a fused silica capillary is negatively charged when it is in contact with standard buffers. Then, the wall is charged with cations. In an electric field, the layer that forms from the cations moves and drags the liquid toward the cathode and this is called the electroosmotic flow (EOF) (Slater et al., 2001). The level of EOF is highly dependent on the electrolyte pH. When pH is below 4, the ionization is small, and the EOF flow rate is not significant. However, when pH is above 9, the silanols are ionized and, therefore, the EOF is much stronger (Altria, 1995a).

The migration time and the sample quantity are the basic parameters obtained from any electropherogram. Migration time is used to estimate the size, and the quantity of product is used to determine the reaction yields and the efficiency of an extraction process (McCord, 2001).

The ionic mobility (μE) is related to the charge-to-mass ratio.

$$\mu E = \left(\frac{q}{6\Pi}\right)\eta r \tag{4.1}$$

where

μE = electrophoretic mobility

q = number of charges

η = solution of viscosity

r = radius of the ion

The speed of the solute ions is related to their mobilities and the magnitude of the applied voltages.

$$v = \mu EE \tag{4.2}$$

where

v = velocity of the ion

E = applied voltage

The extent of flow is related to the charge on the capillary, the buffer viscosity, and the dielectric constant of the buffer.

$$\mu EOF = \left(\frac{\varepsilon\zeta}{\eta}\right) \tag{4.3}$$

where

μEOF = EOF mobility

η = viscosity

ζ = zeta potential

The overall migration time is related to the mobility of solute and the EOF. The term apparent mobility (μA) is measured from the migration time and is a sum of μE and μEOF.

$$\left[\mu A = \mu E + \mu EOF = \left(\frac{lL}{tV}\right)\right] \tag{4.4}$$

where

l = length along the capillary to the detector

V = voltage

L = total length of the capillary (Altria, 1995a)

4.3 INSTRUMENTS

The electrophoresis method was described in 1937. This method has presented significant advances, and with increased availability of commercial CE instruments in the 1990s, this method has been applied as an important alternative in separation science (de Oliveira et al., 2016).

CE is an instrument that performs high-efficiency separations of sample ions in a narrow bore (25–100 μm) capillary tube that is filled with an electrolyte solution. The principal components of CE are a high-voltage power supply, a capillary, a sample introduction system, and an autosampler (Altria, 1995b). CE instrumentation consists of two electrolyte chambers linked by a thin capillary. In spite of the utilization of large electric fields, the temperature gradients and distortions are virtually not found in CE as the microcapillary facilitates rapid heat dissipation. Data on fractionated molecules is acquired automatically by an on-line detector (Mitchelson, 2001).

Studies related to capillary electrophoresis have involved the separation of target analytes in standard solutions. However, this can be explained considering the complexities of the real world samples, which cause depletion of the separation efficiency and sensitivity. Thus, a sample preparation step would be mandatory when food matrices are involved. In other words, the isolation and quantitation of the individual components of a mixture is the fundamental role of any separation process (de Oliveria et al., 2016).

First, the capillary is filled with the required buffer solution. Then, sample (typically 1–20 mL) is introduced at the end of the capillary, away from the detector (usually the anode), and the capillary ends are plunged into reservoirs containing high-voltage electrodes and the buffer solution. Electrodes are composed of inert material, such as platinum. Application of a voltage (for example, 10–30 kV) across the capillary causes electrophoretic and electro end osmotic movements (Altria, 1995c).

The movement of sample ions under the influence of an applied voltage results in their separation when the ions go to the appropriate electrodes and then to the detector. The migration rate is related to the ion size and the number of ionic charges. A smaller ion moves faster than a larger ion with the same number of ionic charges, or an ion with only one charge moves slower than an ion with two charges. Moreover, the sample ions are separated in another direction in the solution during electrophoresis according to the significance of the mass-to-charge ratio (m/z), and when each particle only has a single charge, the absolute value of the force on each particle is the same (Whatley, 2001).

4.3.1 Sample Introduction

The sample can be introduced into the capillary by three mechanisms: hydrodynamic, gravity, and electrokinetic. The principles of operation for the three methods are detailed below.

4.3.1.1 Hydrodynamic Mechanism

The sampling end of the capillary is inserted in the sample solution and a pressure difference is applied.

4.3.1.2 Gravity Mechanism

The sampling end of the capillary is dipped into the sample vial and the capillary is mechanically raised above the height of the detector electrolyte vial.

4.3.1.3 Electrokinetic

The sampling end of the capillary and the high-voltage electrode are immersed into the sample solution (Altria, 1995a).

4.3.2 Capillaries

In the construction of CE capillaries, different materials have been tested: fused silica, borosilicate glass, polytetrafluoroethylene (teflon), etc. Fused silica is the most preferred material in the construction of capillaries due to its high purity. However, handling must be done carefully to prevent breakage (Kuhn and Hoffstetter-Kuhn, 1993).

The inner surface of a capillary is in contact with the separation chemical and the samples, so it is an important factor in the capillary electrophoresis. The inner surface of the capillary participates in the separation process, and the interaction between the analytes and the inner surface of the capillary is prevented by simultaneously controlling EOF. The inner surface of a fused silica capillary depends on the treatment it has experienced (Lambert and Middleton, 1990). The surface is usually formed by siloxane and after this surface has been treated with a base (NaOH or KOH), the siloxane groups are hydrolyzed and a new surface, which is rich in silanol groups, is formed. Silane derivatization reactions are commonly employed to link other groups to silica to separate other compounds. In this way, many types of coatings have been created including polyacrylamide, polyvinyl alcohol, polyethylene glycol, and polyvinylpyrolidone (Bruin et al., 1992; Cobb et al., 1990; Hjerten, 1988).

The buffer choice is important since the buffer used in the capillary modifies the surface. Buffer quality is independent of the instrument system used, so the buffer system can either be made or purchased. The small changes in buffer pH have great impact on the separation with CE. The most commonly employed buffers are phosphate, borate, citrate, acetate, and tris(hydroxymethyl)aminomethane. The selection of a buffer for CE is based on several factors, such as the desired pH value, operating temperature, charge of the buffer relative to the analytes and the capillary inner surface, and the effects on detection.

The buffer concentration is as important as the buffer selection. Increasing the concentration of a buffer will increase the buffering capacity of the system. The addition of different reagents to a buffer used in CE makes it easier to change the separation properties of the system. The most common reagents are: detergents, such as sodium dodecyl sulfate (SDS); the viscosity modifiers, such as linear polyacrylamide; the organic solvents, such as acetonitrile; the denaturants, such as urea; and combinations of these additives, etc. (Ferguson et al., 1980; Van Orman et al., 1991).

4.3.3 Detection

In order to get results from the separation technique, analytes are detected and measured. Detection may be qualitative and/or quantitative.

Absorbance detectors are the most commonly encountered detectors in CE instruments. They measure the absorbance of light energy by the analytes. This absorbance creates a shadow since the analytes pass between the light source and the light detector. The intensity of the shadow is calculated as it is proportional to the amount of material present (Kobayashi et al., 1989).

Fluorescence detectors are one of the detection equipment that do not rely on the measurement of shadows. In these systems, an external energy source is used to excite the analyte molecules to a higher energy state. When these excited molecules return to the normal state they emit energy of a lower wavelength. Then, this wavelength can be detected and recorded as evidence of the passage of the analytes (Wu and Dovichi, 1992).

Amperometry and conductivity detectors can offer a high sensitivity and can thus be applied to several analytes and they require the use of sensing electrodes. In amperometric detection, an electroactive analyte undergoes an electrochemical reaction inside a detector (Wallingford and Ewing, 1987). In conductivity detector, the electrodes are placed opposite to each other in the capillary and so there is no drop of voltage between the two electrodes (Huang et al., 1987; Huang et al., 1991).

Mass spectrometry offers the additional data of molecular weight and, using collision dissociation and MS-MS systems, it can also provide structural information in addition to helping in detection (Banks and Gulcicek 1997).

In all these direct detection methods, the detection is done in the presence of the sample in the detector cell or by increasing the output signal. However, there are certain applications, which are related to the decrease of the background signal, and this signal provides evidence of the passage of analytes through a detector. An indirect detector would not detect the analytes in a direct mode.

4.4 MODES OF CAPILLARY ELECTROPHORESIS (CE)

The main modes of CE are capillary zone electrophoresis (CZE), micellar electrokinetic capillary chromatography (MEKC), capillary electrokinetic chromatography (CEC), capillary isoelectric focusing (CIEF), capillary gel electrophoresis (CGE), and capillary isotachophoresis (CITP) (Bolshakov and Amelin, 2016).

4.4.1 Capillary Zone Electrophoresis (CZE)

CZE is the most universal technique, having shown its usefulness in the separation of a diverse array of analytes varying in size and character (Blatny and Kvasnička, 1999).

In CZE, the sample is introduced into a buffer or gel-filled capillary. Under the influence of an external field, this yields discrete zones, and it passes an in-line detector. Solutes are separated in this technique on the basis of differences in their charge-to-mass ratio (Jenkins, 1999) and they move from one end of the capillary to the other according to the vector sum of electrophoresis and electroosmotic mobility. CZE has been used for analytes as diverse as sodium ions, drugs, and protein molecules.

4.4.2 Capillary Gel Electrophoresis (CGE)

In CGE, separation is based on viscous drag and the capillary is filled with a gel or viscous solution. The separation is effectively based on the molecular size but not on mass-to-charge ratio (Zhu et al., 2012).

4.4.3 Capillary Isotachophoresis (CITP)

In CITP, the sample is introduced between two different buffers. One of these is called to be the leading electrolyte that has the highest mobility in the separation. The second is the trailing electrolyte that has a low mobility. The charges on the analytes and the buffers have to be the same. CITP of peptides and proteins have been used with MS detection (Mala et al., 2013).

4.4.4 Micellar Electrokinetic Capillary Chromatography (MECC)

Electrophoresis is not possible for analytes that are not charged. In order to analyze such analytes, it is necessary to employ some agent. MECC is a modification of CZE. Inside the capillary tube, there are two phases: a pseudostationary phase and an aqueous phase (Jenkins, 1999). A suitable charged detergent, such as SDS, is added to the separation buffer in a concentration that is sufficiently high to allow the formation of micelles. MECC is useful for a wide range of small molecules such as drugs, pesticides, and food additives that are not charged and are sufficiently hydrophobic to associate with the micelles.

4.4.5 Capillary Isoelectric Focusing (CIEF)

Molecules carrying both positively and negatively charged groups exhibit an equal number of positive and negative charges at the isoelectric pH or pI. The molecule has no tendency to migrate in an electrical field and the analyte, having no net charge, ceases to migrate. It remains in this position as long as the pH gradient is stable, typically as long as the voltage is applied. CIEF can separate proteins and it utilizes ampholytes that are mixtures of buffers with a range of pK_a values. These ampholytes facilitate high resolution separation of protein and peptide mixtures. CIEF usually uses a coated capillary; however, if the electroosmotic flow is sufficiently reduced by the use of methylcellulose or hydroxypropylmethylcellulose, then CIEF can be carried out in a fused silica capillary (Jenkins, 1999; Silvertand et al., 2008).

4.4.6 Chiral Electrophoresis

Chiral molecules can exist in two stereospecific forms that are identical in molecular weights and chemical formulae but differ in the arrangement of the atoms in space. Separation of these chiral forms depends on their tendencies to associate differentially with other chiral molecules known as chiral selectors. By incorporating a chiral selector into the CE buffer, it is often possible to separate enantiomers of a chiral molecule. The most commonly employed chiral selector in CE is cyclodextrin. Cyclodextrins may be chemically modified to alter their hydrophobicity or charge. Uncharged cyclodextrins are not suitable for the analysis of uncharged analytes. Chiral CE can be used to separate the enantiomeric forms of pharmaceuticals as well as natural substances, such as amino acids (Whatley, 2001).

4.4.7 Nonaqueous Electrophoresis

Electrophoresis can be applied using acetonitrile, methanol, formamide, dimethylformamide, etc. When two analytes have the same charge-to-mass ratio, this kind of electrophoresis is used.

Moreover, nonaqueous CE has alternatives for analytes that do not separate well under aqueous conditions. Nonaqueous CE has been used for drugs, dyes, preservatives, surfactants, and inorganic ions (Whatley, 2001).

4.4.8 Capillary Electrochromatography (CEC)

CEC is a technique that has high separation efficiency and selectivity. CEC is required for some solvents and samples. CEC is used in the fields of pharmaceutics, biomedicine analysis, and food analysis. In addition to high selectivity, the presence of EOF reduces the solute dispersion in the column (D'orazio et al., 2016).

4.5 APPLICATIONS IN FOOD

The analysis of published works using electrophoresis indicated that only a few studies are dealing with the analysis of food matrices.

Table 4.1 summarizes the modes used for the analysis of various classes of analytes.

Soga and Ross (1999) developed a method for the simultaneous determination of 32 carbohydrates in soy sauce samples, isotonic drinks, and pineapples without derivatization using CZE with indirect UV detection (CZE-UV-ID).

Jager et al. (2007) proposed an extraction method for the determination of carbohydrates such as fructose, glucose, maltose, maltotriose, lactose, galactose, and sucrose in dairy samples and cereals using CZE-UV-ID.

Mateos-Vivas et al. (2017) described a method for the simultaneous determination of unmodified nucleosides and nucleotide mono-, di-, and tri-phosphates in baby food

TABLE 4.1 The Modes Used for the Analysis of Different Compounds

CZE	MEKC	CEC	CIEF	CGE
Ions	Small molecules	Small molecules	Peptides	Nucleic acids
Small molecules	Peptides	Peptides	Proteins	
Peptides		Proteins		
Proteins		Carbohydrates		
Lipids				
Amino acids				
Carbohydrates				

by capillary electrophoresis coupled to mass spectrometry (CE-MS). Then, as in the validated method, a variety of baby food samples (16 in total) such as fish, meat, fruits, and baby dairy desserts that may endogenously contain these analytes were analyzed.

Sun et al. (2017) determined four *Lycoris radiata* alkaloids: galanthamine, homolycorine, lycorenine, and tazettine using capillary electrophoresis electrochemiluminescence after ultrasound-assisted extraction. Under optimal experimental conditions, the four alkaloids were separated within 16 mins. The method validation was applied for each alkaloid.

Nikovaev et al. (2015) have developed a CE microchip with electrochemical detection for the determination of polyphenols in red wine.

Mei et al. (2011) stated that BPA-molecularly imprinted polymeric microspheres were used as the SPE sorbent for the extraction of BPA and it was detected with CE.

Zeng et al. (2013) used the method of CE time-of-flight/MS (CE-TOF/MS) to analyze the urine metabolic profiles of Sprague–Dawley rats fed with four dose levels of Bisphenol A (0, 1, 10, and, 100 μg/kg body weight) for 45 days.

Wu et al. (2015) developed an on-line preconcentration strategy combining field-amplified sample stacking and reversed-polarity stacking for the analysis of vitamin B3, niacinamide, and amino acids including lysine and taurine in functional drinks by microchip electrophoresis.

Rang et al. (2015) developed a quantitative CE method combined with inkjet injection for finding the quantitative concentration and determining theobromine, caffeine, and theophylline in green tea.

Li et al. (2015) used a rapid desalination MEKC method for the determination of five characteristic amino acids of collagen and 15 common amino acids in turtle jelly.

Sander et al. (1999) was using a pressurized CEC (pCEC) gradient mode elution to allow the complete separation of carotenoid isomers (lutein, zeaxantrin, echinenone, cryptoxanthin, carotene, lycopene) in baby food containing 30% creamed corn, 30% macaroni/tomato/beef, 20% orange juice concentrate, 15% powdered infant formula, 3% papaya juice concentrate, and 2% creamed spinach.

Lacná et al. (2017) developed a sensitive capillary electrophoresis method with laser induced fluorescence detection for the determination of malondialdehyde in various biological fluids. The developed method offers sensitivity.

Abidi (2004) tested CEC using silica-C18 and silica-C30 to evaluate the composition of 17 sterols and related esters in soybean, canola, and rice bran oils. They observed that this study is the best stationary phase study depending on the compounds.

Henry et al. (2001) performed a comparison study of four stationary phases for the determination of Vitamin E (α-, β-, γ-tocopherol, α-tocopherol acetate) in the commercial tocopherol supplement. The separation of the isomeric forms of tocopherols (TOHs) is possible by CEC.

Domínguez-Vega et al. (2011) have separated L- and D- citrulline in dietary supplements using the CEC method. The results showed that L- and D-citrulline in commercial food supplements can be detected down to 0.15% of the enantiomeric impurity in the commercial samples manufactured based on L-citrulline.

Medina et al. (2001) determined triglyceride profiles in sunflower, olive, and corn oils using a laboratory-made CEC system. The separation of triglycerides in olive oil after solvent extraction (ACN/50 mM ammonium acetate) was done with nonaqueous CEC.

Feng and Zhao (2009) have extracted (-)-epigallocatechin gallate, (-)-epicatechin gallate, and caffeine in green tea using the preparative CEC apparatus.

Zhang et al. (2014) have determined nitrate and nitrite in ham sausages using CEC.

Jiang et al. (2016) proposed to determine the phthalic acid esters in beverage samples using a surfactant-free microemulsion reinforced hollow-fiber liquid phase microextraction procedure followed by FESI (field enhanced sample injection) with reverse migrating micelles separation.

Scampicchio et al. (2004) have developed a microchip CE procedure for the measurements of some phenolic acids such as chlorogenic, gentisic, ferulic, and vanillic acids present in food with end column amperometric detection.

Masár et al. (2001) presented a study that involves a pioneer work on the analysis of organic acids in wines by chip based CE.

Pobozy et al. (2014) studied the effects of different types of capillaries on the protein profile by CZE-UV involving albumin, globulin, and zein fractions of native and genetically modified maize.

Zhao et al. (1999) and Klamp et al. (1988) reported the different methods of amino acid analysis with CZE-UV detector without the need for prior derivatization. Zhao et al. (1999) analyzed β-N-oxalyl-L-α, β-diaminopropionic acid and homoarginine in grass pea, while Klamp et al. (1988) analyzed 13 free amino acids such as Lys, Arg, His, Gly, Ala, Ser, Asn, Trp, Glu, Phe, Tyr, Pro, and Cys in fresh and canned orange juice and in different types of beers.

Strickland et al. (1996) used two different methods for the determination of amino acids in cheddar cheese. CZE was used in one of the methods, while MEKC was used in the other. Both methods were coupled with UV detection.

Herrero-Martinez et al. (2000) also used CZE-UV to detect the presence of cow milk in other milks and cheeses.

Porto et al. (2015) illustrated the separation of unsaturated fatty acids and the quantification of trans fatty acids using CZE-UV in processed foods such as butter toffee, cake mix, stuffed wafers, chocolate, and mix for Brazilian cheese bread.

REFERENCES

Abidi, S.L., 2004. Capillary electrochromatography of sterols and related steryl esters derived from vegetable oils. *J. Chromatogr. A*, 1059, 199–208.

Altria, K.D., 1995a. Fundamentals of Capillary Electrophoresis Theory, In *Methods in Molecular Biology*, vol. 52, Capillary Electrophoresis Guidebook, Principles, Operation and Applications; Edited by K.D. Altria. The Humana Press, Totowa, NJ, 3–13.

Altria, K.D., 1995b. Standard Commercial Instrument Description, In *Methods in Molecular Biology*, vol. 52, Capillary Electrophoresis Guidebook, Principles, Operation and Applications; Edited by K.D. Altria. The Humana Press, Totowa, NJ, 15–19.

Altria, K.D., 1995c. Typical Operating Procedures, In *Methods in Molecular Biology*, vol. 52, Capillary Electrophoresis Guidebook, Principles, Operation and Applications; Edited by K.D. Altria. The Humana Press, Totowa, NJ, 21–28

Banks, J.F., Gulcicek, E.E., 1997. Rapid peptide mapping by reversed-phase liquid chromatography on nonporous silica with on-line electrospray time -of-flight mass spectrometry. *Anal. Chem.* 69, 3973–3978.

Blatny, P., Kvasnička, F., 1999. Application of capillary isotachophoresis and capillary zone electrophoresis to the determination of inorganic ions in food and feed samples. *J. Chromatogr. A*, 834, 419–431.

Bolshakov, D.S., Amelin, V.G., 2016. Determination of pesticides in environmental materials and food products by capillary electrophoresis. *J. Anal. Chem.*, 71, 965–1013.

Bruin, G.M., Chang, J.P., Kuhlmann, R.H., Zegers, K., Kraak, J.C., Poppe, H., 1992. Capillary zone electrophoretic separations of proteins in poly- ethylene glycol-modified capillaries. *J. Chromatogr.* 471, 429–436.

Cobb, K.A., Dolnik, V., Novotny, M., 1990. Electrophoretic separations of proteins in capillaries with hydrodynamically stable surface structures. *Anal. Chem.*, 61, 2478–2483.

D'orazio, G., Asensio-Ramos, M., Fanali, C., Hernández-Borges, J., Fanali, S., 2016. Capillary electrochromatography in food analysis. *Trends Anal. Chem.*, 82, 250–267.

de Oliveira, M.A.L., Porto, B.L.S., Bastos, C.A., Sabarense, C.M., Vaz, F.A.S., Neves, L.N.O., Duarte, L.M., Campos, N.S., Chellini, P.R., da Silva, P.H.F., de Sousa, R.A., Marques, R., Sato, R.T., Grazul, R.M., Lisboa, T.P., Mendes, T.O., Rios, V.C., 2016. Analysis of amino acids, proteins, carbohydrates and lipids in food by capillary electromigration methods: a review. *Anal. Methods*, 8, 3649–3680.

Domínguez-Vega, E., Crego, A.L., Lomsadze, K., Chankvetadze, B., Marina, M.L., 2011. Enantiomeric separation of FMOC-amino acids by nano-LC and CEC using a new chiral stationary phase, cellulose tris (3-chloro-4-methylphenylcarbamate). *Electrophoresis*, 32, 2700–2707.

Feng, L., Zhao, F., 2009. Preparative electrochromatography of tea polyphenols and caffeine. *Chromatographia*, 69, 1325–1332.

Ferguson, N.J., Braunschweiger, K.I., Braunschweiger, W.R., Smith, J.R., McCormick, J.J., Wasmann, C.C., 1980. Hydrogen ion buffers for bio- logical research. *Anal. Biochem.*, 104, 300–310.

Henry, C.W., Fortier, C.A., Warner, I.M., 2001. Separation of tocopherol isomers using capillary electrochromatography: comparison of monomeric and polymeric C30 stationary phases. *Anal. Chem.*, 73, 6077–6082.

Herrero-Martinez, J.M., Simo-Alfonso, E.F., Ramis- Ramos, G., Gelfi, C., Righetti, P.G., 2000. Determination of cow's milk and ripening time in nonbovine cheese by capillary electrophoresis of the ethanol-water protein fraction. *Electrophoresis*, 21, 633–640.

Hjerten, S., 1988. High performance electrophoresis: elimination of electro end osmosis and solute absorption. *J. Chromatogr.*, 347, 191–198.

Huang, X., Pang, T.K., Gordon, M.J., Zare, R.N., 1987. On column conductivity detector for capillary zone electrophoresis. *Anal. Chem.* 59, 2747–2749.

Huang, X., Zare, R.N., Sloss, S., Ewing, A.G., 1991. End column detection for capillary zone electrophoresis. *Anal. Chem.*, 63, 189–192.

Jager, A.V., Tonin, F.G., Tavares, M.F.M., 2007. Comparative evaluation of extraction procedures and method validation for determination of carbohydrates in cereals and dairy products by capillary electrophoresis. *J. Sep. Sci.*, 30, 586–594.

Jenkins, M.A., 1999. Clinical Applications of Capillary Electrophoresis, In *Methods in Molecular Medicine*, vol. 27, Clinical Applications of Capillary Electrophoresis; Edited by S.M. Palfrey. Humana Press Inc., Totowa, NJ, 1–9

Jiang, T.F., Liang, T.T., Yue, M.E., Wang, Y.H., Lv, Z.H., 2016. Surfactant-free microemulsion reinforced hollow-fiber liquid-phase microextraction combined with micellar electrokinetic capillary chromatography for detection of phthalic acid esters in beverage and urine. *Food Anal. Methods*, 9, 7–15.

Klamp, C.W., Buchberger, W., Turner, M., Fritz, J.S., 1998. Determination of underivatized amino acids in beverage samples by capillary electrophoresis. *J. Chromatogr. A*, 804, 349–355.

Kobayashi, S., Ueda, T., Kikumoto, J., 1989. Photodiode array detection in high-performance capillary electrophoresis. *J. Chromatogr.*, 480, 179–184.

Kuhn, R., Hoffstetter-Kuhn, S., 1993. *Capillary Electrophoresis Principle and Practice.* Springer Verlag, Berlin, p. 153.

Lacná, J., Foret, F., Kubán, P., 2017. Sensitive determination of malondialdehyde in exhaled breath condensate and biological fluids by capillary electrophoresis with laser induced fluorescence detection. *Talanta*, 169, 85–90.

Lambert, W.J., Middleton, D.L., 1990. pH hysteresis effect with silica capillaries in capillary zone electrophoresis. *Anal. Chem.*, 62, 1585–1587.

Landers, J.P., 2008. Introduction to Capillary Electrophoresis, In *Handbook of Capillary and Microchip Electrophoresis and Associated Microtechniques*; Edited by J.P. Landers. CRC Press, Boca Raton, FL, 4.

Li, L.Q., Cai, Y., Yang, M., Shen, Q., Yu, K.M., Cheung, H.Y., 2015. Improvement of reproducibility and sensitivity by reducing matrix effect in micellar electrokinetic chromatography for determination of amino acids in turtle jelly. *Electrophoresis*, 36, 1186–1195.

Mala, Z., Gebauer, P., Bocek, P., 2013. Recent progress in analytical capillary isotachophoresis. *Electrophoresis*, 34, 19–28.

Masár, M., Kaniansky, D., Bodor, R., Jöhnck, M., Stanislawski, B., 2001. Determination of organic acids and inorganic anions in wine by isotachophoresis on a planar chip. *J. Chromatogr. A*, 916, 167–74.

Mateos-Vivas, M., Domínguez-Álvarez, J., Rodríguez-Gonzalo, E., Carabias-Martínez, R., 2017. Capillary electrophoresis coupled to mass spectrometry employing hexafluoro-2-propanol for the determination of nucleosides and nucleotide mono-, di- and tri-phosphates in baby foods. *Food Chem.*, 233, 38–44.

McCord, B., 2001. Quantitative Measurements, In *Methods in Molecular Biology*, vol. 162, Capillary Electrophoresis of Nucleic Acids; vol. 1, Introduction to the Capillary Electrophoresis of Nucleic Acids; Edited by K.R. Mitchelson, J. Cheng. Humana Press Inc., Totowa, NJ, 51–56.

Medina, J.C., Alonso, M.C., Barcelo, D., Lee, M.L., 2001. Simple gradient system for capillary electrochromatography. *J. Microcolumn Sep.*, 13, 351–360.

Mei, S., Wu, D., Jiang, M., Lu, B., Lim, J.-M., Zhou, Y.-K., Lee, Y.-I., 2011. Determination of trace bisphenol A in complex samples using selective molecularly imprinted solid-phase extraction coupled with capillary electrophoresis. *Microchem. J.*, 98, 150–155.

Mitchelson, K.R., 2001. The Application of Capillary Electrophoresis for DNA Polymorphism Analysis, In *Methods in Molecular Biology*, vol. 162, Capillary Electrophoresis of Nucleic Acids; vol. 1, Introduction to the Capillary Electrophoresis of Nucleic Acids; Edited by K.R. Mitchelson, J. Cheng. Humana Press Inc., Totowa, NJ, 3–26.

Nikovaev, A.V., Kartsova, L.A., Filimonov, V.V., 2015. A microfluidic chip for the determination of polyphenolic antioxidants. *J. Anal. Chem.*, 70, 983–988.

Pobozy, E., Sentkowska, A., Piskor, A., 2014. Comparison of three modifications of fused-silica capillaries and untreated capillaries for protein profiling of maize extracts by capillary electrophoresis. *J. Sep. Sci.*, 421, 37, 2388–2394.

Porto, B.L.S., Faria, I.D.L., de Oliveira Mendes, T., de Oliveira, M.A.L., 2015. Fast screening method for the analysis of trans fatty acids in processed food by CZE-UV with direct detection. *Food Control*, 55, 230–235.

Rang, Y., Zeng, H.L., Nakajima, H., Kato, S., Uchiyama, K., 2015. Quantitative on-line concentration for capillary electrophoresis with inkjet sample introduction technique. *J. Sep. Sci.*, 38, 2722–2728.

Sander, L.C., Pursch, M., Märker, B., Wise, S.A., 1999. Separation of carotenoid isomers by capillary electrochromatography with C30 stationary phases. *Anal. Chem.*, 71, 3477–3483.

Scampicchio, M., Wang, J., Mannino, S., Chatrathi, M.P., 2004. Microchip capillary electrophoresis with amperometric detection for rapid separation and detection of phenolic acids. *J. Chromatogr. A*, 1049, 189–94.

Silvertand, L.H.H., Torano, J.S., Bennekom, W.P., Jong, G.J., 2008. Recent developments in capillary isoelectric focusing. *J. Chromatogr. A*, 1204, 157–170.

Slater, G.W., Desruisseaux, C., Hubert, S.J., 2001. DNA Separation Mechanisms during Electrophoresis, In *Methods in Molecular Biology*, vol. 162, Capillary Electrophoresis of Nucleic Acids; vol. 1, Introduction to the Capillary Electrophoresis of Nucleic Acids; Edited by K.R. Mitchelson, J. Cheng. Humana Press Inc., Totowa, NJ, 27–41.

Soga, T., Ross, G.A., 1999. Simultaneous determination of inorganic anions, organic acids, amino acids and carbohydrates by capillary electrophoresis. *J. Chromatogr. A*, 837, 231–239.

Strickland, M., Weimer, B.C., Broadbent, J.R., 1996. Capillary electrophoresis of Cheddar cheese. *J. Chromatogr. A*, 731, 305–313.

Sun, X., Lin, L., Liu, X., Zhang, F., Chi, L., Xia, Q., and Linhardt, R.J., 2016. Capillary electrophoresis–mass spectrometry for the analysis of Heparin Oligosaccharides and low molecular weight Heparin, *Anal. Chem.*, 88(3), 1937–1943.

Sun, S., Wei, Y., Cao, Y., Deng, B., 2017. Simultaneous electrochemiluminescence determination of galanthamine, homolycorine, lycorenine, and tazettine in Lycoris radiata by capillary electrophoresis with ultrasonic-assisted extraction. *J. Chromatogr. B*, 1055–1056, 15–19.

Van Orman, B.B., Liversidge, G.G., McIntire, G.L., 1991. Effects of buffer composition on electroosmotic flow in capillary electrophoresis. *J. Microcol. Sep.*, 2, 176.

Wallingford, R.A. Ewing, A.G., 1987, Characterization of a micro injector for capillary zone electrophoresis. *Anal. Chem.*, 59, 678–681.

Whatley, H., 2001. Basic Principles and Modes of Capillary Electrophoresis, In *Clinical and Forensic Applications of Capillary Electrophoresis*; Edited by J. Peterson, A.A. Mohammad. The Humana Press, Springer Science+Business Media, New York, 21–58.

Wu, M.L., Gao, F., Zhang, Y., Wang, Q.J., Li, H., 2015. Sensitive analysis of amino acids and vitamin B3 in functional drinks via field-amplified stacking with reversed-field stacking in microchip electrophoresis. *Talanta*, 131, 624–631.

Wu, S., Dovichi, N.J., 1992. Capillary zone electrophoresis separation and laser-induced fluorescence detection of zeptomole quantities of fluorescein thiohydantoin derivatives of amino acids. *Talanta*, 39, 173.

Zeng, J., Kuang, H., Hu, C., Shi, X., Yan, M., Xu, L., Wang, L., Xu, C., Xu, G., 2013. Effect of bisphenol A on rat metabolic profiling studied by using capillary electrophoresis time-of-flight mass spectrometry. *Environ. Sci. Technol.*, 47, 7457–7465.

Zhang, Y., Tian, X., Guo, Y., Li, H., Yu, A., Deng, Z., 2014. Analysis of nitrites and nitrates in hams and sausages by open-tubular capillary electrochromatography with a nanolatex-coated capillary column. *J. Agric. Food Chem.*, 62, 3400–3404.

Zhao, L., Chen, X., Hu, Z., Li, Q., Chenb, Q., Li, Z., 1999. Analysis of β-N-oxalyl-l-α, β-diaminopropionic acid and homoarginine in Lathyrus sativus by capillary zone electrophoresis. *J. Chromatogr. A*, 857, 295–302.

Zhu, Z., Lu, J.J., Liu, S. 2012. Protein separation by capillary gel electrophoresis: A review. *Anal. Chim. Acta*, 709, 21–31.

CHAPTER 5

Flow Injection Analysis—Tandem Mass Spectrometry

Claudia Ruiz-Capillas, Ana M. Herrero, and Francisco Jiménez-Colmenero
Instituto de Ciencia y Tecnología de Alimentos y Nutrición (ICTAN-CSIC)

CONTENTS

5.1 INTRODUCTION

Authentication and traceability are essential tools for controlling food quality and safety. Promoting food safety is an essential aspect of public health, which is why there are increasing numbers of controls on products throughout the supply chain, "from farm to fork". The need is thus, on one hand, to verify that a product meets the descriptions on the label so as to detect adulterations and fraud and, on the other hand, to track the product through every stage in its production, from reception of the raw materials to the delivery of the final product—all further geared toward the export and import markets, consumer demands, and government regulations.

Also, the improvements in product quality and safety, traceability controls, and authentication of food benefit the producers by reducing costs and enhancing the competitiveness of those applying them; they also benefit the authorities by assuring the quality of the products and also by guaranteeing and protecting consumer interests (Opara, 2003).

All the components of a food product can be identified and traced, but this requires the right analytical techniques. Finding analytical techniques for the detection of adulteration and for the traceability of food is a challenge for the food industry and the health authorities in every country. The point is to acquire fast, reliable techniques to assure adequate detection, identification, and authentication of the marketed products (and their components) and to verify their compliance with the labeling regulations.

Several technologies for identification and/or verification already exist, particularly for implementing traceable supply chains. These include physicochemical techniques, chromatography, spectroscopy, electronic noses, biosensors, flow injection analysis, molecular biology methods, etc. (Lüdi et al., 1990; Valcarcel and Luque de Castro, 1994; Aarnisalo et al., 2007; Peres et al., 2007; Peris et al., 2009; Aung and Chang, 2014; Ruiz-Capillas and Nollet, 2016).

However, many of these techniques are costly and time consuming and, hence, their utility as routine techniques is limited. Recent advances in techniques like fingerprinting, particularly coupling or tandem flow injection analysis mass spectrometry (FIA-MS), have made it possible to increase their use in the field of food quality and safety, and have opened up new opportunities for enhancing the efficiency and compatibility of these techniques for the verification of food traceability systems and claims to food origin, production, and authenticity.

5.1.1 Flow Injection Analysis (FIA)

The FIA system offers interesting advantages such as versatility, flexibility, accuracy, and suitability for the fast, routine analysis of a large number of samples with high precision. This methodology responds to a growing demand for the mechanization and automation of analysis. FIA has made it possible to automate the most routine analyses, increasing their efficiency and improving the precision of determinations. For that reason, it is used in practically all the fields/areas of science, for medical, environmental and agriculture analysis, food quality control, and also for process analysis in various branches of the industry (Campanella et al., 1996; Ruiz-Capillas and Horner, 1999; Ruzicka and Hansen, 2008; Ruiz-Capillas and Jiménez-Colmenero, 2008; Yebra-Biurrun, 2009; Ruiz-Capillas and Jiménez-Colmenero, 2009; Ruiz-Capillas and Nollet, 2016; Trojanowicz and Kołacińska, 2016). The basic instrumental elements of the typical FIA systems are a pumping device, an injection valve, and a detector which appropriately and continuously records measured signals (Figure 5.1). Small volumes of reagent solutions are injected into the stream of aspirated sample to the system (Ruzicka and Hansen, 1975; Ruiz-Capillas and Horner, 1999; Ruiz-Capillas and Jiménez-Colmenero, 2008; Ruzicka and Hansen, 2008; Cerdà et al., 2016; Trojanowicz, 2016; Campuzano et al., 2016). The most common device used for the pumping of solutions in FIA systems is a single-line or multi-line peristaltic pump (Figure 5.1). In a typical flow system, selectivity of determination is achieved in different ways. If possible, one can use a specific detector responding to particular species only, or the instrumental parameters of detector functioning can be used to provide a specific response to the analyte. The most commonly used detection method is spectrophotometry in the UV/Vis range; it has been used in other detector systems also, such as potentiometry, amperometry, fluorimetry, luminescence, infrared spectroscopy, etc. (Trojanowicz and Matuszewski, 1983; Trojanowicz et al., 1991; Fang, 1994; Sanz-Medel, 1999; Trojanowicz, 2000; Fletcher et al., 2001; Gallignani and Brunetto, 2004; Rodriguez-Mozaz et al., 2004; O'Toole et al., 2005; Trojanowicz, 2008; Ruzicka, 2009; Zagatto et al., 2011; Pokrzywnicka et al., 2012; Strzelak and Koncki, 2013; Trojanowicz and Kołacińska, 2016; Campuzano et al., 2016). Appropriate ion-selective electrodes, selective biosensors, or molecular/atomic spectroscopy can also be applied to spectral parameters to provide a response that is not affected by the other components of the sample (Trojanowicz and Matuszewski, 1983; Trojanowicz et al., 1991; Ferreira and Lima, 1994; O'Toole et al., 2005; Ruzicka, 2009; Pascoa et al., 2012;

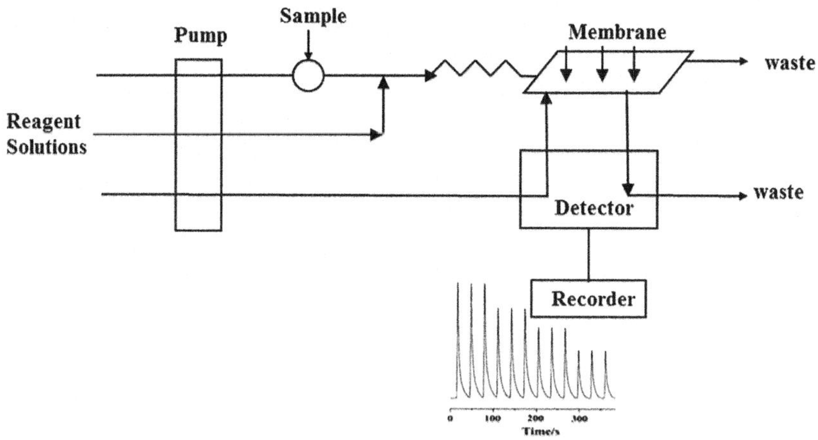

FIGURE 5.1 Diagram of the flow injection/gas diffusion for the determination of trimethylamine. (Adapted from Ruiz-Capillas and Horner, 1999).

Pokrzywnicka et al., 2012; Strzelak and Koncki, 2013; Campuzano et al., 2016; Segundo et al., 2016). Another especially effective way of achieving selective response is by incorporating an appropriate sample processing step into the flow system. In this way, the analyte can be extracted from the sample matrix (Trojanowicz, 2008; Kolev and McKelvie, 2008; Yebra-Biurrun, 2009; Zagatto et al., 2011; Cerdà et al., 2016).

Besides the optimization of hydrodynamic conditions for the determination in FIA systems, on-line sample processing or on-line sample cleanup are essential factors for proper selection and optimization of chemical conditions of determinations (Trojanowicz, 2016; Cerdà et al., 2016).

FIA has been successfully used as an initial step in other instrumentations that are more complex; in general, practically all modern instrumental methods of chemical analysis for measurements in liquids have been employed in FIA measurements, from simple photometry to nuclear magnetic resonance (NMR). One field of such application that is widely investigated in research laboratories is the hyphenation of typical flow analysis systems with instruments for high performance separation, such as liquid and gas chromatography, capillary electrophoresis, mass spectroscopy, etc. (Benkhedda et al., 2002; Keifer, 2007; Almeida et al., 2011; Baker et al., 2012; Clavijo et al., 2015; Trojanowicz, 2016; Campuzano et al., 2016; Segundo et al., 2016). FIA has given the chemical industries a new dimension. As Cristian and Townshend defined it, "flow injection analysis is more than analytical technique. It is a platform for the majority of analytical methods" (Trojanowicz, 2016).

5.1.2 Mass Spectrometry (MS)

MS is a laboratory technique for chemical analyses based on the detection and quantification of ions according to the mass-to-charge ratio (m/z) of compounds in a mixture. The main components of the mass spectrometer are: sample input system, ion source, mass analyzer, and detector (Figure 5.2). MS determination entails the following steps: (1). Generation of molecules (molecular fragments and/or atoms) in gaseous phase; (2) Ionization of these molecules; (3) Separation as determined by the ratio m/z; (4) Detection of ions. The molecules are vaporized and ionized (in the ionization source) and the ions thus generated are accelerated up to an analyzer and separated according to their (m/z) ratios by means of electric or magnetic fields, or simply by determining the time taken to reach a detector. The ions that reach the detector produce an electrical signal that is processed, amplified, and sent to a computer to calculate the mass spectrum (Figure 5.2). This indicates the ion abundances that are calculated on the basis of the (m/z) ratios of the ions detected. This sequence makes it possible to analyze different chemical elements and atomic isotopes with a high degree of precision since each compound is unique and each compound will be ionized and fragmented in a particular way. This is the principle by which mass spectrometry identifies each analyte. This technique can be applied to any kind of analyte that can be vaporized and ionized without decomposition (Hoffman and Stroobant, 2001; Glish and Vachet, 2003; Watson and Sparkman, 2007).

MS is a powerful analytical tool for the identification of compounds. It is very widely used and is very common in analytical laboratories as it makes it possible to study a large variety of compounds—organic, inorganic, or biological (including biopolymers and natural or artificial macromolecules). MS can be used to determine the isotopic content of different chemical elements in a single compound, to obtain information on the molecular mass of the target compound, and also to obtain structural information about it, or simply to detect its presence and/or quantify its concentration. These MS methods have the further advantage of high sensitivity and excellent specificity, which greatly reduces the background interference. Although MS can identify any pure substance almost unequivocally, it cannot normally identify the individual components of a mixture without prior separation of its components owing to the extreme complexity of the spectrum that is obtained by superimposition of the particular signals of each component. Thus, for the analysis of complex samples, hyphenations are used with powerful separation techniques, such as gas chromatography mass spectrometry (GC-MS), liquid chromatography mass spectrometry (LC-MS), capillary electrophoresis mass spectrometry (CE-MS), ion mobility spectrometry mass spectrometry (IMS-MS), triple modalities such as LC-IMS-MS, inductively coupled plasma mass spectrometry (ICP- MS), and FIA-MS, etc. (Fenselau, 1992; Niessen, 1998; McCooeye and Mester, 2006; Schug and Fan, 2012; Di Stefano et al., 2012; Gao et al., 2014; Maher et al., 2015; Nanita and Kaldon, 2016; Ray et al., 2017). Some of these hyphenation techniques, such as LC-MS, have been widely used in

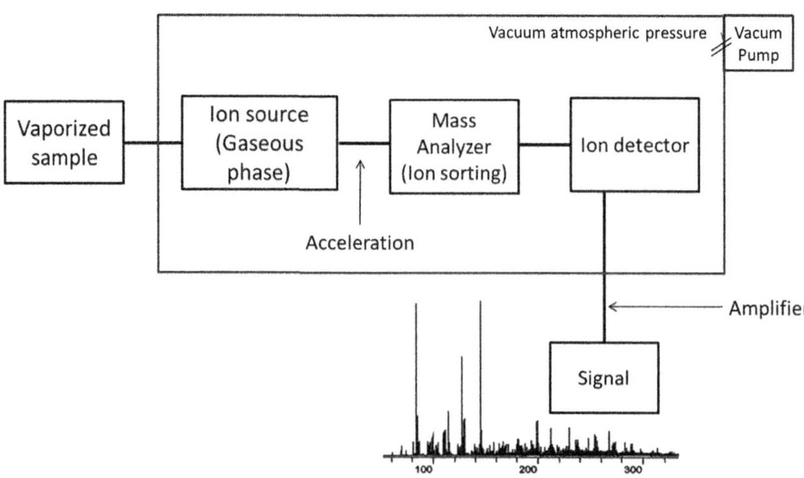

FIGURE 5.2 Schematic of mass spectrometer.

food analysis for the qualification and/or quantification of residual substances or different compounds (Kmellár et al., 2010; MaLik et al., 2010; Di Stefano et al., 2012). However, chromatographic separation is a time -consuming process and the cost of analysis can be increased by the use of expensive columns, the cost of solvents consumed, their disposal, etc. Therefore, development of a faster and more cost effective analytical technique is of significant interest to the food industry, which moreover seeks high-throughput monitoring of compounds during food processing, packaging, storage, or verification of regulatory compliance prior to market delivery. For all these reasons, several FIA-MS methods have recently been developed for the fingerprinting and screening of many compounds in most areas/fields (Vaidyanathan, et al., 2002; Gomez-Ariza, et al., 2006; Beckmann, et al., 2008; Henderickx, et al., 2009; Sun and Chen, 2011; Chen et al., 2011; Fuhrer et al., 2011; Schug and Fan, 2012; Lim et al., 2012; Bhandari et al., 2013; Zhao et al., 2013; Gao et al., 2014; Lu et al., 2015; Nanita and Kaldon, 2016; among others).

5.2 FLOW INJECTION ANALYSIS COUPLED WITH MASS SPECTROMETRY (FIA-MS)

The coupling of FIA and MS is a combined technique (FIA-MS) which can be used to separate and identify complex mixtures, and to mechanize sampling and on-line sample processing, which makes for very simple, fast, easy, and high performance analysis.

In the 1990s, FIA-MS systems were introduced in many laboratories, thanks to the inclusion of electrospray ionization (ESI) in commercial mass spectrometers (Niessen, 1998). Several MS models had a built-in loop injector coupled to liquid chromatography autosamplers and injectors that allowed operation of the mass spectrometer in FIA-ESI-MS mode. These systems were especially used for monitoring non-covalent interactions (Niessen, 1998; Nanita and Kaldon, 2016). At that time, FIA was also used with ICP-MS systems for chemical speciation by reversed phase liquid chromatography, where it was used to estimate the dispersion in sample delivery systems (Thompson and Houk, 1986). However, many of these models were limited to relatively simple matrixes, such as water, and in many cases the thresholds of detection and the quantification limits were not low enough (Geerdink et al., 1993). Major advances have been achieved in this technique to solve these problems. These advances have focused primarily on gaining a better understanding of ion formation and manipulation, and on innovation and advances in instrumentation engineering, injection systems, sample preparation, design, data analysis, etc., with a view to enhancing their selectivity and sensitivity (Van Oudenhove and Devreese, 2013; Nanita and Kaldon, 2016). A major improvement has been achieved in the accuracy and precision of FIA-MS methods, thanks to the optimization of injection systems, which can now deliver microlitre-volume injections reliably and repeatedly (Oellig and Schwack, 2014). The more common FIA-MS methods simply use autosamplers and injectors along with a carrier fluid to insert samples in the ion source for mass spectrometric analysis (Figure 5.3). A less commonly used system, but one typical for FIA, is to mix fluids in the system's manifold to produce a chemical reaction or to create an analyte concentration gradient followed by MS detection (Schug et al., 2010; Nanita, 2013; Santos et al., 2015).

On the other hand, sample manipulation (extraction, purification, and/or preconcentration) is another critical step in FIA-MS determinations and is one of the decisive factors in improving the performance and selectivity of this technique. FIA-MS was designed only for the analysis of liquids similar to LC-MS, and therefore any sample that is not in the liquid state will require some form of preparation (e.g., extraction) prior to FIA-MS

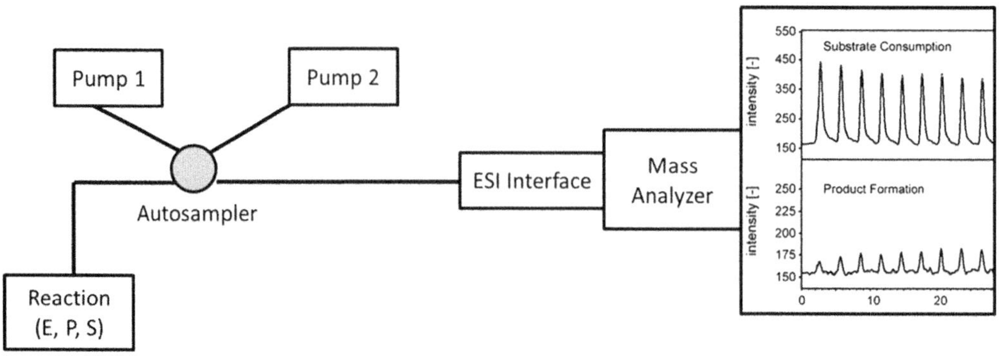

FIGURE 5.3 Schematic setup of the flow injection electrospray ionization mass spectrometry (FIA-ESI-MS) bioassay. Pumps 1 and 2 deliver a constant carrier stream; the reaction mixture consisting of enzyme (E), substrates (S), and respective products (P) is injected by the autosampler (Adapted from Liesener and Karst, 2005).

or LC-MS analysis (Benkhedda et al., 2002; Cerdà and Estela, 2005; Nanita and Kaldon, 2016). Moreover, FIA-MS methods require sample preparation for increased selectivity and compatibility with the technique, as they do not include chromatography with separation of sample components, purification of analytes, removal of matrix effects, etc. Therefore, researchers have reported using various sample extraction/preparation or pretreatment techniques or cleanup steps coupled with FIA-MS/ MS to purify the analytes and reduce the complexity of the sample (i.e., remove potential interferences) prior to instrumental analysis. All this improves their selectivity and ruggedness. There are several sample preparation methods compatible with FIA-MS, such as classic liquid extractions with water immiscible solvents, such as ethyl acetate and hexane (Ito et al., 2003). This quick, easy, cheap, effective, rugged, and safe (QuEChERS) method is also one of the most popular methods for sample extraction. It is based on salting out acetonitrile/water with the addition of salts (NaCl and MgSO4) for extract dehydration and partition of acetonitrile/aqueous layers, and for purification by dispersive solid phase extraction (SPE) (Anastassiades et al., 2003; Nanita et al., 2009; Kmellár et al., 2010; Oellig and Schwack, 2014; Bruzzoniti et al., 2015). However, the presence of significant levels of nonvolatile salts in these extracts may not be compatible with FI-MS for routine chemical residue analysis based on current acquisition methods, since their ions are known to induce adduct formation and reduce the signal intensities of protonated molecular ions. Frequent injection of samples containing these salts, without the possibility of separation/diversion prior to insertion in the mass spectrometer (as in HPLC-MS methods), may also require more frequent ion source maintenance (Nanita and Kaldon, 2016). Another popular method is one known as SALLE (salting out assisted liquid-liquid extraction) in which mass spectrometry-friendly ammonium salts are used; these are volatile and do not accumulate in the ion source or the mass spectrometer atmospheric interface (Wu et al., 2004; Zhang et al., 2010; Nanita and Padivitage, 2013; Nanita and Kaldon, 2015; Yang et al., 2015). Nanita (2011) and Nanita et al. (2011) also presented fast extraction and dilution flow injection mass spectrometry (FED-FI-MS) as a technique to increase the throughput for quantitative multiresidue screening in complex matrixes. This is based on the hypothesis that by the dilution of extracts, direct flow injection MS multiresidue analysis is possible without chromatography, dispersive SPE, or organic/aqueous layer partition. Dilute-and-shoot sample preparation with FIA-MS-MS can provide faster analysis than SPE-FIA-MS-MS analysis (Nanita, 2011;Nanita et al., 2011). FED-FI-MS bridges the gap that currently exists in methods for multiresidue analysis, i.e., between slower sample extraction/cleanup/HPLC/MS methods and the faster ambient ionization/MS alternative. A recent new and fast FED method has been developed by Nanita and Kaldon (2015). This incorporates an ammonium salting out cleanup and was especially designed for use with FIA-MS. This extraction and purification procedure also provides rugged sample analysis when coupled with FIA-MS-MS for analysis of complex pesticide matrixes and allows high-throughput part-per-billion analysis. It is a procedure whereby FIA-MS-MS methods can be expanded to analyze food from plant (sorghum, rice, grapefruit, canola) and animal origin (milk, eggs, beef), as well as body fluids (blood plasma and urine) with increased sensitivity and ruggedness as compared to the earlier method without cleanup (Mol et al., 2008; Zhang et al., 2010).

Along with optimization of the sample composition at the point of injection, it is also important to optimize the carrier solvent used (Nanita and Kaldon, 2016). The solvent must be compatible with the MS ionization method. The most common MS ionization methods include ESI and inductively coupled plasma (ICP) matrix assisted laser desorption/ionization (MALDI). Nanita et al. (2011) assessed the collective sensitivity of 12 pesticides where the extracts were diluted with pure methanol or with methanol containing 1% ammonium acetate, 1% formic acid, or 1% ammonium hydroxide in a triple quadrupole mass spectrometer with an electrospray ion source. When ammonium hydroxide was added, ionization was three times more efficient than when only methanol was used. In case of some analytes, there was some instability under high pH conditions (1% ammonium hydroxide in methanol solution), but the rapid instrumental analysis achieved with FIA-MS yielded accurate results even for relatively large sample sets (Nanita et al., 2011). Ito et al. (2003) employed liquid extraction with ethyl acetate to extract fungicides from citrus fruit prior to FIA-MS-MS instrumental analysis. They were thus able to quantify thiabendazole and imazalil with a limit of quantitation (LOQ) of 1 mg/kg, and o-phenylphenol with a LOQ of 5 mg/kg using a Quattro II triple quadrupole mass spectrometer and FI-ESI.

Then again, advances in different types of mass analyzers (quadrupole, ion traps, etc.) have provided a wide range of analyte selectivity (Bui and Cooks, 1998; Ito et al., 2003; Gomez-Ariza et al., 2006; Nanita et al., 2009; Nanita, 2011; Du et al., 2017). For example, in case of ion trap mass, greater precision has been achieved with accurate mass measurements and higher number of cycles of tandem mass analysis in time. Several (as many as five) ion transitions are performed to ensure adequate method selectivity and to minimize the possibility of false positives (Song et al., 2014). Nanita and Kaldon (2016) proposed analyte fragmentation reactions (ion transitions) that are highly specific in FIA-MS-MS methods using triple quadrupole instruments. The system of FIA-MS-MS analysis includes ion transitions for complex samples that have compounds with similar chemical structures to avoid analyte-to-analyte interferences (Nanita et al., 2009). These systems have been used, for example, in the analysis of structurally similar pesticide residues (e.g., isomers) and yield fragment ions of the same *m/z* and they may share the same ion transitions.

5.3 FIA-MS APPLICATIONS IN FOOD

FIA-MS is increasingly being recognized as a suitable technique for successful applications in many fields—clinical diagnostics, metabolomics, agricultural and environmental sciences, toxicology, pharmaceutical research, detection of adulterated/counterfeit food, etc.—since it makes it possible to determine and quantify a wide variety of analytes at levels ranging from parts-per-billion to parts-per-million in very complex matrixes (such as blood, urine, and a variety of food from plant and animal origin) (Hayward et al., 1990; Kristiansen et al., 1994; Vaidyanathan, et al., 2002; Ito et al., 2003; Wade and Miller, 2005; Beckmann et al., 2008; Enot et al., 2008; Fang et al., 2008; Nanita et al., 2009; Tu et al., 2010; Nanita et al., 2011; Nanita, 2011; Gao et al., 2012; Lu et al., 2014ab; Perez-Gregorio et al., 2014; Gonzalez-Dominguez et al., 2015; etc.).

ESI-MS-MS has been successfully used for rapid quantification and high-throughput mass spectral analysis of vitamins in nutrition formulations, fortified food, and dietary supplements. The principal analytes have been B vitamins, ascorbic acid, and folic acid in multivitamin/multielement tablets. Bhandari and Van Berkel (2012) showed that solid-liquid extraction followed by FI-ESI-MS-MS can quantitatively detect B vitamins with milligram-per-gram concentrations in nutritional supplements without any chromatographic separation. Automated flow injection has been performed using an Agilent 1100 series high performance liquid chromatography (HPLC) system. The HPLC system was coupled to a Thermo Scientific (LTQ XL) linear ion trap mass spectrometer equipped with an ESI source operated in positive ion mode. Sample extracts were directly inserted in the ion source of a mass spectrometer without chromatographic separation. The same authors (Bhandari et al., 2013) also reported the development of a FI-ESI-MS-MS analysis scheme (in negative ion mode) for the quantification of ascorbic acid and folic acid in multivitamin/multielement tablets, which is similar to ID-LC-MS (isotope dilution liquid chromatography/mass spectrometry) analyses of the same materials for high-throughput sample analysis.

FI-MS has also been used to screen for adulterated drugs and dietary supplements, such as lovastatin and sildenafil in red yeast rice products (Song et al., 2012a), or sibutramine in dietary supplements sold for weight loss (Song et al., 2012b). FI-MS-MS offers improved sensitivity, selectivity, and wider detection ranges (Song et al., 2014). Sun and Chen (2011) also tested a FI/MS fingerprinting method in combination with principal component analysis (PCA) to detect *Scutellaria lateriflora* (skullcap) in adulterated dietary supplements. The study was carried out in a FI-MS system consisting of a LCQ classic ion trap mass spectrometer (IT-MS) with an Agilent 1100 LC system (a quaternary pump with a vacuum degasser, a thermostatted column compartment, an autosampler, and a diode array detector). A guard column was used to minimize the potential contamination of the MS system, but this guard column did not provide meaningful separation. Shin et al. (2014) also used a combination of ESI-MS-MS and the Mass Bank database to identify an adulterated, unknown compound in a commercial galactooligosaccharide product. The analyses were carried out in an Acquity ultra performance liquid chromatography (UPLC) system coupled to a Quattro premier XE mass spectrometer equipped with a Z-Spray ESI source as described by Shin et al. (2011) in their study of identification and structure elucidation of a p-phenoxybenzaldehyde in bamboo shoots by HPLC-ESI-MS-MS and NMR. Each sample solution was directly inserted in the electrospray ion source of a mass spectrometer without column separation.

FI-MS has also been used for rapid detection of milk adulteration with soybean, pea, and whey protein (Du et al., 2017). An Acquity UPLC combined with a Xevo G2 quadrupole time-of-flight mass spectrometer (UPLC-Q-TOF-MS, Waters, Milford, Massachusetts, USA) was used for FI-MS analysis. This system did not use a chromatographic column, and a Protein BEH C4 VanGuard precolumn was connected instead as an additional on-line filter. In this study, FI-MS was also combined with chemometric techniques. Specifically, PCA was applied to determine whether there were significant differences between the pure and adulterated FI-MS fingerprints. Flow injection mass spectral fingerprinting has also been used to explore the origin and cultivar (conventionally and organically) of different fruits and vegetables. Chen et al. (2010) and Chen et al. (2011) used FIA-ESI-IT-MS and FI-ESI-TOF-MS to study the chemical composition of Rio red grapefruits from conventionally and organically grown trees and for the differentiation of three *Panax* species, respectively. In this study, the spectral fingerprints were analyzed using analysis of variance and PCA (ANOVA/PCA). Lu et al. (2014b) also used FIA-MS fingerprinting techniques combined with PCA to differentiate organic and conventional sweet basil leaf samples. The same authors (Lu et al., 2014a) also applied partial least squares discriminant analysis to differentiate the Chinese wolfberries by FIA-MS fingerprinting. Similarly, HPLC and FIA-MS fingerprinting have also been tested for their potential in differentiating organic and conventional peppermint, oregano, and sage samples (Gao et al., 2012).

An atmospheric pressure photoionization (APPI) source coupled with quadrupole time-of-flight mass spectrometry (Qq-TOF-MS) has been used to discriminate among different edible oils, such as olive, extra virgin olive, olive pomace, hazelnut, sunflower, corn, and several mixed oils, with high throughput (approximately 1 min-per-sample). It could thus be used to test for olive oil adulteration with other oils, even in case of hazelnut oil, which exhibit a high chemical similarity with olive oil (Gomez-Ariza et al., 2006).

5.4 CONCLUSIONS

FIA-MS is recognized as a suitable technique for applications where quantitative fingerprinting and screening of chemicals need to be performed rapidly and reliably. Moreover, FIA-MS is a simple method that offers high throughput without compromising on sensitivity, precision, and accuracy.

FIA and mass spectroscopy combined with chemometric techniques have been successful methodologies for the qualitative and quantitative analysis of a large variety of analytes in many complex matrixes at levels ranging from parts-per-billion to parts-per-million, and so they are applicable in practically every area of science, including foodstuffs, and they provide a useful tool for the authentication and traceability of foods.

This technique promises much for the future, and so work will undoubtedly go ahead on improving it in terms of on-line sample processing, innovation in instrumentation, accessories, miniaturization, etc., all aimed at reducing costs and making the technique simpler and faster by enhancing automation without reducing, but rather improving, sensitivity.

ACKNOWLEDGMENTS

This research was supported by projects MEDGAN-CM S2013/ABI2913, 2014470E073 and 201470E056 financially by the Comunidad de Madrid (Spain), European Structural, and CSIC.

REFERENCES

Aarnisalo, K., S. Heiskanen, K. Jaakkola, E. Landor, and L. Raaska. 2007. *Traceability of Foods and Foodborne Hazards*. VTT Technical Research Centre of Finland Research Notes, Finland, 2396.

Almeida, M.I.G.S., J.M. Estela, and V. Cerda. 2011. Multisyringe flow injection potentialities for hyphenation with different types of separation techniques. *Anal. Lett.* 44:360–373.

Anastassiades, M., S.J. Lehotay, D. Stajnbaher, and F.J. Schenck. 2003. Fast and easy multiresidue method employing acetonitrile extraction/partitioning and dispersive solid-phase extraction for the determination of pesticide residues in produce. *J. AOAC Int.* 86:412–431.

Aung, M.M., and Y.S. Chang. 2014. Traceability in a food supply chain: Safety and quality perspectives. *Food Control* 39:172–184.

Baker, J.M., J.L. Ward, and M.H. Beale. 2012. Combined NMR and flow injection ESI-MS for Brassicaceae metabolomics. *Methods Mol. Biol.* 860:177–191.

Beckmann, M., D. Parker, D.P. Enot, E. Duval, and J. Draper. 2008. High-throughput, nontargeted metabolite fingerprinting using nominal mass flow injection electrospray mass spectrometry. *Nat. Protoc.* 3(3):486–504.

Benkhedda, K., H.G. Infante, and F.C. Adams. 2002. Inductively coupled plasma mass spectrometry for trace analysis using flow injection on-line preconcentration and time-of-flight mass analyzer. *Trends Anal. Chem.* 21(5):332–342.

Bhandari D., V. Kertesz, and G.J. Van Berkel. 2013. Rapid quantitation of ascorbic and folic acids in SRM 3280 multivitamin/multi-element tablets using flow-injection tandem mass spectrometry. *Rapid Commun. Mass Spectrom.* 27:163–168.

Bhandari, D., and G.J. Van Berkel. 2012. Evaluation of flow-injection tandem mass spectrometry for rapid and high-throughput quantitative determination of b vitamins in nutritional supplements. *J. Agric. Food Chem.* 60:8356–8362.

Bruzzoniti, M.C., L. Checchini, R.M. De Carlo, S. Orlandini, L. Rivoira, and M. Del Bubba 2015. QuEChERS sample preparation for the determination of pesticides and other organic residues in environmental matrices: A critical review. *Anal. Bioanal. Chem.* 406:4089–4116.

Bui, H.A., and R.G. Cooks. 1998. Windows version of the ion trap simulation program ITSIM: A powerful heuristic and predictive tool in ion trap mass spectrometry. *J. Mass Spectrom.* 33:297–304

Campanella, L., K. Pyrzyńska, and M. Trojanowicz. 1996. Chemical speciation by flow-injection analysis. *Talanta* 43(6):825–38.

Campuzano, S., F. Conzuelo, J.A. Reviejo, and J.M. Pingarrón. 2016. Combination of Flow Injection Analysis and Other Analytical Systems Multianalyte Approaches. In *Flow Injection Analysis of Food Additives*. Eds by Claudia Ruiz-Capillas and Leo M.L. Nollet. CRC Press, Taylor and Francis Group, Boca Raton, FL, Pp 101–120.

Cerdà V., J.M. Estela, and J. Avivar. 2016. Multicommutated Flow Injection Analysis, Multisyringe Flow Injection Analysis, and Multipumping Flow Systems Theory and Trends. In *Flow Injection Analysis of Food Additives*. Eds by Claudia Ruiz-Capillas and Leo M.L. Nollet. CRC Press, Taylor and Francis Group, Boca Raton, FL, Pp 79–99.

Cerdà, V., and J.M. Estela. 2005. Automatic pre-concentration and treatment for the analysis of environmental samples using non-chromatographic flow techniques. *Int. J. Environ. Anal. Chem.* 85:231–253.

Chen, P., J.M. Harnly, and G.E. Lester. 2010. Flow injection mass spectral fingerprints demonstrate chemical differences in Rio Red grapefruit with respect to year, harvest time, and conventional versus organic farming. *J. Agric. Food Chem.* 58(8):4545–4553.

Chen, P., J.M. Harnly, and P.B. Harrington. 2011. Flow injection mass spectroscopic fingerprinting and multivariate analysis for differentiation of three panax species. *J. AOAC Int.* 94(1):90–99.

Clavijo, S., J. Avivar, R. Suárez, and V. Cerdà. 2015. Analytical strategies for coupling separation and flow-injection techniques. *Trends Anal. Chem.* 67:26–33.

Di Stefano, V., G. Avellone, D. Bongiorno, V. Cunsolo, V. Muccilli, S. Sforza, A. Dossena, L. Drahos, and K. Vékey. 2012. Applications of liquid chromatography-mass spectrometry for food analysis. *J. Chromatogr. A* 1259:74–85.

Du L., W. Lu, Z.J. Cai, L. Bao, and L.L. Yu. 2017. Rapid detection of milk adulteration using intact protein flow injection mass spectrometric fingerprints combined with chemometrics. *Food Chem.* doi:10.1016/j.foodchem.2017.07.107.

Enot, D.P., W. Lin, M. Beckmann, D. Parker, D.P. Overy, and J. Draper. 2008. Preprocessing, classification modeling, and feature selection using flow injection electrospray mass spectrometry metabolite fingerprint data. *Nat. Protoc.* 3:446–470.

Fang, A.S., X. Miao, P.W. Tidswell, M.H. Towle, W.K. Goetzinger, and J.N. Kyranos. 2008. Mass spectrometry analysis of new chemical entities for pharmaceutical discovery. *Mass Spectrom. Rev.* 27(1):20–34.

Fang, Z.L. 1994. *Flow Injection Atomic Spectrometry.* Wiley, Chichester, UK.

Fenselau, C. 1992. Tandem mass spectrometry: The competitive edge for pharmacology. *Annu. Rev. Pharmacol. Toxicol.* 32:555–78.

Ferreira, I.M.P.L.V., and J.L.F.C. Lima. 1994. Tubular potentiometric detector for flow-injection based on homogeneous crystalline membranes sensitive to copper, cadmium and lead. *Analyst* 119:209–212.

Fletcher, P., K.N. Andrew, A.C. Calocerinos, S. Forbes, and P.J. Worsfold. 2001. Analytical applications of flow injection with chemiluminescence detection-a review. *Luminescence* 16:1–23.

Fuhrer, T., D. Heer, B. Begemann, and N. Zamboni. 2011. High-throughput, accurate mass metabolome profiling of cellular extracts by flow injection-time-of-flight mass spectrometry. *Anal. Chem.* 83:7074–7080.

Gallignani, M., and M.R. Brunetto. 2004. Infrared detection in flow analysis—developments and trends (review). *Talanta* 64:1137–1146.

Gao, B., Y. Lu, F. Qin, P. Chen, H. Shi, D. Charles, and L. Yu. 2012. Differentiating organic from conventional peppermints using chromatographic and flow injection mass spectrometric (FIMS) fingerprints. *J. Agric. Food Chem.* 60(48):11987–11994.

Gao, B., F. Qin, T. Ding, Y. Chen, W. Lu, and L. Yu. 2014. Differentiating organically and conventionally grown oregano using ultraperformance liquid chromatography mass spectrometry (UPLC-MS), headspace gas chromatography with flame ionization detection (Headspace-GC-FID), and flow injection mass spectrum (FIMS) fingerprints combined with multivariate data analysis. *J. Agric. Food Chem.* 62(32):8075–8084.

Geerdink, R.B., P.G.M. Kienhuis, and U.A.T. Brinkman. 1993. Fast screening method for eight phenoxyacid herbicides and bentazone in water optimization procedures for flow-injection analysis-thermospray tandem mass spectrometry. *J. Chromatogr. A* 647:329–339.

Glish G.L., and R.W. Vachet. 2003. The basics of mass spectrometry in the twenty-first century. *Nat. Rev. Drug Discov.* 2(2):140–150.

Gomez-Ariza, J.L., A. Arias-Borrego, and T. Garcia-Barrera. 2006. Use of flow injection atmospheric pressure photoionization quadrupole time-of-flight mass spectrometry for fast olive fingerprinting. *Rapid Commun. Mass Spectrom.* 20:1181–1186.

Gonzalez-Dominguez, R., T. Garcia-Barrera, and J.L. Gomez-Ariza. 2015. Application of a novel metabolomic approach based on atmospheric pressure photoionization mass spectrometry using flow injection analysis for the study of Alzheimer's disease. *Talanta* 131:480–489.

Hayward, M.J., T. Kotiaho, A.K. Lister, R.G. Cooks, and G.D. Austin, R. Narayan, and G.T. Tsao, 1990. On-line monitoring of bioreactions of *Bacillus polymyxa* and *Klabsiella oxytoca* be membrane introduction tandem mass spectrometry with flow injection analysis sampling. *Anal. Chem.* 62:1798–1804.

Henderickx, H.J.W., P.C. Raemakers-Franken, S. van der Wal, C.G. de Koster, A.L.L. Duchateau, and T. Sonke. 2009. Flow injection analysis electrospray ionization mass spectrometry for high-throughput screening of a gene library for amidase activity. *Anal. Biochem.* 394:159–163.

Hoffman, E., and V. Stroobant. 2001. *Mass Spectrometry: Principles and Applications* (2nd ed.). John Wiley and Sons, Hoboken, New Jersey.

Ito, Y., T. Goto, H. Oka, H. Matsumoto, and Y. Miyazaki. 2003. Simple and rapid determination of thiabendazole, imazalil, and o-phenylphenol in citrus fruit using flow-injection electrospray ionization tandem mass spectrometry. *J. Agric. Food Chem.* 51:861–866.

Keifer, P.A. 2007. Flow techniques in NMR spectroscopy. *Annu. Rep. NMR Spectro.* 62:1–47.

Kmellár B., L. Abrankó, P. Fodor, and S.J. Lehotay. 2010. Routine approach to qualitatively screening 300 pesticides and quantification of those frequently detected in fruit and vegetables using liquid chromatography tandem mass spectrometry (LC-MS/MS). *Food Addit. Contam. A* 27(10):1415–1430.

Kolev, S.D., and I.D. McKelvie. 2008. *Advances in Flow Injection Analysis and Related Technique*. Elsevier, Amsterdam, The Netherlands.

Kristiansen, G.K., R. Brock, and G. Bojesen. 1994. Comparison of flow injection/thermospray MS/MS and LC/thermospray MS/MS methods for determination of sulfonamides in meat and blood. *Anal. Chem.* 66:3253–3258.

Liesener, A., Karst, U., 2005. Monitoring enzymatic conversions by mass spectrometry: a critical review. *Anal. Bioanal. Chem.* 382(7):1451–1464.

Lim, C.W., S.H. Tai, and S.H. Chan. 2012. A flow-injection mass spectrometry fingerprinting scaffold for feature selection and quantitation of *Cordyceps* and *Ganoderma* extracts in beverage: A predictive artificial neural network modelling strategy. *AMB Express* 2:43.

Lu, W., Q. Jiang, H. Shi, Y. Niu, B. Gao, and L. Yu. 2014a. Partial least-squares-discriminant analysis differentiating Chinese wolfberries by UPLC–MS and flow injection mass spectrometric (FIMS) fingerprints. *J. Agric. Food Chem.* 62(37):9073–9080.

Lu, Y., B. Gao, P. Chen, D. Charles, and L. Yu. 2014b. Characterization of organic and conventional sweet basil leaves using chromatographic and flow-injection mass spectrometric (FIMS) fingerprints combined with principal component analysis. *Food Chem.* 154:262–268.

Lu, W., H. Shi, and L. Yu. 2015. *Exploratory Chemometric Analysis on Source of Origin and Cultivar of Chinese Wolfberries Using Liquid Chromatograph–Mass Spectrometric and Flow Injection Mass Spectrometric (FIMS) Fingerprints*. Chapter 20, pp 279–292. doi:10.1021/bk-2015-1191.ch020.

Lüdi, H., M.B. Gran, P. Bataillard, and H.M. Widmer. 1990. Flow injection analysis and biosensors: Applications for biotechnology and environmental control. *J. Biotechnol.* 14:71–79.

Maher, S., F.P. Jjunju, and S. Taylor. 2015. 100 years of mass spectrometry: Perspectives and future trends. *Rev. Mod. Phys.* 87(1):113–135.

Malik, A.K., C. Blasco, and Y. Picó. 2010. Liquid chromatography-mass spectrometry in food safety. *J. Chromatogr. A* 1217(25):4018–4040.

McCooeye, M., and Z. Mester. 2006. Comparison of flow injection analysis electrospray mass spectrometry and tandem mass spectrometry and electrospray high-field asymmetric waveform ion mobility mass spectrometry and tandem mass spectrometry for the determination of underivatized amino acids. *Rapid Commun. Mass Spectrom.* 20(11):1801–1808.

Mol, H.G.J., P. Plaza-Bolaños, P. Zomer, T.C. de Rijk, A.A.M. Stolker, P.P.J. Mulder. 2008. Toward a generic extraction method for simultaneous determination of pesticides, mycotoxins, plant toxins, and veterinary drugs in feed and food matrixes. *Anal. Chem.* 80:9450–9459.

Nanita, S.C., J.J. Stry, A.M. Pentz, J.P. McClory, and J.H. May. 2011. Fast extraction and dilution flow injection mass spectrometry method for quantitative chemical residue screening in food. *J. Agric. Food Chem.* 59:7557–7568.

Nanita, S.C., A.M. Pentz, and F.Q. Bramble. 2009. High-throughput pesticide residue quantitative analysis achieved by tandem mass spectrometry with automated flow injection. *Anal Chem.* 81:3134–3142.

Nanita S.C., and N.L.T. Padivitage. 2013. Ammonium chloride salting out extraction/cleanup for trace-level quantitative analysis in food and biological matrices by flow injection tandem mass spectrometry. *Anal. Chim. Acta* 768:1–11.

Nanita S.C., and L.G. Kaldon. 2015. Ammonium salting out extraction with analyte preconcentration for sub-part per billion quantitative analysis in surface, ground and drinking water by flow injection tandem mass spectrometry. *Anal. Methods* 7:2300–2312.

Nanita, S.C., and L.G. Kaldon. 2016. Emerging flow injection mass spectrometry methods for high-throughput quantitative analysis. *Anal. Bioanal. Chem.* 408:23–33.

Nanita, S.C. 2011. High-throughput chemical residue analysis by fast extraction and dilution flow injection mass spectrometry. *Analyst* 136:285–287.

Nanita, S.C. 2013. Quantitative mass spectrometry independence from matrix effects and detector saturation achieved by flow injection analysis with real-time infinite dilution. *Anal. Chem.* 85:11866–11875.

Niessen, W.M.A. 1998. Advances in instrumentation in liquid chromatography-mass spectrometry and related liquid-introduction techniques. *J. Chromatogr. A* 794:407–435.

O'Toole, M., K.T. Lau, and D. Diamond. 2005. Photometric detection in flow analysis systems using integrated PEDDs. *Talanta* 66:1340–1344.

Oellig, C., and W. Schwack. 2014. Planar solid phase extraction clean-up and microliter-flow injection analysis-time-of-flight mass spectrometry for multi-residue screening of pesticides in food. *J. Chromatogr. A* 1351:1–11.

Opara, L.U. 2003. Traceability in agriculture and food supply chain: A review of basic concepts, technological implications, and future prospects. *Food Agric. Environ.* 1(1):101–106.

Pascoa, R.N.M.J., I.V. Toth, and A.O.S.S. Rangel. 2012. Review on recent applications of the liquid waveguide capillary cell in flow based analysis techniques to enhance the sensitivity of spectroscopic detection methods. *Anal. Chim. Acta* 739:1–13.

Peres, B., N. Barlet, G. Loiseau, and D. Montet. 2007. Review of the current methods of analytical traceability allowing determination of the origin of foodstuffs. *Food Control* 18, 228–235.

Perez-Gregorio, M.R., N. Mateus, and V. de Freitas. 2014. Rapid screening and identification of new soluble tannin–salivary protein aggregates in saliva by mass spectrometry (MALDI-TOF-TOF and FIA-ESI-MS). *Langmuir* 30(28): 8528–8537.

Peris, M., and L. Escuder-Gilabert. 2009. A 21st century technique for food control: Electronic noses. Anal. *Chim. Acta* 638(1):1–15.

Pokrzywnicka, M., M. Fiedoruk, and R. Koncki. 2012. Compact optoelectronic flow-through device for fluorimetric determination of calcium ions. *Talanta* 93:106–110.

Ray A., T. Bristow, C. Whitmore, and J. Mosely. 2017. On-line reaction monitoring by mass spectrometry, modern approaches for the analysis of chemical reactions. doi:10.1002/mas.21539.

Rodriguez-Mozaz, S., S. Reder, M. Lopez de Alda, G. Gauglitz, and D. Barcelo. 2004. Simultaneous multi-analyte determination of estrone, isoproturon and atrazine in natural waters by the River ANAlyser (RIANA), and optical immunosensor. *Biosens. Bioelectron.* 19:633–640.

Ruiz-Capillas, C., and F. Jiménez-Colmenero. 2008. Determination of preservatives in meat products by flow injection analysis (FIA). *Food Addit. Contam. A* 25:1167–1178.

Ruiz-Capillas, C., and F. Jiménez-Colmenero. 2009. Application of flow injection analysis for determining sulphites in food and beverages: A review. *Food Chem.* 112:487–493.

Ruiz-Capillas, C., and L. Nollet. 2016. *Flow Injection Analysis of Food Additives*. CRC Press, Member of Francis & Taylor Group, Boca Raton, FL.

Ruiz-Capillas, C., and W.F.A. Horner. 1999. Determination of trimethylamine-nitrogen and total volatile basic-nitrogen in fresh fish by flow injection analysis. *J. Sci. Food Agric.* 79:1982–1986.

Ruzicka, J., and E.H. Hansen. 1975. Flow injection analyses. 1. New concept of fast continuous-flow analysis. *Anal. Chim. Acta* 78:145–157.

Ruzicka, J., and E.H. Hansen. 2008. Retro-review of flow-injection analysis. *Trends Anal. Chem.* 27(5):390–393.

Ruzicka, J. 2009. Tutorial on flow injection analysis. 4th Ed., From beaker to microfluidics, www.flowinjection.com.

Santos, I.C., V.B. Waybright, H. Fan, S. Ramirez, R.B.R. Mesquita, A.O.S.S., Rangel, P. Fryčák, and K.A. Schug. 2015. Determination of noncovalent binding using a continuous stirred tank reactor as a flow injection device coupled to electrospray ionization mass spectrometry. *J. Am. Soc. Mass Spectrom.* 26:1204–1212.

Sanz-Medel, A. 1999. *Flow Analysis with Spectrometric Detectors*, ed. Elsevier, Amsterdam, The Netherlands.

Schug, K.A., Serrano, C., and Fryčák, P.,2010. Controlled band dispersion for quantitative binding determination and analysis with electrospray ionization-mass spectrometry. *Mass Spectrom. Rev.* 29(5):806–829.

Schug, K.A., and Fan, H. 2012. Hyphenation of flow injection analysis with Mass Spectrometry: Versatile and high throughput technique. *LCGC* 10(2):26–33.

Segundo, M.A., L. Barreiros, I.V. Tóth, and L.M. Magalhães. 2016. Flow Injection Analysis. Recent Developments and Future Trends. In *Flow injection Analysis of Food Additives*. Eds by Claudia Ruiz-Capillas and Leo M.L. Nollet. CRC Press, Taylor and Francis Group, Boca Raton, FL, Pp 121–131.

Shin, C., J. Hyun, Y. Lim, J.S. Kim, Y.M. Jang, and S.J. Kang. 2011. Identification and structure elucidation of a p-phenoxybenzaldehyde in bamboo shoots by HPLC-ESI/MS/MS and NMR. *J. Agric. Food Chem.* 59:11579–11584.

Shin, Ch., Y., Lim, E., Yoon, and S.H. Kim. 2014. Detection of adulteration in commercial galactooligosaccharide by flow-injection electrospray ionization tandem mass spectrometry. *Food Control* 36(1):280–286.

Song, F., D. Monroe, A. El-Demerdash, and C. Palmer. 2014. Screening for multiple weight loss and related drugs in dietary supplement materials by flow injection tandem mass spectrometry and their confirmation by liquid chromatography tandem mass spectrometry. *J. Pharm. Biomed.* 88:136–143.

Song, F., A. El-Demerdash, S.J. Lee, and R.E. Smith. 2012a. Fast screening of lovastatin in red yeast rice products by flow injection tandem mass spectrometry. *J. Pharm. Biomed.* 57:76–81.

Song, F., A. El-Demerdash, and S.J. Lee. 2012b. Screening for multiple phosphodiesterase type 5 inhibitor drugs in dietary supplement materials by flow injection mass spectrometry and their quantification by liquid chromatography tandem mass spectrometry. *J. Pharm. Biomed.* 70:40–46.

Strzelak, K., and R. Koncki. 2013. Nephelometry and turbidimetry with paired emitter detector diodes and their application for determination of total urinary protein. *Anal. Chim. Acta* 788:68–73.

Sun, J., and P. Chen. 2011. A flow-injection mass spectrometry fingerprinting method for authentication and quality assessment of *Scutellaria lateriflora*-based dietary supplements. Anal. *Bioanal. Chem.* 401(5):1577–1584.

Thompson, J.J., and R.S. Houk. 1986. Inductively coupled plasma mass spectrometric detection for multi element flow injection analysis and elemental speciation by reversed-phase liquid chromatography. *Anal. Chem.* 58(12):2541–2548.

Trojanowicz, M., J. Szpunar-Lobńska, and Z. Michalski. 1991. Multicomponent analysis with a computerized flow injection system using LED photometric detection. *Mikrochim. Acta* 1:159–169.

Trojanowicz, M., and W. Matuszewski. 1983. Potentiometric flow-injection determination of chloride. *Anal. Chim. Acta* 151:77–84.

Trojanowicz, M., and Kołacińska, K. 2016. Recent advances in flow injection analysis. *Analyst* 141:2085–2139.

Trojanowicz, M. 2016. Flow-Injection Analysis—Origin and Development. In *Flow Injection Analysis of Food Additives*. Eds by Claudia Ruiz-Capillas and Leo M.L. Nollet. CRC Press, Taylor and Francis Group, Boca Raton, FL, Pp 3–34.

Trojanowicz, M. (Ed.). 2000. *Flow Injection Analysis. Instrumentation and Applications.* World Scientific, Singapore.

Trojanowicz, M. (Ed.). 2008. *Advances in Flow Analysis.* Wiley-VCH, Weinheim, Germany.

Tu, Q., T. Wang, and C.J. Welch. 2010. High-throughput metal screening in pharmaceutical samples by ICP-MS with automated flow injection using a modified HPLC configuration. *J. Pharm. Biomed. Anal.* 51:90–95.

Vaidyanathan, S., D.B. Kelland, R. Goodacre. 2002. Flow-injection electrospray ionization mass spectrometry of crude cell extracts for high-throughput bacterial identification. *J. Am. Soc. Mass Spectrom.* 13:118–128.

Valcarcel, M., and M.D. Luque de Castro. 1994. *Flow-through (Bio) Chemical Sensors.* Elsevier, Amsterdam, The Netherlands.

Van Oudenhove, L., and B. Devreese. 2013. A review on recent developments in mass spectrometry instrumentation and quantitative tools advancing bacterial proteomics. *Appl. Microbiol. Biotechnol.* 97:4749–4762.

Wade, N., and K. Miller. 2005. Determination of active ingredient within pharmaceutical preparations using flow injection mass spectrometry. *J. Pharm. Biomed. Anal.* 37(4):669–678.

Watson, J.T., and O.D. Sparkman. 2007. *Introduction to Mass Spectrometry: Instrumentation, Applications, and Strategies for Data Interpretation*, (4th ed.). Jonh Wiley & Sons, Chichester, UK.

Wu, L., E.C. Meurer, and R.G. Cooks. 2004. Chiral morphing and enantiomeric quantification in mixtures by mass spectrometry. *Anal. Chem.* 76:663–671.

Yang, Y., Q. Xu, L. Zhou, D. Zhong, and X. Chen. 2015. High-throughput salting-out-assisted liquid-liquid extraction for the simultaneous determination of atorvastatin, ortho-hydroxyatorvastatin, and para-hydroxyatorvastatin in human plasma using ultrafast liquid chromatography with tandem mass spectrometry. *J. Sep. Sci.* 38:1026–1034.

Yebra-Biurrun, M.C. 2009. *Flow Injection Analysis of Marine Samples.* Nova Science Publishers, Hauppauge, NY.

Zagatto, E.A.G., C.C. Oliveira, A. Townshend, and P. Worsfold. 2011. *Flow Analysis with Spectrophotometric and Luminometric Detection.* Elsevier, Amsterdam, The Netherlands.

Zhang J., R. Rodila, E. Gage, M. Hautman, L. Fan, L.L. King, H. Wu, and T.A. El-Shourbagy. 2010. High-throughput salting-out assisted liquid/liquid extraction with acetonitrile for the simultaneous determination of simvastatin and simvastatin acid in human plasma with liquid chromatography. *Anal. Chim. Acta* 661:167–172.

Zhao, Y., Y. Niu, Z. Xie, H. Shi, P. Chen, and L.L. Yu. 2013. Differentiating leaf and whole-plant samples of di-and tetraploid *Gynostemma pentaphyllum* (Thunb.) makino using flow injection mass spectrometric fingerprinting method. *J. Funct. Foods* 5(3):1288–1297.

CHAPTER 6

Ambient Ionization MS

Leo M.L. Nollet
University College Ghent

CONTENTS

Ambient ionization mass spectrometry (MS) techniques, such as direct analysis in real time (DART), desorption electrospray ionization (DESI), and extractive electrospray ionization (EESI), and similar related techniques have emerged in the past few years. One of the merits of these methods is that they permit rapid and direct measurements of compounds present in the condensed sample phase. Samples are just exposed to the ionization medium under ambient conditions and do not usually need extraction and separation steps. That makes food analysis extraordinarily easy and efficient.

Different types of desorption/ionization sources are classified according to their basic methods of operation, namely spray-based, laser-based, or plasma-based. Albert et al. (1) discusses a number of plasma-based techniques. Preparation, introduction, quantification, and matrix effects are reviewed.

A large number of approaches of ambient ionization techniques exist. A great part of it is discussed in the next sections. DART is explored in more detail.

6.1 DART

DART is a rapid, noncontact analysis of materials at ambient temperature and at ground potential. It is directly able to detect chemicals on surfaces without the need for any sample preparation. For the background and principle of operation see Cody et al. (2). In DART, helium or nitrogen is usually used as the working gas to produce an excited, metastable plasma by a corona discharge and that reacts with the ambient water and air molecules to produce the reactive ionizing species. These species are subsequently transferred toward the sample by a stream of heated working gas, where they ionize the analyte molecules on the surface.

The processes taking place in DART-on source are (3):

- Thermo-desorption of condensed-phase analytes by a stream of a hot gas, which carries active species derived from a plasma discharge.
- Atmospheric pressure chemical ionization (APCI)-like ionization, enabling the acquisition of respective mass spectra.

DART produces relatively simple mass spectra dominated by protonated molecules in the positive ion mode or deprotonated molecules in the negative ion mode.

Some of the following aspects are relevant to the use of DART-MS in food analysis (2):

- **Gas temperature**- The temperature should be carefully optimized to obtain the required intensity of analyte ions. The final result for a particular compound depends on its physicochemical properties.
- **Position** of the sample in the DART ionization region.
- **Dopants**- The presence of dopant vapors in the sampling region may significantly improve the ionization efficiencies of some compounds and, in some cases, enables the formation from analytes that do not directly ionize with DART technology.

DART is an ambient ionization technique undergoing rapid development. With minimal sample pretreatment, the ionization of analyte molecules is feasible outside the MS instrument in the ordinary atmosphere (4). This ionization approach relies upon the fundamental principles of APCI.

This review highlights and critically assesses the applications of DART (and some related desorption/ionization techniques) coupled to various types of MS analyzers for both target and nontarget analyses of complex food matrices. Based on existing studies, DART-MS is presented as a simple, high-throughput tool for:

i Qualitative confirmation of chemical identity.
ii Metabolomic fingerprinting/profiling.
iii Quantification of low-molecular-weight food components, including some trace organic contaminants.

Ibáñez et al. gives an overview of some of the most recent trends in MS-based approaches and their applications in food metabolomics (5). MS is a powerful analytical technique with excellent capabilities, and most emphasis is on descriptions of novel MS-based approaches, such as direct MS analysis using ambient ionization techniques or MS imaging. These authors also highlight the recent applications of multidimensional chromatographic techniques and the analytical multi-platforms.

Cajka et al. (6,7) used SPME-DART-TOF-MS (solid-phase microextraction time-of-flight mass spectrometer) for the authentication of Trappist beers. The preconcentration of analytes was done using a fiber in the headspace (HS) or direct immersion (DI) SPME modes (6). The components were thermally desorbed by the heated helium gas stream of the DART directly. The major advantage of the DART-TOF-MS technique with DI-SPME sampling was the possibility of sorption/desorption/ionization of the relatively polar compounds originated from the hops that were not volatile enough for HS-SPME sampling prior to GC (gas chromatography)-MS.

Metabolomic fingerprinting by DART-TOF-MS was used as a tool for differentiation between chickens fed with feed containing 5%–8% (w/w) chicken bone meal (a banned component) and those representing a reference group, i.e., grown otherwise under the same conditions (8). This differentiation was feasible. Additional experiments conducted after 6 months confirmed the applicability of this approach.

Bentayeb et al. (9) used DART-TOF-MS for the detection and identification of invisible setoff in food packaging.

Because of the high salt/detergent tolerance of the DART ionization source, a new interface for the on-line coupling of capillary electrophoresis (CE) to DART-MS was developed by Chang et al. (10) in 2013, allowing the use of nonvolatile salts and detergents in CE separations to avoid source contamination and ion suppression in MS. This on-line coupled CE-DART-MS was applied to determine endogenous caffeine in Chinese white tea, with the aim of evaluating the applicability of this method in the detection of target molecules in real samples.

DART was also recently developed to be coupled to an Orbitrap or a linear ion trap (LIT)-Orbitrap, and it was used as DART-HRMS (high resolution mass spectrometry) for food safety and quality analysis. For example, Shen et al. (11) proposed a DART-HRMS procedure to distinguish between toxic Japanese star anise and nontoxic Chinese star anise fruits.

A novel approach for the authentication of olive oil samples representing different quality grades has been developed (12). A new type of ion source, DART coupled to a high resolution TOF-MS (HR-TOF-MS), was employed for the comprehensive profiling of triacylglycerols (TAGs) and/or polar compounds extracted with a methanol-water mixture. The main parameters influencing the ionization efficiency of TAGs were the type of sample solvent, degree of sample dilution, ion beam temperature, and presence of a dopant (ammonia vapors). The ionization yield of polar compounds depended mainly on the water content in the extract and on the ion beam temperature. Using DART-TOF-MS, not only the differentiation among extra virgin olive oil (EVOO), olive pomace oil (OPO), and olive oil (OO) could be easily achieved, but also the detection of EVOO adulteration with commonly used adulterant, hazelnut oil (HO), was feasible. Based on linear discriminant analysis (LDA), the introduced method allowed the detection of HO additions of 6% and 15% (v/v) when assessing the DART-TOF-MS mass profiles of polar compounds and TAGs, respectively.

Sophisticated statistical tools are required to extract the full analytical power from high performance thin layer chromatography (HPTLC) (13). The instantaneous DART-MS method is suitable for the sample characterization and differentiation by multivariate data analysis (MVA). The results obtained by the chemometric evaluation of HPTLC and DART-MS data provided complementary information. The complexity, expense, and analysis time were significantly reduced due to the use of statistical tools in the evaluation of fingerprints. The approach allowed the categorization of 91 propolis samples from Germany and other locations based on their phenolic compound profiles. A high level of confidence was thus obtained when combining the orthogonal approaches (HPTLC and DART-MS) for ultrafast sample characterization.

The cubeb berry has been used as a spice and medicinal herb since medieval times (14). In China, *Litsea cubeba* (*Lauraceae*) has been designated as the cubeb tree; in other countries, *Piper cubeba* (*Piperaceae*) is considered as the cubeb. Since most cubeb berries in the South Korean market are imported from China, there is a need to develop an efficient analytical method for the origin detection of cubeb fruit. 32 commercial

TABLE 6.1 Ambient MS Techniques in Food Analysis

Technique	Desorption/ Ionization Principle	Objective	Analytes/Matrix	Sample Preparation	Detection MS System	Performance Characteristics	Ref.
DART	Metastable thermal/ indirect plasma	QUAL	Isoprop/ithioxanthone/ Milk, yoghurt, fat	Thin Layer Chromatography	TOF	N/A	(64)
		QUAL	Melanine/pet food	None	TOF	LOD 100 mg/kg	(65)
		QUAL	Additives, sugars/flavored mineral water	None	TOF	N/A	(66)
		QUAL	Strobilurin pesticides/ wheat	Extraction with ethylacetate	TOF	LOQs 5–30 μg/kg	(67)
		QUAL	Thiabendazole/orange peel	None	TOF	N/A	(68)
		AUTH	Triacylglycerols, phenolic compounds/olive oil	Dilution of oil with toluene, extraction of polar compounds with methanol/water mixture (80:20 *v/v*)	TOF	Discrimation among qulaity grades, detection of 6% adulterant (hazelnut oil)	(12)
		QUAL/ QUAN	Curcumoids/functional drink, curry powder	None, thin layer chromatography	TOF	N/A	(69)
		QUAL	Melamine/milk powder	None	TOF	LOD 1 mg/kg	(70)
		QUAL	Sulfur compounds/garlic	Extraction with methanol/ water/hydrochloric acid mixture (90/9/1, *v/v/v*)	TOF	N/A	(71)
		QUAL	Reactive sulfur compounds/garlic, garlic powder	Extraction from aqueous homogenate by shaking with diethyl ether	TOF	N/A	(72)

(*Continued*)

TABLE 6.1 (*Continued*) Ambient MS Techniques in Food Analysis

Technique	Desorption/ Ionization Principle	Objective	Analytes/Matrix	Sample Preparation	Detection MS System	Performance Characteristics	Ref.
		AUTH	Volatiles, phenolic compounds/beer	Head space and direct immersion solid phase microextraction	TOF	Discrimination among Tr beer brandsappist and non-Trappist	(6)
		QUAN	Melamine, cyanuric acid/milk powder, dried cheese, condensed milk	Extraction with methanol-5% aqueous formic acid (50/50, *v/v*)	TOF	LOQs 450 µg/kg (melamine), 1200 µg/kg (cyanuric acid)	(73)
		QUAN	Selected trichothecenes, *Alternaria* toxins, zearalenone, and sterigmatocystin	Modified QuEChERS extraction and dispersive solid phase extraction clean up	Orbitrap	Lowest calibration levels ranging from 50 to 150 µg/kg	(74)
Desorption atmospheric pressure chemical ionizaton/ DAPCI	Electric plasma species	QUAL	Sudan dyes/tomato sauce; atrazine/pumpkin surface; putrescine, cadaverine/fish meat	None	LIT	N/A	(75)
		AUTH	Nonvolatile and semi-volatile compounds/tea	None	LIT	Discrimination among various types of tea samples	(59)
		QUAL/ QUAN	Melamine, cyanuric acid, melamine-cyanurate complex/milk powder, liquid milk	None	LIT	LODs 1.6 x 10^{-11} g/mm^2 (milk powder), 1.3 x 10^{-12} g/mm^2 (liquid milk)	(20)

(*Continued*)

TABLE 6.1 (*Continued*) Ambient MS Techniques in Food Analysis

Technique	Desorption/ Ionization Principle	Objective	Analytes/Matrix	Sample Preparation	Detection MS System	Performance Characteristics	Ref.
		AUTH	Low molecular compounds/dried sea cucumber	None	LIT	Discrimation among dries sea-cucumber products from different geographical origin	(58)
Low temperature plasma ionization/ LTP	Electric plasma species	QUAN	Melamine/milk powder, liquid milk	None	LIT	LOD 6–15 μg/kg	(76)
		Qual	Volatiles, fatty acids, phenolic compounds/ olive oil	None	LIT	N/A	(55)

Source: Reprinted from (3) with permission.

cubeb samples were collected from China, South Korea, and Indonesia, and DART-MS was employed to obtain a chemical fingerprint of each sample for multivariate analysis. Cubebene was identified as a marker molecule for *P. cubeba*, and all cubeb samples from the South Korean markets were identified as the fruit of *L. cubeba*.

The objective of the study by Vaclavik et al. (15) is the evaluation of the potential of high-throughput DART-HRMS fingerprinting and multivariate regression analysis in the prediction of the extent of acrylamide formation in biscuit samples prepared with various recipes and baking conditions. Information-rich mass spectral fingerprints were obtained by the analysis of biscuit extracts, for the preparation of which aqueous methanol was used. The principal component analysis (PCA) of the acquired data revealed an apparent clustering of samples depending on the extent of heat treatment applied during the baking of biscuits. The regression model for the prediction of acrylamide in biscuits was obtained by partial least square regression (PLSR) analysis of the data matrix representing combined positive and negative ionization mode fingerprints. The model provided a least root mean square error of cross validation (RMSECV) that was equivalent to an acrylamide concentration of 5.4 $\mu g\ kg^{-1}$ and a standard error of prediction (SEP) of 14.8 $\mu g\ kg^{-1}$. According to the authors these results indicate that this strategy can be used to accurately predict the amounts of acrylamide formed during the baking of biscuits. Such rapid estimation of acrylamide concentration can become a useful tool in evaluation of the efficiencies of the processes that are aimed at mitigating this food processing contaminant. However, the robustness of this approach should be tested further with respect to variabilities in the chemical compositions of the ingredients used for the preparation of biscuits.

Factors such as fermentation methods, geographical origin, and season can affect the biochemical composition of tea leaves (*Camellia sinensis* L.) (16). The biochemical composition of oolong tea during the manufacturing and fermentation processes was studied with a nontargeted method utilising ambient ionisation with a DART ion source and MS. Caffeine dominated the positive ionisation spectra throughout the manufacturing process, while the negative ion spectra collected during manufacturing were rich in ions that were likely to be surface lipids. Correlation analyses of the spectra revealed two volatile compounds tentatively identified as indole and geranic acid, along with ammonium and caffeine clusters/adducts with geranic acid that increased in concentration during the fermentation stages of the process. The tentative identifications were assigned using a combination of DART ion trap tandem mass spectrometry (MS^n) and DART accurate mass MS1 and MS2 on tea samples and standard compounds. DART-MS has the potential to rapidly monitor the progress of complex manufacturing processes, such as tea fermentation. In Table 6.1, a number of applications are enumerated.

6.2 DESI

DESI-MS was one of the first reported ambient ionization methods. DESI occurs by a charged droplet pickup process. DESI has been applied to various food analyses in the past couple of years, including qualitative analysis and tissue imaging. A review article (17) with respect to DESI developments in the analysis of chemical contaminants of food summarized the applications of DESI-MS in a broad range of food analytes, and discussed the features of DESI-MS, such as strengths, weaknesses, and desorption/ionization process.

Gerbig et al. (18) used DESI-MS to characterize the triglyceride composition and oxidation behavior of edible oil and margarine samples. After a simple sample preparation protocol for the deposition of samples onto a glass carrier surface, DESI-MS/MS was used to determine the structure of the triglycerides.

DESI has the potential to detect pesticides, natural toxins, veterinary drugs, food additives, adulteration, and packaging migrants contamination in food (19).

Melamine was detected in milk by DESI. The limit of detection (LOD) was found to be 3.4×10^{-15} g mm^{-2} (S/N = 3) for the detection of pure melamine deposited on the paper surface, which was much lower than that for the detection of melamine in powdered milk (1.6×10^{-11} g mm^{-2}, S/N = 3) or liquid milk (1.3×10^{-12} g mm^{-2}, S/N = 3) (20). The significant difference in LOD was ascribed to the relatively strong molecular interactions between melamine and the matrix, such as proteins in the milk products. As demonstrated by the use of DESI for melamine detection, weakening of the molecular interactions between analytes and proteins is proposed as a general strategy to improve the sensitivity of ambient mass spectrometry for the direct detection of analytes bound in protein matrices.

A modification of DESI, termed desorption nano-electrospray (nano-DESI), was tested for the analysis of anthocyanins (21). Acidification of samples and an acidic spray liquid (methanol: water = 75:25 with 0.2% HCOOH) were essential for obtaining good quality spectra. Profiles of the main anthocyanins in wine samples, two vintages (2005 and 2007) of three cultivars (Alibernet, Neronet, and Rubinet), were successfully acquired.

They were in agreement with the results of LC/MS experiments (anthocyanins isolated by solid phase extraction were separated by μ-HPLC with gradient elution and were detected by ESI-MS). Nano-DESI-MS data also allowed the determination of the ratio in which the two cultivars (Neronet and Rubinet) are present in their mixture and the detection of wine coloring by tenturier or elderberry extract. Main anthocyanins in the slices of wine grape, chokeberries and elderberries, or in a wine stain on cotton fabric were also detected.

6.3 EESI

EESI is an ambient ionization source similar to DESI, but it has two colliding aerosols: a sample spray and an electrospray. In a standard EESI, the analytes are neutrally evaporated to collide with an electrospray plume and subsequently ionized by the charged droplets generated by the ESI. Because the ion-suppression effects are reduced by EESI, EESI-MS is capable of directly analyzing complex food samples without significant loss of sensitivity. Li et al. (22) proposed an MS protocol by using a neutral desorption (ND)-EESI-MS approach for the viscous sample analysis of some food samples, such as beer, edible oil, honey, cheese. These samples were treated with a ND sampling gas beam before being characterized by EESI-MS.

A novel method based on ultrasound-assisted EESI-MS has been developed and applied to rapidly detect the presence of melamine in raw milk, wheat gluten, and milk powder with no or minor sample pretreatment (23). The high sample throughput and figures of merit make it especially useful for screening melamine levels well below the current safety limit in various food matrices.

Microjet sampling in combination with EESI-MS was applied for the rapid characterization and classification of EVOO without any sample pretreatment (24). When modifying the composition of the primary ESI spray solvent, mass spectra of an identical EVOO sample showed differences. This demonstrates the capability of this technique to extract molecules with varying polarities and generates rich molecular information of the EVOO. Moreover, with the aid of microjet sampling, compounds with different volatilities (e.g., *E*-2-hexenal, *trans-trans*-2,4-heptadienal, tyrosol and caffeic acid) could be sampled simultaneously. EVOO data was also compared with that of other edible oils. PCA was performed to discriminate between pure EVOO and EVOO adulterated with edible oils. Microjet sampling EESI-MS was found to be a simple, rapid (less than 2 min of analysis time per sample), and powerful method to obtain MS fingerprints of EVOO without requiring any complicated sample pretreatment steps.

Mass spectral fingerprints of various cheese products were rapidly recorded in the mass range of *m/z* 50–300 Da without any sample pretreatment using ND-EESI-MS in negative ion mode (25). The results demonstrate that both volatile and nonvolatile analytes on greasy cheese surfaces can be directly sampled by a ND gas beam. The influence of the ND gas flow on the analyte signal was systematically investigated. Under optimized experimental conditions, reproducible results were obtained using ND-EESI-MS. PCA was applied to differentiate a total of 49 individual cheese samples (four different types), which were purchased from three different supermarkets. All samples were successfully classified according to their types; but the distributors and sensory properties were not distinguishable from the spectra data. The principal components 2, 3, and 4 scores showed excellent capacities of distinguishing the types of cheeses. Molecular markers of interest can be identified by using tandem mass spectrometry and by matching the data with those from reference compounds.

By gently bubbling nitrogen gas through beer—an effervescent beverage—both volatile and nonvolatile compounds can be simultaneously sampled in the form of aerosol (26). This allows fast (within seconds) fingerprinting by EESI-MS in both negative and positive ion modes without the need for any sample pretreatment, such as degassing and dilution. Trace analytes such as volatile esters (e.g., ethyl acetate and isoamyl acetate), free fatty acids (FFA) (e.g., caproic acid, caprylic acid, and capric acid), semi/nonvolatile organic/inorganic acids (e.g., lactic acid), and various amino acids, commonly present in beer at low parts-per-million or at sub-ppm levels, were detected and identified based on tandem MS data. Furthermore, the appearance of solvent cluster ions in the mass spectra gives insight into the sampling and ionization mechanisms: aerosol droplets containing semi/nonvolatile substances are thought to be generated via bubble bursting at the surface of the liquid; these neutral aerosol droplets then collide with the charged primary electrospray ionization droplets, which is followed by analyte extraction, desolvation, ionization, and MS detection. With PCA, several beer samples were successfully differentiated.

6.4 DAPPI

Desorption atmospheric pressure photoionization (DAPPI) is the DESI analog of atmospheric pressure photoionization (APPI). DESI works better with polar analytes that are easy to protonate or deprotonate, whereas DAPPI is often used to ionize nonpolar or neutral compounds. Similar to DESI, DAPPI uses a standard atmospheric pressure ionization interface to collect the produced ions and introduce them into a mass analyzer.

The spray-solvent effect on the analysis of lipids selected from pharmaceutical and food products was studied by two different ambient ionization MS methods—DESI-MS and DAPPI-MS (27). This work demonstrated that DAPPI provided efficient desorption and ionization for less polar, neutral, and ionic lipids, but might cause extensive fragmentation for large, labile compounds; whereas DESI had better performance in the analysis of compounds with sufficient polarity.

6.5 ELDI

Electrospray laser desorption ionization MS (ELDI-MS) is an ambient ionization technique that combines the features of ESI and laser desorption. In ELDI, sample deposits are desorbed by a UV pulse laser and ionized by the ESI-generated droplets under ambient conditions. The use of a laser pulse in ELDI-MS for sampling over a relatively defined area can help improve the spatial resolution of the solid sample, making it suitable for molecular imaging applications. Huang et al. (28) developed an ELDI-MS instrument for the application of molecular imaging on the dry surfaces of two edible fungi (*Ganoderma lucidum* and *Antrodia camphorata*) to characterize the active components. Molecular profiling of the surfaces of the two fungi and spatial distribution of the particular analyte corresponding to the representative m/z signal were obtained by this method.

An electrospray-assisted laser desorption/ionization source with an infrared OPO laser (IR-ELDI) was constructed and optimized for peptide and protein mass spectrometry analysis (29). Similar to ELDI with an ultraviolet laser, IR-ELDI generates multiple charged molecules for the peptides and proteins measured under ambient sampling conditions. Samples in both dried state and in analyte solutions can be directly measured by IR-ELDI even without the presence of a conventional MALDI matrix. However, the analysis of a sample solution is shown to enhance the sensitivity of the MS measurement greatly, as a 100-fold sensitivity gain was achieved for peptide measurements. The LOD of IR-ELDI was determined to be 250 fmol for bradykinin (1.1 kDa), 100 fmol for ubiquitin (8.6 kDa), and 500 fmol for carbonic anhydrase (29 kDa). IR-ELDI is amenable for the MS and MS^n analyses for proteins with up to 80 kDa transferrin.

6.6 STELDI

Olive oil has been analyzed by sorptive tape-like extraction in combination with laser desorption ionization mass spectrometry (STELDI-MS) (30). This powerful combination has some great advantages: no separation steps, solvent-free, matrix-free, and no sample preparation. The olive oil compounds are analyzed by LDI-MS directly from the sample spot on a thin layer chromatography (TLC) plate. The chosen samples represent products commonly used as adulterants in EVOO, and the main monitored ions are lipid adulteration markers. Analytical procedures consisted of profiling the main fatty acids (m/z 255 for palmitic acid; 279 for linoleic acid; 281 for oleic acid; and 283 for stearic acid), triacylglycerols (m/z 901 for LLL and 907 for OOO), and some phenolic compounds (m/z 169 for gallic acid; 193 for ferulic acid; and 195 for 2(4-hydroxyphenyl) ethyl acetate) in EVOO, olive oil (OO), hazelnut oil (HO), and soybean oil (SO). Compound identification was confirmed by the analysis of collision induced dissociation (CID) products in positive (ion [M+Na]+) and negative modes (ion [M–H]–).

6.7 EASI

Porcari et al. (31) reviewed EASI-MS (easy ambient sonic spray ionization mass spectrometry) and several other sister techniques, like Venturi (V-EASI), thermal imprinting (TI-EASI), and Spartan EASI (S-EASI) MS for food quality control and authentication.

These authors discuss the fundamentals, mechanisms, devices, parameters, strategies, and applications of EASI-MS and the related techniques.

EASI is a spray-based technique that employs the action of a polar solvent to promote both desorption and ionization of analytes. EASI does not use voltage, electrical discharges, or high temperatures for ion production.

Chemical profiles for a representative set of 49 propolis ethanolic extracts collected worldwide (North and South America, Europe, Asia, and Oceania) were obtained via EASI-MS (32). This simple and easily implemented fingerprinting technique directly analyzes (without any pre-separation or sample manipulation) a tiny droplet of the ethanolic extract placed on an inert surface under ambient conditions. PCA of the EASI-MS data is shown for the group samples according to the plant sources of their resins, which characterizes their geographical origin.

EASI can be applied in the analysis of:

- **Vegetable oils**: oil characterization; oil authenticity; certification of geographical origin; edible oil oxidation (33)
- **Emulsifiers**
- **Muscle products**
- **Propolis**
- **Coffee beans**
- **Beverages**

EASI fingerprinting has also been used to characterize the quality of other oils that are obtained from andiroba, açai, urucum, and buriti (34–36).

Seed oils from nine native plants of the north and northeast regions of Brazil have been characterized by EASI (37).

The oil of *Jatropha curcas* L. and its biodiesel were monitored with the help of EASI (38,39).

The profiles of TAG of soybean oil obtained with EASI were similar for the others techniques such as MALDI-MS, ESI-MS and GC-FID (34, 40).

Direct EASI fingerprinting allows the differentiation of hazelnut oil and olive oil (41). EASI may also be a tool to inspect the geographical origins of olive oils based on the characteristic phenolic constituents (42).

A further application of EASI is the characterization of Brazil nut oil (43).

EASI-MS efficiently desorbs and ionizes the main oil constituents from an inert surface under ambient conditions and provides the comprehensive TAG and FFA profiles detected mainly as either $[TAG + Na]^+$ or $[FFA - H]^-$ ions (200). EASI (±)-MS analysis is simple, easily implemented, requires just a tiny droplet of the oil, and is performed without any pre-separation or chemical manipulation. It also causes no fragmentation of the TAG ions, hence the diacylglyceride (DAG) and monoacylglyceride (MAG) profiles and contents can also be measured. The EASI (±)-MS profiles of TAG and FFA permit authentication and quality control and can be used, for instance, to access the levels of adulteration, acidity, oxidation, or hydrolysis of vegetable oils in general.

Eberlin and co-authors applied TI-EASI to access the traceability and authenticity of dry-cured hams (44) and to spicify the different fish hand meats (45).

TI-EASI has been used to monitor degradation as a function of the conservation protocol and compositional changes from different storage conditions in caviar samples of Atlantic sturgeon (46).

Propolis has been characterized by EASI-MS (47).

EASI-MS and DESI-MS were applied directly to the surfaces of green Arabica coffee beans. Both techniques were found to function as fast and simple protocols to differentiate the beans treated with dry, semidry, and wet postharvest methods (48).

Several beverages, such as whiskey, wines, *cachaça*, and rum have been characterized via direct infusion ESI-MS, V-EASI, and S-EASI for authenticity purposes (49–51).

6.8 LTP

Recently, a new ambient mass spectrometer, namely the low temperature plasma desorption/ionization MS (LTP-MS), was constructed by Cooks and his colleagues. The LTP ionization source was based on dielectric barrier discharge that created a low-power plasma. The advantages of LTP-MS over some conventional plasma-based techniques are low temperature (no damage to sample), low consumption of power and gas, and capability of coupling to portable MS instruments. Recent studies by Cooks and colleagues reported the LTP-MS analysis of edible oil, the detection of melamine in milk, and the screening of various agrochemicals in different food matrices (52,53).

Martinez-Jarquin and Winkler (54) designed a LTP probe with an adjustable output temperature and a variable beam diameter for the direct detection of organic molecules. A direct analysis of roasted coffee beans is also possible.

Huang et al. (52) developed an LTP probe coupled to a portable mass spectrometer and applied it for the determination and quantitation of melamine in whole milk, fish, milk powder, and other complex food matrices. This work showed the potential of the LTP ambient ionization and tandem MS method for the direct analysis of complex food mixtures.

An ambient ionization technique, the LTP probe that is based on dielectric barrier discharge is used to detect both minor and trace components (FFA, phenolics, and volatiles) in raw, untreated olive oil (55). This method allows the measurement of FFA (the main quality control parameter used to grade olive oil according to quality classes), selected bioactive phenolic compounds, and volatiles.

6.9 PADI

PADI (plasma assisted desorption ionisation) is a recent development by McCoustra et al. It uses a cool plasma to ionize and desorb molecules from a surface (56,57).

PADI has also been successfully applied for the analysis of nicotine in tobacco and thiosulfates in garlic.

6.10 OTHER TECHNIQUES

Wu et al. (58) used DAPCI-LIT-MS (surface desorption APCI mass spectrometry) for the differentiation of sea cucumbers (*Apostichopus japonicus*) according to their habitats. In a different study by Chen et al. (59), DAPCI-MS enabled the rapid analysis and discrimination of tea products without any sample pre-treatment.

PESI (probe electrospray ionization) is an ESI-based technique in which the sample solution transference is performed by a solid needle with a sharp tip (60). PESI features high salt tolerance, direct sampling, and low sample consumption.

α-solanine in potato was analyzed with PESI-MS (61).

SDAPCI (surface desorption ablation flowing atmospheric pressure afterglow) is a combination of corona discharge and ND (60). The distribution of endogenous melamine in a boiled chicken egg in China was imaged by SDAPCI-MS (62).

A rapid analytical method based on nanolitre-level sampling technique and SDAPCI-MS has been developed for the determination of trace dimethoate in juices without any sample preparation (63).

LA-FAPA (laser ablation flowing atmospheric pressure afterglow) combines the laser ablation of analyte molecules with post-ionization and atmospheric pressure plasma. The ion source is a combination of an ablation chamber and an ionization chamber connected with a Teflon tube (60). LA-FAPA is used in imaging experiments of food.

REFERENCES

1. A. Albert, J. T. Shelley, C. Engelhard. Plasma-based ambient desorption/ionization mass spectrometry: State-of-the-art in qualitative and quantitative analysis. Analytical and Bioanalytical Chemistry 406 (2014), 6111–6127.
2. R. B. Cody, J. A. Laramée, J. M. Nilles, H. D. Durst. Direct analysis in real time (Dart^{tm}) mass spectrometry. JEOL News 40 (1) (2005), 8–10.
3. J. Hajslova, T. Cajka, L. Vaclavik. Challenging applications offered by direct analysis in real time (DART) in food-quality and safety analysis. Trends in Analytical Chemistry 30 (2) (2011), 204–218.
4. J. H. Gross. Direct analysis in real-time—A critical review on DART-MS. Analytical and Bioanalytical Chemistry 406 (2014), 63–80.
5. C. Ibáñez, V. García-Cañas, A. Valdés, C. Simó. Novel MS-based approaches and applications in food metabolomics. TrAC Trends in Analytical Chemistry 52 (2013), 100–111.
6. T. Cajka, K. Riddellova, M. Tomaniova, J. Hajslova. Recognition of beer brand based on multivariate analysis of volatile fingerprint. Journal of Chromatography A 1217 (2010), 4195–4203.
7. T. Cajka, K. Riddellova, M. Tomaniova, J. Hajslova. Ambient mass spectrometry employing a DART ion source for metabolomic fingerprinting/profiling: A powerful too for beer origin recognition. Metabolomics 7 (2011), 500–508.
8. T. Cajak, H. Danhelova, M. Zachariasova, K. Riddellova, J. Hajslova. Application of direct analysis in real time ionzation-mass spectrometry (DART-MS) in chicken meat metabolomics aiming at the retrospective control of feed fraud. Metabolomics 9 (2013), 545–557.
9. K. Bentayeb, L. K. Ackerman, T. Lord, T. H. Begley. Non-visible print set-off of photoinitiators in food packaging: Detection by ambient ionisation mass spectrometry. Food Additives & Contaminants: Part A 30 (4) (2013), 750–759.
10. C. Chang, G. Xu, Y. Bai, C. Zhang, X. Li, M. Li, Y. Liu, H. Liu. Online coupling of capillary electrophoresis with direct analysis in real time mass spectrometry. Analytical Chemistry 85 (2013), 170–176.
11. Y. Shen, T. A. van Beek, F. W. Claassen, H. Zuilhof, B. Chen, M. W. F. Nielen. Rapid control of Chinese star anise fruits and teas for neurotoxic anisatin by direct analysis in real time high resolution mass spectrometry. Journal of Chromatography A 1259 (2012), 179–186.
12. L. Vaclavik, T. Cajka, V. Hrbek, J. Hajslova. Ambient mass spectrometry employing direct analysis in real time (DART) ion source for olive oil quality and authenticity assessment. Analytica Chimica Acta 645 (1–2) (2009), 56–63.

13. G. E. Morlock, P. Ristivojevic, E. S. Chernetsova. Combined multivariate data analysis of high-performance thin-layer chromatography fingerprints and direct analysis in real time mass spectra for profiling of natural products like propolis. Journal of Chromatography A 1328 (2014), 104–112.
14. H. Jin Kim, W. Sook Baek, Y. Pyo Jang. Identification of ambiguous cubeb fruit by DART-MS-based fingerprinting combined with principal component analysis. Food Chemistry 129 (3) (2011), 1305–1310.
15. L. Vaclavik, E. Capuano, V. Gökmen, J. Hajslova. Prediction of acrylamide formation in biscuits based on fingerprint data generated by ambient ionization mass spectrometry employing direct analysis in real time (DART) ion source. Food Chemistry 173 (15) (2015), 290–297.
16. K. Fraser, G. A. Lane, D. E. Otter, S. J. Harrison, S.-Y. Quek, Y. Hemar, S. Rasmussen. Monitoring tea fermentation/manufacturing by direct analysis in real time (DART) mass spectrometry. Food Chemistry 141 (3) (2013), 2060–2065.
17. M. W. F. Nielen, H. Hooijerink, P. Zomer, J. G. J. Mol. Desorption electrospray ionization mass spectrometry in the analysis of chemical food contaminants in food. Trends in Analytical Chemistry 30 (2011), 165–180.
18. S. Gerbig, Z. Takats. Analysis of triglycerides in food items by desorption electrospray ionization mass spectrometry. Rapid Communications in Mass Spectrometry 24 (2010), 2186–2192.
19. D. I. Ellis, V. L. Brewster, W. B. Dunn, J. W. Allwood, A. P. Golovanov, R. Goodacre. Fingerprinting food: Current technologies for the detection of food adulteration and contamination. Chemical Society Reviews 41 (2012), 5706–5727.
20. S. Yang, J. Ding, J. Zheng, B. Hu, J. Li, H. Chen, Z. Zhou, X. Qiao. Detection of Melamine in milk products by surface desorption atmospheric pressure chemical ionization mass spectrometry. Analytical Chemistry 81 (7) (2009), 2426–2436.
21. L. Hartmanova, V. Ranc, B. Papouskova, P. Bednar, V. Havlicek, K. Lemr. Fast profiling of anthocyanins in wine by desorption nano-electrospray ionization mass spectrometry. Journal of Chromatography A 1217 (25) (2010), 4223–4228.
22. X. Li, B. Hu, J. Ding, H. Chen. Rapid characterization of complex viscous samples at molecular levels by neutral desorption extractive electrospray ionization mass spectrometry. Nature Protocols 6 (2011), 1010–1025.
23. L. Zhu, G. Gamez, H. Chen, K. Chingin, R. Zenobi. Rapid detection of melamine in untreated milk and wheat gluten by ultrasound-assisted extractive electrospray ionization mass spectrometry (EESI-MS). Chemical Communications (5) (2009), 559–561.
24. W. S. Law, H. W. Chen, R. Balabin, C. Berchtold, L. Meier, R. Zenobi. Rapid fingerprinting and classification of extra virgin olive oil by microjet sampling and extractive electrospray ionization mass spectrometry. Analyst 135 (4) (2010), 773–778.
25. Z. Wu, K. Chingin, H. Chen, L. Zhu, B. Jin, R. Zenobi. Sampling analytes from cheese products for fast detection using neutral desorption extractive electrospray ionization mass spectrometry. Analytical and Bioanalytical Chemistry 397 (4) (2010), 1549–1556.
26. L. Zhu, Z. Hu, G. Gomez, W. S. Law, H. Chen, S. Yang, K. Chingin, R. M. Balabin, R. Wang, T. Zhang, R. Zenobi. Simultaneous sampling of volatile and non-volatile analytes in beer for fast fingerprinting by extractive electrospray ionization mass spectrometry. Analytical and Bioanalytical Chemistry 398 (1) (2010), 405–413.

27. N. M. Suni, H. Aalto, T. J. Kauppila, T. Kotiaho, R. Kostiainen. Analysis of Lipids with desorption atmospheric pressure photoionization-mass spectrometry (DAPPI-MS) and desorption electrospray ionization-mass spectrometry (DESI-MS). Mass Spectrometry 47 (2012), 611–619.
28. M.-Z. Huang, S.-C. Cheng, S.-S. Jhang, C.-C. Chou, C. Chu-Nian, J. Shiea? I. A. Popov, E. N. Nikolaev. Ambient molecular imaging of dry fungus surface by electrospray laser desorption ionization mass spectrometry. International Journal of Mass Spectrometry 325–327 (212) (2012), 172–182.
29. I. X. Peng, R. R. Ogorzalek Loo, E. Margalith, M. W. Little, J. A. Loo. Electrospray-assisted laser desorption ionizationmass spectrometry (ELDI-MS) with an infrared laser for characterizing peptides and proteins. Analyst 135 (2010), 767–772.
30. D. N. de Oliveira, R. R. Catharino, Direct metabolic fingerprinting of olive oils using STELDI-MS. Journal of Food Composition and Analysis 38 (2015), 131–134.
31. A. M. Porcari, G. D. Fernandes, D. Barrera-Arellano, M. N. Eberlin, R. M. Alberici. Food quality and authenticity screening via easy ambient sonic-spray ionization mass spectrometry. Analyst 141 (2016), 1172–1184.
32. A. C. H. F. Sawaya, P. Verardi Abdelnur, M. N. Eberlin, S. Kumazawa, M.-R. Ahn, K.-S. Bang, N. Nagaraja, V. S. Bankova, H. Afrouzan. Fingerprinting of propolis by easy ambient sonic-spray ionization mass spectrometry. Talanta 81 (1–2) (2010), 100–108.
33. R. C. Simas, R. R. Catharino, I. B. S. Cunha, E. C. Cabral, D. Barrera-Arellano, M. N. Eberlin, R. M. Alberici. Instantaneous characterization of vegetable oils *via* TAG and FFA profiles by easy ambient sonic-spray ionization mass spectrometry. Analyst 135 (2010), 738–744.
34. R. C. Sima, R. R. Catharino, I. B. Cunha, E. C. Cabral, D. Barrera-Arellano, M. N. Eberlin. Instantaneous characterization of vegetable oils via TAG and FFA profiles by easy ambient sonic-spray ionization mass spectrometry. Analyst 135 (2010), 738–744.
35. E. C. Cabral, G. F. da Cruz, R. C. Simas, G. B. Sanvido, L. de V. Goncalves, R. V. P. Leal, R. C. F. da Silva, J. C. T. da Silva, L. E. S. Barata, V. S. da Cunha, L. F. de Franca, R. J. Daroda, G. F. de Sa, N. M. Eberlin. Typification and quality control of the Andiroba (Carapa guianensis) oil via mass spectrometry fingerprinting. Analytical Methods 5 (2013), 1385–1391.
36. G. A. Bataglion, F. M. A. da Silva, J. M. Santos, F. N. dos Santos, M. T. Barcia, C. C. de Lourenço, M. J. Salvador, H. T. Godoy, M. N. Eberlin, H. H. F. Kollen. Comprehensive characterization of lipids from Amazonian vegetable oils by mass spectrometry techniques. Food Research International 64 (2014), 472–481.
37. F. N. Santos, J. M. Santos, P. R. R. Mesquita, K. B. Oliveira, F. M. Rodrigues, W. A. Lopes, M. N. Eberlin. Immediate differentiation of unusual seed oils by easy ambient sonic-spray ionization mass spectrometry and chemometric analysis. Analytical Methods, 8 (2016), 3681–3690.
38. K. C. Cardoso, M. J. Da Silva, R. Grimaldi, M. Stahl, R. C. Simas, I. B. S. Cunha, M. N. Eberlin, R. M. Alberici. TAG profiles of *Jatropha curcas* L. Seed oil by easy ambient sonic-spray ionization mass spectrometry. Journal of the American Oil Chemists' Society 89 (2012), 67–71.
39. A. M. A. P. Fernandes, S. El-Khatib, I. B. S. Cunha, A. M. Porcari, M. N. Eberlin, M. J. Silva, P. R. Silva, V. S. Cunha, R. J. Dorado, R. M. Alberici. Chemical characterization of *Jatropha curcas* L. seed oil and its biodiesel by ambient desorption/ionization mass spectrometry. Energy Fuels 29 (5) (2015), 3096–3103.

40. A. M. A. P. Fernandes, G. D. Fernandes, R. C. Simas, D. Barrera-Arellano, M. N. Eberlin, R. M. Alberici. Quantitation of triacylglycerols in vegetable oils and fats by easy ambient sonic-spray ionization mass spectrometry. Analytical Methods 5 (2013), 6969–6975.
41. M. F. Riccio, S. A. Saraiva, L. A. Marques, R. Alberici, R. Haddad, J. C. Moller, M. N. Eberlin, R. R. Catharino. Instantaneous characterization of vegetable oils via TAG and FFA profiles by easy ambient sonic-spray ionization mass spectrometry. European Journal of Lipid Science and Technology 112 (2010), 434–438.
42. M. F. Riccio, A. C. H. F. Sawaya, P. V. Abdelnur, S. A. Saraiva, R. Haddad, M. N. Eberlin, R. R. Catharino. Easy ambient sonic-spray ionization mass spectrometric of olive oils: Quality control and certification of geographical origin. Analytical Letters 44 (2011), 1489–1497.
43. J. Yang. Antioxidant and antiproliferative activities of common edible nut seeds. LWT—Food Science and Technology 42 (2009), 1573–1580.
44. G. D. Fernandes, W. Moreda, D. Barrera-Arellano, G. C. N. Z. Ruiz, P. L. Ferreira, M. N. Eberlin, R. M. Alberici. TAG, DAG and FFA profiles of dry-cured ham by easy ambient sonic-spray ionization mass spectrometry after thermal imprinting. Journal of Brazilian Chemical Society 25 (2014), 1565–1570.
45. A. M. Porcari, N. V. Schwab, R. M. Alberici, E. C. Cabral, D. R. de Moraes, P. F. Montanher, C. R. Ferreira, M. N. Eberlin, J. V. Visentainer. Intact triacylglycerol profiles of fats and meats *via* thermal imprinting easy ambient sonic-spray ionization mass spectrometry. Analytical Methods 4 (2012), 3551–3557.
46. A. M. Porcati, G. D. Fernandes, K. R. A. Belaz, N. V. Schwab, V. G. Santos, R. M. Alberici, V. A. Gromova, N. M. Eberlin, A. T. Lebedev, A. Tata. High throughput MS techniques for caviar lipidomics. Analytical Methods 6 (2014), 2436–2443.
47. A. C. H. F. Sawaya, P. V. Abdelnur, M. N. Eberlin, S. Kumazawa, M. R. Ahn, K. S. Bang, N. Nagaraja, V. S. Bankova, H. Afrouzan. Fingerprinting of propolis by easy ambient sonic-spray ionization mass spectrometry. Talanta 81 (2010), 100–108.
48. R. Garrett, N. V. Schwab, E. C. Cabral, B. V. M. Henrique, D. R. Ifa, M. N. Eberlin, C. M. Rezende. Ambient mass spectrometry employed for direct analysis of intact arabica coffee beans. Journal of the Brazilian Chemical Society 25 (2014), 1172–1177.
49. V. G. Santos, T. Regiani, F. F. G. Dias, W. Romão, J. L. P. Jara, C. F. Klitzke, F. Coelho, M. N. Eberlin. Venturi easy ambient sonic-spray ionization. Analytical Chemistry 83 (2011), 1375–1380.
50. N. V. Schwab, A. M. Porcari, M. B. Coelho, E. M. Schmidt, J. L. Jara, J. V. Visentainer, N. M. Eberlin. Easy dual-mode ambient mass spectrometry with Venturi self-pumping, canned air, disposable parts and voltage-free sonic-spray ionization. Analyst 137 (2012), 2537–2540.
51. J. S. Garcia, B. G. Vaz, Y. E. Corilo, C. F. Ramires, S. A. Saraiva, G. B. Sanvido, E. M. Schmidt, D. R. J. Maia, R. G. Cosso, J. J. Zacca, M. N. Eberlin. Whisky analysis by electrospray ionization-Fourier transform mass spectrometry. Food Research International 51 (2013), 98–106.
52. G. Huang, W. Xu, M. A. Visbal-Onufrak, Z. Ouyang, R. G. Cooks. Direct analysis of melamine in complex matrices using a handheld mass spectrometer. Analyst 135 (2010), 705–711.
53. J. S. Wiley, J. F. Garcia-Reyes, J. D. Harper, N. A. Charipar, Z. Ouyang, R. G. Cooks. Screening of agrochemicals in foodstuffs using low-temperature plasma (LTP) ambinet ionization mass spectrometry. Analyst 135 (2010), 971–979.

54. S. Martinez-Jarquin, R. Winkler. Design of a low-temperature plasma (LTP) probe with adjustable output temperature of organic molecules. Rapid Communications in Mass Spectrometry 27 (2013), 629–634.
55. J. F. Garcia-Reyes, F. Mazzoto J. D. Harper, N. A. Charipar, S. Oradu, Z. Ouyang, G. Sindona, R. G. Cook. Direct olive oil analysis by low-temperature plasma (LTP) ambient ionization mass spectrometry. Rapic Communications in Mass Spectrometry 23 (19) (2009), 3057–3062.
56. Ratcliffe LV, Rutten FJM, Barrett DA, Whitmore T., Seymour D., Greenwood C., Aranda-Gonzalvo Y, Robinson S., McCoustra M. Surface analysis under ambient conditions using plasma-assisted desorption/ionization mass spectrometry. Analytical Chemistry 79 (16) (2007), 6094–6101.
57. T. D. Whitmore, Y. Aranda Gonzalvo, C. L. Greenwood, J. A. Rees, L. V. Ratcliffe, D. A. Barrett, F. J. M. Rutten, M. R. S. McCoustra. Surface analysis by plasma assisted desorption ionisation mass spectrometry (PADI-MS28th ICPIG), July 15–20, 2007, Prague, Czech Republik.
58. Z. Wu, H. Chen, W. Wang, B. Jiu, T. Yang, Z. Zhang, J. Ding, X. Xiao. Differentiation of dried sea cucumber products from different geographical areas by surface desorption atmospheric pressure chemical ionization mass spectrometry. Journal of Agricultural and Food Chemistry 57 (2009), 9356–9364.
59. H. Chen, H. Liang, J. Ding, J. Lai, Y. Huan, X. Qiao. Rapid differentiation of tea products by surface desorption atmospheric pressure chemical ionization mass spectrometry. Journal of Agricultural and Food Chemistry 55 (2007), 10093–10100.
60. E. Handberg, K. Chingin, N. Wang, X. Dai, H. Chen. Mass spectrometry imaging for visualising organic analytes in food. Mass Spectrometry Reviews 2014, 1–18.
61. K. Hiraoka. Probe eelctrospray ionization as a FD. *60th ASMS Conference on Mass Spectrometry and Allied Topics.*
62. S. Yang, H. Chen, Y. Yang, B. Hu, X. Zhang, Y. Zhou, L. Zhang. Imaging melamine in egg samples by ssurface desorption atmospheric pressure chemical ionization mass spectrometry. Chinese Journal of Analytical Chemistry 37 (2009), 315–318.
63. J. Wang, S.-P. Yang, F.-Y. Yan, Y. Liu, M. Li, Y.-H. Song, Y.-B. Zhan, H.-W. Chen. Rapid determination of dimethoate in nanoliter of juice using surface desorption atmospheric pressure chemical ionization mass spectrometry. Chinese Journal of Analytical Chemistry 38 (4) (2010), 453–457.
64. G. Morlock, W. Schwack. Determination of isopropylthioxanthone (ITX) in milk, yoghurt and fat by HPTLC-FLD, HPTLC-ESI/MS and HPTLC-DART/MS. Analytical and Bioanalytical Chemistry 385 (3) (2006), 586–595.
65. T. M. Vail, P. R. Jones, O. D. Sparkman. Rapid and unambiguous identification of melamine in contaminated pet food based on mass spectrometry with four degrees of confirmation. Journal of Analytical Toxicology 3 (6) (2007), 304–312.
66. T. Cajka, L. Vaclavik, K. Riddellova, J. Hajslova. GC-TOF-MS and DART-TOF-MS: Challenges in the analysis of soft drink. LC GC Europe 21 (2008), 250–255.
67. J. Schurek, L. Vaclavik, D. Hooijerink, O. Lacina, J. Poutska, M. Sharman, M. Caldow, M. Nielen, J. Hajslova. Control of strobilurin fungicides in wheat using direct analysis in real time accurate time-of-flight and desorption electrospray ionization linear ion trap mass spectrometry. Analytical Chemistry 80 (24) (2008), 9567–9575.
68. JEOL AccuTOF mass spectrometer with DART application data. Rapid detection of fungicide in orange peel. www.jeol.co.jp/en/applications/pdf/ms/ms_note_en002.pdf 2017 (accessed 1 April 2017).

69. H. J. Kim, Y. P. Yang. Direct analysis of curcumin in turmeric by DART-MS. Phytochemical Analysis 20 (5) (2009), 372–377.
70. A. J. Dane, R. B. Cody. Selective ionization of melamine in powdered milk by using argon direct analysis in real time (DART) mass spectrometry. Analyst (Cambridge, UK) 135 (2010), 696–699.
71. R. Kubec, R. B. Cody, A. J. Dane, R. A. Musah, J. Schraml, A. Vattekkatte, E. Block, Applications of direct analysis in real time–mass spectrometry (DART-MS) in Allium chemistry. (Z)-Butanethial S-Oxide and 1-Butenyl Thiosulfinates and Their S-(E)-1-Butenylcysteine S-Oxide Precursor from Allium siculum. Journal of Agricultural and Food Chemistry 58 (2) (2010), 1121–1128.
72. E. Block, A. J. Dane, S. Thomas, R. B. Cody. Applications of direct analysis in real time mass spectrometry (DART-MS) in Allium chemistry. 2-Propenesulfenic and 2-Propenesulfinic Acids, Diallyl Trisulfane S-Oxide, and other reactive sulfur compounds from crushed garlic and other alliums. Journal of Agricultural and Food Chemistry 58 (8) (2010), 4617–4625.
73. L. Vaclavik, J. Rosmus, B. Popping, J. Hajslova. Rapid determination of melamine and cyanuric acid in milk powder using direct analysis in real time-time-of-flight mass spectrometry. Journal of Chromatography A 1217 (25) (2010), 4204–4211.
74. L. Vaclavik, M. Zachariasova, V. Hrbek, J. Hajslova. Analysis of multiple mycotoxins in cereals under ambient conditions using direct analysis in real time (DART) ionization coupled to high resolution mass spectrometry. Talanta 82 (5) (2010), 1950–1957.
75. H. Chen, J. Zheng, X. Zhang, M. Luo, Z. Wang, X. Qiao. Surface desorption atmospheric pressure chemical ionization mass spectrometry for direct ambient sample analysis without toxic chemical contamination. Journal of Mass Spectrometry 42 (8) (2007), 1045–1056.
76. M. Huang, Z. Ouyang, R. G. Cooks. High-throughput trace melamine analysis in complex mixtures. Chemical Communications 5 (2009), 556–558.

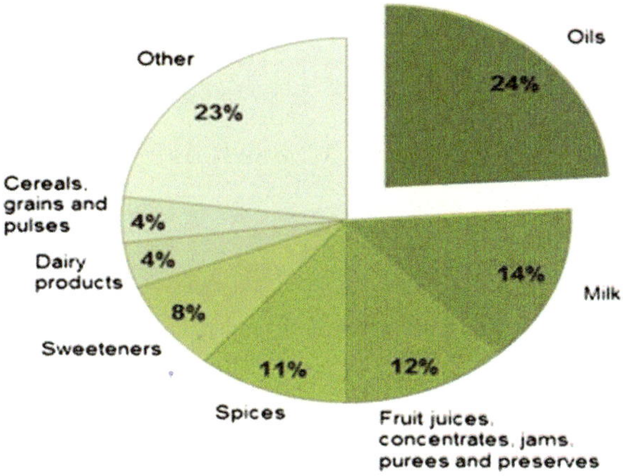

FIGURE 2.1 Percentage of food fraud cases from 1980 until 2010. (Reproduced with permission from Moore et al., Journal of Food Science, 77 (2012), pg 118–126. With permission.)

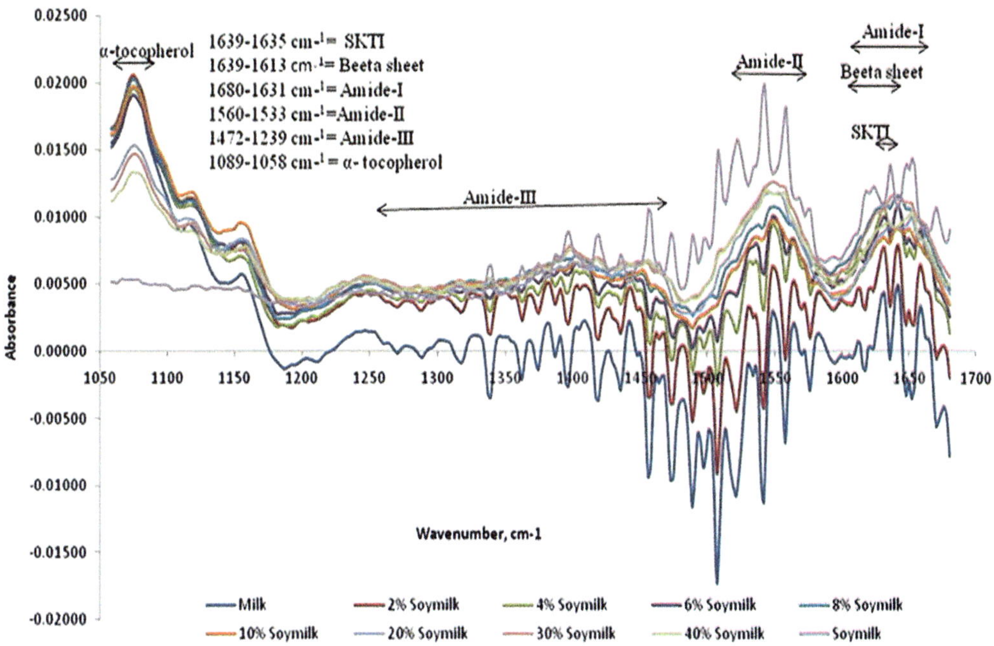

FIGURE 2.4 Typical spectra of milk, soymilk, and milk adulterated with soymilk at different levels in the spectral range 1680–1058 cm^{-1}. (Reproduced with permission form Botelho et al., Food Chemistry, 181 (2015), pg 31–37. With permission.)

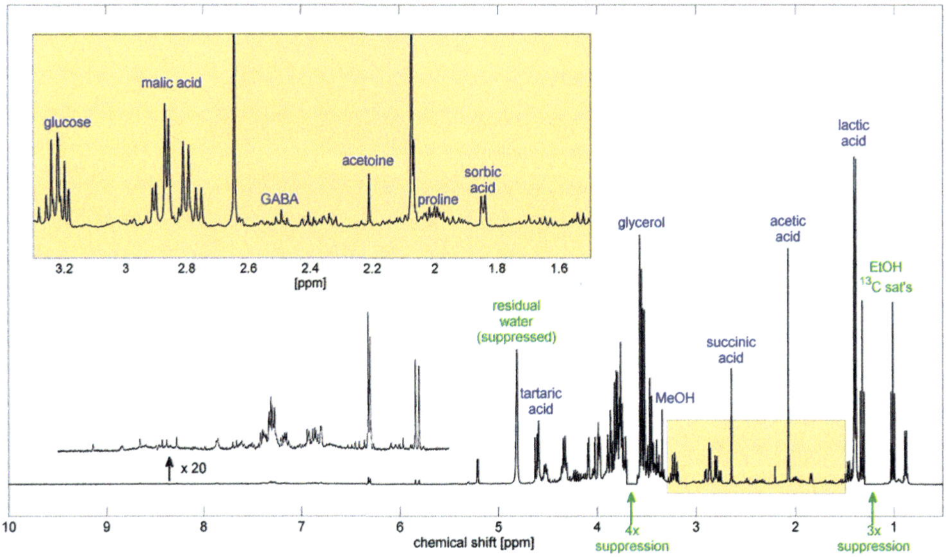

FIGURE 3.5 ^{1}H NMR spectrum of wine with suppression of the water and ethanol peaks, identifying several low molecular weight metabolites of wine. Reprinted with permission from (Godelmann et al., 2013). Copyright 2013 American Chemical Society.

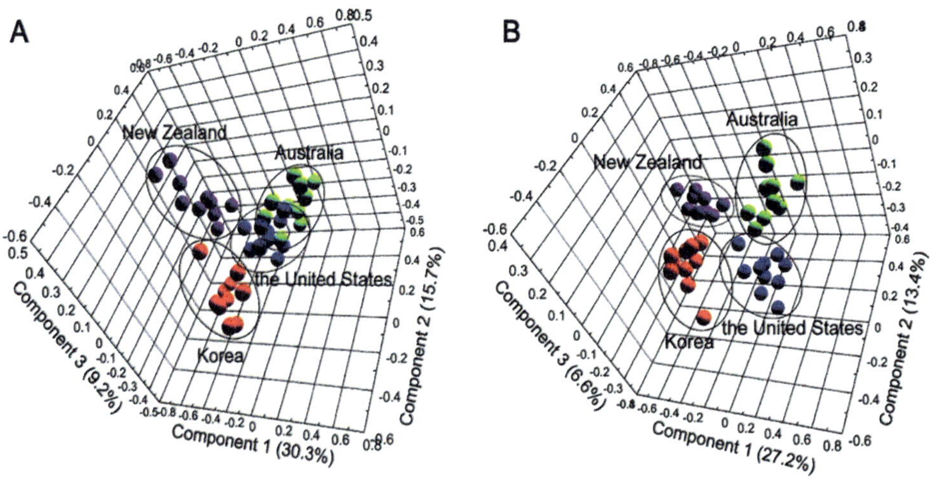

FIGURE 3.6 PCA (A) and OPLS-DA (B) 3D score plots derived from the ^{1}H NMR spectra of beef sirloin (or chuck) extracts obtained from Australia, South Korea, New Zealand, and the United States. Reprinted with permission from (Jung et al., 2010). Copyright 2010 American Chemical Society.

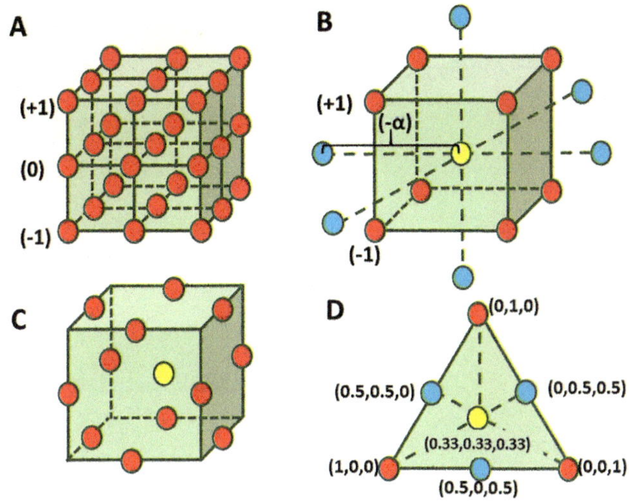

FIGURE 10.3 Schematic representation of (A) a full factorial 3^3 design, (B) a central composite design, (C) a Box-Behnken design, and (D) a mixture simplex centroid design. All of them are for three factors.

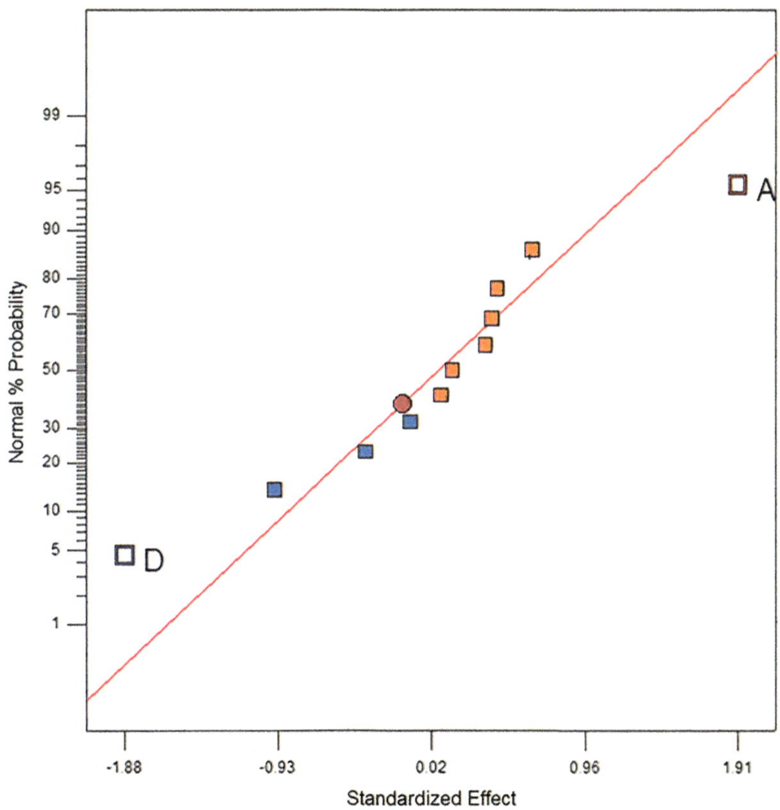

FIGURE 10.4 Normal probability plot of the effects on the phenolic compounds extraction experiment [25].

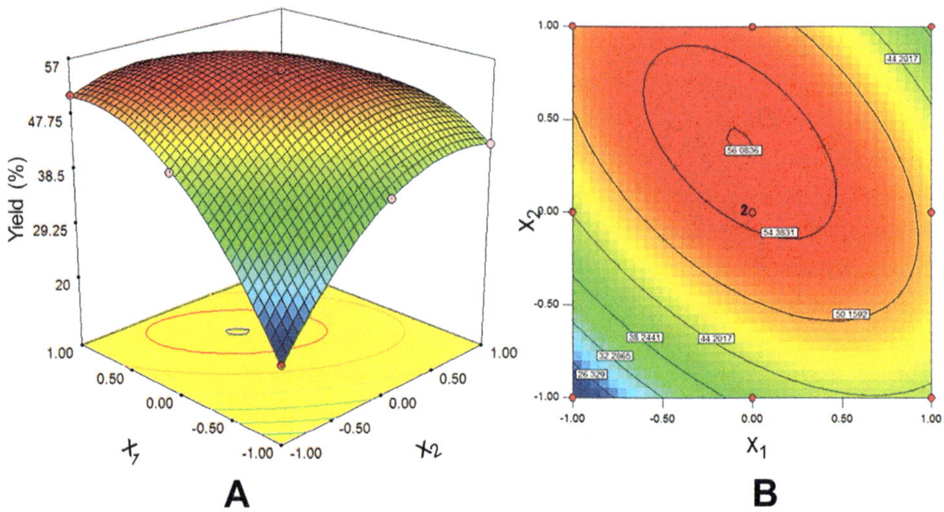

FIGURE 10.5 Surface of response (A) and the contour plot (B) corresponding to the response (%) of yield in a chemical reaction [22].

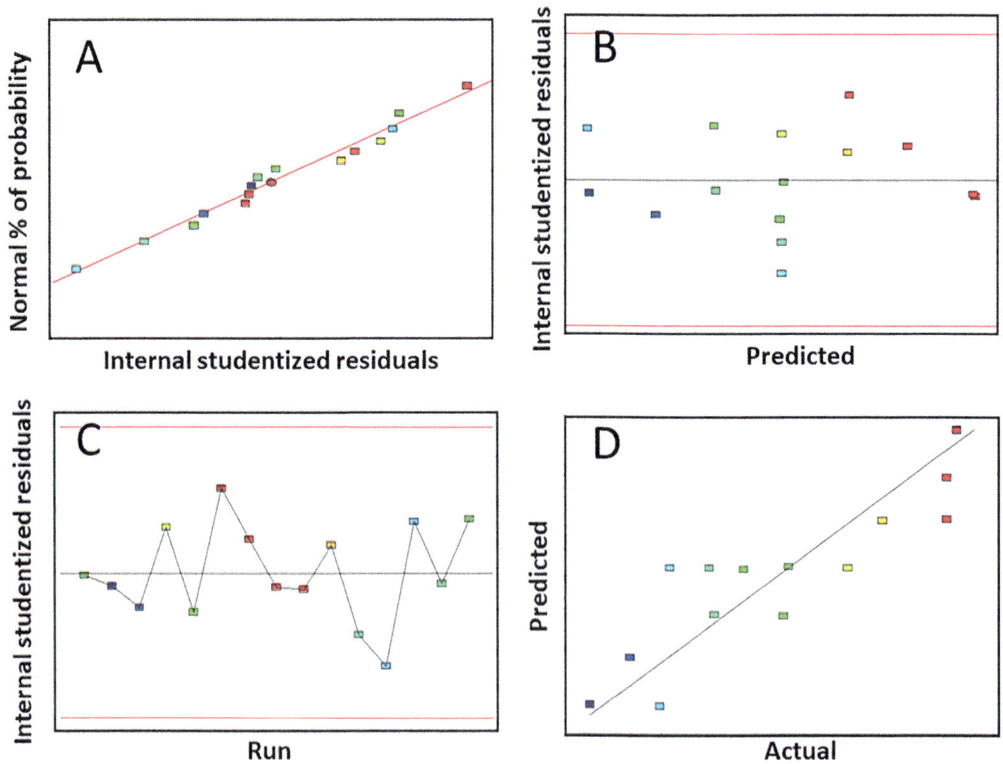

FIGURE 10.7 Diagnosis of normality distribution assumption plots for response pH. (A) Normal plot of residuals, (B) residuals vs. predicted values, (C) residuals vs. run, and (D) predicted vs. actual values.

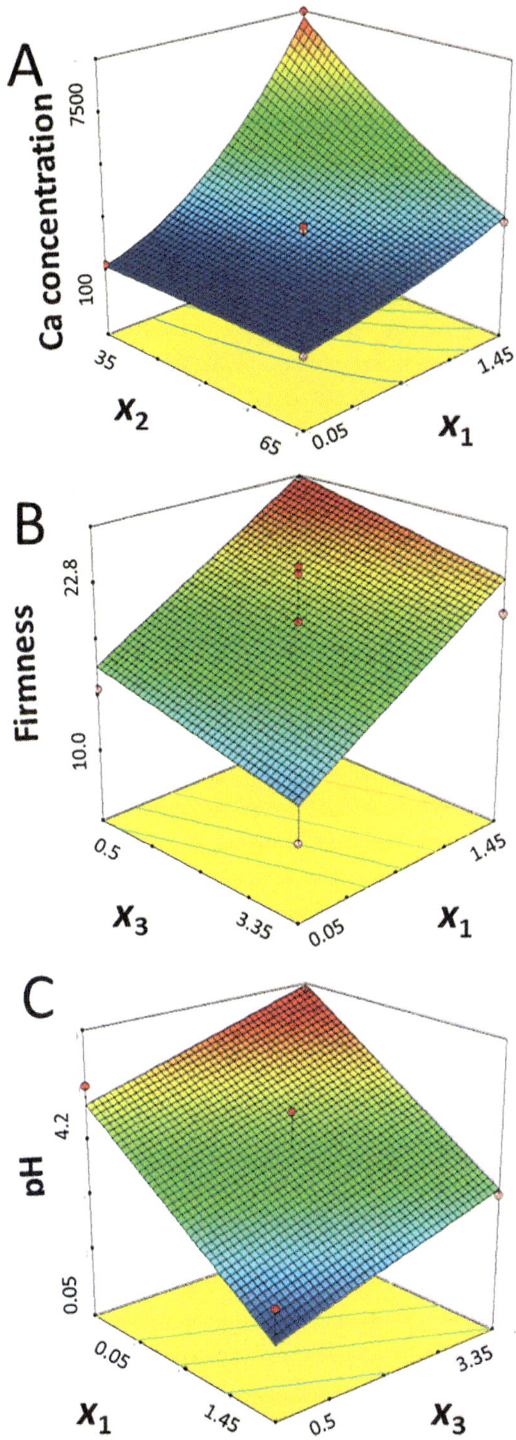

FIGURE 10.8 Response surfaces plotted following the models represented in Eqs. (10.6–10.8). In all cases, just two factors are represented, maintaining the other at a constant value equal to the center of the design. The two most important factors are plotted.

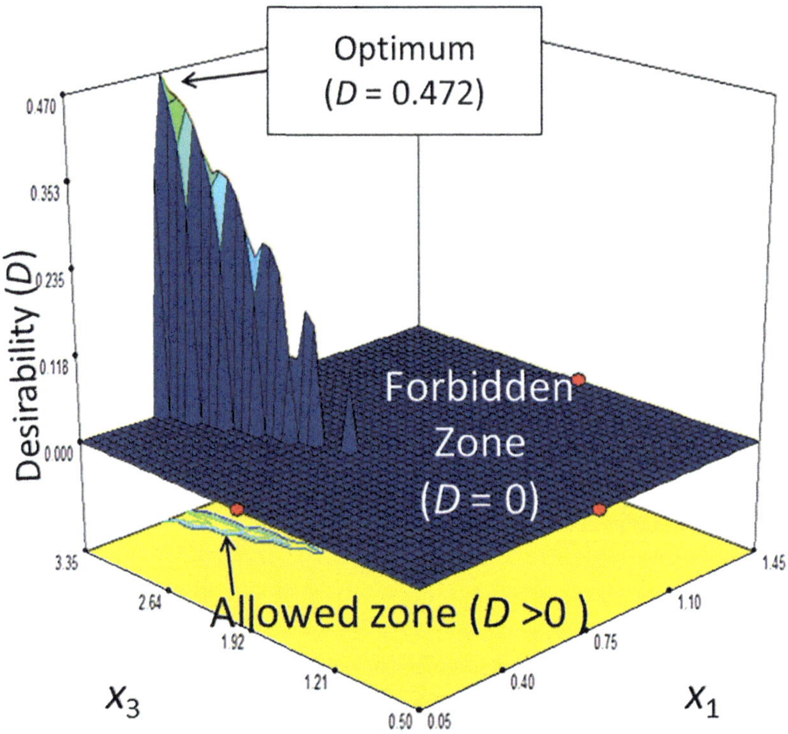

FIGURE 10.10 Surface for the desirability (D) for the combination x_1 vs. x_3.

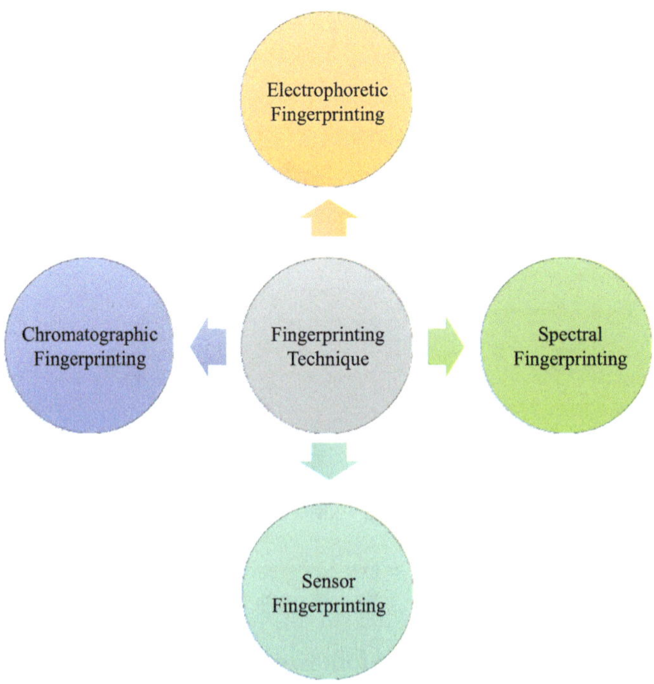

FIGURE 20.1 Classification of fingerprinting techniques.

CHAPTER 7

DNA-Based Methodologies

Marta Muñoz-Colmenero
University of the Basque Country UPV/EHU

CONTENTS

7.1 DNA-BASED MARKERS IN FOOD TRACEABILITY

The identification of species present in food is essential for providing the consumers with complete and faithful information, through labels, which offers them the possibility to make decisions according to their preferences, ethics, religious beliefs, or specific health problems (Jacquet & Pauly, 2008; Demirhan et al., 2012). In a globalized food market, traceability is important not only to ensure safety of food but also to avoid origin and quality fraud (Aung & Chang, 2014). The edible products can be sold in different forms, from fresh and whole, for example many vegetables or some fish, to fully processed products, such as pre-cooked products, gelatins, etc. with different modifications of the original species. The different manipulations of food hinder the identification of the species present in the products, as well as their origin since the morphological characteristics are lost during manufacturing process (Lago et al., 2014).

In spite of the large number of forms in which food is available, from processed to unprocessed, the molecular tools based on deoxyribonucleic acid molecule (DNA) have been described to be the most effective for the unequivocal identification of species (Céspedes et al., 1998a; Kochzius et al., 2010; Doosti et al., 2014; Pegels et al., 2014; Safdar & Junejo, 2015; Muñoz-Colmenero et al., 2017). A clear advantage of DNA-based techniques over other molecular techniques is the high resistance and stability of the DNA molecule, which remains poorly degraded even when the food products are cooked or highly processed by strong abrasive methods (Ardura et al., 2012; Arun et al., 2013; Bornatowski et al., 2013; Torelli et al., 2014), maintaining its effectiveness even when other molecules, such as proteins, get highly modified or degraded (Woolfe & Primrose, 2004). In this sense, DNA-based methods can be applied to different matrices (Mafra et al., 2008) as it is possible to recover DNA even from small traces of initial organic material (Hellberg & Morrissey, 2011; Galimberti et al., 2013). In addition, DNA contains more information than proteins since it has coding and noncoding regions (Espiñeira and Santaclara, 2016), the last ones being strongly variable (Taberlet et al., 1991). Besides the nuclear DNA (or bacterial chromosome in prokaryotes), other DNA molecules also exist in organisms such as the chloroplast and mitochondrial DNAs that are present in more than one copy, offering a higher probability of recovery even in the most degraded

TABLE 7.1 Summary of Main Applications, Advantages, and Disadvantages of the Different DNA-Based Molecular Methods Used in Food Traceability

Method	Applications	Advantages	Disadvantages
PCR-GE	• Species detection (presence/ absence) • Label authenticity • Adulterations • GMO detection	• Cheap and rapid • No complex requirements • Minimum effort • Can be used as a preparative technique	• Toxicity of acrylamide gels • Discriminates among species only on the basis of amplicon size • Can't discriminate species with the same amplicon size and different sequences • Not a quantification method
PCR-DGGE/ TGGE	• Establishment of the product origin ("fingerprinting") • Discrimination among closely related species or populations (less used)	• High resolution to discriminate different sequences (mutations/ nucleotide variation) with same amplicon size • It is possible to analyze different patterns at the same time	• Toxicity of acrylamide gels • Needs a high quality of DNA • Needs software to analyze the resulting images (band pattern)
PCR-RFLP	• Identification of species present in food • Used in mixes, fresh, and processed samples • Detects adulterations	• High discrimination among species • Cheap, quick, and repeatable • Can be used in processed food • Cost effective even in developing countries for routine systems	• No sequence information • Can give similar band patterns using universal primers for some taxonomic groups

(Continued)

TABLE 7.1 (*Continued*) Summary of Main Applications, Advantages, and Disadvantages of the Different DNA-Based Molecular Methods Used in Food Traceability

Method	Applications	Advantages	Disadvantages
AFLP	• Finds new species-specific/ population-specific markers • Classifies and identifies cultivars • Stock identification and population assignment • Detects adulterations (less used)	• Analyzes several fragments of genome at the same time • High resolution and sensitivity • High reproducibility	• Laborious and time-consuming • Requires high-quality DNA
PCR-sequencing / DNA-Barcoding	• Discriminates among closer species • Identifies unknown samples • Assays the accuracy of the labels • Checks frauds	• Provides information about the nucleotide sequence in a target species • Very resolute in discriminating among closer species • Can be applied in fresh, modified, and highly processed samples • DNA barcoding and mini-barcoding offer an extra database and a standardized method to identify samples	• More expensive and time-consuming than other methods • Needs some prior information about the sequence of target species • The databases are not always reliable • The databases are not complete for all species
SNP	• Identification of a specific set of species/ organisms • Checks the accuracy and reliability of labeling	• High abundance of SNPs in all genomes • Rapid, efficient, and accurate species identification • Low cost and high reproducibility • Can be applied in modified products and degraded DNA • Adequate for automatized analyses • Easy implementation and interpretation of results	• High cost of SNP discovery • Necessary to know the DNA sequence of target species • It is implemented in a specific set of organisms

(*Continued*)

TABLE 7.1 (*Continued*) Summary of Main Applications, Advantages, and Disadvantages of the Different DNA-Based Molecular Methods Used in Food Traceability

Method	Applications	Advantages	Disadvantages
Microsatellites (SSR)	• Identification of varieties, cultivars, and animal stocks • Assignment of individual samples to a putative source • Authentication of food components	• High variability and genetic codominance. • Fast and lower cost • Possibility to perform in multiplex PCR • Possibility of automatization • Useful in cooked and some processed foods	• Needs good reference databases
PCR-cloning-sequencing	• Detection of several species at the same time • Analyses of mixtures	• Can separate the different species present in a mixture or complex food.	• Expensive • Complex process in the laboratory • Very dependent on the chosen primers
DNA-Chips	• Detection of several species at same time (GMOs, cultivars, etc.) • Analyses of mixtures	• High sensitivity and accuracy • Fast, adaptable, and comparable	• It is very dependent on the data banks of available DNA sequences or markers
RT-PCR	• To identify and quantify minute traces of species in feed and foodstuffs • GMO testing	• High sensitivity • It is able to quantify • Monitoring the early stages of PCR	• High cost of reagents, equipment, and trained professionals
HRM	• Very commonly used to identify closely related species, cultivars, or PDO • Other uses: GMO screening, allergen detection, or detection of food pathogens	Allows the discrimination of species with minimal variations in the sequences (SNP or small indels)	• High cost of the instrument and reagents
NGS	• Food microbiology • Microbial profile for specific biogeographical region or "terroir" • Identifies all the species present in a food product (also mixtures and processed products)	• High capacity of analysis (in the order of millions of DNA sequences) • Analyzes several products (samples) at the same time • High resolution in mixed, highly degraded samples, and in samples that are genetically very similar	• High cost of the equipment and the analysis, but that cost is decreasing

samples (Rooney et al., 2015). The mitochondrial genome has a mutation rate higher than nuclear DNA, thus presenting a greater nucleotide variation, which makes it suitable for differentiation between species, even when they are closely related (Galal-Khallaf et al., 2016a). The genes contained in the DNA of these organelles have no introns; they have a few spacers and repetitive sequences or pseudogenes and, in most organisms, they have uniparental inheritance (Birki, 1995; Hoekstra, 2000), which simplifies their use because heterozygotes will not occur in these cases. All the above-mentioned traits make them widely applicable as target genes in food traceability.

7.2 POLYMERASE CHAIN REACTION (PCR)

DNA-based techniques used for food traceability and food safety are based on PCR (Costa Leal et al., 2015), which results in the amplification of a specific region of the genome (amplicon) delimited by previously selected primers. This technique presents a high potential for application due to its speed, simplicity, sensibility, and specificity in obtaining a large number of copies of target sequences from a minimal amount of sample tissue (Bottero & Dalmasso, 2011; Horreo et al., 2013). In addition, the prices of the equipment and reagents needed to perform PCR are continuously decreasing, thus making it a more economical process (Costa Leal et al., 2015). These techniques can offer different levels of identification—individual traceability to ensure food safety, and traceability of individuals to their sources, breeds, or species—and can detect possible labeling fraud (Dalvit et al., 2007).

For a successful PCR, the DNA extraction method is an important step because it cleans all the potential inhibitors of the posterior PCR. The selection of amplified molecular markers is also crucial since in highly processed food, such as canned products, DNA may be highly degraded and broken due to which the shorter fragments will have more probabilities of being amplified (Fajardo et al., 2010). Furthermore, a balance must exist between the specificity/universality of the primers with respect to the template DNA, anchoring well and equally with all targeted species in order to recover all the species required for a specific study (Wilcox et al., 2013; Muñoz-Colmenero et al., 2017), for instance, only a specific species among several species in the same taxonomical group or all the different species present in a mixture like gelatin or feed. Likewise, the resulting amplicon should be able to discriminate between all targeted species, presenting enough variability in its sequence. Nuclear, mitochondrial, and chloroplast genes have been used for the identification of species in food, using one or the other depending on the DNA-based methodology applied and on the taxonomic groups to be amplified.

Also, the identification or detection of several components by only one PCR is possible by using different pairs of primers in the same reaction, thus amplifying different markers at the same time. This reaction is called multiplex PCR and it reduces the unit cost of assay and saves time (Natonek-Wiśniewska et al., 2013). However, sometimes, the standardization of this method is complicated due to uneven amplification efficiencies of the different primer couples or due to the low sensitivity of the test, which may result in false negative results (Ballin et al., 2009; Zhang, 2013).

Once the PCR is carried out, the PCR product can be visualized and examined by several methodologies. Some of the most commonly used methods in food traceability are explained in this chapter, offering a review of their advantages and disadvantages, as well as their principal uses (summarized in the Table 7.1).

7.3 GEL ELECTROPHORESIS (GE)

One of the simplest and cheapest methods to visualize the PCR product (from a simple or multiplex PCR) is by using a conventional gel electrophoresis (Galimberti et al., 2013). This method is based on loading the PCR product in a matrix and the application of an electric field to separate the macromolecules (or fragments of them) by their sizes and charges. The molecules will run from the cathode to the anode, moving through the pores of the matrix, migrating to a greater distance than the smaller particles do, since they cross the matrix much easily and with greater speed, thus moving farther away from the loading zone. Currently, the most commonly used matrix to separate the DNA molecules or their fragments is 0.7%–2% agarose, but more agarose should be added to separate the shorter fragments (Brody et al., 2004). In some cases, polyacrylamide instead of agarose is used to separate very small DNA fragments (PAGE), but it is necessary to manipulate it with care because the acrylamide is a potent neurotoxin in both liquid and powdered forms (Roach et al., 2003). To visualize the DNA molecules in the gel, they can be stained with different substances, like ethidium bromide that intercalates into DNA and fluoresces with ultraviolet light (UV)—this compound is a strong mutagen—or with *SimplySafe* that emits green fluorescence under UV light and is less dangerous. In addition, DNA size markers and/or mass ladders may be loaded with the samples for checking the size or amount of PCR products obtained.

In food traceability, this methodology is frequently applied as an analytical technique to detect the presence/absence of some species in feedstuffs and foodstuffs since it is economical, rapid, and does not require complex equipment in the laboratory, and it is thus an ideal, simple, and cost effective approach for commercial screening. With a simple reaction and electrophoresis gel we can validate the label authenticity, check adulterations, or even detect potential genetically modified organisms (GMOs) by choosing specific primers.

Several examples can be found in literature about the multiple applications of this method: in the meat industry, for analyzing the species contained in the commercial samples of Portuguese game meat *Alheira* sausages (Amaral et al., 2014), or detecting the presence of chicken and donkey meats as adulterants in raw ground meats (Mousavi et al., 2015); in the seafood industry, for identifying the most important species of scallops in processed products (Marín et al., 2013); a rapid authentication and traceability of Yellow Drum (*N. albiflora*) from *surimi* products (Zhao et al., 2013); in the feed industry, for testing the presence of ruminant ingredients in feedstuff (Martin et al., 2007), or applying a multiplex PCR-GE to discriminate the different types of meats in pet food and ruminant feeds (Safdar & Junejo, 2015); etc. One drawback of this method is the low resolution used to discriminate the species that have different sequences but amplicons of the same size. In that case, other techniques such as DGGE (denaturing gradient gel electrophoresis) or RFLP (restriction fragment length polymorphism) are used to discriminate sequences with different base compositions.

Finally, GE after PCR can also be used in food traceability as a preparative technique to check the PCR product before continuing the analyses with other methodologies, such as RFLP, cloning, DNA sequencing, etc.

7.4 DENATURING GRADIENT GEL ELECTROPHORESIS (DGGE) AND TEMPERATURE GRADIENT GEL ELECTROPHORESIS (TGGE)

These approaches use polyacrylamide gels that include a denaturing agent, usually urea or formamide, for creating a denaturing gradient (DGGE)—the most commonly used method— or increase the temperature during the process (TGGE). Both are based on the separation of PCR products of the same size but having different sequences, which are equivalent to different points of denaturation (melting profile) (Ercolini, 2004). DNA or RNA molecules are denatured and separated according to the sequence of nucleotides they contain. The result is a pattern of bands that represents the molecular fingerprint of the analyzed sample. The gels are stained with different stains that are reactive to UV radiation, such as SYBR Gold. Then the images of the band patterns are analyzed through different software.

This methodology has been most applied in the field of microbial ecology. In fact, in food traceability, one of its main uses has been to establish the geographical origin of the product through the analyses of the bacterial communities present in the food, assuming that different bacterial communities are associated with organisms from different geographical locations (Costa Leal et al., 2015). This methodology is also called "fingerprinting" and it has been applied successfully to determinate the geographical origin of: fish species—such as Pangasius fish (Le Nguyen et al., 2008a) or Tilapia (Tatsadjieu et al., 2010); of fruits—such as clementine (Le Nguyen et al., 2008b), nectarines, and peaches (Bigot et al., 2015); in cheese (Arcuri et al., 2013); etc. Although less commonly, this technique is also used for the direct discrimination among and within species, for example, in the assay of the commercial ivory shell (*Babylonia areolata*), and it results in a robust method without sequencing (Chiu et al., 2015).

The PCR-DGGE method is less time-consuming, costly, and labor-intensive than PCR sequencing, since it is able to detect almost all the mutations present in the target samples. It has proved to be a good tool for monitoring fresh products, but this methodology has not demonstrated the same results for all kinds of processed products, being highly dependent on the DNA quality (Hovda et al., 2007).

7.5 RESTRICTION FRAGMENT LENGTH POLYMORPHISM (RFLP)

A specific region of DNA is amplified by PCR with universal primers and the PCR product is digested by restrictions enzymes (RE), resulting in a specific band pattern of restriction fragments when the digested PCR product is run and visualized in gel electrophoresis or acrylamide, depending on the size and number of the fragments (Kumar et al., 2014). Different samples or species can be identified from different number of fragments and/or from different lengths of such fragments, resulting in a particular band pattern that can be sometimes called "fingerprinting" as in the DGGE method (Partis et al., 2000). When this methodology is used over one locus, it is called single locus probes (SLP), and when it is used over more than one locus, it is called multi locus probes (MLP) (Chakraborty & Jin, 1993). A critical step for successful RFLP application is the careful choice of the target gene/s as well as the REs applied to cut the PCR product. Such selection has to ensure the maximal differentiation between the investigated species and the complete, not partial, cut of the sequences for preventing ambiguous results caused due to intraspecies polymorphism (Wolf et al., 1999). The visualization of band pattern may be done with stains reactive to UV, and it can be analyzed directly from the gel or through an image analysis software.

RFLPs are widely employed for the typing of species with several target genes, especially within the mitochondrion (Céspedes et al., 1998b; Hold et al., 2001; Zhang et al., 2006). It has been one of the most commonly used technologies for identifying the species present in food, being considered as a highly discriminatory, cheap, quick, reliable, and repeatable method that can be used to analyze fresh and processed food and even mixes (Meyer et al., 1995; Partis et al., 2000; Girish et al., 2005; Lanzilao et al., 2005). It is turning out to be cost effective for a long food traceability chain, especially in developing countries that have less access to sequencing technologies in routine systems (Murugaiah et al., 2009). One big advantage of using PCR-RFLP is that one universal PCR-primer system in combination with a few REs can be sufficient for species identification (Meyer et al., 1995), but it is important to note that using universal primers, when there are many species in a determined taxonomic group, can produce the same restriction pattern for the studied and not-studied species, and a false diagnosis might be obtained (Santaclara et al., 2007). This possibility may be decreased by the use of specific primers that are more driven toward only a few taxonomic groups (Armani et al., 2012).

In the last few years, the use of RFLP has decreased due to the rise of cheaper DNA sequencing technologies, which are considered to be more precise (Kumar et al., 2014). However, this methodology continues to be applied in the food industry for identification of species present in commercial products and to detect the adulteration in food. It is commonly applied to know the composition of meats (Ong et al., 2007; Chen et al., 2010; Stamoulis et al., 2010; Kumar et al., 2014), to check the reliability of meat for halal populations (Murugaiah et al., 2009), to detect any mislabeling in seafood (Santaclara et al., 2007; Rea et al., 2009; Armani et al., 2012; Sumathi et al., 2015), and to identify the origin species of milk (Abdel-Rahman & Ahmed, 2007), etc.

7.6 AMPLIFIED FRAGMENT LENGTH POLYMORPHISM (AFLP)

This technique is also known as "fingerprinting" and it was described by Zabeau & Vos (1993) and Vos et al. (1995). It is based on the digestion of DNA present in the cells using several REs. A selection of the amplified fragments is made and specific adapters are attached at the sticky ends of the restriction sites. Through primers complementary to the adapters, these fragments are finally amplified and run in a DGGE to visualize the band pattern by autoradiography or fluorescence or are run through capillary electrophoresis gel to visualize the profile. The obtained pattern is analyzed based on the presence/absence of bands or peaks corresponding to the different restriction fragments, thus distinguishing the polymorphisms present in the genome of the target species. The principal anonymous markers used largely consist of noncoding DNA (Wong et al., 2001).

This methodology has several advantages as it has the ability of analyzing numerous fragments of genome at the same time, without having any prior information about the sequences in the genome of the target species (Meudt & Clarke 2007). AFLP has a high probability of detecting species-specific or population-specific polymorphisms because it generates hundreds of genetic markers (Gomez-Uchida et al., 2003). In addition, this technique is considered to have high reproducibility, resolution, and sensitivity along the entire length of the genome (Mueller & Wolfenbarger, 1999; Shirasawa et al., 2004). However, this method also has some drawbacks, as it is a laborious and time-consuming procedure and it requires a high-quality DNA to be amplified by PCR and to obtain reproducible results (Kalaitzis & El-Zein, 2016). Hence, in many cases, the need exists to convert the AFLP markers into sequence-specific PCR markers (Parker & Langridge, 2000; von Stackelberg et al., 2003).

Despite the obligatory step of PCR amplification and its potential disadvantages, the AFLP method has been considered as one of the most reliable and informative fingerprinting procedures for classifying and identifying olive cultivars to establish the origin of oil (Busconi et al., 2003; Pafundo et al., 2005). Furthermore, the AFLP technique has been used to develop efficient methods to identify cultivars of rice (Sardesai et al., 2002; Shirasawa et al., 2004) and potato (McGregor et al., 2000), as well as to detect the adulteration in ginseng products (Ha et al., 2002). This technology has already demonstrated a good ability to stock identification and population assignments, even for populations with weak differentiation (Campbell et al., 2003). This ability may be used to assess species identification and authenticity, for example, in fish and seafood samples, differentiating even fish and crustacean domestic strains (Maldini et al., 2006).

7.7 MICROSATELLITES OR SIMPLE SEQUENCE REPEATS (SSR)

The simple or single sequence repeats (SSR), also known as microsatellites (Mafra et al., 2008), are repeated genomic motifs usually ranging in length from 2 to 13 base pairs, which are abundantly distributed across the organism's genome and have high levels of allele polymorphism due to variations in the number of repeats, leading to high genetic diversity (Chistiakov et al., 2006; Pasqualone et al., 2007). They can be amplified easily by PCR and their detection can be automatized, labeling one primer with a fluorescent dye and analyzing it with capillary electrophoresis-based genetic analyzers according to the length and repeats of the marker in each sample. (Alves et al., 2009; Rodriguez-Ramirez et al., 2011). This process requires less manual intervention and it is faster and of lower cost (Koumi et al., 2004). An internal standard, containing DNA fragments of known sizes and labeled with a different colored dye, is electrophoresed with each sample to calibrate sizes from run to run. In addition, using different colored dyes for the known size ranges of SSRs, it is possible to amplify several microsatellites in the same multiplex PCR (Butler et al., 2004).

Due to their great variability, high reproducibility, and adequate automation, the microsatellites are considered a good option in the traceability field. One application of SSR is the variety and cultivar identification, because of their properties of genetic codominance, high reproducibility, and multiallelic variation, for instance, in grapes (Vignani et al., 1996; Faria et al., 2000), rice (Zeng et al., 2004), tomato cultivars (Savo Sardaro et al., 2012, Iquebal et al., 2013), olive oil cultivars (Cipriani et al., 2002), and even in the identification of cultivars of complex olive oil (Pasqualone et al., 2007). The microsatellites have also been successfully applied for individual identification in natural or brood stocks and in breeding programs based on divergent allele frequencies in different populations, which along with adequate statistical methodologies are able to enable a probabilistic assignment of individual samples to putative sources (Ogden et al., 2008). This assignation methodology is applied in different research fields like fisheries—for example cod, trout, salmon, or mussels (Hansen et al., 2001; Chistiakov et al., 2006; Larrain et al., 2014)—or the meat industry (Vazquez et al., 2004; Orrù et al., 2006).

Other common uses of these molecular markers are in effective authentication of food components, such as sugarcane (Parida et al., 2009) or *Hinai-Jidori* chicken (Rikimaru & Takahashi, 2007). In addition, this methodology can be used for cooked or processed food as cooked meats or oils (Rodriguez-Ramirez et al., 2011; Pasqualone et al., 2007) that cannot be tested by all the molecular techniques.

The main disadvantage of this methodology is that for an effective implementation in traceability, with successful identification of species, varieties, or cultivars, the availability of a comprehensive DNA database with reference samples for the SSR panels is necessary (Alves et al., 2009). For some food, the databases are still incomplete.

7.8 PCR SEQUENCING AND DNA BARCODING

The main advantage of applying PCR amplification followed by the sequencing reaction is that it allows the detection of the nucleotide sequence present in the molecular region of interest in the species for which it is applied. The most common sequencing method is "DNA sequencing with chain-terminating inhibitors" developed by Frederick Sanger in 1977 (Sanger et al., 1977), extending one strand of DNA from a primer which is complementary to the other template strand that we want to examine. Each time an inhibitor (modified deoxynucleotides = di-deoxynucleotidetriphosphates (ddNTPs)) is added to the chain, the polymerase stops the amplification and the chain ends. At the end of the reaction, there will be chains with all possible lengths, each one ending with a labeled ddNTPs. For their automated detection by sequencing machines, the ddNTPs are i.e., fluorescently labeled, and they are excited by a laser when they run through capillary electrophoresis, detecting the ddNTPs at the end position of each sequence length and giving a complete chromatogram of the target marker.

This method is more expensive and time-consuming than the simple visualization of the PCR product in a gel (Teletchea, 2009). This analytical method is one of the most commonly used for routine inspection of the composition of food or feed (Natonek-Wiśniewska et al., 2013). Currently, sequencing doesn't require very expensive equipment since there have been decreases in price, and it can be applied for fresh, modified, or highly processed samples, provided that in the latter case the chosen marker is of a small size to ensure its PCR amplification (Horreo et al., 2013). This method allows us to know the complete nucleotide sequence of the molecular marker in a target species, as it is possible to detect the difference between even one nucleotide of close species, offering an unambiguous identification of species in food. However, for choosing the most suitable molecular marker for discriminating among the target species, it is necessary to have some information about their sequences. In case of unidentified samples, the availability of public sequence databases such as GenBank (NCBI: www.ncbi.nlm.nih.gov/genbank/), European Nucleotide Archive (EMBL-EBI: www.ebi.ac.uk/ena), or DNA Data Bank of Japan (DDBJ: www.ddbj.nig.ac.jp/) makes it possible to compare the sequences in the unidentified products with the reference sequences submitted to such databases, and to identify the unknown sample as long as sequences in the present species have been previously uploaded to the databases. Several studies based on sequencing have been performed to identify species present in foodstuff and feedstuff (Colgan et al., 2001; Quinteiro et al., 2001; Girish et al., 2004; Rojas et al., 2009; Lago et al., 2011a and 2011b; Espiñeira & Vieites, 2012a; Horreo et al., 2013; Ali et al., 2014; Abuzinadah et al., 2015; Armani et al., 2015a and 2016; Wen et al., 2015). Despite the great utility of this technique and its wide use in the field of food traceability and identification of unknown samples using the BLAST tool or with phylogenetical analyses like "forensically informative nucleotide sequencing (FINS)", the reliability of the sequences included in the public databases has been discussed. Some failures or inaccuracies can be found in the databases, so the reference sequences used in a study must be strongly checked before making conclusions (Vilgalys, 2003; Galimberti et al., 2013). In addition, when little known or low internationalization species are assayed, there may not be any sequences for the species loaded in the databases, which should be checked before drawing conclusions to avoid false negatives (Vilgalys, 2003; Galimberti et al., 2013).

With the aim of standardizing the identification of organisms through DNA-based molecular tools, in 2003, the researcher Paul Hebert (University of Guelph, Canada) proposed to start an international project—"Barcode of Life" (BOL: www.barcodeoflife.

org/)—based on the use of short genetic sequences. Barcodes, from selected fragments of the genome, showed enough differences in their nucleotide sequences to distinguish unambiguously the different species (Hebert et al, 2003a). With the generated information was created a public database that can be used to identify unknown samples by comparing them against the preexisting classification (BOL database (BOLD)). In the field of food traceability, the availability of standardized molecular markers and an extra database for checking the species present in a food and feed products was a clear advantage, since several food products lose the taxonomic features of the original species due to processing, allowing for their identification only through molecular tools (Hellberg & Morrissey, 2011). The barcode most commonly used for the identification of majority of the animal species is a fragment of the mitochondrial cytochrome oxidase subunit I gene (COI), with several studies supporting its discrimination ability (Hebert et al., 2003b; Kochzius et al., 2010; Carvalho et al., 2015; Muñoz-Colmenero et al., 2015 and 2016a; Asis et al., 2016; Chang et al., 2016; Kane & Hellberg, 2016; Quinto et al., 2016). This gene shows sections conserved within the species to anchor the primers, but they exhibit enough divergence among the species to enable the identification at species level, in most cases (Hebert & Gregory, 2005). In case of plants, two chloroplast genes, Maturase K (MatK) and Ribulose-1,5-bisphosphate carboxylase/oxygenase (rbcL), have been determined as barcodes and are widely used (Theodoridis et al., 2012; Federici et al., 2014), although discrimination of plants is a more complex process and less references are available for this. In some cases, a combination of both the barcode genes is needed because rbcL has high universality but less species resolution and matK affords high resolution but less universality (Techen et al., 2014; Bruni et al., 2015). In some other cases, an extra marker is applied, such as internal transcribed spacer 2 (ITS2) or the ribosomal genes (rRNA) (Xin et al., 2013; Selvaraj et al., 2014), which is also used for the identification of the fungi species (Khaund & Joshi, 2015). The 16S rRNA gene has been frequently used in identifying the communities of protists in the last few years (www.earthmicrobiome.org/).

One disadvantage of DNA barcoding identification is that, as in the other databases, if one particular species is not submitted to BOLD then it is not possible to obtain a species-level identification (Milton et al., 2013).

In samples that are highly processed, where the DNA can be fragmented and degraded, shorter sequences have been proposed, such as mini-barcodes, providing a perfect tool adapted to low quality DNA and expanding the sample types where DNA barcoding can be applied (Meusnier et al., 2008; Hajibabaei & McKenna, 2012; Armani et al., 2015b; Fields et al., 2015; Shokralla et al., 2015; Janjua et al., 2017; Mitchell & Hellberg, 2016).

7.9 SINGLE NUCLEOTIDE POLYMORPHISM (SNP)

Single nucleotide polymorphisms (SNPs) are the fundamental units of genetic variation based on the changes in a single nucleotide of a DNA sequence within a locus. Such variations must occur at a frequency of at least 1% in the studied population for it to be considered as a SNP and not as a single nucleotide variant (SNV) or a point mutation (Karki et al., 2015). Some authors also consider the changes of a few nucleotides or the small insertions or deletions in the genome sequences, most commonly called indels, as SNPs (Smith et al., 2005). Several sequencing and bioinformatics tools can be used to develop and identify SNPs, for example, PCR amplicon targeted sequencing, sequencing of AFLPs bands, RNA sequencing, NGS methods (shotgun, amplicon sequencing, etc.), bioinformatics checking within expressed sequence tags (ESTs) databases, etc. (Liu & Cordes, 2004; Smith et al., 2005). This part of discovery and development is the most expensive step in case of this molecular method, and this fact must be considered when its potential utility is evaluated (Schlötterer, 2004). There is a diverse range of approaches to genotype SNPs including TaqMan probe, which needs expensive specific probes, SNP microarray panel, or with a multiple amplification system, such as SnapShot, which is based on the determination of fragment sizes and on fluorescent label identification at multiple SNP sites at the same time by capillary electrophoresis with high resolution melting detection (HRM). Besides, the aforementioned methods like AFLPs and RFLPs can also be used, but they require more time for implementation (Ogden, 2008; Wu et al., 2017).

Their high abundance in the genome together with the fact that they have proved to be a rapid, efficient, and accurate methodology for the identification of species, makes them suitable for automated and routine analyses with high reproducibility (Lindblad-Toh et al., 2000; Garvin & Gharrett, 2007; Kumar et al., 2009); thus, they have become popular as DNA-based markers that are implemented in food traceability and quality control. Many studies have applied SNP methodology to identify the species present in a large variety of food, such as salmonids (Smith et al., 2005), hakes (Machado-Schiaffino et al., 2008), Atlantic albacore and Bay of Biscay anchovy (Montes et al., 2017), eels (Itoi et al., 2005), lobster (Villacorta-Rath et al., 2016), olive oil (Agrimonti et al., 2011), beef products with a special certification of “protected geographic indication” (PGI) (Negrini et al., 2008), breed label fraud in the Chinese meat market (Rogberg-Muñoz et al., 2016), and sheep meat (Wu et al., 2017).

The SNPs are robust when they are applied in modified samples, being able to trace degraded DNA extracted from a complex matrix as they can be detected within a range of low-sized amplicons (Agrimonti et al., 2011). Furthermore, the implementation and interpretation of the results are easy and it is not necessary to have advanced knowledge or technical expertise to apply this technique (Machado-Schiaffino et al., 2008). Nevertheless, despite its wide use in the detection and identification of species, this methodology shows a clear disadvantage since it requires access to the correct DNA sequence of the organisms (e.g., strains/varieties or ecotypes, species), so its application is often limited to just a few taxa in a specific set of target organisms (Ogden, 2008; Galimberti et al., 2013).

7.10 PCR-CLONING-SEQUENCING AND DNA-CHIPS OR DNA MACRO/MICROARRAY

The simultaneous detection of several species is certainly one of the greatest challenges in the field of food traceability. Two methods offer convincing and reliable solutions for the examination of complex mixtures of PCR products and, potentially, for the identification of hundreds of thousands of species simultaneously, and they are cloning-sequencing and DNA-chips (also known as DNA macroarray or DNA microarray) (Teletchea et al., 2005).

The PCR-cloning-sequencing methodology can be used when the food contains a mixture of different species and the objective is to know all the components, and not just a specific one (Teletchea et al., 2005). This technique is based on the separation of each amplified molecule of DNA through the cloning process, including each DNA molecule in a vector (ligation) and the vector in a bacteria (transformation). The bacteria are spread in a growing medium on a plate and each bacteria (colony) will have one specific sequence from the PCR product inside its vector. If the molecular marker is amplified using primers that are anchored in the vector by PCR in each bacterial colony (PCR-colony) and if the PCR product is sequenced, this will be only one element of the mixture. For recovering the cloned sequences, the target fragment is amplified from the selected colonies, and by amplifying and sequencing enough bacterial colonies, it will be possible to determine all the different elements present in the analyzed food. This methodology has been used to analyze some complex mixtures of food, such as *surimi* (Galal-Khallaf et al., 2016a), candies (Muñoz-Colmenero et al., 2016b), honey (Bruni et al., 2015), etc.

There are two main constrains in this method: the choice of primers and the expense of the method. When the goal is to figure out all the species present in a food product, the chosen molecular marker must be able to amplify all the potential species present in it. In this sense, universal primers such as barcode primers are a good option, but their specificity, discrimination, and amplification effectiveness capacities should be checked beforehand (Galimberti et al., 2013). The cost of this methodology is high, so cloning many samples and sequencing a large number of clones could be unaffordable for some traceability projects (Muñoz-Colmenero et al., 2017). Thus, it is more advisable to apply this methodology in food where a large number of species will not be expected (Galal-Khallaf et al., 2016b).

DNA-chips are based on the hybridization of several strands of a specific molecular marker located in known positions on a surface, with the sequences present in a food sample (Peres et al., 2007). The sequences are amplified in order to increase the amount of available DNA and are marked with fluorescence before putting them on the chip. There is almost no limit to the number of capture probes that can be introduced into a chip and the hybridization positions are measured through image analyses of the fluorescence (Leimanis et al., 2006). This methodology is very sensitive, accurate, fast, and highly adaptable, allowing a screening throughput that is easily comparable with that of other laboratory techniques (Leimanis et al., 2006; Peres et al., 2007; Teletchea et al., 2008). DNA-chips have been often used to detect GMOs (Leimanis et al., 2006; Xu et al., 2006; Elenis et al., 2008; Nikolic et al., 2009), cultivars (Xie et al., 2006; Consolandi et al., 2008; Agrimonti et al., 2011), as well as the main vertebrates present in food (Teletchea et al., 2008).

DNA-chips have some drawbacks that makes it necessary to create data banks of molecular markers that are able to discriminate among the species of interest (Peres et al., 2007).

7.11 REAL-TIME PCR (RT-PCR) AND HIGH RESOLUTION MELTING (HRM)

In contrast to other PCR techniques, RT-PCR is a very sensible methodology that is able to identify and quantify minute traces of species in feed and foodstuffs (Pegels et al., 2014), even in molecules that are heavily degraded, for example, in highly processed food (Natonek-Wiśniewska et al., 2013). Besides monitoring the amplification, this useful tool is also able to quantify at the first stages of PCR, giving more accurate results than what would be obtained after PCR (Fajardo et al., 2010). Two chemicals are mostly used: SYBR Green and TaqMan. The first one is a dye that turns fluorescent when it binds to the double strands of DNA, so its fluorescence would be proportional to the amplification; the second one binds directly to the specific target amplicon, between the forward and reverse primers, so it is more specific, but it is also a more expensive technique (Ballin et al., 2009).

The barriers in the laboratory application of this methodology are the high cost of reagents, equipment, and trained professionals (Soares et al., 2010; Safdar & Junejo, 2015). However, despite the costs, many laboratories have opted for this technique. It is used in the analysis of different food products and for different purposes. RT-PCR is very useful in GMO testing due to its ability to monitor the amplification and quantification in only one step (Weighardt, 2003; Miraglia et al., 2004; Elenis et al., 2008). Other applications have been focused on the detection of different species that are used as ingredients in feedstuff, for instance, avian species in farm animal and pet feeds (Pegels et al., 2014); poultry and pork in pet food (Safdar & Abasiyanik, 2013a); bovine and caprine species in ruminant feeds (Safdar et al., 2014a); beef and soybean species in processed sausages (Safdar & Abasiyanik, 2013b); soybean, poultry, horse, and pork species in sausages (Safdar et al., 2014b); *Gadus morhua* (Herrero et al., 2010), *Scomber scombrus* (Velasco et al., 2013), *Mullus* sp. (Espiñeira & Vieites, 2015), *Molva molva* (Taboada et al., 2017), Anglerfish species (Herrero et al., 2011), and cephalopods (Espiñeira & Vieites, 2012b) in seafood products; and in foodstuff as bovine DNA in meats, milk, and cheeses (Zhang et al., 2007); etc.

HRM is a more recent method that was developed after a collaboration of researchers from the University of Utah (UT, USA) and the industry (Idaho Technology, UT, USA) in 2002 (Reed et al., 2007). This method is based on the post-PCR melting curve analysis, for instance, SYBR Green, but HRM allows the discrimination among amplicons with minimal variations in their sequences, even single-base variations or small indels (Druml & Cichna-Markl, 2014). The temperature is controlled very accurately with increments of the temperature in the order of decimals, and the melting is monitored with fluorescence that is sent out by saturation dyes—that do not inhibit the DNA polymerase at high concentrations—and is received by high resolution instruments (Erali et al., 2008). In addition, this technique is not a destructive one, so it is possible to perform gel electrophoresis or sequencing later. One drawback of RT-PCR is the high cost of the instrument (Vossen et al., 2009).

The application of HRM in food traceability and authenticity purposes is more recent, and it plays an important role in the identification of closely related species, cultivars, or denomination of origin designations (protected denomination origin: PDO) usually by using microsatellites or DNA barcodes as molecular markers (Mackay et al., 2008; Ganopoulos et al., 2011; Bosmali et al., 2012; Madesis et al., 2012; Sakaridis et al., 2013; Pereira et al., 2017). Other fields where this method is applied with good results are the screening of GMOs (Kluga et al., 2013), allergen detection (Costa et al., 2012), and identification of pathogenic species related to food even at a strain level (Lilliebridge et al., 2011).

7.12 NEXT GENERATION SEQUENCING TECHNIQUES (NGS) OR HIGH-THROUGHPUT SEQUENCING

In recent years, NGS techniques have captivated the field of molecular biology. These new techniques provide a high capacity for analysis and make it possible to generate a huge amount of data that can reach the order of millions of DNA sequences. This allows for the identification of all the species present in one food product (large number of sequences and analyses of entire genomes) and the analysis of several products at the same time using different barcode sequences to unambiguously label the different samples (barcoding amplicon sequencing) (Madesis et al., 2014). These features are greatly useful for food typing, especially in mixed or highly degraded samples, as well as in genetically similar ones (Weber-Lehmann et al., 2014).

These techniques have been mainly applied in microbiology because of their ability to identify the species that are present in a sample without cultivating them, thus allowing the identification of a larger number of species (Wang et al., 2012). In the field of food traceability, the use of these techniques is being proposed for some years (Diaz-Sanchez et al., 2013). Despite their greater precision and volume of analysis, NGS technologies have been applied only in limited cases due to the high cost equipment and analyses, but those costs are decreasing (Madesis et al. 2014). Different sequencing systems are currently available such as: the Roche 454 sequencer whose methodology detects the emission of light during the release of a pyrophosphate molecule during nucleotide incorporation (Zaiko et al., 2015) and it was one of the first to be used; the Illumina MiSeq sequencer (the smallest from the brand, Life Technologies) based on the fluorescence generated during nucleotide incorporation, which is captured by a camera, and it is one of the most famous and the primarily chosen method in small laboratories (Quail et al. 2012); and the ion torrent personal genome machine (PGM), which detects the pH changes produced by the release of protons during the incorporation of nucleotides (Diaz-Sanchez et al., 2013), and it is one of the best methods for the screening of amplicons (Diaz-Sanchez et al., 2013). For all of them, the costs of both machine and analyses are still high, but they are becoming more affordable each year (Grada & Weinbrecht, 2013).

One of the first studies that used these methodologies was performed by Ortola-Vidal et al. (2007) for the identification of fruit species in yogurts. However, their use has been related more to food microbiology (Ercolini, 2013). Besides, the microbial profile detected by these methodologies had been already tested with the aim of relating a sample to a specific biogeographical region or "terroir" for different wine varieties (Gilbert et al., 2014). During the last decade, NGS has also been applied to identify the species present in food, especially for their ability to detect a large number of species with a single assay (Tillmar et al., 2013). Some examples are: the detection of animal species in candies (Muñoz-Colmenero et al., 2017), fish traceability (Galal-khallaf et al., 2016a), seafood traceability (Giusti et al., 2017), discrimination of plant species present in honey (Valentini et al., 2010), and plant and animal DNA present in traditional Chinese medicines (Coghlan et al., 2012).

Despite the fact that NGS analyses are more expensive than the other DNA-based methodologies, these are finding a place in the field of food traceability with an increase in the number of studies that use them. However, NGS technologies need to be correctly evaluated, validated, and standardized before being considered for routine food analyses in laboratories (Muñoz-Colmenero et al., 2017). A parallel improvement in bioinformatic tools will be necessary for these techniques to become successful. Currently, DNA analyses are performed independently, but may be in the future analyses related to the biological contaminants, authenticity of raw materials, undeclared mixtures, misbranding, GMO contamination, and others can be performed in a single run (Corrado, 2016).

REFERENCES

Abdel-Rahman SM, Ahmed MMM. 2007. Rapid and sensitive identification of buffalo's, cattle's and sheep's milk using species-specific PCR and PCR–RFLP techniques. Food Control 18: 1246–1249.

Abuzinadah OHA, Yacoub HA, El Ashmaoui HM, Ramadan HAI. 2015. Molecular detection of adulteration in chicken products based on mitochondrial 12S rRNA gene. Mitochondrial DNA 26: 337–340.

Agrimonti C, Vietina M, Pafundo S, Marmiroli N. 2011. The use of food genomics to ensure the traceability of olive oil. Trends in Food Science & Technology 22: 237–244.

Ali ME, Rahman MM, Abd Hamid S, Mustafa S, Bhassu S, Hashim U. 2014. Canine-specific PCR assay targeting cytochrome b gene for the detection of dog meat adulteration in commercial frankfurters. Food Analytical Methods 7: 234–241.

Alves E, Fernandez AI, Fernandez-Rodrıguez A, Perez-Montarelo D, Benitez R, Ovilo C, Rodrıguez C Silio L. 2009. Identification of mitochondrial markers for genetic traceability of European wild boars and Iberian and Duroc pigs. Animal 3: 1216–1223.

Amaral JS, Santos CG, Melo VS, Oliveira MBPP, Mafra I. 2014. Authentication of a traditional game meat sausage (*Alheira*) by species-specific PCR assays to detect hare, rabbit, red deer, pork and cow meats. Food Research International 60: 140–145.

Arcuri EF, El Sheikha AF, Rychlik T, Piro-Métayer I, Montet D. 2013. Determination of cheese origin by using 16S rDNA fingerprinting of bacteria communities by PCR-DGGE: Preliminary application to traditional Minas cheese. Food Control 30: 1–6.

Ardura A, Horreo JL, Hernandez E, Jardon A, Pola IG, Martinez JL, Garcia-Vazquez E. 2012. Forensic DNA analysis reveals use of high trophic level marine fish in commercial aquaculture fish meals. Fisheries Research 115–116: 115–120.

Armani A, Castigliego L, Tinacci L, Gandini G, Gianfaldoni D, Guidi A. 2012. A rapid PCR-RFLP method for the identification of *Lophius* species. European Food Research and Technology 235: 253–263.

Armani A, Tinacci L, Xiong X, Castigliego L, Gianfaldoni D, Guidi A. 2015a. Fish species identification in canned pet food by BLAST and Forensically Informative Nucleotide Sequencing (FINS) analysis of short fragments of the mitochondrial 16s ribosomal RNA gene (16S rRNA). Food Control 50: 821–830.

Armani A, Guardone L, Castigliego L, D'Amico P, Messina A, Malandra R, Gianfaldoni D, Guidi A. 2015b. DNA and Mini-DNA barcoding for the identification of Porgies species (family *Sparidae*) of commercial interest on the international market. Food Control 50: 589–596.

Armani A, Giusti A, Guardone L, Castigliego L, Gianfaldoni D, Guid A. 2016. Universal primers used for species identification of foodstuff of animal origin: Effects of oligonucleotide tails on PCR amplification and sequencing performance. Food Analytical Methods 9: 1199–1209.

Arun OO, Yılmaz F, Muratoglu K. 2013. PCR detection of genetically modified maize and soy in mildly and highly processed foods. Food Control 32: 525–531.

Asis AM, Lacsamana JK, Santos MD. 2016. Illegal trade of regulated and protected aquatic species in the Philippines detected by DNA barcoding. Mitochondrial DNA 27: 659–666.

Aung MM, Chang YS. 2014. Traceability in a food supply chain: Safety and quality perspectives. Food Control 39: 172–184.

Ballin N, Vogensen F, Karlsson A. 2009. Species determination we can we detect and quantify meat adulteration? Meat Science 83: 165–174.

Bigot C, Meile J-C, Kapitan A, Montet D. 2015. Discriminating organic and conventional foods by analysis of their microbial ecology: An application on fruits. Food Control 48: 123–129.

Birki Jr. CW. 1995. Uniparental inheritance of mitochondrial and chloroplast genes: Mechanisms and evolution. Proceedings of the National Academy of Sciences 92: 11331–11338.

Bornatowski H, Braga RR, Simões Vitule JR. 2013. Shark mislabeling threatens biodiversity. Science (letters) 340: 923.

Bottero MT, Dalmasso A. 2011. Animal species identification in food products: Evolution of biomolecular methods. The Veterinary Journal 190: 34–38.

Bosmali I, Ganopoulos I, Madesis P, Tsaftaris A. 2012. Microsatellite and DNA-barcode regions typing combined with High Resolution Melting (HRM) analysis for food forensic uses: A case study on lentils (Lens *culinaris*). Food Research International 46: 141–147.

Brody JR, Calhoun ES, Gallmeier E, Creavalle TD, Kern SE. 2004. Ultra-fast high-resolution agarose electrophoresis of DNA and RNA using low-molarity conductive media. Biotechniques 37: 598, 600, 602.

Bruni I, Galimberti A, Caridi L, Scaccabarozzi D, De Mattia F, Casiraghi M, Labra M. 2015. A DNA barcoding approach to identify plant species in multiflower honey. Food Chemistry 170: 308–315.

Busconi M, Foroni C, Corradi M, Bongiorni C, Cattapan F, Fogher C. 2003. DNA extraction from olive oil and its use in the identification of the production cultivar. Food Chemistry 83: 127–134.

Butler JM, Federica Buef E, Crivellente F, McCord BR. 2004. Forensic DNA typing by capillary electrophoresis using the ABI Prism 310 and 3100 genetic analyzers for STR analysis. Electrophoresis 25: 1397–1412.

Campbell D, Duchesne P, Bernatchez L. 2003. AFLP utility for population assignment studies: Analytical investigation and empirical comparison with microsatellites. Molecular Ecology 12: 1979–1991.

Carvalho DC, Palhares RM, Drummond MG, Frigo TB. 2015. DNA barcoding identification of commercialized seafood in South Brazil: A governmental regulation forensic program. Food Control 50: 784–788.

Céspedes A, García T, Carrera E, González I, Sanz B, Hernández PE, Martín R. 1998a. Identification of flatfish species using polymerase chain reaction (PCR) amplification and restriction analysis of the cytochrome b gene. Journal of Food Science 63: 206–209.

Céspedes A, García T, Carrera E, González I, Sanz B, Hernández PE, Martín R. 1998b. Polymerase Chain Reaction–Restriction Fragment Length Polymorphism Analysis of a Short Fragment of the Cytochrome b Gene for Identification of Flatfish Species. Journal of Food Protection 61(12): 1684–1685.

Chakraborty R, Jin L. 1993. Determination of relatedness between individuals using DNA fingerprinting. Human Biology 65: 875–895.

Chang C-H, Lin H-Y, Ren Q, Lin Y-S, Shao K-T. 2016. DNA barcode identification of fish products in Taiwan: Government-commissioned authentication cases. Food Control 66: 38–43.

Chen S-Y, Liu Y-P, Yao Y-G. 2010. Species authentication of commercial beef jerky based on PCR-RFLP analysis of the mitochondrial 12S rRNA gene. Journal of Genetics and Genomics 37: 763–769.

Chistiakov DA, Hellemans B, Volckaert FAM. 2006. Microsatellites and their genomic distribution, evolution, function and applications: A review with special reference to fish genetics. Aquaculture 255: 1–29.

Chiu T-H, Kuo C-W, Lin H-C, Huang D-S, Wu P-L. 2015. Genetic diversity of ivory shell (*Babylonia areolata*) in Taiwan and identification of species using DNA-based assays. Food Control 48: 108–116.

Cipriani G, Marrazzo MT, Marconi R, Cimato A, Testolin R. 2002. Microsatellite markers isolated in olive (Olea europaea L.) are suitable for individual fingerprinting and reveal polymorphism within ancient cultivars. Theoretical and Applied Genetics 104: 223–228.

Coghlan ML, Haile J, Houston J, Murray DC, White NE, Moolhuijzen P, Bellgard MI, Bunce M. 2012. Deep sequencing of plant and animal DNA contained within traditional Chinese medicines reveals legality issues and health safety concerns. PLoS Genetics 8: e1002657.

Colgan S, O'Brien L, Maher M, Shilton N, McDonnell K, Ward S. 2001. Development of a DNA-based assay for species identification in meat and bone meal. Food Research International 34: 409–414.

Consolandi C, Palmieri L, Severgnini M, Maestri E, Marmiroli N, Agrimonti C, Baldoni L, Donini P, De Bellis G, Castiglioni B. 2008. A procedure for olive oil traceability and authenticity: DNA extraction, multiplex PCR and LDR–universal array analysis. European Food Research and Technology 227: 1429–1438.

Corrado G. 2016. Advances in DNA typing in the agro-food supply chain. Trends in Food Science & Technology 52: 80–89.

Costa J, Mafra I, Oliveira MBPP. 2012. High resolution melting analysis as a new approach to detect almond DNA encoding for Pru du 5 allergen in foods. Food Chemistry 133: 1062–1069.

Costa Leal M, Pimentel T, Ricardo F, Rosa R, Calado R. 2015. Seafood traceability: Current needs, available tools, and biotechnological challenges for origin certification. Trends in Biotechnology 33: 331–336.

Dalvit C, De Marchi M, Cassandro M. 2007. Genetic traceability of livestock products: A review. Meat Science 77: 437–449.

Demirhan Y, Ulca P, Senyuva HZ. 2012. Detection of porcine DNA in gelatine and gelatine-containing processed food products—Halal/Kosher authentication. Meat Science 90: 686–689.

Diaz-Sanchez S, Hanning I, Pendleton S, D'Souza D. 2013. Next-generation sequencing: The future of molecular genetics in poultry production and food safety. Poultry Science 92: 562–572.

Doosti A, Dehkordi PG, Rahimi E. 2014. Molecular assay to fraud identification of meat products. Journal of Food Science and Technology 51: 148–152.

Druml B, Cichna-Markl M. 2014. High resolution melting (HRM) analysis of DNA – Its role and potential in food analysis. Food Chemistry 158: 245–254.

Elenis DS, Kalogianni DP, Glynou K, Ioannou PC, Christopoulos TK. 2008. Advances in molecular techniques for the detection and quantification of genetically modified organisms. Analytical and Bioanalytical Chemistry 392: 347–354.

Erali M, Voelkerding KV, Wittwer CT. 2008. High resolution melting applications for clinical laboratory medicine. Experimental and Molecular Pathology 85: 50–58.

Ercolini D. 2004. PCR-DGGE fingerprinting: novel strategies for detection of microbes in food. Journal of Microbiological Methods 56: 297–314.

Ercolini D. 2013. High-throughput sequencing and metagenomics: moving forward in the culture-independent analysis of food microbial ecology. Applied and Environmental Microbiology 79: 3148–3155.

Espiñeira M, Vieites JM. 2012a. Authentication of the most important species of rockfish by means of fins. European Food Research and Technology 235: 929–937.

Espiñeira M, Vieites JM. 2012b. Rapid method for controlling the correct labeling of products containing common octopus (*Octopus vulgaris*) and main substitute species (*Eledone cirrhosa* and *Dosidicus gigas*) by fast real-time PCR. Food Chemistry 135: 2439–2444.

Espiñeira M, Vieites JM. 2015. Genetic system for traceability of goatfishes by FINS methodology and authentication of mullets (*Mullus barbatus* and *Mullus surmuletus*) by RT-PCR. European Food Research and Technology 240: 423–429.

Espiñeira M Santaclara FJ. 2016. The use of molecular biology techniques in food traceability. In Advances in food traceability techniques and technologies: Improving quality throughout the food chain, pp. 91–118, Espiñeira M, Santaclara FJ (Eds.), WP Woodhead Publishing series in Food Science, Technology and Nutrition, ELSEIVER.

Fajardo V, González I, Martín I, Rojas M, García T, Martín R. 2010. A review of current PCR-based methodologies for the authentication of meats from game animal species. Trends in Food Science and Technology 21: 408–421.

Faria MA, Magalhaes R, Ferreira MA, Meredith CP, Ferreira Monteiro F. 2000. *Vitis vinifera* must varietal authentication using microsatellite DNA analysis (SSR). Journal of Agricultural and Food Chemistry 48: 1096–1100.

Federici S, Fontana D, Galimberti A, Bruni I, de Mattia F, Cortis P, Galasso G et al. 2014. A rapid diagnostic approach to identify poisonous plants using DNA barcoding data. Plant Biosystems 149: 537–545.

Fields AT, Abercrombie DL, Eng R, Feldheim K, Chapman DD. 2015. A novel mini-DNA barcoding assay to identify processed fins from internationally protected shark species. PLoS ONE 10: e0114844.

Galal-Khallaf A, Osman AGM, Carleos CE, Garcia-Vazquez E, Borrell YJ. 2016a. A case study for assessing fish traceability in Egyptian aquafeed formulations using pyrosequencing and metabarcoding. Fisheries Research 174: 143–150.

Galal-Khallaf A, Ardura A, Borrell YJ, Garcia-Vazquez E. 2016b. Towards more sustainable surimi? PCR-cloning approach for DNA barcoding reveals the use of species of low trophic level and aquaculture in Asian *surimi*. Food Control 61: 62–69.

Galimberti A, De Mattia F, Losa A, Bruni I, Federici S, Casiraghi M, Martellos S, et al. 2013. DNA barcoding as a new tool for food traceability. Food Research International 50: 55–63.

Garvin MR, Gharrett AJ. 2007. DEco-TILLING: an inexpensive method for single nucleotide polymorphism discovery that reduces ascertainment bias. Molecular Ecology Notes 7: 735–746.

Ganopoulos I, Argiriou A, Tsaftaris A. 2011. Adulterations in Basmati rice detected quantitatively by combined use of microsatellite and fragrance typing with high resolution melting (HRM) analysis. Food Chemistry 129: 652–659.

Gilbert JA, van der Lelie D, Zarraonaindia I. 2014. Microbial terroir for wine grapes. PNAS 111: 5–6.

Girish PS, Anjaneyulu ASR, Viswas KN, Anand M, Rajkumar N, Shivakumar BM, Bhaskar S. 2004. Sequence analysis of mitochondrial 12S rRNA gene can identify meat species. Meat Science 66: 551–556.

Girish PS, Anjaneyulu ASR, Viswas KN, Shivakumar BM, Anand M, Patel M, Sharma B. 2005. Meat species identification by polymerase chain reaction-restriction fragment length polymorphism (PCR-RFLP) of mitochondrial 12S rRNA gene. Meat Science 70: 107–112.

Giusti A, Tinacci L, Sotelo CG, Marchetti M, Guidi A, Zheng W, Armani A. 2017. Seafood identification in multispecies products: assessment of 16srRNA, cytb and COI universal primer efficiency as preliminary analytical step for setting up metabarcoding Next Generation Sequencing (NGS) techniques. Journal of Agricultural and Food Chemistry 65: 2902–2912.

Gomez-Uchida D, Weetman D, Hauser L, Galleguillos R, Retamal M, 2003. Allozyme and AFLP analyses of genetic population structure in the hairy edible crab *Cancer setosus* from the Chilean coast. Journal of Crustacean Biology 23: 486–494.

Grada A, Weinbrecht K. 2013. Next-generation sequencing: Methodology and application. Journal of Investigative Dermatology 133: e11.

Ha WY, Shaw PC, Liu J, Yau FCF, Wang J. 2002. Authentication of *Panax ginseng* and *Panax quinquefolius* using amplified fragment length polymorphism (AFLP) and directed amplification of minisatellite region DNA (DAMD). Journal of Agricultural and Food Chemistry 50: 1871–1875.

Hajibabaei M, McKenna C. 2012. DNA mini-barcodes. In DNA barcodes: Methods and protocols (part of the Methods in Molecular Biology book series (MIMB, volume 858)), pp. 339–353, Kress WJ, Erickson DL (Eds.), Springer, USA.

Hansen MM, Kenchington E, Nielsen EE. 2001. Assigning individual fish to populations using microsatellite DNA markers. Fish and Fisheries 2: 93–112.

Hebert PDN, Cywinska A, Ball SL, deWaard JR. 2003a. Biological identifications through DNA barcodes. Proceedings of the Royal Society 270: 313–321.

Hebert PDN, Ratnasingham S, deWaard JR. 2003b. Barcoding animal life: cytochrome c oxidase subunit 1 divergences among closely related species. Proceedings of the Royal Society B-Biological Sciences 270: S96–S99.

Hebert PDN, Gregory TR. 2005. The promise of DNA barcoding for taxonomy. Systematic Biology 54:852–859.

Hellberg RSR, Morrissey MT. 2011. Advances in DNA-based techniques for the detection of seafood species substitution on the commercial market. Journal of Laboratory Automation: JALA 16: 308–321.

Herrero B, Madrinan M, Vieites JM, Espineira M. 2010. Authentication of Atlantic cod (*Gadus morhua*) using real time PCR. Journal of Agricultural and Food Chemistry 58: 4794–4799.

Herrero B, Vieites JM, Espineira M. 2011. Duplex real-time PCR for authentication of anglerfish species. European Food Research and Technology 233: 817–823.

Hoekstra RF. 2000. Evolutionary origin and consequences of uniparental mitochondrial inheritance. Human Reproduction 15(Suppl. 2): 102–111.

Hold GL, Russell VJ, Pryde SE, Rehbein H, Quinteiro J, Vidal R, Rey-Mendez M, et al. 2001. Development of a DNA-based method aimed at identifying the fish species present in food products. Journal of Agricultural and Food Chemistry 49: 1175–1179.

Horreo JL, Ardura A, Pola IG, Martinez JL, Garcia-Vazquez E. 2013. Universal primers for species authentication of animal foodstuff in a single polymerase chain reaction. Journal of the Science of Food and Agriculture 93: 354–361.

Hovda MB, Sivertsvika M, Lunestad BT, Lorentzen G, Rosnes JT. 2007. Characterization of the dominant bacterial population in modified atmosphere packaged farmed halibut (*Hippoglossus hippoglossus*) based on 16S rDNA-DGGE. Food Microbiology 24: 362–371.

Iquebal MA, Sarika, AV, Verma N, Rai A, Kumar D. 2013. First whole genome based microsatellite DNA marker database of tomato for mapping and variety identification. BMC Plant Biology 13: 197.

Itoi S, Nakaya M, Kaneko G, Kondo H, Sezaki K, Watabe S. 2005. Rapid identification of eels *Anguilla japonica* and *Anguilla anguilla* by polymerase chain reaction with single nucleotide polymorphism-based specific probes. Fisheries Science 71: 1356–1364.

Jacquet JL, Pauly D. 2008. Trade secrets: Renaming and mislabeling of seafood. Marine Policy 32: 309–318.

Janjua S, Abbas FI, William K, Ullah Malik I, Mehr J. 2017. DNA Mini-barcoding for wildlife trade control: A case study on identification of highly processed animal materials. Mitochondrial DNA Part A: DNA Mapping, Sequencing, and Analysis 28: 544–546.

Kalaitzis P, El-Zein Z. 2016. Olive oil authentication, traceability and adulteration detection using DNA-based approaches. Lipid Technology 28: 173–176.

Kane DE, Hellberg RS. 2016. Identification of species in ground meat products sold on the U.S. commercial market using DNA-based methods. Food Control 59: 158–163.

Karki R, Pandya D, Elston RC, Ferlini C. 2015. Defining "mutation" and "polymorphism" in the era of personal genomics. BMC Medical Genomics 8: 37.

Khaund P, Joshi SR. 2015. DNA barcoding of wild edible mushrooms consumed by the ethnic tribes of India. Gene 550: 123–130.

Kluga L, Folloni S, Kagkli DM, Bogni A, Foti N, Savini C, et al. 2013. Combinatory SYBRÒ Green real-time PCR screening approach for tracing materials derived from genetically modified rice. Food Analytical Methods 6: 361–369.

Kochzius M, Seidel C, Antoniou A, Botla SK, Campo D, Cariani A, Garcia-Vazquez E, et al. 2010. Identifying fishes through DNA barcodes and microarrays. PLoS ONE 5: e12620.

Kumar P, Gupta VK, Misra AK, Modi DR, Pandey BK. 2009. Potential of molecular markers in plant biotechnology. Plant Omics 2: 141–162.

Kumar D, Singh SP, Karabasanavar NS, Singh R, Umapathi V. 2014. Authentication of beef, carabeef, chevon, mutton and pork by a PCR-RFLP assay of mitochondrial cytb gene. Journal of Food Science and Technology 51: 3458–3463.

Koumi P, Green HE, Hartley S, Jordan D, Lahec S, Livett RJ, Tsang KW, Ward DM. 2004. Evaluation and validation of the ABI 3700, ABI 3100, and the MegaBACE 1000 capillary array electrophoresis instruments for use with short tandem repeat microsatellite typing in a forensic environment. Electrophoresis 25: 2227–2241.

Lago FC, Herrero B, Madriñán M, Vieites JM, Espineira M. 2011a. Authentication of species in meat products by genetic techniques. European Food Research and Technology 232: 509–515.

Lago FC, Herrero B, Vieites JM, Espiñeira M. 2011b. Genetic identification of horse mackerel and related species in seafood products by means of forensically informative nucleotide sequencing methodology. Journal of Agricultural and Food Chemistry 59: 2223–2228.

Lago FC, Alonso M, Vieites JM Espiñeira M. 2014. 16-Fish and seafood authenticity – Species identification. In Seafood processing: Technology, quality, and safety, pp. 419–452, Boziaris IS (Ed.), John Wiley & Sons, Ltd, UK.

Lanzilao I, Burgalassi F, Fancelli S, Settimelli M, Fani R. 2005. Polymerase chain reaction-restriction fragment length polymorphism analysis of mitochondrial cytb gene from species of dairy interest. Journal of Association of Official Analytical Chemists International 88: 128–135.

Larrain MA, Díaz NF, Lamas C, Uribe C, Araneda C. 2014. Traceability of mussel (Mytilus chilensis) in southern Chile using microsatellite molecular markers and assignment algorithms. Exploratory survey. Food Research International 62: 104–110.

Leimanis S, Hernandez M, Fernandez S, Boyer F, Burns M, Bruderer S, Glouden T, Harris N, Kaeppeli O, Philipp P, Pla M, Puigdomenech P, Vaitilingom M, Bertheau Y, Remacle J. 2006. A microarray-based detection system for genetically modified (GM) food ingredients. Plant Molecular Biology 61: 123–139.

Le Nguyen DD, Ngoc HH, Dijoux D, Loiseau G, Montet D. 2008a. Determination of fish origin by using 16S rDNA fingerprinting of bacterial communities by PCR-DGGE: An application on Pangasius fish from Viet Nam. Food Control 19: 454–460.

Le Nguyen DD, Gemrot E, Loiseau G, Montet D. 2008b. Determination of citrus fruit origin by using 16S rDNA fingerprinting of bacterial communities by PCR-DGGE: An application to clementine from Morocco and Spain. Fruits 63: 75–84.

Lilliebridge RA, Tong SYC, Giffard PM, Holt DC. 2011. The utility of high-resolution melting analysis of SNP nucleated PCR amplicons – An MLST based Staphylococcus aureus typing scheme. PLoS ONE 6: e19749.

Lindblad-Toh K, Winchester E, Daly MJ, Wang DG, Hirschhorn JN, Laviolette JP, Ardlie K, et al. 2000. Large-scale discovery and genotyping of single-nucleotide polymorphisms in the mouse. Nature Genetics 24: 381–386.

Liu Z, Cordes J. 2004. DNA marker technologies and their applications in aquaculture genetics. Aquaculture 238: 1–37.

Machado-Schiaffino G, Martinez JL, Garcia-Vazquez E. 2008. Detection of mislabeling in hake seafood employing mtSNPs-based methodology with identification of eleven hake species of the genus *Merluccius*. Journal of Agricultural and Food Chemistry 56: 5091–5095.

Mackay JF, Wright CD, Bonfiglioli RG. 2008. A new approach to varietal identification in plants by microsatellite high resolution melting analysis: Application to the verification of grapevine and olive cultivars. Plant Methods 4: 8.

Madesis P, Ganopoulos I, Anagnostis A, Tsaftaris A. 2012. The application of Bar-HRM (barcode DNA-high resolution melting) analysis for authenticity testing and quantitative detection of bean crops (Leguminosae) without prior DNA purification. Food Control 25: 576–582.

Madesis P, Ganopoulos I, Sakaridis I, Argiriou A, Tsaftaris A. 2014. Advances of DNA-based methods for tracing the botanical origin of food products. Food Research International 60: 163–172.

Mafra I, Ferreira IMPLVO, Oliveira MBPP. 2008. Food authentication by PCR-based methods. European Food Research and Technology 227: 649–665.

Maldini M, Nonnis Marzano F, González FortesG, Papa R, Gandolfi G. 2006. Fish and seafood traceability based on AFLP markers: Elaboration of a species database. Aquaculture 261: 487–494.

Marín A, Fujimoto T, Arai K. 2013. Rapid species identification of fresh and processed scallops by multiplex PCR. Food Control 32: 472–476.

Martın I, Garcıa T, Fajardo V, Lopez-Calleja I, Hernandez PE, Gonzalez I, Martın R. Species-specific PCR for the identification of ruminant species in feedstuffs. 2007. Meat Science 75: 120–127.

McGregor CE, Lambert CA, Greyling MM, Louw JH, Warnich L. 2000. A comparative assessment of DNA fingerprinting techniques (RAPD, ISSR, AFLP and SSR) in tetraploid potato (Solanum tuberosum L.) germplasm. Euphytica 113: 135–144.

Meudt HM, Clarke AC. 2007. Almost forgotten or latest practice? AFLP applications, analyses and advances. TRENDS in Plant Science 12: 106–117.

Meusnier I, Singer GAC, Landry JF, Hickey DA, Hebert PDN, Hajibabaei M. 2008. A universal DNA mini-barcode for biodiversity analysis. BMC Genomics 9: 214.

Meyer R, Hofelein C, Luthy J, Candrian U. 1995. Polymerase chain reaction-restriction fragment length polymorphyism analysis: A simple method for species identification in food. Journal of Association of Official Analytical Chemists International 78: 1542–1551.

Miraglia M, Berdal KG, Brera C, Corbisier P, Holst-Jensen A, Kok EJ, Marvin HJP, Schimmel H, Rentsch J, van Rie JPPF, Zagon J. 2004. Detection and traceability of genetically modified organisms in the food production chain. Food and Chemical Toxicology 42: 1157–1180.

Mitchell JK, Hellberg RS. 2016. Use of the mitochondrial control region as a potential DNA mini-barcoding target for the identification of canned tuna species. Food Analytical Methods 9: 2711–2720.

Milton M, Pierossi P, Ratnasingham S. 2013. Barcode of life data systems handbook, Biodiversity Institute of Ontatio, Guelph, Ontario, Canada.

Montes I, Laconcha U, Iriondo M, Manzano C, Arrizabalaga H, Estonba A. 2017. Reduced SNPs panels for assigning Atlantic albacore and Bay of Biscay anchovy individuals to their geographic origin: Toward sustainable fisheries management. Journal of Agricultural and Food Chemistry 65: 4351–4358.

Mousavi SM, Khaniki GJ, Eskandari S, Rabiei M, Samiee SM, Mehdizadeh M. 2015. Applicability of species-specific polymerase chain reaction for fraud identification in raw ground meat commercially sold in Iran. Journal of Food Composition and Analysis 40: 47–51.

Mueller UG, Wolfenbarger LL. 1999. AFLP genotyping and fingerprinting. Trends in Ecology & Evolution 14: 389–394.

Muñoz-Colmenero M, Klett-Mingo M, Díaz E, Blanco O, Martinez JL, Garcia-Vazquez E. 2015. Evolution of hake mislabeling niches in commercial markets. Food Control 54: 267–274.

Muñoz-Colmenero M, Blanco O, Arias V, Martinez JL, Garcia-Vazquez E. 2016a. DNA authentication of fish products reveals mislabeling associated with seafood processing. Fisheries 41: 128–138.

Muñoz-Colmenero M, Martínez JL, Roca A, García-Vázquez E. 2016b. Authentication of commercial candy ingredients using DNA PCR-cloning methodology. Journal of the Science of Food and Agriculture 96: 859–867.

Muñoz-Colmenero M, Martinez JL, Roca A, Garcia-Vazquez E. 2017. NGS tools for traceability in candies as high processed food products: Ion Torrent PGM versus conventional PCR-cloning. Food Chemistry 214: 631–636.

Murugaiah C, Mohd Noor Z, Mastakim M, Bilung LM, Selamat J, Radu S. 2009. Meat species identification and Halal authentication analysis using mitochondrial DNA. Meat Science 83: 57–61.

Natonek-Wiśniewska M, Krzyścin P, Piestrzyńska-Kajtoch A. 2013. The species identification of bovine, porcine, ovine and chicken components in animal meals, feeds and their ingredients, based on COX I analysis and ribosomal DNA sequences. Food Control 34: 69–78.

Negrini R, Nicoloso L, Crepaldi P, Milanesi E, Marino R, Perini D, Pariset L, et al. 2008. Traceability of four European protected geographic indication (PGI) beef products using single nucleotide polymorphisms (SNP) and Bayesian statistics. Meat Science 80: 1212–1217.

Nikolic Z, Taški-Ajdukovic K, Jevtic A, Marinkovic D. 2009. Detection of GM soybean in food products by simultaneous employment of three pairs of PCR primers. Food Research International 42: 349–352.

Ogden R. 2008. Fisheries forensics: The use of DNA tools for improving compliance, traceability and enforcement in the fishing industry. Fish and Fisheries 9: 462–472.

Ong SB, Zuraini MI, Jurin WG, Cheah YK, Tunung R,Chai LC, Haryani Y, Ghazali FM, Son R. 2007. Meat molecular detection: Sensitivity of polymerase chain reaction-restriction fragment length polymorphism in species differentiation of meat from animal origin. ASEAN Food Journal 14: 51–59.

Orrù L, Napolitano F, Catillo G, Moioli B. 2006. Meat molecular traceability: How to choose the best set of microsatellites? Meat Science 72: 312–317.

Ortola-Vidal A, Schnerr H, Rojmyr M, Lysholm F, Knight A. 2007. Quantitative identification of plant genera in food products using PCR and Pyrosequencing ((R)) technology. Food Control 18: 921–927.

Pafundo S, Agrimonti C, Marmiroli N. 2005. Traceability of plant contribution in olive oil by amplified fragment length polymorphisms. Journal of Agricultural and Food Chemistry 53: 6995–7002.

Parida SK, Kalia SK, Kaul S, Dalal V, Hemaprabha G, Selvi A, Pandit A, Singh A, Gaikwad K Sharma TR, Srivastava, PS, Singh NK, Mohapatra T. 2009. Informative genomic microsatellite markers for efficient genotyping applications in sugarcane. Theoretical and Applied Genetics 118: 327–338.

Parker GD, Langridge P. 2000. Development of a STS marker linked to a major locus controlling flour colour in wheat (Triticum aestivum L.). Molecular Breeding 6: 169–174.

Partis L, Croan D, Gua Z, Clark R, Coldham T, Murby J. 2000. Evaluation of a DNA fingerprinting method for determining the species origin of meats. Meat Science 54: 369–376.

Pasqualone A, Montemurro C, Summo C, Sabetta W, Caponio F, Blanco A. 2007. Effectiveness of microsatellite DNA markers in checking the identity of protected designation of origin extra virgin olive oil. Journal of Agricultural and Food Chemistry 55: 3857–3862.

Pegels N, González I, García T, Martín R. 2014. Avian-specific real-time PCR assay for authenticity control in farm animal feeds and pet foods. Food Chemistry 142: 39–47.

Pereira L, Gomes S, Castro C, Eiras-Dias JE, Brazão J, Graça A, Fernandes JR, Martins-Lopes P. 2017. High resolution melting (HRM) applied to wine authenticity. Food Chemistry 216: 80–86.

Peres B, Barlet N, Loiseau G, Montet D. 2007. Review of the current methods of analytical traceability allowing determination of the origin of foodstuffs. Food Control 18: 228–235.

Quail MA, Smith M, Coupland P, Otto TD, Harris SR, Connor TR, et al. 2012. A tale of three next generation sequencing platforms: Comparison of Ion Torrent, Pacific Biosciences and Illumina MiSeq sequencers. BMC Genomics 13: 341.

Quinteiro J, Vidal R, Izquierdo M, Sotelo CG, Chapela MJ, Perez-Martın RI, Rehbein H, et al. 2001. Identification of hake species (*Merluccius* genus) using sequencing and PCR-RFLP analysis of mitochondrial DNA control region sequences. Journal of Agricultural and Food Chemistry 49: 5108–5114.

Quinto CA, Tinoco R, Hellberg RS. 2016. DNA barcoding reveals mislabeling of game meat species on the U.S. commercial market. Food Control 59: 386–392.

Rea S, Storani G, Mascaro N, Stocchi R, Loschi AR. 2009.Species identification in anchovy pastes from the market by PCR-RFLP technique. Food Control 20: 515–520.

Reed GH, Kent JO, Wittwer CT. 2007. High-resolution DNA melting analysis for simple and efficient molecular diagnostics. Pharmacogenomics 8: 597–608.

Rikimaru K, Takahashi H. 2007. A method for discriminating a Japanese brand of chicken, the Hinai-jidori, using microsatellite markers. Poultry Science 86: 1881–1886.

Roach JAG, Andrzejewski D, Gay ML, Nortrup D, Musser SM. 2003. Rugged LC-MS/MS survey analysis for acrylamide in foods. Journal of Agricultural and Food Chemistry 51: 7547–7554.

Rodriguez-Ramirez R, González-Córdova AF, Vallejo-Cordoba B. 2011. Review: Authentication and traceability of foods from animal origin by polymerase chain reaction-based capillary electrophoresis. Analytica Chimica Acta 685: 120–126.

Rogberg-Muñoz A, Wei S, Ripoli MV, Guo BL, Carino MH, Lirón JP, Prando AJ, et al. 2016. Effectiveness of a 95 SNP panel for the screening of breed label fraud in the Chinese meat market. Meat Science 111: 47–52.

Rojas M, González I, Fajardo V, Martín I, Hernández PE, García T, Martín R. 2009. Authentication of meats from quail (Coturnix coturnix), pheasant (Phasianus colchicus), partridge (Alectoris spp.), and guinea fowl (Numida meleagris) using polymerase chain reaction targeting specific sequences from the mitochondrial 12S rRNA gene. Food Control 20: 896–902.

Rooney JP, Ryde IT, Sanders LH, Howlett EH, Colton MD, Germ KE, Mayer GD, et al. 2015. PCR based determination of mitochondrial DNA copy number in multiple species. Methods in Molecular Biology 1241: 23–38.

Safdar M, Abasıyanık MF. 2013a. Simultaneous identification of pork and poultry origins in pet foods by a quick multiplex real-time PCR assay using EvaGreen florescence dye. Applied Biochemistry and Biotechnology 171: 1855–1864.

Safdar M, Abasıyanık MF. 2013b. Development of fast multiplex real-time PCR assays based on EvaGreen florescence dye for identification of beef and soybean origins in processed sausages. Food Research International 54: 1652–1656.

Safdar M, Junejo Y, Arman K, Abasıyanık MF. 2014a. Rapid bovine and caprine species identification in ruminant feeds by duplex real-time PCR melting curve analysis using EvaGreen fluorescence dye. Molecular Biotechnology 56: 770–776.

Safdar M, Junejo Y, Arman K, Abasıyanık MF. 2014b. A highly sensitive and specific tetraplex PCR assay for soybean, poultry, horse and pork species identification in sausages: Development and validation. Meat Science 98: 296–300.

Safdar M, Junejo Y. 2015. A multiplex-conventional PCR assay for bovine, ovine, caprine and fish species identification in feedstuffs: Highly sensitive and specific. Food Control 50: 190–194.

Sakaridis I, Ganopoulos I, Argiriou A, Tsaftaris A. 2013. High resolution melting analysis for quantitative detection of bovine milk in pure water buffalo mozzarella and other buffalo dairy products. International Dairy Journal 28: 32–35.

Sanger F, Nicklen S, Coulson AR. 1977. DNA sequencing with chain-terminating inhibitors. Proceedings of the National Academy of Sciences 74: 5463–5467.

Santaclara FJ, Espiñeira M, Vieites JM. 2007. Genetic identification of squids (families ommastrephidae and loliginidae) by PCR–RFLP and FINS methodologies. Journal of Agricultural and Food Chemistry 55: 9913–9920.

Sardesai N, Kumar A, Rajyashri KR, Nair S, Mohan M. 2002. Identification and mapping of an AFLPs marker linked to Gm7, a gall midge resistance gene and its conversion to a SCAR marker for its utility in marker aided selection in rice. Theoretical and Applied Genetics 105: 691–698.

Savo Sardaro ML, Marmiroli M, Maestri E, Marmiroli N. 2012. Genetic characterization of Italian tomato varieties and their traceability in tomato food products. Food Science & Nutrition 1: 54–62.

Schlötterer C. 2004. The evolution of molecular markers—just a matter of fashion? Nature Reviews Genetics 5: 63–69.

Selvaraj D, Sarma RK, Shanmughanandhan D, Srinivasan R, Ramalingam S. 2014. Evaluation of DNA barcode candidates for the discrimination of the large plant family Apocynaceae. Plant Systematics and Evolution 301: 1263–1273.

Shirasawa K, Kishitani S, Nishio T. 2004. Conversion of AFLP markers to sequence-specific markers for closely related lines in rice by use of the rice genome sequence. Molecular Breeding 14: 283–292.

Shokralla S, Hellberg RS, Handy SM, King I, Hajibabaei M. 2015. A DNA mini-barcoding system for authentication of processed fish products. Scientific Reports 5: 15894.

Smith CT, Elfstrom CM, Seeb LW, Seeb JE. 2005. Use of sequence data from rainbow trout and Atlantic salmon for SNP detection in Pacific salmon. Molecular Ecology 14: 4193–4203.

Soares S, Amaral JS, Mafra I, Oliveira MBPP. 2010. Quantitative detection of poultry meat adulteration with pork by a duplex PCR assay. Meat Science 85: 531–536.

Stamoulis P, Stamatis C, Sarafidou T, Mamuris Z. 2010. Development and application of molecular markers for poultry meat identification in food chain. Food Control 21: 1061–1065.

Sumathi G, Jeyasekaran G, Jeya Shakila R, Sivaraman B, Arunkumar G, Manimaran U, Sukumar D. 2015. Molecular identification of grouper species using PCR-RFLP technique. Food Control 51: 300–306.

Taberlet P, Gielly L, Pautou G, Bouvet J. 1991. Universal primers for amplification of three non-coding regions of chloroplast DNA. Plant Molecular Biology 17: 1105–1109.

Taboada L, Sánchez A, Sotelo CG. 2017. A new real-time PCR method for rapid and specific detection of Ling (*Molva molva*). Food Chemistry 228: 469–475.

Tatsadjieu NL, Maïworé J, Hadjia MB, Loiseau G, Montet D, Mbofung CMF. 2010. Study of the microbial diversity of *Oreochromis niloticus* of three lakes of Cameroon by PCR-DGGE: Application to the determination of the geographical origin. Food Control 21: 673–678.

Techen N, Parveen I, Pan Z, Khan IA. 2014. DNA barcoding of medicinal plant material for identification. Current Opinion in Biotechnology 25: 103–110.

Teletchea F, Maudet C, Hanni C. 2005. Food and forensic molecular identification: Update and challenges. TRENDS in Biotechnology 23: 359–366.

Teletchea F, Bernillon J, Duffraisse M, Laudet V, Hänni C. 2008. Molecular identification of vertebrate species by oligonucleotide microarray in food and forensic samples. Journal of Applied Ecology 45: 967–975.

Teletchea F. 2009. Molecular identification methods of fish species: Reassessment and possible applications. Reviews in Fish Biology and Fisheries 19: 265–293.

Theodoridis S, Stefanaki A, Tezcan M, Aki C, Kokkini S, Vlachonasios KE. 2012. DNA barcoding in native plants of the Labiatae (Lamiaceae) family from Chios Island (Greece) and the adjacent Cesme-Karaburun Peninsula (Turkey). Molecular Ecology Resources 12: 620–633.

Tillmar AO, Dell'Amico B, Welander J, Holmlund G. 2013. A universal method for species identification of mammals utilizing next generation sequencing for the analysis of DNA mixtures. PLoS One 8: e83761.

Torelli A, Marieschi M, Bruni R. 2014. Authentication of saffron (Crocus sativus L.) in different processed, retail products by means of SCAR markers. Food Control 36: 126–131.

Valentini A, Miquel C, Taberlet P. 2010. DNA barcoding for honey biodiversity. Diversity 2: 610–617.

Vazquez JF, Perez T, Ureña F, Gudin E, AlbornoZ J, Dominguez A. 2004. Practical application of DNA fingerprinting to trace beef. Journal of Food Protection 67: 972–979.

Velasco A, Sánchez A, Martínez I, Santaclara FJ, Pérez-Martín RI, Sotelo CG. 2013. Development of a real-time PCR method for the identification of Atlantic mackerel (*Scomber scombrus*). Food Chemistry 141: 2006–2010.

Vignani R, Bowers JE, Meredith CP. 1996. Microsatellite DNA polymorphism analysis of clones of Vitis vinifera 'Sangiovese'. Scientia Horticulturae 65: 163–169.

Vilgalys R. 2003. Taxonomic misidentification in public DNA databases. New Phytologist 160: 4–5.

Villacorta-Rath C, Ilyushkina I, Strugnell JM, Green BS, Murphy NP, Doyle SR, Hall NE, et al. 2016. Outlier SNPs enable food traceability of the southern rock lobster, *Jasus edwardsii*. Marine Biology 163: 223.

von Stackelberg M, Lindemann S, Menke M, Riesselmann S, Jacobsen H-J. 2003. Identification of AFLP and STS markers closely linked to the def locus in pea. Theoretical and Applied Genetics 106: 1293–1299.

Vos P, Hogers R, Bleeker M, Reijans M, van de Lee T, Hornes M, Frijters A, et al. 1995. AFLP: A new technique for DNA fingerprinting. Nucleic Acids Research 23: 4407–4414.

Vossen RHAM, Aten E, Roos A, Den Dunnen JT. 2009. High-resolution melting analysis (HRMA) – More than just sequence variant screening. Human Mutation 30: 860–866.

Wang Y, Leung HCM, Yiu SM, Chin FYL. 2012. MetaCluster 4.0: A novel binning algorithm for NGS reads and huge number of species. Journal of Computational Biology 19: 241–249.

Weber-Lehmann J, Schilling E, Gradl G, Richter DC, Wiehler J, Rolf B. 2014. Finding the needle in the haystack: Differentiating "identical" twins in paternity testing and forensics by ultra-deep next generation sequencing. Forensic Science International: Genetics 9: 42–46.

Weighardt F. 2003. Quantitative PCR for the detection of GMOs. In The analysis of food samples for the presence of genetically modified organisms (Special publication No. I.03.114) (Session 10, pp. 1–20), Querci M, Jermini M, Van den Eede G (Eds.), European C European Commission DG-JRC.

Wen J, Zeng L, Sun Y, Chen D, Xu Y, Luo P, Zhao Z, Yu Z, Fan S. 2015. Authentication and traceability of fish maw products from the market using DNA sequencing. Food Control 55: 185–189.

Wilcox TM, McKelvey KS, Young MK, Jane SF, Lowe WH, Whiteley AR, Schwartz MK. 2013. Robust detection of rare species using environmental DNA: The importance of primer specificity. PLoS ONE 8: e59520.

Wolf C, Rentsch J, Hubner P. 1999. PCR-FRLP analysis of mitochondrial DNA: A reliable method for species identification. Journal of Agricultural and Food Chemistry 47: 1350–1355.

Wong A, Forbes MR, Smith ML. 2001. Characterization of AFLP markers in damselflies: Prevalence of codominant markers and implications for population genetic applications. Genome 44: 677–684.

Woolfe M, Primrose S. 2004. Food forensics: Using DNA technology to combat misdescription and fraud. Trends in Biotechnology 22: 222–226.

Wu Q, Zhou G, Yang S, Tur Abulikemu B, Luo R, Zhang Y, Li X, et al. 2017. SNP genotyping in sheep from northwest and east China for meat traceability. Journal of Consumer Protection and Food Safety 12: 125–130.

Xie Y, McNally K, Li C-Y, Leung H, Zhu Y-Y. 2006. A high-throughput genomic tool: Diversity array technology complementary for rice genotyping. Journal of Integrative Plant Biology 48: 1069–1076.

Xin T, Yao H, Gao H, Zhou X, Ma X, Xu C, Chen J, et al. 2013. Super food *Lycium barbarum* (Solanaceae) traceability via an internal transcribed spacer 2 barcode. Food Research International 54: 1699–1704.

Xu J, Miao H, Wu H, Huang W, Tang R, Qiu M, Wen J, Zhu S, Li Y. 2006. Screening genetically modified organisms using multiplex-PCR coupled with oligonucleotide microarray. Biosensors and Bioelectronics 22: 71–77.

Zaiko A, Martínez JL, Ardura A, Clusa L, Borrel YJ, Samuiloviene A, et al. 2015. Detecting nuisance species using NGST: Methodology shortcomings and possible application in ballast water monitoring. Marine Environmental Research 112: 64–72.

Zabeau M, Vos P. 1993. Selective restriction fragment amplification: A general method for DNA fingerprinting. European Patent Office. Publication 0 534 858 A1 bulletin 93/13.

Zeng L, Kwon T-R, Liu X, Wilson C, Grieve CM, Gregorio GB. 2004. Genetic diversity analyzed by microsatellite markers among rice (Oryza sativa L.) genotypes with different adaptations to saline soils. Plant Science 166: 1275–1285.

Zhang J, Huang H, Cai Z, Huang L. 2006. Species identification in salted products of red snappers by semi-nested PCR-RFLP based on the mitochondrial 12S rRNA gene sequence. Food Control 17: 557–563.

Zhang C-L, Fowler MR, Scott NW, Lawson G, Slater A. 2007. A TaqMan real-time PCR system for the identification and quantification of bovine DNA in meats, milks and cheeses. Food Control 18: 1149–1158.

Zhang C. 2013. Semi-nested multiplex PCR enhanced method sensitivity of species detection in further-processed meats. Food Control 31: 326–330.

Zhao W, Zhao Y, Pan Y, Wang X, Wang Z, Xie J. 2013. Authentication and traceability of *Nibea albiflora* from surimi products by species-specific polymerase chain reaction. Food Control 31: 97–101.

SECTION 2

Authentication and Traceability

CHAPTER 8

Introduction to Food Authentication Based on Fingerprinting Techniques

Md. Rizwanullah, Nausheen Khan, and Saima Amin
Jamia Hamdard

Javed Ahmad
Najran University

Syed Sarim Imam
Glocal University

Khalid Umar Fakhri and Mohd. Moshahid Alam Rizvi
Jamia Millia Islamia

CONTENTS

8.1 INTRODUCTION

Food authentication is the process that verifies that a food is in compliance with its label description. This may include, among others, the origin (species, geographical, or genetic), production method (conventional, organic, traditional procedures, free range), or processing technologies (irradiation, freezing, microwave heating). The declaration of specific quality attributes in high-value products is of particular interest since these products are often the target of fraudulent labeling. Proof of provenance is an important topic for food safety, food quality, and consumer protection, as well as the compliance with national legislation, international standards, and guidelines (Aung and Chang, 2014). Due to the globalization of food markets and the resulting increase in variability and availability of food products from other countries, consumers are increasingly interested in knowing the geographical origin along with the assumed quality of the products they eat and drink. The quality assurance and the methods used to authenticate foodstuffs are of great interest both from commercial and legal points of view (Drivelos and Georgiou, 2012).

Authenticity has been a major concern for consumers, producers, and regulators since ancient times (Posudin et al., 2015). Modern instrumentation and advances in basic sciences and in information and communication technologies provide the means for precise measurement and elucidation of food origin (Georgiou and Danezis, 2015). Since the beginning of the 20th century, organizations that set standards and control the origin of ingredients and the production processes have appeared all over the world, e.g., the French "Institut National des Appellations d'Origine (INAO)", Italy's "Denominazione di Origine Controllata", Spain's "Denominación de Origen", South Africa's "Wine of Origin", or the United States' "American Viticultural Areas". The production of consumer goods according to the standardized procedures normally results in better products and is rewarded with higher prices at the point of sale. Unfortunately, these financial benefits attract the production of counterfeit food and illegal food trades (Danezis et al., 2016).

In Europe, origin is one of the main authenticity issues concerning food. European Union legislations that reserves specific names for food and beverages of particular qualities or reputations have been abundant since the dawn of the European integration process. These legislations introduced regulatory frameworks for wines and spirits and quality schemes for food products, including PDO (protected designation of origin) that links products to the defined geographical area where they are produced, PGI (protected geographical indication) that links products to a geographical area where at least one production step occurs, and TSG (traditional specialties guaranteed) that protects traditional methods of production. Furthermore, recently defined optional quality terms (OQT) such as, "mountain product" and "product of island farming" were defined. The purpose of these EU schemes is to protect the reputation of regional food products and to promote good practices in rural and agricultural activities. Such practices help the producers obtain premium prices for authentic products and minimize the unfair and misleading competition from non-genuine products, which are usually of inferior quality or of different flavors (1151/2012 EU Regulation). The information includes a characterization of the geographical region and reinforces the consumer perception of the special qualities attributed to the mountain and island products. In case of cultivated species, EU indicates that a reference should be made to the country in which the food undergoes the final production stage. Vigorous research activities in EU are supported by the coordinating actions of the European Union, including "Food Integrity", "MoniQa", and "TRACE" within HORIZON 2020 (Danezis et al., 2016).

Determination of food authenticity is an important issue in quality control and food safety. Authenticity testing is a quality criterion for food and food ingredients, which is increasingly becoming a result of legislative protection for regional food. Thus, there is a pressing need for accurate, standardized food authentication techniques (Posudin et al., 2015; Georgiou and Danezis, 2015; Cubero-Leon et al., 2014; Reinholds et al., 2015). Food authenticity testing does not serve only consumers but the stakeholders also include the food industries that are seeking the opportunities to assure their food products' labeling compliance and branding. Regulatory authorities are asking for an extended and updated list of the analytical methods for the confirmation of authentic food products and to support law enforcement.

8.2 IMPORTANCE OF FOOD AUTHENTICATION (Dennis, 1998)

- Incorrect labeling can be a commercial fraud, as species of higher commercial value can be substituted with lower value species, e.g., pork (*Sus scrofa*) substitution in beef (*Bos taurus*).
- Misleading labeling can even cause health issues in sensitive consumers due to undisclosed allergens. Food allergies are an emerging public health issue. Regulatory bodies (WHO/EC) enlisted the prevalent and severe allergens (milk, egg, fish, peanut, soybean, whey, and gluten) that industries and producers should label on packaged food.
- Food labeling also helps consumers choose certain food products over others as per their lifestyles (vegetarian or nonvegetarian) or religious sentiments (e.g., pork meat should be absent for Muslims and Jews).

8.3 CLASSIFICATION OF AUTHENTICITY ISSUES (Dennis, 1998)

8.3.1 Species of Origin

Generally, species used in food products are misdescribed, in the form of substitution of one species for another. Thus, in a survey of battered fried fish that were sold to the consumers, the species of fish was found to be misdescribed in 5% of the cases. In this instance, there was no evidence of a fraud being carried out for economic benefit since, as often as not, an equally expensive fish was substituted. Irrespective of whether a commercial fraud is intended or not, the consumer gets a product of a different genetic makeup. A consumer may as well prefer certain species with an expectation that she would not contract certain diseases. For example, the claims that beef has come from Aberdeen Angus may potentially require testing. Such claims confer a commercial advantage because some consumers consider that pure bred beef herds are less likely to have contracted bovine spongiform encephalopathy (BSE).

8.3.2 Geographical Region of Origin

The product quality is usually associated with its geographical origin. For example, the pricing of wines is largely based on the region, pertaining to the quality, just like in the cases of Assam tea, sausages, and olive oil. European legislation has been developed for the protection of geographical indications and designations of origin for agricultural produce and foodstuffs.

8.3.3 Commercial Treatment

The food products are usually subjected to a number of commercial treatments, either necessary or desirable, for example, pasteurization (not adding any value to the product) or value adding processes or quality assurance processes. An example of quality assurance process is cold pressed (i.e., virgin) olive oil that should demonstrate the absence of refined oil. Value adding procedures include food irradiation or the use of genetically modified organisms (GMOs), which seek to benefit the consumer through the retardation of spoilage or through the production of cheaper food.

8.3.4 Water

Water is the earliest and cheapest adulterant of milk and beer, which may not be harmful but alleges a commercial fraud. It is commonly used to expand the quantity of foodstuffs in liquid products such as wine and also in meat products.

8.3.5 Brands

Most businesses are keen to protect the authenticity of their brand. Authenticity is already questionable for non-food businesses at large and, increasingly, it is becoming important for the food brands as well; for example, the Scotch whisky industry goes to considerable lengths to protect its brands from being counterfeited, particularly in some third-world countries where Scotch can command premium prices and where counterfeiting of brand is rife.

8.4 EFFECTS OF FOOD FRAUD

There can be no argument that consumers have the right to an accurate and informative labeling. Studies linking sensory perceptions and chemical compositions are therefore helpful in ensuring that molecules which have a sensory effect are present in premium quality food. A unique study investigated the biological effects of mixing peppermint oil with corn-mint oil (Lis-Balchin et al., 1997). Both materials exhibited a wide range of activities against different species of bacteria, but both showed consistent spasmolytic activities on guinea pig ileum suggesting that they both are equally effective in treating conditions associated with smooth muscles (e.g., irritable bowel syndrome). Thus, a mixture of these oils would not necessarily disadvantage consumers who took the oils for medical purpose. In most cases, the materials used for extending food are innocuous (e.g., sugar or water). However, if honest labeling is not enforced through legislation then there is a possibility that more harmful practices would arise. The addition of ethylene glycol to Austrian wine some 10 years ago is a good example. What is less known is that the incorporation of this material in "elixir sulfanilamide" in 1938 led to 105 deaths in the U.S. and forced fundamental changes in the operation of the Food and Drug Administration (Wax, 1995). The adulteration of food may lead to long term health effects in survivors. A follow-up study on survivors of the Spanish toxic oil syndrome found that 58% still suffered from symptoms 12 years after the poisoning event (Kaufman et al., 1995). There can therefore be no doubt about the need for continued vigilance in the determination of food authenticity.

8.5 FOOD AUTHENTICATION BY FINGERPRINTING TECHNIQUES

In a way similar to the use of human fingerprints in people recognition, food authentication by instrumental fingerprinting is based on verifying that the signal is consistent with a reference signal within the sample and with instrumental variability. Therefore, the practical application is not immediate, but it necessarily requires the attainment of a previous stage that involves the establishment of an instrumental fingerprint that is characteristic of the authenticity.

Food authentication and food fraud detection are two sides of the same coin. They are similar problems and their solutions require parallel analytical strategies. In principle, three analytical approaches could be distinguished for obtaining the target information in the structure of the matter: (i) the information is obtained from the food chemical composition (chromatographic techniques); (ii) works at biomolecular level, fundamentally on DNA (DNA-based fingerprinting techniques); and (iii) these are focused on the stable isotopic compositions of certain atoms (spectroscopic technique).

8.5.1 Spectroscopic Fingerprinting Techniques

Modern spectroscopic techniques such as IR, Raman, NMR, mass, and UV are all capable of high sample throughput and automated analysis. Born as tools for the structural assignment of a compound's inorganic chemistry in the first half of 20th century, their potentiality was initially either masked by poor knowledge of their theoretical mechanisms or by their diffusion that was limited by the instrumental complexity and cost. However, they expanded rapidly over the remaining half of the century, after the introduction of relatively low cost but reliable bench top instruments in the 1990s and after the development of software that were able to manage data from complex component mixtures. One of the greatest advantages of these techniques comes from their capability to deliver fast quantitative data, thus providing a superior alternative to the long and tedious procedure of the wet chemistry (Gallo and Ferranti, 2016).

8.5.1.1 Fourier Transform Infrared (FT-IR) Spectroscopy

FT-IR spectroscopy allows for the extremely rapid, high-throughput, and nondestructive analysis of a wide range of sample types, in both research and industrial environments. This technique is based on the principle that when a sample is interrogated with an IR beam (usually in the mid-infrared 4,000–400 cm^{-1} or near-infrared 14,000–4,000 cm^{-1}), functional groups within the sample will absorb this radiation and vibrate in one of a number of recognized ways (such as stretching or bending vibrations) and these vibrations/absorptions can then be correlated directly with (bio)chemical species. Therefore, the obtained IR spectra can be reported as a chemical 'fingerprint', because it is characteristic of the specific chemical sample under analysis and, hence, each and every (bio) chemical will have its own unique IR fingerprint (Dunn and Ellis, 2005; Ellis et al., 2007).

8.5.1.1.1 Mid-Infrared (MIR) Spectroscopy

A significant proportion of the work undertaken in recent years has concentrated on the mid-IR part of the electromagnetic (EM) spectrum, as this part of the spectrum is where the fundamental vibration is seen (as opposed to overtones or harmonics in near-IR) and is particularly information rich. With relevance to biological applications, the MIR can be further broken down into specific regions, or windows, of biological interest such as: C–Hx stretching vibrations from fatty acids (3050–2800 cm^{-1}); CQO, N–H and C–N from proteins and peptides (1750–1500 cm^{-1}); C–O and C–O–C from polysaccharides (1200–900 cm^{-1}); and P–O vibrations from nucleic acids (1245–1220 and 1090–1085 cm^{-1}) (Gallo and Ferranti, 2016).

The rapidity and reproducibility of FT-IR spectroscopy is demonstrated through the large body of research published using this technology and, indeed, its rapidity cannot be overstressed as thousands of spectra can be collected within a 24-h period (Ellis et al., 2007). Moreover, due to its holistic nature, FT-IR has been recognized as a valuable tool for metabolic fingerprinting, as it is able to analyze carbohydrates, amino acids, fatty acids, lipids, proteins, nucleic acids, and polysaccharides rapidly and simultaneously with a minimum amount of sample preparation or background training (Ellis et al., 2007; Ellis and Goodacre, 2006).

Applications of FT-IR spectroscopic technique are broad in scope, and this was accommodated in the 1990s within the biosciences, as the technique was established to be mostly useful for the discrimination of axenically cultured bacteria (Goodacre et al., 1998), and work on food-related bacteria continues (Lu et al., 2011). The work undertaken during the '90s spawned a considerable amount of interest which led to other areas of research, such as metabolomics, more specifically referred to as metabolic fingerprinting. Concomitant with these multidisciplinary applications, FT-IR and, particularly, NIR spectroscopy have been viewed as exceptionally useful spectroscopic tools for the analysis of foodstuffs (Wishart, 2008; Winder et al., 2006).

8.5.1.1.2 Near-Infrared (NIR) Spectroscopy

For many decades, much of the food-related IR research was almost exclusively reserved to NIR spectroscopy, perhaps due to the robustness and simplicity of the instrumentation, its highly precise and efficient nature, and its portability. Though in terms of portability, more recently, handheld portable FT-IR (Birkel and Rodriguez-Saona, 2011) and Raman (Robinson et al., 2012) instruments are becoming more widely available, though these are not yet as mature as NIR. Indeed, in regards to the maturity of this technique, it has been stated that on/in-line NIR instrumentation has a proven record of 40 years in the food processing industry (Cozzolino, 2011), particularly with respect to continuous monitoring and control of process and product quality (Huang et al., 2008). The rapidity of this method also allows for frequent measurements at all stages, from production and collection of raw ingredients and process monitoring to quality control of finished products (Huang et al., 2008). These instruments usually have a very high signal-to-noise ratio (Buning-Pfaue, 2003) and, like MIR, can measure multiple components in a food sample simultaneously, and they also have remote sampling capability allowing for the collection of real-time information in a process stream (Woodcock et al., 2008).

For food and feedstuff analyses, this method clearly has many advantages similar to those of MIR (both being IR methods), though two main advantages that NIR has over MIR analysis are that: (i) NIR light enters much deeper into an intact food sample (>10 mm), as well as through different packaging materials that are transparent to NIR light; (ii) the water absorbance in the NIR region is not as strong as in the MIR so analysis is more routine and no extra accessories are required. However, the main disadvantage of NIR is the low sensitivity to some minor chemical constituents (Buning-Pfaue, 2003). Food science related NIR technique represents a large research community and a long history of applications in the food industry, and neither its inclusion here nor its continuing contribution to food science are not to be underestimated (Reid et al., 2006; Wang et al., 2004).

8.5.1.2 Raman Spectroscopy

Raman spectroscopy is a versatile, nondestructive analytical technique, which, like IR spectroscopies, provides unique spectral fingerprints for many analytes. The technique is seemingly ideally suited to the analysis of foodstuffs, largely due to the diversity of analytes which can be probed; ranging from macro-food components such as lipids, proteins, and carbohydrates to the minor components such as dyes, pigments, and flavorings. Also, of interest are the microorganisms which play a significant role in food spoilage (Herrero, 2008; He et al., 2008). The use of Raman spectroscopic technique is significantly increased by its ability to couple to microscopes, resulting in 1 mm resolution that is perfect for single microbe identification and for in situ or on-line analysis when interfaced with fiber optic probes (Qin et al., 2010; Sowoidnich et al., 2010). Like FT-IR spectroscopic techniques, minimal or no sample preparation is required. Furthermore, due to the confocal nature, Raman spectroscopy permits the samples to be profiled through different varieties of transparent materials and allows the analysis of food products through glass or plastic packaging. Moreover, Raman spectroscopic technique is able to detect analytes in solutions with minimal interference from water. The combination of the above properties with the prospect of portable instrumentation make Raman spectroscopy a promising fingerprinting technique for food science (Li-Chan, 1996).

8.5.1.3 Nuclear Magnetic Resonance (NMR) Spectroscopy

NMR spectroscopy is a fingerprinting technique that takes advantage of the magnetic properties of certain nuclei (Bothwell and Griffin, 2011). As with the vibrational approaches (FT-IR and Raman spectroscopy) described above, NMR spectroscopy can also be used to identify functional groups and it produces important structural information and determines the physical and chemical properties of atoms within a molecule or chemical. NMR spectroscopy can be used across a wide range of sample types, including solids and solutions, as well as complex mixtures (Heinzmann et al., 2010). A vast majority of elements possess a magnetic spin number greater than zero that is necessary to produce the NMR effect, and the common nuclei observed by NMR spectroscopy have a spin quantum number of 1/2 resulting in two spin states. The transitions between these two spin states are detected when the nucleus is placed in a magnetic field and induced by radiation in the radiofrequency region of the electromagnetic spectrum (Bothwell and Griffin, 2011). The radiation is able to move through a variety of materials with ease with minimal heating effects and is sensitive to different chemical environments (Gil et al., 2003).

Moreover, an NMR spectrum can contain resonances that provide significant quantities of structural information; this further enables the identification of individual constituents through the interpretation of chemical shifts, for example (Dunn et al., 2011). However, as with all the approaches mentioned thus far, NMR does suffer from relatively low sensitivity, especially so if compared directly to mass spectrometry-based techniques (vide infra) (Biais et al., 2009; Dunn et al., 2008). Whilst this is the case at the moment, there is momentum within the NMR research community to move toward ever higher magnetic fields in order to improve sensitivity and to explore hyperpolarizability effects (Golman et al., 2006). NMR spectroscopy therefore presents itself as another potential tool of choice for the analysis of food and feedstuffs and, indeed, it has a substantial record of food-based applications over the last three decades (Guillou et al., 1992).

8.5.1.4 Mass Spectrometry

Mass spectroscopy is a powerful analytical technique that is conducted by the formation of positively or negatively charged analytes (ions) and subsequent detection and measurement of the mass-to-charge (m/z) ratio of these ions (Dunn, 2008). Mass spectrometers operate through introduction of sample, formation of ions in the ionization source, separation of ions based on their m/z ratio in a mass analyzer, and finally detection of ions. Sample introduction can involve electrophoretic or chromatographic separation of analytes prior to detection (capillary electrophoresis, liquid chromatography, gas chromatography) or direct introduction to the mass spectrometer that includes direct infusion mass spectrometry (DIMS) (Smedsgaard and Nielsen, 2005), desorption electrospray ionisation (DESI) (Nielen et al., 2011), and matrix assisted laser desorption ionisation (MALDI) (Nicolaou et al., 2011). The development of charged analytes in the ionization source is an essential requirement as subsequent mass analysis applies electrical and magnetic fields that will only separate ions of different m/z ratios if the analytes are charged (either with a positive or negative charge). Different ionization sources include DESI, MALDI, electron impact ionization (EI) in combination with gas chromatography (GC) as well as atmospheric pressure ionization (API) sources including electrospray ionization (ESI) and atmospheric pressure chemical ionization (APCI) often combined with liquid chromatography (LC). Mass analyzers involve electrical or magnetic fields to separate ions of different m/z ratios and include quadrupole, triple-quadrupole, ion trap, time-of-flight (TOF), magnetic sector, Orbitrap, and Fourier transform ion cyclotron resonance (FTICR) mass analyzers. When combined with chromatographic separation, it provides the ability to analyze complex samples containing analytes of similar or identical masses and physicochemical properties, with specificity provided by mass and chromatographic separation.

Mass spectrometers offer a number of advantages and limitations for application in the food industry. Targeted analytical methods of high specificity and sensitivity are available to quantify analytes of interest accurately. Quantification of analytes is typically done with liquid chromatography coupled to triple quadrupole mass spectrometers (LC-MS/MS) that provides sufficiently high levels of sensitivity and specificity (Li et al., 2010; Lutter et al., 2011). GC-MS platforms are also applied for analyte quantification (Fantoni and Simoneau, 2003; Kempf et al., 2008). The coupling of chromatography and mass spectrometry platforms also enables the detection of analytes in complex samples at sub-μg ml^{-1} or sub-μg g^{-1} concentrations. Indeed, LC-MS/MS platforms are the golden standards for performing quantification in the food and many other industries including the pharmaceutical industry for regulatory purposes. However, these techniques need large sample preparation and, therefore, it needs time to go from sampling to achievement of results. This can be extensively longer (minutes to hours) than for other spectroscopic fingerprinting techniques such as FT-IR, Raman, and NMR. Other mass spectrometry-based fingerprinting techniques can be applied with minimal (e.g., MALDI) or no (e.g., DESI) sample preparation and allow quick analysis of sample, though generally with lower sensitivity and specificity.

8.5.1.5 UV Spectrometry

The corresponding electromagnetic wave of UV absorption spectra has short wavelength and high energy that reflects the transition of valence electron energy levels of elements. It is sensitive to the conjugate structure (conjugated olefin and unsaturated carbonyl compounds) and aromatic compounds. The UV spectrum is comparatively simpler with broad peaks, as the electronic energy levels change with the transition of vibration levels. The UV fingerprinting technique is mainly used in the detection of main ingredients in food, additives, pesticides, and veterinary drugs residues (Zhang et al., 2011). The recent invention of a portable UV spectroscopic testing device for authentication of scotch whisky highlights the application potential of UV for industrial use, as it can be used in the field and cuts screening time down to less than 1 min (MacKenzie et al., 2004). Whereas, the signal of UV absorption in qualitative analysis is less, so it is difficult to analyze complex components. The penetrating power of UV light is weak, so it requires good transparency of the samples. And the sample pretreatment for UV spectrum is relatively cumbersome. Future research in these techniques can yield faster and more economic methods of identification. Introduction of sufficiently robust and reliable instrumentation design shall also lead to the expansion of its use in industrial applications.

8.5.2 Chromatographic Fingerprinting Techniques

The essential aim of the chromatographic fingerprinting technique is the separation of chemical species and not obtaining instrumental signals related to the chemical composition of the test material under study. This last key role is played by the detector that is the true analytical measurement instrument and is defined as a device that measures the change in the composition of the eluent by measuring the physical or chemical properties (Viaene et al., 2015). The most common modalities of chromatography are GC and LC.

8.5.2.1 Gas Chromatography

GC is an appropriate technique for the detection of volatile liquids or solids and gas mixture. Advantages of this technique include high resolution, high sensitivity, and short separation time. Due to the evolution of gas chromatographic technique with the emergence of different types of sensitive detectors, the detecting sensitivity of this technique is generally higher than most high performance liquid chromatography (HPLC) detectors (Zhang et al., 2011).

GC has been widely used for testing food additives. In the research by Yingliang et al. (2006), clenbuterol hydrochloride and ractopamine hydrochloride residues in animal liver were detected by solid phase extraction (SPE) with GC-MS. The detection limit was 0.30–1.00 mg/kg that proved it to be an effective identification method for these two residues. Zhan et al. (2008) studied the detection of commonly used additives: methylparaben, ethylparaben, propylparaben, butylparaben, BHA, BHT, and TBHQ using GC-MS. The detection limit was 0.01–0.5 mg/kg and the recovery rate was between 94.4%–105.1%. This technique had been utilized by Tian et al. (2006) for the detection of volatile flavor constituents in Jinhua ham, which confirmed that the main content of volatile flavor constituents in the Jinhua ham were acid, aldehyde, and ketone compounds with minor quantities of sulfur and heterocyclic compounds.

There are a few drawbacks of using gas chromatographic techniques in food analysis. Samples of the complex food usually require deep cleanup and fractionation, removal of the nonvolatile constituents or other disrupting chemicals, and preparation of relatively simple mixtures so that the samples can be inserted into the analysis system without causing too much damage to the chromatography column. In addition, gas chromatography methods can only solve about 20% organic analysis; it needs esterification and other pretreatment to reduce the boiling points of materials with high boiling points. Improper handling of any of the steps would result in the loss or degradation of the samples (Zhang et al., 2011).

8.5.2.2 Liquid Chromatography

LC is widely used in the detection of food additives, toxic substances, pesticides, and veterinary drug residues. This technique has rapid analysis ability, high separation efficiency, and good sensitivity. LC could analyze macromolecules with high boiling points and thermal stability. Due to the continuous development of a large variety of workstation software and the combination with MS, the applications of LC have been greatly broadened. The data acquired from a chromatographic method include values for retention time, peak intensity (area or height) and, if MS is coupled, peak m/z. In GC-MS-based approaches, the spectra comprise several m/z fragments from a single analyte, whereas in LC-MS there is less fragmentation and isotopic peaks can be detected. In addition, the use of comprehensive two-dimensional chromatography opens up the possibility of having an additional chromatographic dimension, as there are two retention times instead of one, thus increasing the potential of profiling and identifies markers that makes data handling even more challenging (Hyotylainen and Oresic, 2013).

The versatility of the chromatographic methods allows the analyst to interact with both the separation and measurement steps in order to acquire an analytical signal with the maximum useful information, as well as to select the data to be treated. As a consequence, these methods are excellent ways of getting proper fingerprints for food “identification” as well as for food authentication (Bosque-Sendra et al., 2012).

Li et al. (2009a) integrated LC with electrospray ionization tandem mass spectrometry method (LC-ESI-MS/MS) to measure 156 kinds of pesticide residues in five concentrated fruit and vegetable juices. The recovery rates for majority of the pesticides were between 70%–120%, while the detection and quantification ranges were 0.10–56.77 mg/kg and 0.33–189.23 mg/kg, respectively. In the research carried out by She et al. (2008), a method has been introduced to determinate 24 kinds of sulfa drug residues in milk by ultra performance liquid chromatography-quadrupole mass spectrometry (UPLC-MS/MS). With this method they could get results within 10 mins with high recovery rates and sensitivity. In a similar way, Zhang et al. (2009) tested the residues of sulfonamides in *Thelitopenaeus vannamei* and *Carassius carassius* with satisfactory results. Li et al. (2009b) developed a fast method to detect the residues of malachite green (MG) and its metabolite leucomalachite malachite green (LMG) in eel by HPLC, which lasted for 14 mins with separate recovery rates of 85.7%–89.7% and 87.6%–91.2%, respectively.

8.5.3 Thermal Fingerprinting Technique - Differential Scanning Calorimetry (DSC)

DSC is a thermoanalytical fingerprinting technique that evaluates the difference between the quantity of heat required to increase the temperature of a sample and a suitable reference material as a function of temperature. This differential heat flow is then recorded as a peak. The enthalpy change is directly proportional to the area under the peak, and this indicates whether the thermal event is exothermic (energy producing) or endothermic (energy consuming). DSC has played an important role in the analysis of the thermal behavior of fats, oils, carbohydrates, and proteins, particularly in the field of food science and technology. Moreover, DSC is a rapid and simple technique as in most cases no specific sample preparation is needed. It is also well known as an environment friendly technique as it does not require any chemical or solvent utilization (Nur Azira and Amin, 2016).

In case of food component interactions, thermal events deduced from DSC may be the elimination or appearance of a new exothermic or endothermic peak. Additionally, any changes in peak onset, peak shape, peak maximum, peak offset temperature, and relative height might be considered. Currently, DSC is used in a large number and variety of applications. It is suitable for analysis of food systems as they are often exposed to cooling or heating during processing; indeed, data obtained can be directly used to examine the thermal transition of a food system. In addition, the availability of a wide range of DSC sample pans enables researchers to study materials either in the solid or liquid forms (Elmer, 2013).

Recently, DSC is employed for the detection of butter adulteration with lard (Nurrulhidayah et al., 2015). Butter has very similar physicochemical characteristics to lard, which makes lard a preferable adulterant in butter. DSC provides a unique thermal fingerprinting technique for lard and butter.

DSC has been successful in the detection of adulteration with lard and beef tallow in canola oil (Marikkar et al., 2002) and with lard in palm oil (Marikkar et al., 2003). Apart from fats and oils, the differentiation of honey samples on the basis of floral origin and adulteration was possible using DSC (Cordella et al., 2003). Although DSC holds much potential for the authentication of food products, reported applications till date are limited, making DSC more intriguing for future research.

8.5.4 DNA-Based Fingerprinting Technique - Polymerase Chain Reaction (PCR)

DNA-based methods are considered more reliable on account of the stability of DNA under environmental conditions, as well as farming and production techniques used during the processing of certain food products (Martellos and Labra, 2013).

The identification of plant species relies on the principle of DNA polymorphisms or genetic variations that occur as a result of naturally arising mutations in the genetic code (Liu and Cordes, 2004). First, the DNA is extracted from the target organism and then the DNA fragment of interest is amplified using PCR in order to identify the species-specific genetic polymorphisms. The amplicons that arise from the PCR assay are then analyzed to expose the characteristic polymorphisms under study.

DNA isolation methods: Detection of the botanical origin of food requires the isolation of DNA that is free of inhibitors, and in sufficient quantity and quality. However, the DNA extraction from food products possesses several problems and limitations and requires special treatment (Martins-Lopes et al., 2013). Plant material is going through a number of procedures that result in DNA degradation. Thus, it may not result in satisfactory PCR amplification because the method requires intact targets. Therefore, PCR products should be shorter than 200 bp (base pairs) for potentiating optimal results (Unseld et al., 1995). Moreover, food products contain a large range of substances including carbohydrates, fat, and chemicals that are often inhibitory to the PCR reaction leading to false results (negative or positive). Thus, optimization procedures should be applied in a case-by-case manner. Different methods have been proposed for DNA extraction (Di Bernardo et al., 2007; Pafundo et al., 2011). Yet, phenol chloroform is the most commonly described (Köchl et al., 2005). In addition, most of the biotechnology companies have developed commercial kits for high-throughput DNA extraction, such as NucleoSpin® Food Kit, NucleoSpin Plant II (Macherey-Nagel GmbH, Germany), Wizard Magnetic DNA Purification for Food (Promega, WI, USA), DNeasy® Plant Mini Kit, DNeasy Tissue Kit, QIAamp DNA Mini Kit (Qiagen GmbH, Germany), and Gene Elute Plant Kit (Sigma-Aldrich, MO, USA), with very good results in terms of quality and quantity.

8.5.4.1 Conventional PCR and Sequencing

Conventional PCR amplification is based on the hybridization of specific oligonucleotides to a target DNA and synthesis, in vitro, of millions of DNA copies flanked by these primers. The amplification of DNA fragments, followed by agarose gel electrophoresis for fragment size verification, is the simplest PCR strategy applied to verify the presence of a species in a plant derived food product. Although conventional PCR has the disadvantage of relying entirely on the end point, there are quite a few researchers who continue to use this method and its variations for identification assays of plant species.

Conventional PCR was also used in assays aiming to confirm the ingredient list of food products. Han et al., (2012) compared three different techniques (PCR, real-time PCR, and denaturing high performance liquid chromatography (DHPLC)) to detect ingredients in fruit juices. The authors managed to develop an accurate and rapid PCR method with the intention to detect ingredients from seven fruit species (grape, strawberry, orange, apple, peach, pear, and mandarin) for labeling regulations on fruit juice. On the contrary, Olivieri et al. (2012) performed the sequence analysis of 18S rDNA PCR products in order to identify six different orders of plants (*Apiales, Fabales, Asterales, Solanales, Brassicales,* and *Sapindales*) in a single honey specimen. "Universal" plant primers were used in this assay that targeted the ribosomal RNA genes and allowed the identification of six plant species and also of microorganism (fungi and bacteria) species.

DNA-based methods have also been developed for the detection of allergens originating from plant species. As described by Dahinden et al. (2001), the application of quantitative competitive PCR with primers specific for the chloroplast trnL gene allowed the detection of wheat, barley, and rye in gluten-free food products. Lately, Mayer et al. (2012) performed a qualitative and indirect quantitative analysis of wheat in spelt products using a PCR-RFLP method that was based on the differences found in the g-gliadin gene.

Marieschi et al. (2011a) developed a method based on SCAR (sequence characterized amplified region) markers in order to identify *Satureja montana* L. and *Origanum majorana* L. in commercial samples of Mediterranean oregano. SCAR markers have also been used by the same team for the identification and detection of *Olea europaea*, which is a new adulterant of Mediterranean oregano. The limit of detection (LOD) of this method in multiple commercial batches was found to be 1% (Marieschi et al., 2011b). SCAR markers, initially developed for the detection and identification of plants, were also used for tracing adulterants in traded ground chili (Dhanya et al., 2011) and saffron (Marieschi et al., 2012).

Soares et al. (2010) developed a PCR assay targeting the soybean lectin gene to detect trace amounts of soybean in meat sausages. Similarly, Sakai et al. (2010) described a specific PCR that targeted the chloroplast rbcL gene for the detection of banana residues in processed food. In another assay, Handy et al. (2011) reported a DNA-based method for identification of the species level of pine nuts in FDA complaint samples by using the barcoding ycf1 chloroplast gene. Following a different approach, Hernández et al. (2010) described a simple random amplified polymorphic DNA (RAPD)-PCR method for the identification of smoked paprika *Pimentón de la Vera* adulteration with paprika elaborated from five foreign varieties.

Despite the fact that conventional PCR techniques have been used for the identification of plant species, the method has a major disadvantage, that of the lack of standardization across species. Moreover, conventional PCR techniques present a few drawbacks that limit their use. The main concern is that these protocols should be restricted to research labs, since the amplification products are detected by gel electrophoresis using UV luminescence and the ethidium bromide double strand DNA intercalating dye. Secondly, the standardization of the PCR procedure (from sampling to result interpretation) is difficult since its execution demands specific skills and cautious handling. Moreover, conventional PCR has the limitation of not providing information on the quantitative analysis of the food product as long as its amplification products do not give quantitative information regarding the initial amount of target molecules, especially in food products (Madesis et al., 2014).

8.5.4.2 Quantitative Real-Time PCR

Information on the quantity of a substituted ingredient is essential for verifying an intended or unintended adulteration, or for evaluating if a limit set by law has been exceeded. In those cases, the application of quantitative real-time PCR (qRT-PCR) is essential. Finally, another crucial application of qRT-PCR is the detection of GMOs and the quantitation of GMO ingredients in commercial food products.

Palmieri et al. (2009) described a qualitative and quantitative qRT-PCR method to detect the presence of berry DNA in various food products. Real-time PCR has also been applied as a tool for the authentication of olive oil (Giménez et al., 2010). The authors also supported that qRT-PCR was useful in quantifying the DNA extracted from oil and, therefore, in evaluating the yields from different methods of extraction. However, they found that the size of the rbcL amplification product used in their method was critical for the success of analysis.

Later, Pafundo et al. (2011) developed a fast SYBR Green DNA melting curve assay for identifying the presence of tomato, carrot, and celery in commercial food. LOD for carrot and tomato DNA was about 1 mg/kg, whereas for celery DNA it was 10 mg/kg.

Recently, a real-time PCR method for the detection of traces of black (*Brassica nigra*) and brown mustard (*Brassica juncea*) in food products has been described (Palle-Reisch et al., 2013). The *B. nigra* partial RT gene was selected for reverse transcriptase from gypsy-like retroelement 13G42-26. With the exception of white mustard, this technique does not demonstrate any cross-reactivity with other *Brassicaceae* species. Low cross-reactivities were observed with cinnamon, ginger, rye, cumin, fenugreek, and turmeric, but can be overlooked since these substances are normally present in very low quantities in food products containing mustard. The LOD of this method reached 0.1 pg when using serially diluted DNA extracts from black and brown mustard. The cross-reactivity might pose a major drawback since the method could detect even 0.1 pg of *Brassicaceae*.

For the detection of wheat, Mujico et al. (2011) selected a fragment of 51 bp from the intergenic region limited by genes that encode the 25S and 18S wheat ribosomal RNAs.

The qRT-PCR applied to this region was capable of quantifying wheat contaminations in raw materials and complex food products. The sensitivity of the method was found to be comparable to the commercially available enzyme-linked immunosorbent assays (ELISAs) for prolamin analysis.

qRT-PCR has also been used for the detection of allergens like nuts in food products. Despite their healthy attributes, some tree nuts have been accused of being responsible for allergenic reactions in individuals. For this purpose, Pafundo et al. (2009) developed two real-time PCR assays, based on melting curve analysis, in order to detect almond species in biscuits. The Pru 1 and rbcL genes were selected and the sensitivity test proved that almond genomic and plasmid DNAs could be detected down to one copy.

8.5.4.3 Real-Time PCR Using TaqMan Probes

TaqMan probes are defined as hydrolysis probes that are developed to improve the specificity of real-time PCR assays (Holland et al., 1991). The key feature of TaqMan probes is that they produce a fluorescent signal only when they are specifically hybridized to the target DNA sequence.

Several qRT-PCR methods based on TaqMan probes have been described so far for the detection and identification of wheat, barley, and/or rye DNA (Hernandez et al., 2005; Zeltner et al., 2009). The use of a dual-labeled probe in a qRT-PCR targeting a specific microsatellite sequence for the D genomic region of wheat has been described (Sonnante et al., 2009). Other researchers illustrated a TaqMan PCR method for the detection of plant species in honey (Laube et al., 2010).

TaqMan probes PCR methods have been excessively used for the identification or detection of nuts in food products. A novel and highly sensitive TaqMan probe PCR assay was described for the detection of Brazil nut (Röder et al., 2010). The LOD of this method was 5 mg/kg of Brazilian nut. Several other assays based on TaqMan probes have been reported for the quantitative detection of tree nuts such as Brazilian nuts, pecan, macadamia, and walnuts in various food products (Bren et al., 2010; Brezna et al., 2006; Brezna and Kuchta, 2008; Brezna et al., 2009).

Recently, López-Calleja et al. (2013) described a TaqMan PCR method for the detection of peanut in 123 commercial food products. Two genetic markers in the Arah2 (125 bp) and ITS1 (90 bp) gene were used in this assay along with TaqMan probes. Detection levels of 10 and 0.1 ppm were obtained with the use of Arah2 and ITS primers, respectively, on a sensitivity test using mixtures spiked with different concentrations. Overall, TaqMan probes proved very sensitive and specific, yet the need for the development and synthesis of different dual-labeled probes for particular target sequences increases the assay setup and costs and limits their use.

8.5.4.4 High Resolution Melting (HRM) Analysis

A new DNA based-method that has been applied for tracing food products is HRM, which measures the rate of double stranded DNA dissociation to single stranded DNA with increasing temperature (Reed and Wittwer, 2004). The high resolution and precision of the instruments used in HRM analysis and the development of saturating DNA dyes are able to produce more accurate melting profiles, permitting its use for genotyping at variety level (single nucleotide polymorphism (SNP), simple sequence repeats (SSR) markers) (Ganopoulos et al., 2012; Mader et al., 2008) and for identification and quantification of adulterants in various food products (Bosmali et al., 2012; Faria et al., 2013; Ganopoulos et al., 2013; Mader et al., 2011; Madesis et al., 2012).

Numerous applications of HRM analysis have been described recently. HRM analysis was used for the first time on SSR fragments by Ganopoulos et al. (2011a) in an authenticity assay for processed sweet cherry products. The same HRM assay demonstrated its potential as a sensitive and reliable diagnostic tool for the simultaneous detection and quantification of rice food product adulterants using microsatellite and fragrance typing with HRM analysis (Ganopoulos et al., 2011b). The use of HRM analysis coupled to the plant DNA barcoding regions (trnL and rpoC1) allowed the detection and authentication of PDO Fava Santorinis (Ganopoulos et al., 2012). This method was proven to be very sensitive, allowing the detection of non-Fava Santorinis adulterants in Fava Santorinis commercial products in ratios as low as 1:100. HRM analysis, based on microsatellites combined with the use of species specific rpoC1 chloroplast region as a marker, was also applied for a quantitative detection of lentil adulterants in species and variety level (Bosmali et al., 2012). Madesis et al. (2012) described the application of the barcode DNA high resolution melting (Bar-HRM) analysis method with the universal plant DNA barcoding region trnL that allowed the identification, adulteration, and quantification of the major Greek and Mediterranean bean species. The LOD of this method was as low as 1%.

HRM is a very promising technique due to its low cost and the decreased time involved in the analysis of the results, especially when compared to the sequence-based species discrimination methods. Moreover, the use of small amplification products allows the analysis of processed food. One minor disadvantage is the need to obtain a melting curve database to serve as a standard when one needs to identify unknown samples. Thus, it is of great importance to develop methods and/or techniques to allow the curves produced by the HRM analysis to be transferred across different laboratories. However, the relatively low cost involved permits the simultaneous running of unknown samples and selected standards in each analysis (Madesis et al., 2014).

8.5.4.5 Capillary Electrophoresis (CE)

Over the last few years, several CE-based methods have been proposed for the detection of GMOs and for the authenticity evaluation of food products. These assays were based on specific DNA markers, and CE analysis was selected for their detection (Madesis et al., 2014).

Rodríguez-Plaza et al. (2006) applied two microsatellite markers (VVMD5 and ZAG79) together with CE analysis and laser induced fluorescence (CGE-LIF) in order to identify the grape (*Vitis vinifera*) variety of musts. Following a similar approach, Ganopoulos et al. (2011a) developed an analytical method for the detection of *Tragana Edessis* sweet cherry, in both fresh and processed fruit products using six microsatellite markers. Vemireddy et al. (2007) described a multiplex microsatellite marker technique for the detection of adulteration in Basmati rice (*Oryza sativa*). The researchers used eight microsatellite loci with labeled primers for amplification by multiplex PCR. These loci were detected by CE with the intention of discriminating among the different Basmati commercial products, evolved Basmati and aromatic non-Basmati long-grain rice varieties in mixtures. The reliability of the method in quantifying adulteration was also tested using data obtained from the co-amplified polymorphic targets in rice mixtures.

Another application of CE in wine authenticity based on the use of SNPs has been proposed for the detection of Cabernet Sauvignon grapevine variety in musts and wines derived from Agiorgitiko variety (Spaniolas et al., 2008a). This method was based on PCR amplification of the sequence containing the polymorphism, followed by the digestion of the PCR product with a specific restriction enzyme and CE analysis. Samples of

Agiorgitiko containing the Cabernet variety down to a level of 10% adulteration could be detected by this method.

In addition, Spaniolas et al. (2008b) developed a CE method that detected chloroplast polymorphisms for the authentication of plant oils. Universal primers were used to amplify different lengths of plastid DNA sequences from each plant species. Two years later, Spaniolas et al. (2010) described another PCR-based method that could differentiate a range of 11 plant species commonly used for oil extraction. This method was also based on polymorphisms of the plastid genome that were detected by the use of CE on the basis of different lengths of their corresponding PCR amplicons. CE method's major disadvantage is the need for gel electrophoresis and, thus, the use of UV luminescence and ethidium bromide. Moreover, CE methods have a high risk of contamination which is also a major problem.

8.5.4.6 Microarrays

DNA microarrays, compared to the other DNA-based methods, have the advantage of identifying simultaneously up to hundreds of thousands of species (Teletchea et al., 2005). Zhu et al. (2008) developed a DNA microarray (PNX array) method for the identification of several *Panax* (ginseng) plant species based on the nucleotide differences among these plants. The authors used 35 kinds of specific oligonucleotides as probes (33 probes corresponded to the species-specific nucleotide substitutions, and two probes were used as positive and negative controls). A partial 18S rRNA gene from *Panax* plants and drugs and also from their derived health foods was used to hybridize to the PNX array. Specific fluorescent patterns were produced for each *Panax* species, which could serve as barcode patterns for quick distinction. Another method for the identification of ginseng species was described by Niu et al. (2011) who proposed a subtracted diversity array (SDA) method for the differentiation of two ginseng species, *Panax ginseng* and *Panax quinquefolius. P. quinquefolius* is frequently used for the adulteration of *P. ginseng,* and Niu et al. (2011) suggested that they could detect an adulteration down to 10% which was not as low as for the PCR methods.

Another application of microarrays was used for the differentiation of several cereal species and varieties in a complex cereal mixture (Prins et al., 2008). The author presented a padlock-microarray system, which targeted a relatively short sequence of approximately 45 nucleotides from the D genome.

8.5.4.7 Next Generation Sequencing (NGS)

Recent advances in sequencing technologies and the continuous development of new sequencing platforms are expected to give a boost, in the near future, to the detection methods for the traceability of food products. The possibility to deep sequence entire genomes provides the scientific community with new DNA markers able to distinguish accurately the raw materials in processed food where DNA or at least small DNA fragments are still present. The high throughput of the NGS technologies permits the preparation of DNA samples from different extracts to be marked with different DNA tags, mixed and processed all together. Thus, it will be possible to analyze, in parallel, a very high number of samples, thus lowering the analysis cost (Madesis et al., 2014).

Despite their advantages, NGS technologies, due to the high cost of equipment and analysis have been applied in limited cases, but as the cost diminishes it will allow for massive use of the method. A NGS (pyrosequencing) method has been described for the identification of fruit species in yogurts (Ortola-Vidal et al., 2007). The presence of 2% w/w rhubarb in raspberry yogurt was detected by this method.

8.6 CONCLUSION

This chapter reviews different fingerprinting techniques used in food authentication. The result shows that the fingerprinting techniques have a wide adoption in food detection, covering most of the food testing aspects. It also suggests that fingerprinting technologies have their advantages in food detection: accurate, good reproducibility, low detection limit, rapid analysis, and so on. Fingerprinting technique has become an important technology in food detection and as it is developing rapidly, the applications of fingerprinting technique would be more extended. Spectroscopic and chromatographic fingerprinting techniques in food authentication are absolutely suitable and valid for working in-house when authentication of food samples is required. However, when transferability of the techniques is needed, for acceptance by the authorities, producers, and consumers, the use of food reference material is essential, for instance; then the fingerprinting signal could be normalized between the equipment. In addition, compositional information on significant chemical markers could be used as a second step for confirmation of the authenticity as well as for transferability. Molecular authentication of fresh and processed food based on DNA-based fingerprinting techniques provides an important contribution to the identification of genetic material since, in principle, DNA can be extracted from any fresh and processed food products. Finally, the application of DNA-based fingerprinting techniques for the authentication of food components can improve the transparency of food production systems, helping honest producers in the protection of their food products.

REFERENCES

Aung, M.M., Chang, Y.S. 2014. Traceability in a food supply chain: safety and quality perspectives. Food Control 39:172–84.

Biais, B., Allwood, J.W., Deborde, C., et al. 2009. ^{1}H NMR, GC-EI-TOFMS, and data set correlation for fruit metabolomics: application to spatial metabolite analysis in melon. Anal Chem 81(8):2884–94.

Birkel, E., Rodriguez-Saona, L. 2011. Application of a portable handheld infrared spectrometer for quantitation of trans fat in edible oils. J Am Oil Chem Soc 88(10):1477–83.

Bosmali, I., Ganopoulos, I., Madesis, P., Tsaftaris, A. 2012. Microsatellite and DNA-barcode regions typing combinedwith High ResolutionMelting (HRM) analysis for food forensic uses: A case study on lentils (Lens culinaris). Food Res Int 46(1):141–7.

Bosque-Sendra, J.M., Cuadros-Rodríguez, L., Ruiz-Samblás, C., de la Mata, A.P. 2012. Combining chromatography and chemometrics for the characterization and authentication of fats and oils from triacylglycerol compositional data-a review. Anal Chim Acta 724:1–11.

Bothwell, J.H., Griffin, J.L. 2011. An introduction to biological nuclear magnetic resonance spectroscopy. Biol Rev Camb Philos Soc 86(2):493–510.

Bren, B., Dudov, H., Kuchta, T. 2010. A novel real-time polymerase chain reaction method for the detection of Brazil nuts in food. J AOAC Int 93(1):197–201.

Brezna, B., Hudecová, L., Kuchta, T. 2006. A novel real-time polymerase chain reaction (PCR) method for the detection of walnuts in food. Eur Food Res Tech 223(3):373–7.

Brezna, B., Kuchta, T. 2008. A novel real-time polymerase chain reaction method for the detection of pecan nuts in food. Eur Food Res Tech 226(5):1113–8.

Brezna, B., Piknová, L., Kuchta, T. 2009. A novel real-time polymerase chain reaction method for the detection of macadamia nuts in food. Eur Food Res Tech 229(3):397–401.

Buning-Pfaue, H. 2003. Analysis of water in food by near infrared spectroscopy. Food Chem 82(1):107–15.

Cordella, C., Faucon, J.P., Cabrol-Bass, D., Sbirrazzuoli, N. 2003. Application of DSC as a tool for honey floral species characterization and adulteration detection. J Therm Anal Calorim 71(1): 279–290.

Cozzolino, D. 2011. Infrared methods for high throughput screening of metabolites: Food and medical applications. Comb Chem High Throughput Screen 14(2):125–31.

Cubero-Leon, E., Peñalver, R., Maquet, A. 2014. Review on metabolomics for food authentication. Food Res Int 60:95–107.

Dahinden, I., von Büren, M., Lüthy, J. 2001. A quantitative competitive PCR system to detect contamination of wheat, barley or rye in gluten-free food for coeliac patients. Eur Food Res Tech 212(2):228–33.

Danezis, G.P., Tsagkaris, A.S., Camin, F., Brusic, V., Georgiou, C.A. 2016. Food authentication: Techniques, trends & emerging approaches. TrAC Trends Anal Chem 85:123–32.

Dennis, M.J. (1998). Recent developments in food authentication. Analyst 123(9):151–56.

Dhanya, K., Syamkumar, S., Siju, S., Sasikumar, B. 2011. SCAR markers for adulterant detection in ground chilli. British Food J 113(5):656–68.

Di Bernardo, G., Del Gaudio, S., Galderisi, U., Cascino, A., Cipollaro, M. 2007. Comparative evaluation of different DNA extraction procedures from food samples. Biotechnol Prog 23(2):297–301.

Drivelos, S.A., Georgiou, C.A. 2012. Multi-element and multi-isotope-ratio analysis to determine the geographical origin of foods in the European Union. Trends Anal Chem 40:38–51.

Dunn, W.B. 2008. Current trends and future requirements for the mass spectrometric investigation of microbial, mammalian and plant metabolomes. Phys Biol 5(1):011001.

Dunn, W.B., Broadhurst, D., Ellis, D.I., et al. 2008. A GC-TOF-MS study of the stability of serum and urine metabolomes during the UK Biobank sample collection and preparation protocols. Int J Epidemiol 37(1):23–30.

Dunn, W.B., Broadhurst, D.I., Atherton, H.J., Goodacre, R., Griffin, J.L. 2011. Systems level studies of mammalian metabolomes: the roles of mass spectrometry and nuclear magnetic resonance spectroscopy. Chem Soc Rev 40(1):387–426.

Dunn, W.B., Ellis, D.I. 2005. Metabolomics: Current analytical platforms and methodologies. TrAC Trends Anal Chem 24(4):285–94.

Ellis, D.I., Dunn, W.B., Griffin, J.L., Allwood, J.W., Goodacre, R. 2007. Metabolic fingerprinting as a diagnostic tool. Pharmacogenomics 8(9):1243–66.

Ellis, D.I., Goodacre, R. 2006. Metabolic fingerprinting in disease diagnosis: biomedical applications of infrared and Raman spectroscopy. Analyst 131(8):875–85.

Elmer, P. 2013. Guide to selection of DSC sample pans. Technical Note. Waltham, MA, USA.

Fantoni, L., Simoneau, C. 2003. European survey of contamination of homogenized baby food by epoxidized soybean oil migration from plasticized PVC gaskets. Food Addit Contam 20(11):1087–96.

Faria, M.A., Magalhaes, A., Nunes, M.E., Oliveira, M. 2013. High resolution melting of trnL amplicons in fruit juices authentication. Food Control 33(1):136–41.

Gallo, M., Ferranti, P. 2016. The evolution of analytical chemistry methods in foodomics. J Chromatogr A 1428: 3–15.

Ganopoulos, I., Argiriou, A., Tsaftaris, A. 2011a. Microsatellite high resolution melting (SSR-HRM) analysis for authenticity testing of protected designation of origin (PDO) sweet cherry products. Food Control 22(3–4):532–41.

Ganopoulos, I., Argiriou, A., Tsaftaris, A. 2011b. Adulterations in Basmati rice detected quantitatively by combined use ofmicrosatellite and fragrance typing with High Resolution Melting (HRM) analysis. Food Chem 129(2):652–9.

Ganopoulos, I., Bazakos, C., Madesis, P., Kalaitzis, P., Tsaftaris, A. 2013. Barcode-DNA High Resolution Melting (Bar-HRM) analysis as a novel closetubed and accurate tool for olive oil forensic use. J Sci Food Agric 93(9):2281–6.

Ganopoulos, I., Bosmali, I., Madesis, P., Tsaftaris, A. 2012. Microsatellite genotyping with HRM (High Resolution Melting) analysis for identification of the PGI common bean variety Plake Megalosperma Prespon. Eur Food Res Tech 234(3):501–8.

Georgiou, C.A., Danezis, G.P. 2015. Elemental and isotopic mass spectrometry. In: Advanced mass spectrometry for food safety and quality, Ed. Y. Pico, Elsevier.

Gil, A.M., Duarte, I.F., Godejohann, M., Braumann, U., Maraschin, M. Spraul, M. 2003. Characterization of the aromatic composition of some liquid foods by nuclear magnetic resonance spectrometry and liquid chromatography with nuclear magnetic resonance and mass spectrometric detection. Anal Chim Acta 488(1):35–51.

Giménez, M.J., Pistón, F., Martína, A., Atienza, S.G. 2010. Application of real-time PCR on the development of molecular markers and to evaluate critical aspects for olive oil authentication. Food Chem 118(2):482–7.

Golman, K., in't Zandt, R., Thaning, M. 2006. Real-time metabolic imaging. Proc Natl Acad Sci USA 103(30):11270–75.

Goodacre, R., Timmins, E.M., Burton, R., et al. 1998. Rapid identification of urinary tract infection bacteria using hyperspectral whole-organism fingerprinting and artificial neural networks. Microbiology 144(5):1157–70.

Guillou, C., Remaud, G., Martin, G.J. 1992. Applications of NMR to the characterization and authentication of foods and beverages. Trends Food Sci Technol 3:197–201.

Han, J., Wu, Y., Huang, W., et al. 2012. PCR and DHPLC methods used to detect juice ingredient from 7 fruits. Food Control 25(2):696–703.

Handy, S.M., Parks, M.B., Deeds, J.R., et al. 2011. Use of the chloroplast gene ycf1 for the genetic differentiation of pine nuts obtained from consumers experiencing dysgeusia. J Agric Food Chem 59(20):10995–11002.

He, L., Kim, N.J., Li, H., Hu, Z., Lin, M. 2008. Use of a fractal-like gold nanostructure in surface-enhanced raman spectroscopy for detection of selected food contaminants. J Agric Food Chem 56(21):9843–7.

Heinzmann, S.S., Brown, I.J., Chan, Q., et al. 2010. Metabolic profiling strategy for discovery of nutritional biomarkers: proline betaine as a marker of citrus consumption. Am J Clin Nutr 92(2):436–43.

Hernández, A., Aranda, E., Martín, A., et al. 2010. Efficiency of DNA typing methods for detection of smoked paprika "Pimenton de la Vera" adulteration used in the elaboration of dry-cured Iberian pork sausages. J Agric Food Chem 58(22):11688–94.

Hernandez, M., Esteve, T., Pla, M. 2005. Real-time polymerase chain reaction based assays for quantitative detection of barley, rice, sunflower, and wheat. J Agric Food Chem 53(18):7003–9.

Herrero, A.M. 2008. Raman spectroscopy a promising technique for quality assessment of meat and fish: A review. Food Chem 107(4):1642–51.

Holland, P.M., Abramson, R.D., Watson, R., Gelfand, D.H. 1991. Detection of specific polymerase chain reaction product by utilizing the 5′–3′ exonuclease activity of Thermus aquaticus DNA polymerase. Proc Natl Acad Sci U S A 88(16):7276–80.

Huang, H., Yu, H., Xu, H., Ying, Y. 2008. Near infrared spectroscopy for on/in-line monitoring of quality in foods and beverages: A review. J Food Eng 87(3):303–13.

Hyotylainen, T., Oresic, M. 2013. Data handling. In: Chromatographics methods in metabolomics, Ed. T. Hyotylainen and S. Wiedmer, 183–94. Royal Society of Chemistry, Cambridge.

Kaufman, L.D., Martinez, M.I., Serrano, J.M., Gomez-Reino, J.J. 1995. 12-year followup study of epidemic Spanish toxic oil syndrome. J Rheumatol 22(2):282–88.

Kempf, M., Beuerle, T., Bühringer, M., et al. 2008. Pyrrolizidine alkaloids in honey: risk analysis by gas chromatography-mass spectrometry. Mol Nutr Food Res 52(10):1193–200.

Köchl, S., Niederstätter, H., Parson, W. 2005. DNA extraction and quantitation of forensic samples using the phenol-chloroform method and real-time PCR. Forensic DNA Typing Protocols. Springer.

Laube, I., Hird, H., Brodmann, P., et al. 2010. Development of primer and probe sets for the detection of plant species in honey. Food Chem 118(4):979–86.

Li, W., Herrman, T.J., Dai, S.Y. 2010. Rapid determination of fumonisins in corn-based products by liquid chromatography/tandem mass spectrometry. J AOAC Int 93(5):1472–81.

Li, Y., Gao, H.P., Yao, A. 2009b. Detection on the residues of malachite green and its metabolite in eel by high performance liquid chromatography. J Anhui Agri Sci 37(1):31–33.

Li, Y., Zheng, F., Wang, M., Pang, G. 2009a. Rapid screening and confirmation of 156 pesticide residues in concentrated fruit and vegetable juices using liquid chromatography-tandem mass spectrometry. Chinese J Chromatography 27(2):127–37.

Li-Chan, E.C.Y. 1996. The applications of Raman spectroscopy in food science. Trends Food Sci Technol 7(11):361–70.

Lis-Balchin, M., Deans, S.G., Hart, S. 1997. A study of the variability of commercial peppermint oils using antimicrobial and pharmacological parameters. Med Sci Res 25:151–52.

Liu, Z.J., Cordes, J.F. 2004. DNA marker technologies and their applications in aquaculture genetics. Aquaculture 238(1–4):1–37.

López-Calleja, I.M., de la Cruz, S., Pegels, N., et al. 2013. Development of a real-time PCR assay for detection of allergenic trace amounts of peanut (Arachis hypogaea) in processed foods. Food Control 30(2):480–90.

Lu, X., Al-Qadiri, H.M., Lin, M., Rasco, B.A. 2011. Application of mid-infrared and raman spectroscopy to the study of bacteria. Food and Bioprocess Technology 4(6): 919–35.

Lutter, P., Savoy-Perroud, M.C., Campos-Gimenez, E., et al. 2011. Screening and confirmatory methods for the determination of melamine in cow's milk and milk-based powdered infant formula: Validation and proficiency-tests of ELISA, HPLC-UV, GC-MS and LC-MS/MS. Food Control 22(6):903–13.

MacKenzie, W.M., Aylott, R.I. 2004. Analytical strategies to confirm Scotch whisky authenticity. Part II: Mobile brand authentication. Analyst 129(7):607–12.

Mader, E., Lukas, B., Novak, J. 2008. A strategy to setup codominant microsatellite analysis for high-resolution-melting-curve-analysis (HRM). BMC Genet 9:69.

Mader, E., Ruzicka, J., Schmiderer, C., Novak, J. 2011. Quantitative high-resolution melting analysis for detecting adulterations. Anal Biochem 409(1):153–5.

Madesis, P., Ganopoulos, I., Anagnostis, A., Tsaftaris, A. 2012. The application of Bar-HRM (Barcode DNA-High Resolution Melting) analysis for authenticity testing and quantitative detection of bean crops (Leguminosae) without prior DNA purification. Food Control 25(2):576–82.

Madesis, P., Ganopoulos, I., Sakaridis, I., Argiriou, A., Tsaftaris, A. 2014. Advances of DNA-based methods for tracing the botanical origin of food products. Food Res Int 60:163–72.

Marieschi, M., Torelli, A., Bianchi, A., Bruni, R. 2011a. Detecting Satureja montana L. and Origanum majorana L. bymeans of SCARePCR in commercial samples of Mediterranean oregano. Food Control 22(3–4):542–48.
Marieschi, M., Torelli, A., Bianchi, A., Bruni, R. 2011b. Development of a SCAR marker for the identification of *Olea europaea* L.: A newly detected adulterant in commercial Mediterranean oregano. Food Chem 126(2):705–9.
Marieschi, M., Torelli, A., Bruni, R. 2012. Quality control of saffron (*Crocus sativus* L.): Development of SCAR markers for the detection of plant adulterants used as bulking agents. J Agric Food Chem 60(44):10998–11004.
Marikkar, J.M.N., Ghazali, H.M., Man, Y.B.C., Lai, O.M. 2002. The use of cooling and heating thermograms for monitoring of tallow, lard and chicken fat adulterations in canola oil. Food Res Int 35(10):1007–14.
Marikkar, J.M.N., Ghazali, H.M., Man, Y.B.C., Lai, O.M. 2003. Differential scanning calorimetric analysis for determination of some animal fats as adulterants in palm olein. J Food Lipids 10(1):63–79.
Martellos, S., Labra, M. 2013. DNA barcoding as a new tool for food traceability. Food Res Int 50:55–63.
Martins-Lopes, P., Gomes, S., Pereira, L., Guedes-Pinto, H. 2013. Molecular markers for food traceability. Extraction 28(29):31–37.
Mayer, F., Haase, I., Graubner, A., et al. 2012. The use of polymorphisms in the gamma-gliadin gene of spelt and wheat as tool for authenticity control. J Agric Food Chem 60(6):1350–7.
Mujico, J.R., Lombardía, M., Mena, M.C., Méndez, E., Albar, J.P. 2011. A highly sensitive real-time PCR system for quantification of wheat contamination in gluten-free food for celiac patients. Food Chem 128(3):795–801.
Nicolaou, N., Xu, Y., Goodacre, R. 2011. MALDI-MS and multivariate analysis for the detection and quantification of different milk species. Anal Bioanal Chem 399(10):3491–502.
Nielen, M.W.F., Hooijerink, H., Zomer, P., Mol, J.G.J. 2011. Desorption electrospray ionization mass spectrometry in the analysis of chemical food contaminants in food. TrAC Trends Anal Chem 30(2):165–80.
Niu, L., Mantri, N., Li, C.G., et al. 2011. Detection of Panax quinquefolius in Panax ginseng using 'subtracted diversity array'. J Sci Food Agric 91(7):1310–15.
Nur Azira, T., Amin, I. 2016. Advances in differential scanning calorimetry for food authenticity testing. In: *Advances in food authenticity testing*. Ed. G. Downey, 311–35. Woodhead Publishing group, Elsevier.
Nurrulhidayah, A.F., Arieff, S.R., Rohman, A., Amin, I., Shuhaimi, M., Khatib, A. 2015. Detection of butter adulteration with lard using differential scanning calorimetry. Int Food Res J 22(2):832–9.
Olivieri, C., Marota, I., Rollo, F., Luciani, S. 2012. Tracking plant, fungal, and bacterial DNA in honey specimens. vol. 57. Blackwell Publishing Ltd.
Ortola-Vidal, A., Schnerr, H., Rojmyr, M., Lysholm, F., Knight, A. 2007. Quantitative identification of plant genera in food products using PCR and Pyrosequencing® technology. Food Control 18(8):921–7.
Pafundo, S., Gullì, M., Marmiroli, N. 2009. SYBR® GreenER™ real-time PCR to detect almond in traces in processed food. Food Chem 116(3):811–5.
Pafundo, S., Gullì, M., Marmiroli, N. 2011. Comparison of DNA extraction methods and development of duplex PCR and real-time PCR to detect tomato, carrot, and celery in food. J Agric Food Chem 59(19):10414–24.

Palle-Reisch, M., Wolny, M., Cichna-Markl, M., Hochegger, R. 2013. Development and validation of a real-time PCR method for the simultaneous detection of black mustard (*Brassica nigra*) and brown mustard (*Brassica juncea*) in food. Food Chem 138(1):348–55.

Palmieri, L., Bozza, E., Giongo, L. 2009. Soft fruit traceability in food matrices using real-time PCR. Nutrients 1(2):316–28.

Posudin, Y.I., Peiris, K.S., Kays, S.J. 2015. Non-destructive detection of food adulteration to guarantee human health and safety. Ukrainian Food J 4(2):207–60.

Prins, T.W., van Dijk, J.P., Angeline Van Hoef, A.M., et al. 2008. Towards amultiplex cereal traceability tool using padlock probe ligation on genomic DNA. Food Chem 118(4):966–73.

Qin, J., Chao, K., Kim, M.S. 2010. Raman chemical imaging system for food safety and quality inspection. Trans ASABE 53(6):1873–82.

Reed, G.H., Wittwer, C.T. 2004. Sensitivity and specificity of single-nucleotide polymorphism scanning by high-resolution melting analysis. Clin Chem 50(10):1748–54.

Reid, L.M., O'Donnell, C.P., Downey, G. 2006. Recent technological advances for the determination of food authenticity. Trends Food Sci Technol 17:344–53.

Reinholds, I., Bartkevics, V., Silvis, I.C.J., van Ruth, S.M., Esslinger, S. 2015. Analytical techniques combined with chemometrics for authentication and determination of contaminants in condiments: A review. J Food Compos Anal 44:56–72.

Robinson, A.M., Harroun, S.G., Bergman, J., Brosseau, C.L. 2012. Portable electrochemical surface-enhanced Raman spectroscopy system for routine spectroelectrochemical analysis. Anal Chem 84(3):1760–64.

Röder, M., Filbert, H., Holzhauser, T. 2010. A novel, sensitive and specific real-time PCR for the detection of traces of allergenic Brazil nut (*Bertholletia excelsa*) in processed foods. Anal Bioanal Chem 398(5):2279–88.

Rodríguez-Plaza, P., González, R., Moreno-Arribas, M.V., et al. 2006. Combining microsatellite markers and capillary gel electrophoresis with laser-induced fluorescence to identify the grape (Vitis vinifera) variety of musts. Eur Food Res Tech 223(5):625–31.

Sakai, Y., Ishihata, K., Nakano, S., et al. 2010. Specific detection of banana residue in processed foods using polymerase chain reaction. J Agric Food Chem 58(14):8145–51.

She, Y.X., Liu, J.J., Wang, J. 2008. Determination of 24 sulfonamides in milk by ultra performance liquid chromatography combined with Quadrupole mass spectrometry. J Instrum Anal 27(12):1313–17.

Smedsgaard, J., Nielsen, J. 2005. Metabolite profiling of fungi and yeast: From phenotype to metabolome by MS and informatics. J Exp Bot 56(410):273–86.

Soares, S., Mafra, I., Amaral, J.S., Oliveira, M.B.P.P. 2010. A PCR assay to detect trace amounts of soybean in meat sausages. Int J Food Sci Tech 45(12):2581–88.

Sonnante, G., Montemurro, C., Morgese, A., et al. 2009. DNA microsatellite region for a reliable quantification of soft wheat adulteration in durum wheat-based foodstuffs by real-time PCR. J Agric Food Chem 57(21):10199–204.

Sowoidnich, K., Schmidt, H., Maiwald, M., Sumpf, B., Kronfeldt, H.D. 2010. Application of diode-laser Raman spectroscopy for in situ investigation of meat spoilage. Food Bioprocess Technol 3:878–82.

Spaniolas, S., Bazakos, C., Awad, M., Kalaitzis, P. 2008b. Exploitation of the chloroplast trnL (UAA) intron polymorphisms for the authentication of plant oils by means of a lab-on-a-chip capillary electrophoresis system. J Agric Food Chem 56(16):6886–91.

Spaniolas, S., Bazakos, C., Spano, T., Zoghby, C., Kalaitzis, P. 2010. The potential of plastid trnL (UAA) intron polymorphisms for the identification of the botanical origin of plant oils. Food Chem 122(3):850–6.

Spaniolas, S., Tsachaki, M., Bennett, M. J., Tucker, G.A. 2008a. Toward the authentication of wines of Nemea denomination of origin through cleaved amplified polymorphic sequence (CAPS)-based assay. J Agric Food Chem 56(17):7667–71.

Teletchea, F., Maudet, C., Hanni, C. 2005. Food and forensic molecular identification: Update and challenges. Trends Biotechnol 23(7):359–66.

Tian, H.X., Wang, Z., Xu, S.Y. 2006. Separation and identification of volatile flavors of Jinhua Ham by gas chromatography-mass spectrometry coupled with head space solid phase microextraction. Chinese J Chromatography 24(2):177–80.

Unseld, M., Beyermann, B., Brandt, P., Hiesel, R. 1995. Identification of the species origin of highly processed meat products by mitochondrial DNA sequences. Genome Res 4(4):241–43.

Vemireddy, L.R., Archak, S., Nagaraju, J. 2007. Capillary electrophoresis is essential for microsatellite marker based detection and quantification of adulteration of Basmati rice (*Oryza sativa*). J Agric Food Chem 55(20):8112–7.

Viaene, J., Goodarzi, M., Dejaegher, B., et al. 2015. Discrimination and classification techniques applied on Mallotus and Phyllanthus high performance liquid chromatography fingerprints. Anal Chim Acta 877:41–50.

Wang, D.J., Zhou, X.Y., Jin, T.M., Hu, X.N., Zhong, J.E., Wu, Q.T. 2004. Application of near-infrared spectroscopy to agriculture and food analysis. Spectrosc Spect Anal 24:447–50.

Wax, P.M. 1995. Elixirs, diluents, and the passage of the 1938 Federal Food, Drug and Cosmetic Act. Ann Intern Med 122:456–60.

Winder, C.L., Gordon, S.V., Dale, J., Hewinson, R.G., Goodacre, R. 2006. Metabolic fingerprints of *Mycobacterium bovis* cluster with molecular type: implications for genotype-phenotype links. Microbiology 152(9):2757–65.

Wishart, D.S. 2008. Metabolomics: applications to food science and nutrition research. Trends Food Sci Technol 19(9):482–93.

Woodcock, T., Downey, G., O'Donnell, C.P. 2008. Better quality food and beverages: The role of near infrared spectroscopy. J Near Infrared Spectrosc 16(1):1–29.

Yingliang, W., Xiaowei, L., Suying, L., Jianzhong, S., Ting, Y. 2006. Determination of residual clenbuterol and ractopamine in animal liver tissues by solid phase extraction with gas chromatography-mass spectrometry. Chinese J Anal Chem 34(8):1083–86.

Zeltner, D., Glomb, M., Maede, D. 2009. Real-time PCR systems for the detection of the gluten-containing cereals wheat, spelt, kamut, rye, barley and oat. Eur Food Res Tech 228(3):321–30.

Zhan, J., Yu, X.J., Huang, W. 2008. A simple method for the simultaneous determination of 7 food additives in foods with complex matrices by GC/MS. J Anal Sci 24(6):729–31.

Zhang, J., Zhang, X., Dediu, L., Victor, C. 2011. Review of the current application of fingerprinting allowing detection of food adulteration and fraud in China. Food Control 22:1126–35.

Zhang, X., Zheng, B., Chen, X.C. 2009. Simultaneous determination of 4 sulfonamide residues in fishery products by high performance liquid chromatography with fluorescence detection. Food Sci 30(8):235–37.

Zhu, S., Fushimi, H., Komatsu, K. 2008. Development of a DNA microarray for authentication of ginseng drugs based on 18S rRNA gene sequence. J Agric Food Chem 56(11):3953–59.

CHAPTER 9

Introduction to Food Traceability Based on Fingerprinting Techniques

Md. Rizwanullah, Nausheen Khan, and Saima Amin
Jamia Hamdard

Javed Ahmad
Najran University

Syed Sarim Imam
Glocal University

Khalid Umar Fakhri and Mohd. Moshahid Alam Rizvi
Jamia Millia Islamia

CONTENTS

9.1 INTRODUCTION

After a detailed investigation of the true definition of traceability, Olsen and Borit (2013) came up with a compendium of all the definitions saying that traceability is "The ability to access any or all information relating to that which is under consideration, throughout its entire life cycle, by means of recorded identifications."

Historically, food scares have been with human beings for many years. Atkins (2008) has discussed that, in Europe, food scares (especially zoonotic hazards) have been with the society of UK for at least 150 years. Saltini and Akkerman (2012) mentioned that only in Europe foodborne illness affects about 1% of the population (approximately 7 million people) each year. Only in 2011, approximately 16.7% of the population (47.8 million people) got sick in America due to food-related illnesses (Resende-Filho & Hurley, 2012).

In the modern livestock production sector, long distance animal transport is increasing. This in turn has not only increased the potential for infection and spread of diseases among the livestock, but it has also exposed the sector to bioterrorism attacks. For example, in the U.S. cattle production sector, a terrorist attack (infection) at a single point can result in a loss of around 23 million cattle within 8 days (Greger, 2007). These challenges have triggered the importance of animal identification and certification processes. Historically, the issue of animal identification (e.g., by marking their bodies) may be traced back to 3,800 years (Smith, Pendell, Tatum, Belk, & Sofos, 2008). However, it is still an important issue within the agriculture and food supply chain (FSC).

Another risk for food such as contamination with radioactive materials disturbs the FSC. After the release of radioactive materials from damaged nuclear plants due to the earthquake in Japan in 2011 (WHO, 2011), many countries have implemented intensive food control measures in concern to their food trade relationships with Japan, while some countries completely suspended transporting food from Japan. A study in Finland (Orre, 2005) has indicated that, as a proactive strategy, effective training on food logistics is important. Such training should include cleaning warehouses and retails, controlling and transporting non-contaminated food items to retail storage facilities after the radioactive materials have passed through a region.

In addition to the public health risks, food crises also lead to economic crises due to direct and indirect (damage to reputation and brand name) costs of product recalls. The indirect cost dominates the recall cost as the loss of market value and reputation can lead to a total bankruptcy of the brand name (Saltini & Akkerman, 2012). Therefore, traceability is an important component of contemporary supply chains in the production industry in general and in the food sector in particular, as the food sector is sensitive from the human and animal health point of view (Olsen & Aschan, 2010).

The terms "tracing" and "tracking" are generally discussed in traceability. Tracing is the backward process where the origin is identified through the histories or records in supply chain and tracking is the forward process where the end users and trading partners are identified through the locations in the supply chain, but both the terms provide visibility to the supply chain (Van Dorp, 2002). Stefansson and Tilanus (2001) point out that the tracking and tracing systems must be connected to the physical transportation system and information system.

The current traceability systems are characterized by the inability to link food chain records, inaccuracies and errors in records, and delays in obtaining essential data, which are fundamental in case of food disease outbreaks (Badia-Melis et al., 2015). These systems should also address the recalls and withdrawals of nonconsumable products, however there are some recent reports covering the implementation of food assurance systems that do not mention the traceability issue although they are highly related to food traceability, e.g., implementation of good manufacturing practices in a mozzarella cheese processing plant (Dias et al., 2012), the assessment of costs for the implementation of food safety systems in a small dairy plant (Cusato et al., 2014), or implementation of a food safety system in a dairy processing plant (Cusato et al., 2013).

Food recalls are associated not only with information problems but also with logistics problems (McCallum, 2012). The lack of coordinated logistics operations definitely affects the product flow from farm to fork and, consequently, the efficiency of food traceability efforts. Again, the development and implementation of IT-supported food traceability systems (FTS) can improve the operational planning and increase the efficiencies of food logistics processes (Bourlakis & Bourlakis, 2006; Heyder, Theuvsen, & Hollmann-Hespos, 2012; Liao, Chang, & Chang, 2011; Manos & Manikas, 2010; Saltini & Akkerman, 2012; Van Dorp, 2003). Bourlakis and Bourlakis (2006) strongly argued that IT operations should be formulated alongside the logistics operations of food retailers, and if continuous maintenance services are in place, these IT operations can increase the operational efficiency of the retailers.

9.2 DRIVING FORCES FOR FOOD TRACEABILITY

Often, a combination of two or more factories influences the development and implementation of FTSs. Many driving forces behind the development and implementation of FTSs have been identified. These drivers have been categorized into five groups: food safety and quality, regulatory, social, economic, and technological concerns.

9.2.1 Regulatory Concerns

New legislations (introduced to address the safety and quality concerns and resolve ownership disputes) are identified as important driving forces. Many food companies implement FTSs mainly to fulfil these regulatory issues and stay in the market. Recent developments indicate that the need to maintain market power (and protect against the loss of consumer confidence) and handle political pressure to protect consumer welfare are emerging as major drivers that push large retailers to invest in food traceability projects (Bertolini et al., 2006; Heyder et al., 2012; Liao et al., 2011; Resende-Filho & Hurley, 2012).

Nowadays, mandatory food traceability laws are being enforced and EU has realized this by introducing the general food law (GFL) (Kher et al., 2010; Schwägele, 2005). Accordingly, traceability data can be mandatory or optional (Folinas et al., 2006). Mandatory data include lot number, product ID, product description, supplier ID, quantity, unit of measure, and buyer ID. Optional data include supplier's name, contact information, receipt date, country of origin, date of pack, trade unit, transportation vehicle ID, logistics service provider ID, buyer's name, and dispatch date.

9.2.2 Safety and Quality Concerns

In the recent two decades, food traceability has become an important issue due to food crises—such as foot-and-mouth disease, bovine spongiform encephalopathy (BSE), dioxin crisis, avian flu, melamine contamination of milk, and other food safety incidents involving aquatic products as well as food counterfeiting—and due to issues of sustainable production including labor issues (Bertolini et al., 2006; Engelseth, 2009; Hobbs, Bailey, Dickinson, & Haghiri, 2005; Hong et al., 2011; Kelepouris et al., 2007; Liao et al., 2011; Liu, Kerr, & Hobbs, 2012; Salampasis et al., 2012; Van Dorp, 2003; Van Rijswijk, Frewer, Menozzi, & Faioli, 2008; Wognum, Bremmers, Trienekens, Van der Vorst, & Bloemhof, 2011). Food quality and safety crises, in turn, cause significant crises for the economic and marketing relationships at the national and international levels. Liu et al. (2012) reported that, in 2002, EU banned the import of aquatic products from China claiming that residues from veterinary medicines, pesticides, and heavy metals detected in the aquatic food products exceeded EU standards. This might have affected the trade market in China, a country that exports about 3.06 million tons of aquatic products per year with the value of US$ 9.74 billion.

9.2.3 Social Concerns

Increasing consumer confidence in their food, changing lifestyles, increasing consumer incomes, increasing awareness in the society about health are some of the social issues that motivate food companies to implement traceability systems. The improvement in food crises management enables the concerned agencies to build capacities to ensure food safety and security, which in turn strengthens the social and political security of a nation. In regard to the contemporary FTSs, companies should not attempt only to comply with the government rules, but they should also provide adequate information that consumers need to know, such as the variety of the food attributes, country of origin, animal welfare, and genetic engineering related issues (Golan et al., 2004).

9.2.4 Economic Concerns

The economic benefit of traceability systems is considered as a relatively less important factor, as an efficient full chain traceability system is capital- and resource-intensive and requires a significant initial investment. However, better market access, better product prices, and potential government funding were identified as driving forces. For example, in the United States, in addition to controlling the spread of animal diseases, economical motives have also influenced the development of traceability systems in the livestock sector (Golan et al., 2004). Such economic factors are: protecting property (animals) from theft; proving (through traceability documentation) that animals possess valuable attributes such as up-to-date vaccinations, proving that animal welfare provisions are in place so that the animals can fetch higher prices.

9.2.5 Technological Concerns

Effective traceability systems require more complex devices and systems that do not attract the attention of food companies due to the complexities and high costs associated with them. However, the emerging new and cheaper technologies are motivating companies to develop full chain traceability systems that can integrate information at all stages of the supply chain. Especially, the decreasing costs and increasing effectiveness of the new traceability systems that are enhanced by the development of nanotechnology-based traceability devices (Chrysochou, Chryssochoids, & Kehagia, 2009; Karippacheril, Rios, & Srivastava, 2011, pp. 285–310) are expected to highly motivate the food companies to actively participate in the development and implementation of FTSs.

9.3 BENEFITS OF FOOD TRACEABILITY

Kher et al. (2010) studied the views of food risk management professionals (38 experts participated in the study) in Europe and pointed out that all experts who participated in answering the questionnaire agreed that the advantages of FTS implementation far outweighs the disadvantages (e.g., initial investment cost and extra workload). The benefits of effective FTS can be categorized broadly into social benefits (Schwägele, 2005; Wilson & Clarke, 1998), authorities' benefits (Donnelly & Olsen, 2012; McMeekin et al., 2006; Smith et al., 2005), and food companies' benefits (Golan et al., 2004; Kher et al., 2010; Wognum et al., 2011). However, for better explanations, the benefits identified in this chapter are increase in customer satisfaction, improvement in food crises management, improvement in FSCM (food supply chain management), competence development (for companies), technological and scientific contribution, and contribution to agricultural sustainability.

9.3.1 Increase in Customer Satisfaction

Consumer satisfaction is reflected by the increase in consumer confidence in the food available in the market and by the availability of adequate information that helps in making the food choice. Such information can be organized or analyzed using FTSs and can be communicated to customers and other stakeholders. The occurrence of food safety incidents that attract media attention globally can erode consumer confidence. Such reduced consumer trust in food in the market should be improved and for this FTS is an important tool (Bosona & Gebresenbet, 2013).

9.3.2 Improvement in Food Crises Management

Regarding the food safety and quality issue, traceability in the FSC helps to: minimize the production and distribution of unsafe or poor quality products; limit the extent of damage by facilitating product recall activities; establish the extent of liability in cases of food safety failures by identifying the evidence (based on documented information) of negligence or improper production practices; and identify the existence and quantities of genetically modified organisms (GMOs) in the food products (Bosona & Gebresenbet, 2013). Traceability also helps to verify the existence of credence attributes through bookkeeping records that establishes their creation and preservation (Golan et al., 2004). The credence attributes can be content attributes (e.g., calcium content in a glass of enriched orange juice cannot be determined by the consumers by just drinking it) or process attributes (e.g., earth friendly and fair trade cannot be detected by the consumers or by the testing equipment) (Bosona & Gebresenbet, 2013).

The traceability systems enable the accessibility to integrated data throughout the production, storage, distribution, quality control, and selling processes (Schwägele, 2005). As food traceability is mainly concerned with providing safe and quality food products to the consumers, the traceability data shall include all the relevant data concerning: (a) the origins and types of food/feed, ingredients, meat animals (species, authenticity, age); (b) processes in all the stages of FSC (primary production, further processing, testing, storage, distribution, retailing, consumption); and (c) resources used (people, machines, transport equipment). These integrated data and information are very important in food crises management (Bosona & Gebresenbet, 2013).

9.3.3 Improvements in FSCM

One of the major benefits of FTS is its importance for FSCM. Traceability helps to increase the efficiency of FSCM by: (Golan et al., 2004) reducing the costs of supply-related activities, mainly the logistics costs; providing information trail (regarding each product type), starting from raw agricultural inputs to products in the retailer; differentiating food with undetectable attributes; and enabling firms to manage their resources efficiently using new traceability information systems. The improvement in FSCM enables the FSC partners to increase cooperation among them and develop their technical and economic competence (Bosona & Gebresenbet, 2013).

9.3.4 Competence Development

Effective FTS can be a source of competitive advantage for the FSC partners (Alfaro & Rábade, 2009; Xiaoshuan, Jian, Feng, Zetian, & Weisong, 2010) because it: (i) enables to solve food safety problems; (ii) provides a good faith legal defense in product liability cases; (iii) enables a company to understand its logistics system well; iv) provides promotional advantages by connecting the manufacturer with the consumers (Hall, 2010); and v) enables to develop products of better quality in the long run using the laboratory based-test results and availability of traceability information. In general, food producers who have efficient production systems and distribution channels are competent in the market.

9.3.5 Technological and Scientific Contributions

The increasing implementation of FTSs promotes researches in FSC, because designing an effective FTS requires adequate knowledge about the production supply channel and addressing the four pillars of food traceability (Regattieri, Gamberi, & Manzini, 2007). These pillars are: product identification (volume, weight, perishability); data to trace (degree of detail, dynamism, data storage); product routing (production activities, equipment, storage systems); and traceability tools (accuracy and reliability of captured data). The new traceability devices used to capture, store, and transmit data significantly encourage the efforts to develop more advanced technologies and promote future research regarding food traceability and FSCM (Bosona & Gebresenbet, 2013).

9.3.6 Contribution in Agricultural Sustainability

Traceability systems increase the quality of food and food production systems as they increase the awareness of workers through their focus on data capturing and documentation processes (Donnelly & Olsen, 2012). Traceability data is very instrumental for transparency of food production and sourcing that in turn helps the implementation of sustainability initiatives, especially at the farm level. For example, preventing the depletion of fish stocks is an environmental concern, and the EU has regulations for ensuring that fish are sourced only from a legal catch or farm (Donnelly & Olsen, 2012). Across Europe and America, about 50% of the seafood could not be traced back to the origin (Donnelly & Olsen, 2012). Fish products can be traced through product labels, external tags, chemical markings (e.g., tattoos, chemical branding with inorganic substances like silver nitrate or potassium nitrate), physical markings (e.g., fin clipping), DNA markers (Hayes, Sonesson, & Gjerde, 2005), and other more advanced technologies. FTS also reduces the food product losses in FSC as traceability activities promote effective packaging technologies. For example, Sonneveld (2000) has pointed out that, due to effective packaging, food losses during distribution have been reduced to levels less than 1%.

9.4 BARRIERS IN IMPLEMENTING EFFECTIVE FOOD TRACEABILITY SYSTEMS

In developing and implementing FTSs, food companies face many barriers. The problems identified during this review have been categorized as resource limitation, information limitation, standard limitation, capacity limitation, and awareness limitation.

9.4.1 Resource Limitation

Developing and implementing traceability systems is an expensive and complicated task that can lead to financial problems. Moreover, allocating the costs and benefits among the partners of FSC needs extra efforts and costs. This can even lead to initial resistance (against the implementation of FTSs) by some partners. It also requires much administration and paper work, especially for companies implementing the traceability system for the first time (Kher et al., 2010). Food recall costs may be influenced by factors such as time taken to discover the fault, ability to identify and locate the affected product batch, and value of the product (Randrup, Wu, & Jørgen, 2012). Traceability is a complex task as its application is associated with some key requirements (Salampasis et al., 2012):

- Traceability should be able to address both internal and chain traceability with information about a product's total lifecycle.
- The traceability has to completely address both backward traceability and forward traceability of the product with adequate related information.
- The traceability has to be cost effective and user friendly for implementation and operation. The traceability system also has to be extensible to easily accommodate new traceability data.

9.4.2 Information Limitation

Traceability in agriculture is associated with an inherent uncertainty that makes it difficult to acquire certain and timely data during all the stages of FSC. According to the European legislation on food traceability EC 178(2002), article 18, it is important to: establish a traceability system during all the stages of food production, processing, storage, and distribution; implement (apply) one-step-backward and one-step-forward approaches; and label adequately the food/feed before placing them in the market. However, the EU regulation (EC 178(2002)) lacks detailed internal traceability requirements (Bosona & Gebresenbet, 2013). Bertolini et al. (2006), Donnelly, Karlsen, and Dreyer (2012) and Hu, Zhang, Moga, and Neculita (2013) argue that addressing both internal traceability (within a company) and chain traceability (between companies in the supply chain) with clear connections between internal and chain traceability data enables the achievement of fast and precise tracing activities. According to Canavari et al. (2010), the traceability information can be classified as "strategic" information (information on product quality, social and environmental ethics, service level) and "operational" information (information on a legal requirement, hygienic-sanitary safety). It is also necessary to have information about the presence of GMO components in the food, because the consumers require such information; there is inadequate knowledge about the long term effects of GMO on health and environment; and there are no real international agreements on the principles, testing methods, and safety evaluations (Regattieri et al., 2007).

Food packing and labeling are the main requirements for implementing (paper or electronic-based) effective FTS as these activities link the material flow with the information flow (Manos & Manikas, 2010). However, in some cases, customers desire to buy loose produce (unpacked), especially in case of fresh produce. Such cases can be considered as extra challenges in realizing traceability information flow.

9.4.3 Standard Limitation

Lack of adequate and standardized data and means of data exchange are the areas that need more efforts and research work to improve FTSs. FTS is usually complicated due to variations in data capturing, inconsistency in types of captured data, variations in sharing data within a facility and among FSC partners, and lack of definitions of key terms such as "lot" or "batch" (McEntire et al., 2010). The major problem that is common for traceability techniques, such as numerical code, bar code, radio frequency identification (RFID) tags, is the lack of standardization that creates compatibility problems among the different solutions introduced by the different actors in a supply chain (Regattieri et al., 2007; Salampasis et al., 2012).

Canavari et al. (2010) argued that every company or supply network must find its own model to fit its purposes and activities. In developed countries, automated data capturing and electronic data coding showed good progress at food company levels but, due to a lack of standardization, data transmission from one actor to another was difficult and introducing sector-specific data terminology was recommended as an effective way to tackle the problem (Thakur & Donnelly, 2010). Structured data lists, vocabularies, and ontology can be considered as appropriate tools to achieve universal data exchange (Thakur et al., 2011). However, the code standardization exercise is costly (Van Dorp, 2003).

9.4.4 Capacity Limitation

The complex nature of FTS requires skilled staff for its development, implementation, and management. Moreover, different partners in the FSC have different objectives of FTS, making full chain traceability more complex. The FSC partners may not have adequate skills and employees to effectively implement and manage the FTS (Xiaoshuan et al., 2010). Even in case of small food producers, FTS can place more work load as they experience capacity limitation (Engelseth, 2009; Schwägele, 2005).

9.4.5 Awareness Limitation

Considering traceability as an extra burden, lack of clear information about the benefits of traceability and an absence of uniform willingness among FSC partners to participate in the implementation of traceability systems are problems related with a lack of awareness. For example, the study by Liao et al. (2011) indicated that introducing FTS in developing countries is not an easy task due to technical limitation and lack of awareness. These authors identified that the Taiwan Agriculture and Food Traceability Program, initiated in 2004 by the Taiwan government, failed mainly due to a lack of awareness among the farmers about the traceability program. This indicates that effective training and education programs are required to increase the technical capacities and the awareness levels of all the participants (Bosona & Gebresenbet, 2013).

9.5 FINGERPRINTING TECHNIQUES IN FOOD TRACEABILITY

9.5.1 DNA-Based Fingerprinting Techniques

Recently, Galimberti et al. (2013) reviewed the usefulness of DNA barcoding to certify food identity by tracking the origin and provenance of raw materials at different levels of their transformation. DNA barcoding permits the discrimination among biological entities by analyzing the variability in a single or in a few standard molecular marker(s) (Casiraghi, Labra, Ferri, Galimberti, & De Mattia, 2010). In this context, DNA sequence(s) identify different food products in the same way a supermarket scanner uses the black stripes of the UPC barcode to identify any purchase. The application of this tool opened up new opportunities to track not only the common crops and breeds but also those minor crops and local products that still lack reference genetic fingerprints (Galimberti et al., 2014). As an example, DNA barcoding was extensively applied in the last decade to verify the origin of seafood (Becker, Hanner, & Steinke, 2011) and to exclude commercial frauds from the production and distribution process (Barbuto et al., 2010; Carvalho, Palhares, Drummond, & Frigo, 2015; Cutarelli et al., 2014). The success of seafood molecular identification allowed the U.S. Food and Drug Administration to propose DNA barcoding as a routine approach for the authentication of fish-based commercial products (Yancy et al., 2008). Both consumers and food producers may take advantage of a DNA barcoding screening, especially for the items distributed as shredded or powered material, which otherwise would be deemed as unidentifiable by a simple morphological analysis (Cornara et al., 2013). Among these, promising results were obtained in studies performed on commercial spices (De Mattia et al., 2011), herbal teas

(Li et al., 2012), and fruit juices (Faria, Magalhães, Nunes, & Oliveira, 2013). Analysis of the case studies suggests that DNA barcoding is a sensitive, fast, and cheap approach, which is able to identify and track a wide panel of raw materials and derived food commodities. The cost and time effectiveness of DNA barcoding and the recent developments in innovative sequencing technologies allow a certain degree of automation in species identification, which is particularly useful in simultaneous monitoring activities for multiple foodstuffs and batches.

As a diagnostic tool, the DNA barcoding approach can be more or less fallacious, and it should be taken into account that failures are mainly in the essence of biological species rather than in the method (Casiraghi et al., 2010). DNA barcoding performance is strongly influenced by the molecular variability of the organisms. As an example, the method still cannot be easily applied for the differentiation of genetically modified raw materials, breeds, and cultivars based on the standard molecular markers. The modified genomic tracts usually do not involve the plastidial or nuclear regions analyzed in a classical DNA barcoding approach. However, given the increasing demand for a fast and reliable traceability system for these kinds of products, a panel of additional markers (i.e., promoters, reporter genes) can be applied in combination with classical DNA barcodes. As an example, very recent studies showed the potential of high resolution melting (HRM) analyses when coupled with the investigation of DNA barcoding markers (bar-HRM) to differentiate cultivars and closely related species and to authenticate protected designation of origin (PDO) of some food products (Druml & Cichna-Markl, 2014; Ganopoulos, Bazakos, Madesis, Kalaitzis, & Tsaftaris, 2013; Ganopoulos, Madesis, Darzentas, Argiriou, & Tsaftaris, 2012; Jaakola, Suokas, & Häggman, 2010).

Following the advent of molecular biology, a plethora of laboratory techniques have been developed and most of these are now extensively adopted in food control activities. Molecular approaches permit the identification of food related microorganisms and the estimation of their relative abundance, thus providing a fast, accurate, and economic detection tool. Most techniques rely on the analysis of genetic DNA markers and become increasingly important in food microbiology. They identify microorganisms rapidly and accurately, complementing or substituting classical methods (Ceuppens et al., 2014; Chakraborty, Doss, Patra, & Bandyopadhyay, 2014).

Denaturing gradient gel electrophoresis (DGGE) is one of the most used fingerprinting techniques in food microbiology. It is based on the separation of polymerase chain reaction (PCR) amplicons of the same size but different sequences. Fragments are separated in a denaturing gradient gel based on their differential denaturation (melting) profile (Ercolini, 2004). In recent years, PCR-DGGE has been largely used to characterize bacteria and yeasts in fermented products (Muyzer, De Waal, & Uitterlinden, 1993; Peres, Barlet, Loiseau, & Montet, 2007) and to define the origin of raw material, starting from the characteristics of its yeast or bacterial communities, as in the case of fruit (El Sheikha, Bouvet, & Montet, 2011; El Sheikha, Durand, Sarter, Okullo, & Montet, 2012; El Sheikha, Métayer, & Montet, 2011) and fish (El Sheikha & Montet, 2014; Le Nguyen, Ngoc, Dijoux, Loiseau, & Montet, 2008). However, it is not always possible to resolve DGGE fragments when the difference in sequence is not wide enough or when different DNA fragments have identical melting behavior (Ercolini, 2004).

Aiming to differentiate microorganisms at the species level, methods based on DNA sequencing are currently the most adopted. In many cases, when a fast and accurate response is needed, a “DNA barcoding-like” approach is the most reliable (Chakraborty et al., 2014). Many scientists used 16S rRNA gene as a universal marker for species-level typing of microorganisms (Bokulich & Mills, 2012; Claesson et al., 2010; Janda &

Abbott, 2007). This genomic region is considered as a 'bacterial barcode' due to its peculiar properties (Patel, 2001): it is present in all the bacterial species; it contains sufficient information (1,500 bp long) to differentiate species and, in some cases, strains (Muñoz-Quezada et al., 2013); and the 16S rRNA relies upon an impressive archive of reference sequences, such as Greengenes (De Santis et al., 2006) and SILVA (Pruesse et al., 2007). Amplicons belonging to whole genomic extraction conducted on the matrices under investigation (e.g., food products) are sequenced and reads are compared to reference databases to identify the operational taxonomic units—OTUs (Sandionigi et al., 2015).

Several studies test analytical approaches for the DNA-based detection of emergent food microbial contaminants in a wide panel of food products (Fusco & Quero, 2014; Velusamy, Arshak, Korostynska, Oliwa, & Adley, 2010). Such techniques allow the detection of specific bacteria and strains in different steps of the FSC as reported, for example, in the cases of seafood and meat manufacturing (Amagliani, Brandi, & Schiavano, 2012; Norhana, Poole, Deeth, & Dykes, 2010; Zbrun et al., 2013). In international trade, major food categories, like those cited above, are commonly shipped for very long distances and are therefore exposed to various contaminants such as *Salmonella*, *Listeria*, and *Campylobacter*. PCR and real-time PCR-based methods are nowadays routinely used for the detection of these pathogens. Primer combinations also permit the simultaneous identification of a panel of foodborne pathogens in a single reaction (Galimberti et al., 2015).

Progresses in sequencing technologies and bioinformatics analysis of data nowadays lead to more complex scenarios in food control activities. Detection approaches targeting one or a few microorganisms are not sufficient for reliable characterizations of quality and safety of foodstuffs. Recent technological advances offer a panel of analytical tools that are able to screen the whole microbial community present in food matrices. The use of universal markers produces several DNA barcode fragments, corresponding to each bacterial species present in a food sample (Galimberti et al., 2015). With the ultimate goal of characterizing the complete spectrum of microorganisms, the traditional Sanger sequencing approach results inadequate. To date, several novel approaches, referred to as "next generation sequencing" (NGS) and, more recently, "high throughput sequencing" (HTS), have been developed (Ercolini, 2013; Mayo et al., 2014; Solieri et al., 2013).

HTS techniques can provide sequence data around a hundred times faster and cheaper than the conventional Sanger approach. Sequencers from 454 Life Sciences/Roche (producing about a million sequences of 800–1,000 base length), Solexa/Illumina, and Applied Biosystems SOLiD technology (producing over a billion sequences of 50–500 base length) were produced as second-generation technologies; some other competitive instruments that appeared in the market are Ion Torrent and PacBio. HTS technologies also permit the preparation of several DNA samples from different extracts and their marking with different DNA tags, mixed and processed at the same time. Thanks to these practical advantages, it is possible to analyze in parallel a very large number of samples and, hence, lower the analysis cost. The reduction in cost and time for generating DNA sequence data has resulted in a range of new successful applications, including food traceability and, especially, food microbiology (Madesis et al., 2014).

9.6 CONCLUSION

Food traceability has new uses now since these new traceability concepts can register not only the origin of the product, but the new parameters integrated in the intelligent logistics can know the quality of the product and, therefore, enhance the chain design and the goods flow. The connection between identifiers or environment/event loggers allows a dynamic traceability of quality and, finally, it has become possible to achieve a homogeneous traceability system. DNA-based technologies are emerging as widely adopted tools for microbial characterization of a large number of matrices and ecosystems, among which food is one category. In the field of food quality and safety assessment, the vast majority of published studies focus on fermented beverages and dairy products because of their relevance and economic value in the global market. Other food categories such as meat and seafood are widely distributed worldwide, but many aspects of their microbial ecology are largely unknown. In recent years, due to the growing accessibility of modern analytical technologies (i.e., HTS), the initial studies on these apparently less complex food matrices are gradually emerging.

The performance of full chain food traceability should be evaluated against its overall goal, which should incorporate important specific goals such as: compliance with rules and legislation; food safety and quality; social and stakeholders' satisfaction; economic benefits; and technological and scientific benefits. The efficiency and effectiveness of the introduced FTS in achieving these specific goals can pinpoint the performance level of the system. The performance of FTS can also be evaluated by its ability to improve the breadth, depth, precision, and access of information. Such an improved information flow will in turn enhance the integration of logistics activities and will improve the FSC management on the whole.

REFERENCES

Alfaro, J.A., and Rábade, L.A. 2009. Traceability as a strategic tool to improve inventory management: a case study in the food industry. *Int J Prod Econ* 118:104–110.

Amagliani, G., Brandi, G., Schiavano, G.F. 2012. Incidence and role of Salmonella in seafood safety. *Food Res Int* 45:780–788.

Atkins, P. 2008. Fear of animal foods: a century of zoonotics. *Appetite* 51:18–21.

Badia-Melis, R., Mishra, P., Ruiz-García, L. 2015. Food traceability: New trends and recent advances. A review. *Food Control* 57:393–401.

Barbuto, M., Galimberti, A., Ferri, E., et al. 2010. DNA barcoding reveals fraudulent substitutions in shark seafood products: The Italian case of "palombo" (Mustelus spp.). *Food Res Int* 43:376–381.

Becker, S., Hanner, R., Steinke, D. 2011. Five years of FISH-BOL: brief status report. *Mitochondrial DNA* 22:3–9.

Bertolini, M., Bevilacqua, M., Massini, R. 2006. FMECA approach to product traceability in food industry. *Food Control* 17:137–145.

Bokulich, N.A., and Mills, D.A. 2012. Next-generation approaches to the microbial ecology of food fermentations. *BMB Rep* 45:377–189.

Bosona, T., and Gebresenbet, G. 2013. Food traceability as an integral part of logistics management in food and agricultural supply chain. *Food Control* 33:32–48.

Bourlakis, M., and Bourlakis, C. 2006. Integrating logistics and information technology strategies for sustainable competitive advantage. *J Enterpr Inform Manag* 19(4):389–402.

Canavari, M., Centonze, R., Hingley, M., Spadoni, R. 2010. Traceability as part of competitive strategy in the fruit supply chain. *Br Food J* 112(2):171–186.

Carvalho, D.C., Palhares, R.M., Drummond, M.G., Frigo, T.B. 2015. DNA barcoding identification of commercialized seafood in South Brazil: A governmental regulatory forensic program. *Food Control* 50:784–788.

Casiraghi, M., Labra, M., Ferri, E., Galimberti, A., De Mattia, F. 2010. DNA barcoding: a six question tour to improve users' awareness about the method. *Brief Bioinform* 11:440–453.

Ceuppens, S., Li, D., Uyttendaele, M., et al. 2014. Molecular methods in food safety microbiology: interpretation and implications of nucleic acid detection. *Compr Rev Food Sci Food Saf* 13:551–577.

Chakraborty, C., Doss, C.G.P., Patra, B.C., Bandyopadhyay, S. 2014. DNA barcoding to map the microbial communities: current advances and future directions. *Appl Microbiol Biotechnol* 98:3425–3436.

Chrysochou, P., Chryssochoids, G., Kehagia, O. 2009. Traceability information carriers. The technology backgrounds and consumers' perceptions of the technological solutions. *Appetite* 53:322–331.

Claesson, M.J., Wang, Q., O'Sullivan, O., et al. 2010. Comparison of two next-generation sequencing technologies for resolving highly complex microbiota composition using tandem variable 16S rRNA gene regions. *Nucleic Acids Res* 38:200.

Cornara, L., Borghesi, B., Canali, C., et al. 2013. Smart drugs: green shuttle or real drug? *Int J Legal Med* 127:1109–1123.

Cusato, S., Gameiro, A.H., Corassin, C.H., et al. 2013. Food safety systems in a small dairy factory: implementation, major challenges, and assessment of systems' performances. *Foodborne Pathog Dis* 10(1):6–12.

Cusato, S., Gameiro, A.H., Sant'Ana, A.S., et al. 2014. Assessing the costs involved in the implementation of GMP and HACCP in a small dairy factory. *Qual Assur Saf Crops Foods* 6(2):135–139.

Cutarelli, A., Amoroso, M.G., De Roma, A., et al. 2014. Italian market fish species identification and commercial frauds revealing by DNA sequencing. *Food Control* 37:46–50.

De Mattia, F., Bruni, I., Galimberti, A., et al. 2011. A comparative study of different DNA barcoding markers for the identification of some members of Lamiacaea. *Food Res Int* 44:693–702.

De Santis, T.Z., Hugenholtz, P., Larsen, N., et al. 2006. Greengenes, a chimera-checked 16S rRNA gene database and workbench compatible with ARB. *Appl Environ Microbiol* 72:5069–5072.

Dias, M.A.C., Sant'Ana, A.S., Cruz, A.G., et al. 2012. On the implementation of good manufacturing practices in a small processing unity of mozzarella cheese in Brazil. *Food Control* 24(1):199–205.

Donnelly, K.A.M., and Olsen, P. 2012. Catch to landing traceability and the effects of implementation – a case study from the Norwegian white fish sector. *Food Control* 27:228–233.

Donnelly, K.A.M., Karlsen, K.M., Dreyer, B. 2012. A simulated recall study in five major food sectors. *Br Food J* 114(7):1016–1031.

Druml, B., and Cichna-Markl, M. 2014. High resolution melting (HRM) analysis of DNA—Its role and potential in food analysis. *Food Chem* 158:245–254.

El Sheikha, A.F., and Montet, D. 2014. Fermented fish and fish products: snapshots on culture and health. In *Microorganisms and Food Fermentation*, eds. R.C. Ray, and D. Montet, 390..; CRC Press, New Hampshire; Boca Raton.

El Sheikha, A.F., Bouvet, J.M., Montet, D. 2011. Biological bar-code for the determination of geographical origin of fruits by using 28S rDNA fingerprinting of fungal communities by PCRDGGE: An application to Shea tree fruits. *Qual Assur Saf Crops Foods* 3:40–47.

El Sheikha, A.F., Métayer, I., Montet D. 2011. A Biological bar-code for determining the geographical origin of fruit by using 28S rDNA fingerprinting of fungi communities by PCRDGGE: An application to *Physalis* fruits from Egypt. *Food Biotechnol* 25:115–129.

El Sheikha, A.F., Durand, N., Sarter, S., Okullo, J.B.L., Montet D. 2012. Study of the microbial discrimination of fruits by PCR-DGGE: Application to the determination of the geographical origin of *Physalis* fruits from Colombia, Egypt, Uganda and Madagascar. *Food Control* 24:57–63.

Engelseth, P. 2009. Food product traceability and supply network integration. *J Business Indust Market* 24(5):421–430.

Ercolini, D. 2004. PCR-DGGE fingerprinting: novel strategies for detection of microbes in food. *J Microbiol Methods* 56:297–314.

Ercolini, D. 2013. High-throughput sequencing and metagenomics: moving forward in the culture independent analysis of food microbial ecology. *Appl Environ Microbiol* 79:3148–3155.

Faria, M.A., Magalhães, A., Nunes, M.E., Oliveira, M.B.P.P. 2013. High resolution melting of *trn*L amplicons in fruit juices authentication. *Food Control* 33:136–141.

Folinas, D., Manikas, I., Manos, B. 2006. Traceability data management for food chains. *Br Food J* 108(8):622–633.

Fusco, V., and Quero, G.M. 2014. Culture-dependent and culture-independent nucleic acid-based methods used in the microbial safety assessment of milk and dairy products. *Compr Rev Food Sci Food Saf* 13:493–537.

Galimberti, A., De Mattia, F., Losa, A., et al. 2013. DNA barcoding as a new tool for food traceability. *Food Res Int* 50:55–63.

Galimberti, A., Labra, M., Sandionigi, A., et al. 2014. DNA barcoding for minor crops and food traceability. *Adv Agric* 2014:831875.

Galimberti, A., Bruno, A., Mezzasalma, V., et al. 2015. Emerging DNA-based technologies to characterize food ecosystems. *Food Res Int* 69:424–433.

Ganopoulos, I., Bazakos, C., Madesis, P., Kalaitzis, P., Tsaftaris, A. 2013. Barcode DNA high-resolution melting (Bar-HRM) analysis as a novel close-tubed and accurate tool for olive oil forensic use. *J Sci Food Agric* 93:2281–2286.

Ganopoulos, I., Madesis, P., Darzentas, N., Argiriou, A., Tsaftaris, A. 2012. Barcode High Resolution Melting (Bar-HRM) analysis for detection and quantification of PDO "Fava Santorinis" (*Lathyrus clymenum*) adulterants. *Food Chem* 133:505–512.

Golan, E., Krissoff, B., Kuchler, F., et al. 2004. Traceability in U.S. food supply: Economic theory and industry studies. Agricultural Economic Report [830]. USDA/Economic Research Service, Available at www.ers.usda.gov. Accessed 15.05.2017.

Greger, M. 2007. The long haul: risks associated with livestock transport. *Biosecur Bioterror* 5(4):301–311.

Hall, D. 2010. Food with a visible face: traceability and the public promotion of private governance in the food system. *Geoforum* 41:826–835.

Hayes, B., Sonesson, A.K., Gjerde, B. 2005. Evaluation of three strategies using DNA markers for traceability in aquaculture species. *Aquaculture* 250:70–81.

Heyder, M., Theuvsen, L., Hollmann-Hespos, T. 2012. Investments in tracking and tracing systems in the food industry: a PLS analysis. *Food Policy* 37:102–113.

Hobbs, J.E., Bailey, D.V., Dickinson, D.L., Haghiri, M. 2005. Traceability in the Canadian red meat sector: do consumers care? *Can J Agric Econ* 53:47–65.

Hong, I.H., Dang, Jr F., Tsai, Y.H., et al. 2011. An RFID application in the food supply chain: a case study of convenience stores in Taiwan. *J Food Eng* 106:119–126.

Hu, J., Zhang, X., Moga, L. M., Neculita, M. 2013. Modeling and implementation of the vegetable supply chain traceability system. *Food Control* 30(1):341–353.

Jaakola, L., Suokas, M., Häggman, H. 2010. Novel approaches based on DNA barcoding and high-resolution melting of amplicons for authenticity analyses of berry species. *Food Chem* 123:494–500.

Janda, J.M., and Abbott, S.L. 2007. 16S rRNA gene sequencing for bacterial identification in the diagnostic laboratory: pluses, perils, and pitfalls. *J Clin Microbiol* 45:2761–2764.

Karippacheril, T.G., Rios, D.L., Srivastava, L. 2011. Global markets, global challenges: Improving food safety and traceability while empowering smallholders through ICT. ICT in Agriculture, Module 12. Available at www.ictinagriculture.org/sites/ictinagriculture.org/files/ICT%20in%20Agriculture%20eSourcebook_0.pdf. Accessed 15.05.2017.

Kelepouris, T., Pramatari, K., Doukidis, G. 2007. RFID-enabled traceability in the food supply chain. *Industr Manag Data Sys* 107(2):183–200.

Kher, S.V., Frewer, L.J., De Jonge, J., et al. 2010. Experts' perspectives on the implementation of traceability in Europe. *Br Food J* 112(2):261–274.

Le Nguyen, D.D., Ngoc, H.H., Dijoux, D., Loiseau, G., Montet, D. 2008. Determination of fish origin by using 16S rDNA fingerprinting of bacterial communities by PCR-DGGE: An application on Pangasius fish from Viet Nam. *Food Control* 19:454–460.

Li, M., Wong, K.L., Chan, W.H., et al. 2012. Establishment of DNA barcodes for the identification of the botanical sources of the Chinese 'cooling' beverage. *Food Control* 25:758–766.

Liao, P.A., Chang, H.H., Chang, C.Y. 2011. Why is the food traceability system unsuccessful in Taiwan? Emprical evidence from a national survey of fruit and vegetable farmers. *Food Policy* 36:686–693.

Liu, H., Kerr, W.A., Hobbs, J.E. 2012. A review of Chinese food safety strategies implemented after several food safety incidents involving export of Chinese aquatic products. *B Food J* 114(3):372–386.

Madesis, P., Ganopoulos, I., Sakaridis, I., Argiriou, A., Tsaftaris, A. 2014. Advances of DNA based methods for tracing the botanical origin of food products. *Food Res Int* 60:163–172.

Manos, B., and Manikas, I. 2010. Traceability in the Greek fresh produce sector: drivers and constraints. *Br Food J* 112(6):640–652.

Mayo, B., Rachid, C., Alegría, Á., et al. 2014. Impact of Next Generation Sequencing techniques in food microbiology. *Curr Genomics* 15:293–309.

McCallum, S. 2012. Protecting the food supply: hi-tech food banks and the safety of food supply chains. *DomPrep J* 8(4):5–6.

McEntire, J.C., Arens, S., Bernstein, M., et al. 2010. Traceability (product tracing) in food systems: an IFT (Institute of Food Technology) report submitted to the FDA (Food and Drug Administration), volume 1: technical aspects and recommendations. *Compr Rev Food Sci Food Saf* 9:92–158.

McMeekin, T.A., Baranyi, J., Bowman, J., et al. 2006. Information systems in food safety management. *Int J Food Microbiol* 112:181–194.

Muñoz-Quezada, S., Chenoll, E., María Vieites, J., et al. 2013. Isolation, identification and characterisation of three novel probiotic strains (*Lactobacillus paracasei* CNCM I-4034, *Bifidobacterium breve* CNCM I-4035 and *Lactobacillus rhamnosus* CNCM I-4036) from the faeces of exclusively breast-fed infants. *Br J Nutr* 109:S51–S62.

Muyzer, G., De Waal, E.C., Uitterlinden, A.G. 1993. Profiling of complex microbial populations by denaturing gradient gel electrophoresis analysis of polymerase chain reactionamplified genes coding for 16S rRNA. *Appl Environ Microbiol* 59:695–700.

Norhana, W., Poole, S.E., Deeth, H.C., Dykes, G.A. 2010. Prevalence, persistence and control of Salmonella and Listeria in shrimp and shrimp products: A review. *Food Control* 21:343–361.

Olsen, P., and Aschan, M. 2010. Reference method for analyzing material flow, information flow and information loss in food supply chains. *Trends Food Sci Tech* 21:313–320.

Olsen, P., and Borit, M. 2013. How to define traceability. *Trends Food Sci Tech* 29:142–150.

Orre, K. 2005. The logistics of food supply following radioactive fallout. *J Environ Radioact* 83:429–432.

Patel, J.B. 2001. 16S rRNA gene sequencing for bacterial pathogen identification in the clinical laboratory. *Mol Diagn* 6:313–321.

Peres, B., Barlet, N., Loiseau, G., Montet, D. 2007. Review of the current methods of analytical traceability allowing determination of the origin of foodstuffs. *Food Control* 18:228–335.

Pruesse, E., Quast, C., Knittel, K., et al. 2007. SILVA: a comprehensive online resource for quality checked and aligned ribosomal RNA sequence data compatible with ARB. *Nucleic Acids Res* 35:7188–7196.

Randrup, M., Wu, H., Jørgen, B.M. 2012. On the track of fish batches in three distribution networks. *Food Control* 26:439–445.

Regattieri, A., Gamberi, M., Manzini, R. 2007. Traceability of food products: general framework and experimental evidence. *J Food Eng* 81:347–356.

Resende-Filho, M.A., and Hurley, T.M. 2012. Information asymmetry and traceability incentives for food safety. *Int J Prod Econ* 139:596–603.

Salampasis, M., Tektonidis, D., Kalogianni, E. 2012. TraceALL: a semantic web framework for food traceability systems. *J Sys Inform Tech* 14(4):302–317.

Saltini, R., and Akkerman, R. 2012. Testing improvements in the chocolate traceability system: Impact on product recalls and production efficiency. *Food Control* 23(1):221–226.

Sandionigi, A., Vicario, S., Prosdocimi, E.M., et al. 2015. Towards a better understanding of Apis mellifera and Varroa destructor microbiomes: introducing 'phyloh'as a novel phylogenetic diversity analysis tool. *Mol Ecol Resour* 15(4):697–710.

Schwägele, F. 2005. Traceability from a European perspective. *Meat Sci* 71:164–173.

Smith, G.C., Pendell, D.L., Tatum, J.D., Belk, K.E., Sofos, J.N. 2008. Post slaughter traceability. *Meat Sci* 80:66–74.

Smith, G.C., Tatum, J.D., Belk, K.E., et al. 2005. Traceability from a US perspective. *Meat Sci* 71:174–193.

Sonneveld, K. 2000. What drives (food) packaging innovation? *Pack Tech Sci* 13:29–35.

Stefansson, G., and Tilanus, B. 2001. Tracking and tracing: principles and practice. *Int J Serv Tech Manag* 2(3–4):187–206.

Solieri, L. Dakal, T.C. and Giudici, P. 2013. "Next-generation sequencing and its potential impact on food microbial genomics,"*Annals of Microbiology* 63(1):21–37.

Thakur, M., and Donnelly, K.A.M. 2010. Modeling traceability information in soybean value chains. *J Food Eng* 99:98–105.

Thakur, M., Sørensen, C.F., Bjørnson, F.O., Forås, E., Hurburgh, C.R. 2011. Managing food traceability information using EPCIS framework. *J Food Eng* 103:417–433.

Van Dorp, K.J. 2002. Tracking and tracing: a structure for development and contemporary practices. *Logist Inform Manag* 15(1):24–33.

Van Dorp, K.J. 2003. Beef labeling: the emergence of transparency. *Supply Chain Manag* 8(1):32–40.

Van Rijswijk, W., Frewer, L.J., Menozzi, D., Faioli, G. 2008. Consumer perceptions of traceability: a cross-national comparison of the associated benefits. *Food Qual Prefer* 19:452–464.

Velusamy, V., Arshak, K., Korostynska, O., Oliwa, K., Adley, C. 2010. An overview of foodborne pathogen detection: in the perspective of biosensors. *Biotechnol Adv* 28:232–254.

Wilson, T.P., and Clarke, W.R. 1998. Food safety and traceability in the agricultural supply chain: using the internet to deliver traceability. *Supply Chain Manag* 3(3):127–133.

Wognum, P.M., Bremmers, H., Trienekens, J.H., Van der Vorst, J.G.A.J., Bloemhof, J.M. 2011. Systems for sustainability and transparency of food supply chains – current status and challenges. *Adv Eng Infor* 25:65–76.

Xiaoshuan, Z., Jian, Z., Feng, L., Zetian, F., Weisong, M. 2010. Strengths and limitations on the operating mechanisms of traceability system in agro food, China. *Food Control* 21:825–829.

Yancy, H.F., Zemlak, T.S., Mason, J.A., et al. 2008. Potential use of DNA barcodes in regulatory science: applications of the Regulatory Fish Encyclopedia. *J Food Prot* 71:210–217.

Zbrun, M.V., Romero-Scharpen, A., Olivero, C., et al. 2013. Occurrence of thermotolerant Campylobacter spp. at different stages of the poultry meat supply chain in Argentina. *N Z Vet J* 61:337–343.

SECTION 3

Multivariate Statistical Techniques

CHAPTER 10

Experimental Design

Héctor C. Goicoechea
CONICET-Universidad Nacional del Litoral

CONTENTS

10.1 INTRODUCTION

Multivariate design of experiments (DOE) and optimization of processes through the well-known response surface methodology (RSM) are issues of paramount importance in real world applications considering that its implementation consumes less time and requires fewer efforts and resources than the use of univariate procedures for the same purpose. DOE and RSM applications enable the gathering of quality information and at the same time they allow the number of experiments to be considerably minimized [1], and they have been proved to be useful for developing, improving, and optimizing a wide variety of processes [2] including food production [3–6], analytical applications [7–10], industrial world [11–15], and bioprocesses [16–19].

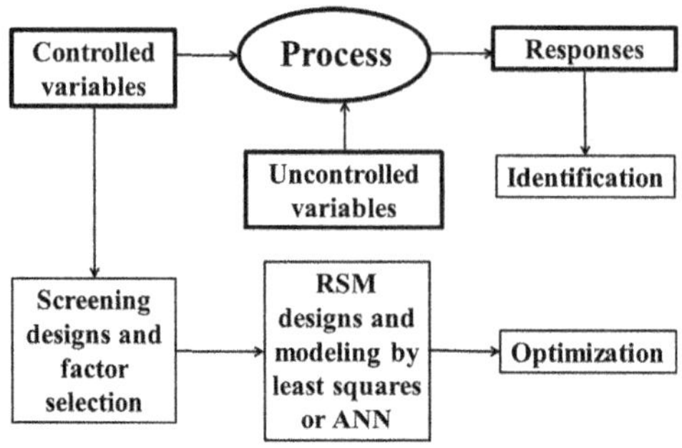

FIGURE 10.1 Scheme showing the steps generally followed for DOE and RSM implementation.

The application of this attractive tool includes a series of consecutive steps that are as follows: (a) identification of the responses being influenced by controlled and uncontrolled variables; (b) screening study in which an initial large number of factors suspected of influencing one or several responses are reduced by the use of screening designs, followed by statistical evaluation (mainly ANOVA, analysis of variance); (c) response surface design should be built, which provides the data that must be generally modeled through least squares (LS) fitting or exceptionally through artificial neural networks (ANN) [20]; and (d) when a large number of responses are involved, they have to be simultaneously optimized following appropriate criteria. In the latter case, the desirability function is the most popular tool to be applied [20]. The scheme displayed in Figure 10.1 shows the steps that are generally followed for DOE and RSM implementations.

A bibliographic revision reveals that two strategies can be implemented for optimization purposes, the univariate and multivariate approaches, which can be described as follows:

a. One variable at time (OVAT) where one selected factor varies, while the others remain constant. This procedure is usually applied, and it does not take into account the interactions between factors, generating a large number of experiments. Usually the experimental domain explored here is smaller compared with that examined by the multivariate approach. Thus, this procedure provides just local information, i.e., knowledge about only those points of the experimental domain at which the experiments were performed.
b. The multivariate strategy in which several factors are simultaneously studied in a predefined number of experiments, varying together the levels of all the factors involved in the process. This strategy has important advantages as compared to the univariate strategy, as discussed by Leardi in his tutorial [21].

In this chapter, the role of DOE and optimization through the multivariate approach will be analyzed and several critical issues will be discussed. Finally, one application will be presented in the context of the food world to illustrate the concepts presented in the chapter.

10.2 RESPONSE VARIABLES

As can be appreciated in Figure 10.1, controlled and uncontrolled variables affect the process, and their effects can be observed by the variations in one or several responses.

Thus, the first step when applying DOE and RSM is the selection of a variable (or several) that can provide the necessary information during the evaluation of the process, which is to be subjected to the optimization procedure. This variable is called response and, according to the objective, it may be necessary to study more than one response. In the latter case, it will be called multiple response optimization.

The following examples extracted from the literature can be mentioned to illustrate different responses analyzed in the food world: (a) isoflavone extraction yield (%) from soybean flour in a pressurized water system [22]; (b) three responses: calcium uptake, firmness, and pH of diced tomatoes evaluated during a calcification process [23]; and (c) optimization of the foaming conditions of tomato juice by studying three responses: expansion volume, foam density, and drainage volume that is an indication of foaming ability and foam stability [24]. The second example will be used to illustrate the whole procedure in the last part of this chapter.

10.3 FACTORS AND THEIR LEVELS - SCREENING DESIGNS

The second step consists of careful detection and examination of all the factors that may affect the process, and on which the experimenter can execute changes, i.e., controlled variables. The experimental domain must be defined for each factor. In addition, a method of control and accurate measurement must be established. The factors can be divided as quantitative, qualitative, and mixture-related (e.g., in the formulation of fish patties with a blend of three kinds of fish—mullet, sheepshead, and croaker). Since the number of factors to be considered can be important, it is necessary to perform screening experiments to determine the experimental variables and interactions that have a significant influence on one or several responses. In screening designs, the factors are usually examined at two levels (–1, +1). The range between the levels is the broadest interval in which the factor can be varied for the system under study, and it is chosen on the basis of literature information or earlier knowledge.

Full factorial, fractional factorial, and Plackett-Burman designs (PBD), all of them at two levels for each factor (k), are the most widely used during the selection of factors as they are economic and efficient. The latter word refers to the number of coefficients that can be adjusted in the modeling step (see Figure 10.2). The factorial fractional design is the most commonly used for screening purposes. This design enables the evaluation of a relatively large number of factors in a small number of experiments by fractioning a full

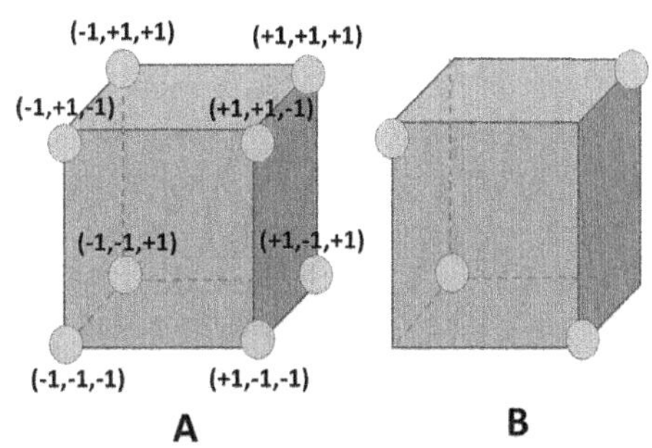

FIGURE 10.2 Schematic representation of (A) full factorial 2^3 design and (B) a 2^{3-1} fractional factorial design, both for three factors.

factorial 2^k design in a 2^{k-p} design, where p represents the number of independent design generators chosen to fractionate the design. For example, when examining five factors, a half-fraction factorial design with $2^{5-1}=16$ experiments or a quarter-fraction factorial design with $2^{5-2}=8$ experiments are possible. However, the latter design does not permit the estimation of all major and interaction effects separately because some of them are estimated together, i.e., they are confounded. Finally, PBD are highly fractioned designs which allow studying a large number of factors with a reduced number of experiments (for example, 11 factors can be studied by running just 12 experiments). Unfortunately, the main effects are aliased with double interactions and, consequently, the experimenter should be extremely careful when implementing PBD.

Figure 10.2 shows the experimental space of both a 2^3 full factorial (A) and a 2^{3-1} fractional factorial (B) designs, involving eight and four experiments, respectively.

10.4 DESIGNS FOR RESPONSE SURFACE METHODOLOGY

Several designs are available for implementation when the objective is to model a second order response surface. Each one has its own advantages and disadvantages, and the experimenter selects the proper design that fits the experiment. Different criteria are taken into account, for example, the need to carry out a reduced number of experiments, prohibited zones in the experimental region, confidence in the parameters to be adjusted, etc. Following are brief descriptions of four of the most commonly used designs in published reports: full factorial, central composite, Box-Behnken, and mixture simplex centroid designs.

10.4.1 Full Factorial Design (FFD)

This design is an extension of the full factorial 2^k presented for screening designs. In this case, the number of experiments is equal to 3^k, i.e., for three factors ($k=3$) at least 27 experiments are required (Figure 10.3A).

10.4.2 Central Composite Design (CCD)

This design is one of the most commonly used. It is built with a two-level (–1 and +1) factorial design points (Fp), axial or "star" points (Sp) and center points (Cp), with all factors set to 0.

All the factors in Sp are set to 0, except one factor with the value $\pm\alpha$. The latter number determines the location of the Sp in the design and usually varies from 1 to $\sqrt{k}$. The former allocates the axial points on the face of the cube or hypercube and is known as face-centered central composite design. $\sqrt{k}$ results in experimental points placed in the domain of a cube or hypercube.

If $\alpha=f^{1/4}$, the design will be rotable (f is the number of Fp). Rotability is an interesting property that ensures a reasonably stable distribution of the prediction variance in the entire design space. In this case, the designs will have five experimental levels, an attractive fact when complex models are to be adjusted.

In Figure 10.3B, the experimental points of a spherical central composite design with three factors (a, b, and c) are represented (full factorial 3^3). The number of experiments is

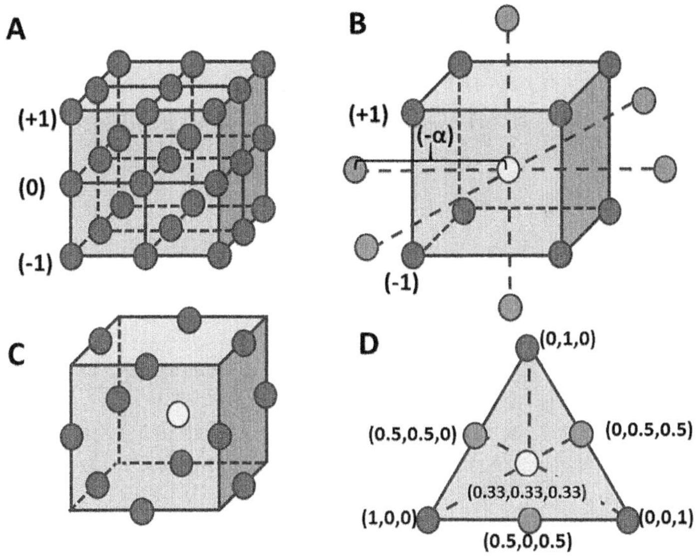

FIGURE 10.3 (**See color insert.**) Schematic representation of (A) a full factorial 3^3 design, (B) a central composite design, (C) a Box-Behnken design, and (D) a mixture simplex centroid design. All of them are for three factors.

equal to $2^k+2k+Cp$, i.e., for three factors ($k=3$) and one central point, at least 15 experiments are required, although replications of the central points are recommended in order to evaluate the lack of fit (see below).

10.4.3 Box-Behnken Design (BBD)

Figure 10.3C shows the schematic representation of a BBD. The number of experiments for this design is equal to $2k\ (k-1)+Cp$, i.e., for three factors ($k=3$) and one central point, at least 13 experiments are required. As can be appreciated, this design requires lesser (or equal) number of experiments than the two designs previously presented. Thus, for six factors ($k=6$), the minimal number of required experiments for the experimental designs will be the following: FFD (729), CCD (77), and BBD (49).

10.4.4 Mixture Simplex Centroid Design

In the special case in which the analyzed factors are the components of a mixture, their levels are not independent of one another. Thus, the use of a mixture design is the option. Mixture designs allow studying the effects of the ratio variations among the variables. The domain is a regular figure having as many vertices as components in a space for which the dimensionality is equal to the number of components minus one [2].

Figure 10.3D displays the graphical representation of a mixture design with three components. It is an equilateral triangle whose vertices correspond to combinations containing 100% of a single component. Each of the three sides represents a binary mixture. The internal points correspond to ternary mixtures. These designs are usually augmented with additional points in the interior of the experimental region. It should be noted that the models used in mixture designs differ from the polynomials used in response surface for independent variables. They are the well-known Scheffé polynomials, which can be linear, quadratic, full cubic, and special cubic [2].

10.5 CONSIDERATIONS FOR COMPLETING THE EXPERIMENTS

The following considerations should be taken into account in this part of the general procedure:

a. The observations and the errors related with the measurements of the responses should be independent random variables.
b. If the number of experiments exceeds the amount that can be done in one day or in a work sequence, these should be performed in blocks, each one corresponding to a different day or work sequence. Thus, it will be necessary to take into account the blocking effect.
c. In order to decide if the effect is significant or not, the standard deviation of the effect should be known and compared with the random error. The first one can be computed from the standard deviation of an experimental measurement (experimental error). On the other hand, a random error is due to common causes of the process and represents the observed variability of the response that cannot be explained through the factors under study. This error may include the effect of the non-studied factors (uncontrolled variables of Figure 10.1) and operator errors during the execution of the experiments. If the latter variabilities are important, it is not possible to distinguish the true effect that the studied factor exerts on the response. Thus, it is very important for the operator errors to remain small or negligible, and to avoid free variations of any factor that has a significant influence on the response.

10.6 MODELING

10.6.1 Least Squares (LS) Modeling

10.6.1.1 Screening

For screening designs and factor selections, the general approach to the statistical analysis for every response being analyzed includes examining factors, signs and magnitudes by building an initial model for the response, performing statistical tests, refining the model by removing any insignificant variable from the initial model, and analyzing residuals in order to check model adequacy and assumptions [2]. The effect of a factor on a response is estimated as the difference between the average response of the experiments with positive signs and the average response of the experiments with negative signs (in a codified design). This value is related with the coefficients of the estimated model. The response (y) in a factorial screening experiment for two factors (x_1 and x_2) can be described by the following linear with interaction regression model:

$$y = b_0 + b_1 x_1 + b_2 x_2 + b_{1-2} x_1 x_2 + e \tag{10.1}$$

where b_0 is the overall mean effect, b_1 and b_2 represent the effects of the factors x_1 and x_2, b_{1-2} is the effect of the interaction between both the factors, and e is a random error component, which represents other sources of variability not accounted for in the model. Thus, e includes effects such as measurement error on the response and other sources of variations inherent in the system like instrumental background noise and the effects of non-studied variables. It is assumed that e has a normal distribution with mean $\mu=0$ and variance σ^2, which can be estimated by performing replicated experiments.

Several statistical tools are usually employed to assess the significance of the effects and to decide which of them should be considered in the final model and which ones should be included in the error. One of them is the well-known normal and half-normal probability plot in which the negligible effects are normally distributed and tend to fall along a straight line on this plot. On the contrary, significant effects will have non-zero means and will not lie along the straight line. Another well-known plot is the Pareto chart where factor values are represented in a descending order by bars, indicating whether they are significant or not by a threshold established through any statistical test (Bonferroni test, for example) [2].

Nevertheless, the selected factors from any graph analysis should be examined by ANOVA [2,20,21]. The latter is a statistical test used to analyze the differences between group means and their associated procedures, such as "variation" among and between groups. ANOVA estimates three sample variances: a total variance based on all the observation deviations from the grand mean, an error variance based on all the observation deviations from their appropriate treatment means, and treatment variances. The test procedure involves partitioning the total sum of squares (SS_{Total}) into a sum of squares due to the model ($SS_{Regression}$) and a sum of squares due to the residual or error (SS_{Error}). To determinate the statistical significance of the model and the coefficients of Eq. 10.1, the F-test is used. To obtain the coefficients, the LS methodology that fits the model is applied to set the experimental data finding coefficient values that minimize the residual term e.

The following example extracted from the literature will illustrate the process of factor selection [25]. Sweet potato peels are usually considered as waste in food service operations, but they have a high phenolic content and antioxidant activity. In this report, the effect of nine extraction variables on phenolic compound extraction was studied by the authors. They are solvent-to-solid ratio, time, pH, peeling cut depth, particle size, temperature, solvent, sample amount, and agitation, and these were combined in a 12-run PBD to screen for key factors. Figure 10.4 shows the normal probability plot of the effects in which all the effects that lie along the line are negligible, whereas two large effects are far from the line. The important effects that emerge from this analysis are the main effects A (solvent-to-solid ratio) and D (peel cut depth). On the other hand, Table 10.1 displays the results of the ANOVA performed for this analysis. As can be concluded, the extraction of phenolic compounds was significantly ($p < 0.05$) affected only by solvent-to-solid ratio (positively) and peel cut depth (negatively) with similar effect magnitudes. The final equation for the model in terms of coded factors is the following: $y=3.34+0.6\times A - 0.94\times D$.

10.6.1.2 Response Surface Modeling

In a similar way, for response surface modeling, once the data corresponding to the optimization stage have been collected, a mathematical model can be built for each response fitting a second (or higher) order polynomial function. The response (y) in a RSM experiment for two factors (x_1 and x_2) can be described by the following regression model:

$$y = b_0 + b_1x_1 + b_2x_2 + b_{1-2}x_1x_2 + b_{1-1}x_1^2 + b_{2-2}x_2^2 + e \qquad (10.2)$$

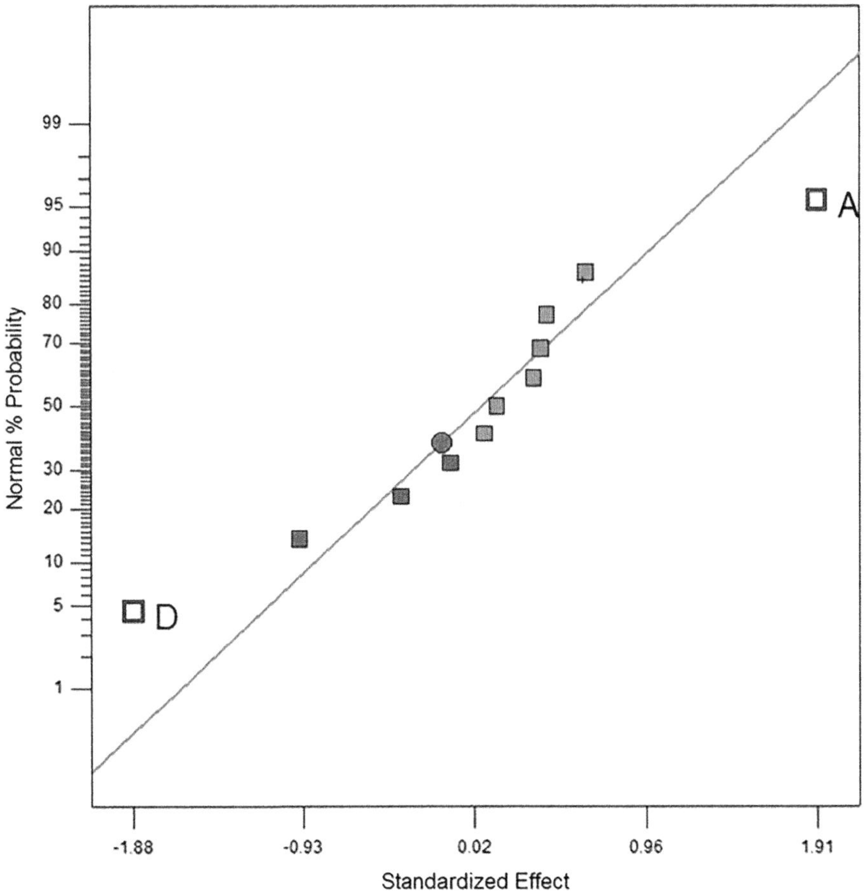

FIGURE 10.4 **(See color insert.)** Normal probability plot of the effects on the phenolic compounds extraction experiment [25].

TABLE 10.1 Results of ANOVA for the Phenolic Compound Extraction Experiment [25]

Source	Sum of Squares	Degree of Freedom	Mean Square	*F*-value	*p*-value
Model	21.55	2	10.77	15.71	0.0012
A	10.96	1	10.96	15.99	0.0031
D	10.58	1	10.58	15.43	0.0035
Residual	6.17	9	0.69	–	–
Total	27.72	11	–	–	–

where b_0, b_1, b_2, $b_{1\text{-}2}$, and e are similar to Eq. 10.1, and $b_{1\text{-}1}$ and $b_{2\text{-}2}$ corresponds to the quadratic terms. Generally, only second order interactions are taken into account because higher order interactions are not significant and may be confused with the main effects. Significance of the coefficients are also investigated through ANOVA, as mentioned before for the screening step.

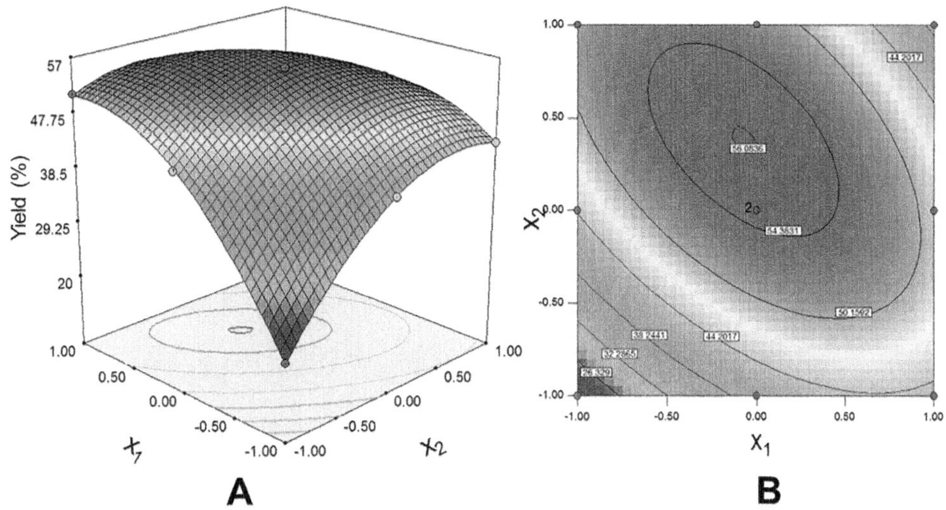

FIGURE 10.5 **(See color insert.)** Surface of response (A) and the contour plot (B) corresponding to the response (%) of yield in a chemical reaction [22].

The fitted model must be able to properly describe the data performance in order to make statistical predictions. In the case when only two factors are optimized, RSM generates a graphical view of the system, since the response can be represented as a solid surface in a three-dimensional space. When more than two factors are being optimized, the graphical representation is made for two of them, maintaining the others at constant values; thus, a small fraction of the surface is shown. The contour plot consists of lines of constant response, corresponding to a specific height of the response surface. Figure 10.5 shows both the surface of response (A) and the contour plot (B) corresponding to the response (%) of yield in a chemical reaction [21] that is affected by two factors described by the following equation in coded values: $y = 55.04 + 2.25\,x_1 + 6.10\,x_2 - 9.10\,x_1\,x_2 - 8.34\,x_1^2 - 8.89\,x_2^2$.

When applying LS regression, it should be assumed that the expected values of the errors are near zero, independent of constant variance and, at least, approximately normally distributed. That is why these assumptions should be checked [2]. As was commented for the fitting of linear models, to determine if the multiple regression fit is significant for the second order model, a test of ANOVA should be applied. The model is considered satisfactory when the regression is significant and an insignificant lack of fit is obtained for the selected confidence level, for example ($p < 0.05$) and ($p > 0.05$), respectively. However, obtaining a significant model does not necessarily mean that it explains the variation in the data correctly. Consequently, it is necessary to evaluate the residual plots, the coefficient of determination (R^2), and the adjusted coefficient of determination (R_{adj}^2), representing the percentage of variance explained by the model. The normal probability plot indicates whether the residuals follow a normal distribution, one of the basic conditions for the validity of ANOVA. The homogeneity of the variance (another requirement for ANOVA validity) can be evaluated by the plot of residual versus the ascending predicted response values. Two additional plots are also important: (a) the residuals versus experimental run order plot, which allows to detect the uncontrolled variables that may have influenced the response during the experiment, and (b) the residuals versus levels of each factor plot, which checks whether the variance not accounted for by the model is different at diverse levels of a factor.

At least three ways to minimize the lack of normality or residuals heteroscedasticity can be found in the literature [2]: (a) to use nonparametric methods such as ANN; this will be briefly discussed in the next section; (b) to implement a generalized linear modeling (GLM), a flexible generalization of ordinary linear regression that allows for response variables without a normal distribution; and (c) to analyze a transformed response that meets the requirements of normality (an example of this option is analyzed at the end of this chapter).

10.6.2 Artificial Neural Networks (ANN) Modeling

The ANN methodology was especially created to model nonlinear information, simulating some properties of the human brain. The well-known multilayer feed-forward networks [26] or multilayer perceptron (MLP) networks are often used for prediction, classification, and optimization. For optimization purposes, the architecture mostly used consists of three layers of neurons or nodes, which are the basic computing units. First is the input layer with the number of active neurons equal to the number of factors being investigated. Then comes one (or two) hidden layer with a variable number of active neurons (which should be optimized). Finally, there is an output layer that has a unit for each response.

The neurons are linked in a hierarchical manner, i.e., the outputs of one layer of nodes are used as inputs for the next layer and so on. In the hidden layer, it is common to use a sigmoid function. The output of the hidden neuron j, O_j, is calculated as:

$$O_j = f\left[\sum_{i=1}^{m}\left(s_i w_{ij} + w_{bj}\right)\right] \tag{10.3}$$

where s_i is the input from neuron i in the layer above to neuron j in the hidden layer, w_{ij} are the connection weights between neurons i and j, w_{bj} is the bias to neuron j, and m is the total number of neurons in the layer above. Linear functions are generally used both in the input and output layers. The number of hidden layers and the number of neurons in each hidden layer must be optimized to achieve a satisfactory fitting ability of the network, which is associated to a satisfactory predictive ability.

It is important to stress that ANNs trained with the latter rule have a remarkable advantage, as there is no need to know the exact form of the function on which the model should be built. Furthermore, neither the functional type nor the number of model parameters need to be given. This is the main difference between modeling by LS regression or ANNs.

An additional sort of ANN, based on the use of radial basis functions (RBF) has been recently introduced for nonlinear multivariate function estimation and regression tasks [27]. RBF networks have a single hidden layer of neurons incorporating Gaussian transfer functions and a linearly activated output layer. In comparison with MLP networks, RBF offers some advantages such as robustness toward noisy data as well as a faster training phase, the latter being of crucial importance. In the field of optimization, RBF was used for the prediction of optimal culture conditions for maximum hairy root biomass yield [28].

10.7 OPTIMIZATION

The graphical representation of the model provides a suitable way to find the optimal location, i.e., a maximum, a minimum, a target value, or a region of interest. As commented before, two types of graphs may prove helpful: the response surface in the three-dimensional space (Figure 10.5A) and the graph of contours, which is the projection of the surface in a plane, represented as lines of constant response, each one corresponding to a specific height of the surface (Figure 10.5B). It should be stressed that in both graphics the response is represented as a function of two factors. According to the established optimization criterion, the optimum can be found by simple visual inspection of the mentioned graphs.

When more than two factors are studied, those which are not plotted must be set at a constant value, thus a limited part of the experimental domain is shown and the optimum is not necessarily seen in the graph. For this reason, the value of the fixed variable must be selected very carefully. The overlaying of contour plots constructed allows a visual search for the best compromise region satisfying response requirements. However, if more than three factors are being analyzed, the superposition of contour plots becomes a difficult task. Therefore, sometimes a more formal analytical approach to the second order surface is necessary, when it is generated in k dimensions. One possible approach is to differentiate the adjusted polynomial with respect to each of the factors, and then allowing the derivative to be set to zero. In this way, the stationary point of the system can be found, which is usually the "optimal candidate" corresponding to a point with a maximum or minimum response in the hyper-surface (k-dimensional), or a saddle point.

10.8 MULTIPLE RESPONSE OPTIMIZATION: THE DESIRABILITY FUNCTION

Usually, the optimization procedure involves more than one response. In the case when three or more responses have to be optimized, it is not possible to do it separately because a number of solutions equal to the variables under study would be attained. In the optimization of a process, the overall solution must be included in an optimal region, leading to a certain degree of compliance with the proposed criteria for each variable of the system; namely, a compromise solution must be found.

One solution to optimize the multiple responses was presented in 1980 by Derringer and Suich: the "desirability function" [29]. Since that moment, this function has been widely used in industry and many other applications [20]. This function is based on the idea that the quality of a product that has many features is completely unacceptable if one of them lies outside a desirable threshold. The goal is to find operating conditions that ensure agreement with the criteria of all the involved responses and, at the same time, to provide the best value of compromise in the desirable joint response [2]. The objective is achieved by converting the multiple responses into a single one. Interestingly, the desirability function allows to find factor levels to simultaneously reach the optimal values for all the evaluated responses, taking into account the researcher's priorities during the optimization procedure.

To implement the strategy suggested by Derringer and Suich, first, an individual desirability function $d_i(\hat{y}i)$ for each response $\hat{y}_i(x)$ must be created by using the fitted models (by LS or ANN) and by establishing the optimization criteria. The individual desirability always takes values between 0–1, where $d_i(\hat{y}_i)$ is equal to 0 for an undesirable response, and 1 represents a completely desirable value (ideal response). Intermediate values of $d_i(\hat{y}_i)$ indicate more or less desirable responses.

Different functions can be built, depending on the optimization criteria adopted, within an acceptable range of response values given by $(U_i–L_i)$, where U_i is the upper acceptable value for the response and L_i is the lower one. As an example, if the response has to be maximized, $d_i(\hat{y}_i)$ is described by the following equation:

$$d_i\left(\hat{y}_i(x)\right) = \begin{bmatrix} 0 & \text{if } \hat{y}_i(x) < L_i \\ \left(\dfrac{\hat{y}_i(x) - L_i}{U_i - L_i}\right)^s & \text{if } Li \le \hat{y}_i(x) \le Ui \\ 1 & \text{if } \hat{y}_i(x) > Ui \end{bmatrix} \quad (10.4)$$

where s is a power value named "weight" set to determine how important it is for $\hat{y}_i$ to be close to the maximum. Other functions should be built for other criteria [29].

It should be noted that factor levels might also be included in the optimization procedure in order to prioritize the use of certain suitable conditions within the experimental region. Figure 10.6 shows a graphical representation of the desirability function for the maximum optimization criteria. As can be seen in this figure, on giving different values to the exponent s in Eq. 10.4, the condition can be more or less restrictive.

Once the n variables (factor and responses levels) are transformed in desirability functions, they are combined in a single function named global desirability (D), which allows to find out the best combined responses. The following equation is used:

$$D = \left(d_1^{r_1} x d_2^{r_2} x \ldots x d_n^{r_n}\right)^{\frac{1}{\sum r_i}} = \left(\prod_{i=1}^{n} d_i^{r_i}\right)^{\frac{1}{\sum r_i}} \quad (10.5)$$

where r_i is the importance of each variable relative to the others. The latter can be used in those cases when one or more responses should be taken into account more carefully than the others.

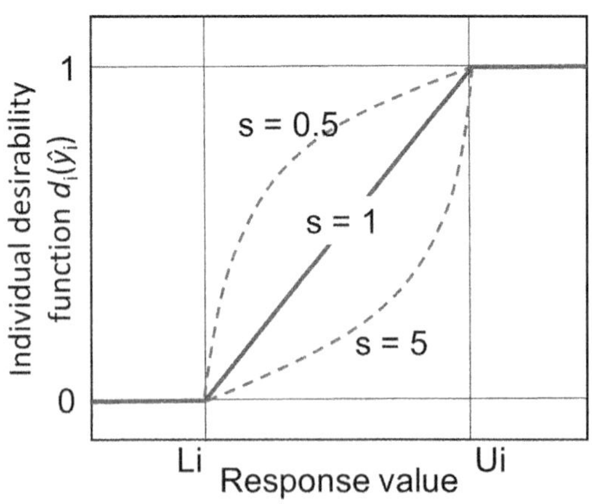

FIGURE 10.6 Graphical representation of the desirability function for the maximum optimization criteria.

Interestingly, because Eq. 10.5 consists of a cumulative product, if D reaches a value different from zero, all the variables which are being simultaneously optimized can be considered to have a desirable value. On the other hand, if one of the responses is completely undesirable, $d_i(\hat{y}_i)=0$, D will be zero. The optimization procedure implies to maximize D or to find a region in which the values of D are different from zero. Whatever the case, it should be kept in mind that the goal of an optimization procedure is to find a good set of conditions that will meet all the goals, but not to get to a D value equal to 1. This value is completely dependent on how closely the L_i and U_i limits are set relative to the actual optimum and on the different weights or priorities assigned to each response.

As showed for the surface of a single response, D can also be graphically represented for the experimental space, calculating several values of D using $d_i(\hat{y}_i)$ for all the responses from a large number of combinations of factor levels (a grid). The function usually finds more than one combination of factor levels where the responses are acceptable and represented as a flattened top on the surface of D plot. In these cases, it can be stated that the optimization has generated a robust solution. The final step in multiple response optimization is to predict the values that the different responses can take at the optimal factor combinations using the fitted models. Finally, the process should be carried out to verify if the predictions match the experimental values.

10.9 DESCRIPTION OF A LITERATURE EXAMPLE

In the following, an example extracted from the literature will be presented and discussed in order to illustrate the most important aspects of the RSM strategy that were introduced in the previous sections. The work deals with the study of quality characteristics of diced tomatoes, i.e., their textural integrity and firmness, carried out by Floros et al. in 1992 [23]. This fact is particularly critical in aseptic processing and packaging, because the diced product is subjected to high shear stress as it gets transferred through pumps, pipes, strainers, pressurized tanks, and fillers. Therefore, actions are needed to secure the textural integrity and improve the firmness of the product.

The objectives of the study were the following: (a) to better understand relationships between factors affecting the calcification process (calcium concentration, solution temperature, and contact time) and the responses determining effectiveness of the process (total calcium content, firmness, and pH of the tomato dices after treatment) through the use of RSM; (b) to determine optimum processing conditions to provide a final product with calcium less than 800 μg g^{-1}, maximum firmness, and pH as low as possible.

To apply the RSM, the authors decided to build a BBD as the one described in Section 10.4.3 and displayed in Figure 10.3C, with 15 experiments including three central points. The experimental combinations of the three factors: calcium concentration (x_1 in % $CaCl_2$), solution temperature (x_2 in °C), and treatment time (x_3 in min), and the corresponding responses: calcium content (y_1 in μg g^{-1}), firmness (y_2 in Ng^{-1}), and pH (y_3) are displayed in Table 10.2. For each experiment, the total calcium content was determined by atomic absorption spectroscopy, firmness was measured using a multiple blade share cell and computed as maximum shear force, and pH was measured using a pH meter.

As commented before, one model should be built for every response, adjusting each response to the factors by LS (or ANN). In the present case, LS was applied, and the ANOVA test applied to the three responses indicated that in both the cases linear models were the best for fitting F (y_2) and pH (y_3) responses, with $p = 0.0007$ and $p = 0.0004$, respectively. For these two responses, the lack of fit was not significant in both cases, with

TABLE 10.2 15-Runs Experimental Design (Box-Behnken) Built to Find the Optimal Conditions on a Calcification Process for Diced Tomatoes [23]

	Factors			Responses		
Experiment	x_1=Calcium Concentration (% $CaCl_2$)	x_2=Solution Temperature (°C)	x_3=Treatment Time (min)	y_1=Calcium Content (μg g^{-1})	y_2=Firmness (N g^{-1})	y_3=pH
1	0.75	65.0	3.35	2890.3	20.79	4.01
2	1.45	50.0	3.35	7490.2	26.48	3.61
3	0.75	35.0	3.35	1698.1	22.85	3.69
4	0.75	50.0	1.93	1274.3	20.00	4.13
5	0.75	35.0	0.50	804.9	17.75	3.92
6	0.05	50.0	3.35	253.5	14.81	4.33
7	0.75	65.0	0.50	1162.5	21.57	4.33
8	0.05	50.0	0.50	152.0	10.10	4.35
9	0.05	65.0	1.93	390.4	15.40	4.35
10	0.05	35.0	1.93	248.1	12.85	4.20
11	0.75	50.0	1.93	1505.9	23.53	3.85
12	0.75	50.0	1.93	1660.3	24.12	3.77
13	1.45	35.0	1.93	5578.8	24.81	3.75
14	1.45	65.0	1.93	3720.0	23.04	3.86
15	1.45	50.0	0.50	1842.9	20.59	4.00

p=0.4927 and p=0.8843, respectively. The R^2_{adj} were 0.716 for F response and 0.785 for pH response. It should be stated that a value superior to 0.70 is expected for this statistical parameter.

When analyzing CC (y_1) response, a significant lack of fit indicated that the model did not appropriately fit the experimental system (p=0.0396). For this reason, a square root transformation was performed on CC (sqy_1) and it resulted in a significant two-factors interaction model ($p < 0.0001$) with no significant lack of fit (p=0.2015) and R^2=0.955. The mathematical expressions for the obtained models are showed in Eqs. 10.6, 10.7, and 10.8:

$$\mathrm{sq}(y_1) = 40.1 + 25.2x_1 + 10.1x_2 + 10.0x_3 - 4.4x_1x_2 + 10.0x_1x_3 \quad (10.6)$$

$$(y_2) = 19.9 + 5.2x_1 + 0.3x_2 + 1.9x_3 \quad (10.7)$$

$$(y_3) = 4.0 - 0.3x_1 + 8.3x_2 - 0.1x_3 \quad (10.8)$$

where x_1, x_2, and x_3 are Ca, Te, and Ti factors, respectively. Only the factors that are significant for the responses have been included in the above equations.

Figures 10.7A–C show three graphs which are commonly used for the diagnosis of fulfillment of normality distribution, homoscedasticity, and independence; in this case corresponding to the response pH. They are the normal plot of residuals, the plot of residuals vs. predicted values, and the plot of residuals vs. experimental runs. In addition, Figure 10.7D corresponds to the plot of predicted vs. actual values, and it gives an idea about how well the model fits the data. As can be appreciated, the assumptions are fulfilled to some degree.

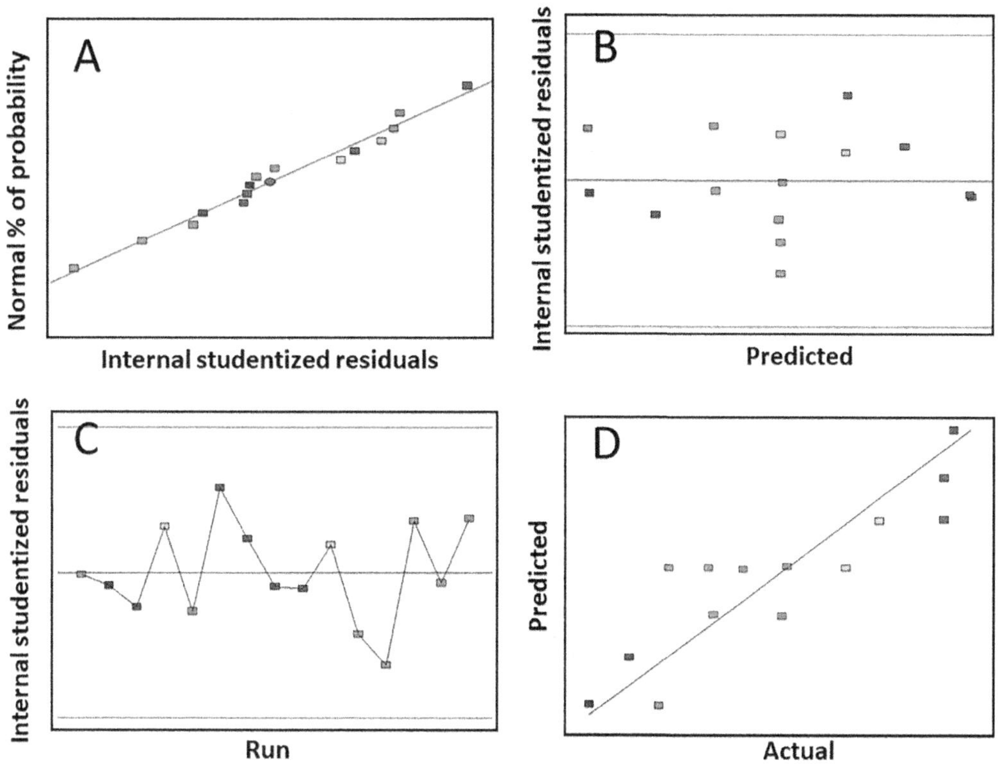

FIGURE 10.7 **(See color insert.)** Diagnosis of normality distribution assumption plots for response pH. (A) Normal plot of residuals, (B) residuals vs. predicted values, (C) residuals vs. run, and (D) predicted vs. actual values.

The response surfaces plotted following the models represented in Eqs. 10.6, 10.7, and 10.8 are represented in Figures 10.8A–C. In all the cases, just two factors can be represented, maintaining the other one at a constant value (the center of the design). Thus, for the three responses, the two most important factors are plotted.

Now, let us go to the optimization step. Considering that the optimization criteria, which are established by the experimenter, are the following: minimize y_1 between 700.0–1100.0 μg g^{-1}, maximize y_2 between 18.50–20.50 N g^{-1}, and minimize pH between 3.92–3.95. Thus, those criteria are used to build the individual desirability functions according to the equations that are similar to Eq. 10.4. Once individual desirability functions are computed using the models presented in Eq. 10.6–10.8, the global desirability can be obtained with Eq. 10.5. In the present case, the following combinations of factors can be selected as optimum: $x_1=0.41$; $x_2=35.00$; and $x_3=3.35$. They generate the following response predictions: $y_1=886$; $y_2=19.9$; and $y_3=3.89$. As can be appreciated, the requirements established a priori are fulfilled. The D value computed is equal to 0.472. Figure 10.9A–C shows the contour plots for the global desirability function corresponding to the three combinations of factors. As can be seen, owing to the strict conditions established by the experimenter, only a reduced region on the plot can be used. In addition, these three plots show a large zone, which is totally forbidden since D reach values equal to zero. Figure 10.10 shows the surface for the combination x_1 vs. x_3.

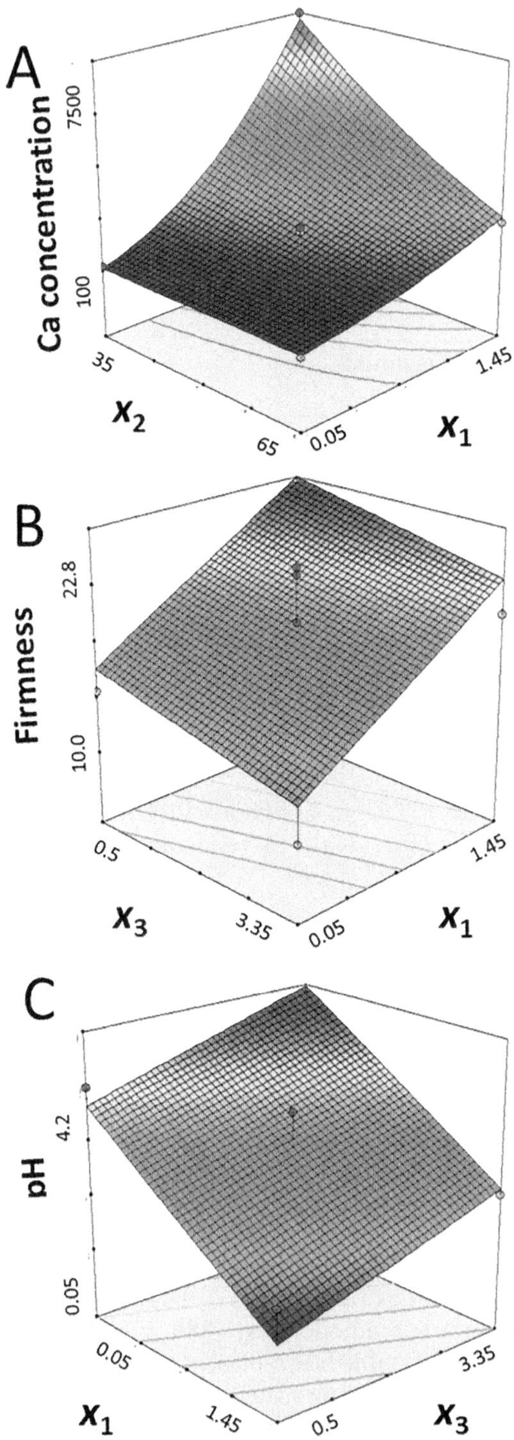

FIGURE 10.8 **(See color insert.)** Response surfaces plotted following the models represented in Eqs. (10.6–10.8). In all cases, just two factors are represented, maintaining the other at a constant value equal to the center of the design. The two most important factors are plotted.

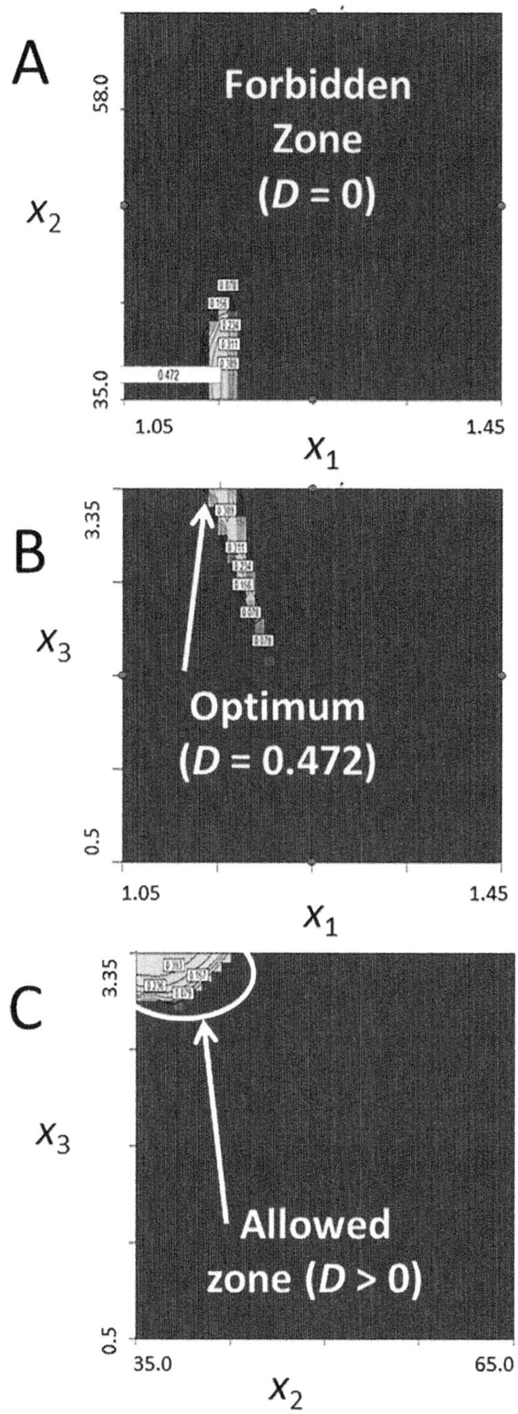

FIGURE 10.9 Contour plots for the global desirability function corresponding to the three combinations of factors.

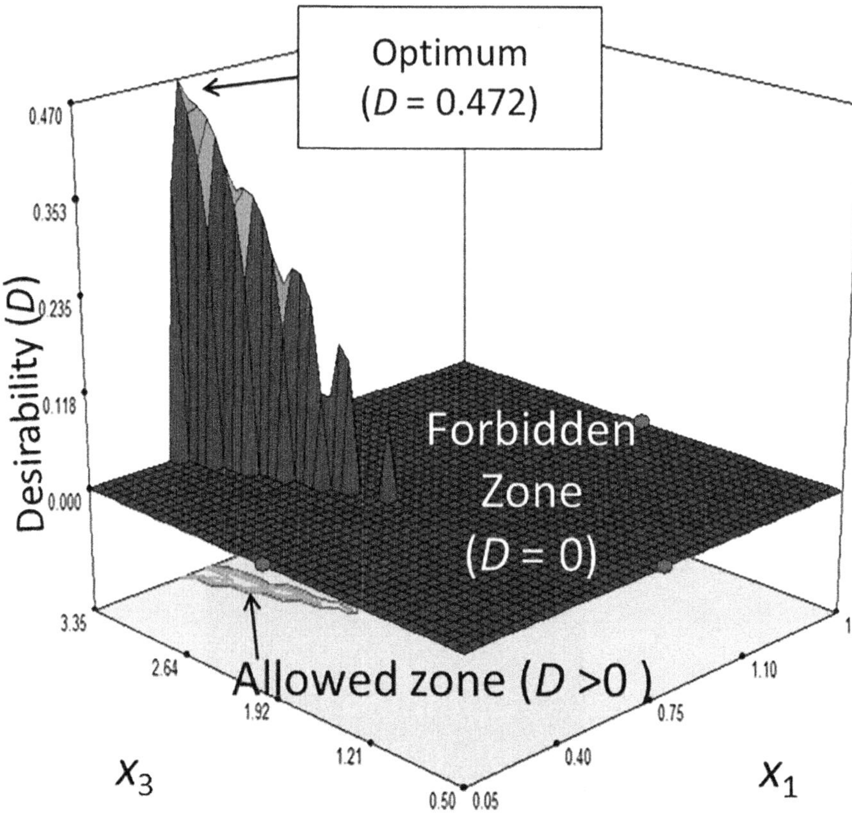

FIGURE 10.10 **(See color insert.)** Surface for the desirability (D) for the combination x_1 vs. x_3.

10.10 SOFTWARE

Several software are available for the implementation of DOE and optimization. All of them allow the implementation of different tools regarding screening assays and RSM, each one with its particular way of use. Among the most commonly used, the following can be mentioned: Design Expert (www.statease.com/), Minitab (www.minitab.com/), SPSS (www.ibm.com/analytics/us/en/technology/spss/), JMP (www.jmp.com/), R: The R Project for Statistical Computing (www.r-project.org/), Six Sigma (www.isixsigma.com/six-sigma-software/), Unscrumbler (www.camo.com/), NCSS (www.ncss.com/), Cornestone (www.camline.com/), and SAS (https://support.sas.com/rnd/app/qc/qc/qcdesign.html).

ACKNOWLEDGMENT

The author is grateful to CONICET (Consejo Nacional de Investigaciones Científicas y Técnicas, Project PIP-2015 N° 0111) and ANPCyT (Agencia Nacional de Promoción Científica y Tecnológica, Project PICT 2014-0347) for financial support.

REFERENCES

1. R. Brereton. *Chemometrics: Data Analysis for the Laboratory and Chemical Plant.* John Wiley & Sons, Chichester, 2003.
2. R.H. Myers, D.C. Montgomery. *Response Surface Methodology: Process and Product Optimization Using Designed Experiments (Wiley Series in Probability and Statistics).* Wiley, New York, 2009.
3. R. Mohammadi, M.A. Mohammadifar, A.M. Mortazavian, M. Rouhi, J.B. Ghasemi, Z. Delshadian. Extraction optimization of pepsin-soluble collagen from eggshell membrane by response surface methodology (RSM). Food Chem. 190 (2016) 186–193.
4. G. Astray, B. Gullón, J. Labidi, P. Gullón. Comparison between developed models using response surface methodology (RSM) and artificial neural networks (ANNs) with the purpose to optimize oligosaccharide mixtures production from sugar beet pulp. Ind. Crops Prod. 92 (2016) 290–299.
5. Z. Šumić, A. Vakula, A. Tepić, J. Čakarević, J. Vitas, B. Pavlić. Modeling and optimization of red currants vacuum drying process by response surface methodology (RSM). Food Chem. 203 (2016) 465–475.
6. D. Turan, E. Capanoglu, F. Altay. Investigating the effect of roasting on functional properties of defatted hazelnut flour by response surface methodology (RSM). LWT—Food Sci. Technol. 63 (2015) 758–765.
7. M.A. Bezerra, R.E. Santelli, E.P. Oliveira, L.S. Villar, L.A. Escaleira. Response surface methodology (RSM) as a tool for optimization in analytical chemistry. Talanta 76 (2008) 965–977.
8. G. Hanrahan, R. Montes, F.A. Gomez. Chemometric experimental design based optimization techniques in capillary electrophoresis: A critical review of modern applications. Anal. Bioanal. Chem. 390 (2008) 169–179.
9. S.L.C. Ferreira, R.E. Bruns, H.S. Ferreira, G.D. Matos, J.M. David, G.C. Brandao, E.G.P. da Silva, L.A. Portugal, P.S. dos Reis, A.S. Souza, W.N.L. dos Santos. Box-Behnken design: An alternative for the optimization of analytical methods. Anal. Chim. Acta 597 (2007) 179–186.
10. B. Dejaegher, Y. Vander Heyden. Experimental designs and their recent advances in set-up, data interpretation, and analytical applications. J. Pharm. Biomed. Anal. 56 (2011) 141–158.
10. I. Noshadi, N.A.S. Amin, R.S. Parnas. Continuous production of biodiesel from waste cooking oil in a reactive distillation column catalyzed by solid heteropolyacid: Optimization using response surface methodology (RSM). Fuel 94 (2012) 156–164.
12. G. Dingstad, F. Westad, T. Naes. Three case studies illustrating the properties of ordinary and partial least squares regression in different mixture models. Chemom. Intell. Lab. Syst. 71 (2004) 33–45.
13. N. Ortega, S.M. Albillos, M.D. Busto. Application of factorial design and response surface methodology to the analysis of bovine caseins by capillary zone electrophoresis. Food Control 14 (2003) 307–315.
14. J.V. Nardi, W. Acchar, D. Hotza. Enhancing the properties of ceramic products through mixture design and response surface analysis. J. Eur. Ceramic Soc. 24 (2004) 375–379.
15. Z.K.Awad, T. Aravinthan, Y. Zhuge, F. Gonzalez. A review of optimization techniques used in the design of fibre composite structures for civil engineering applications. Mater. Design 33 (2012) 534–544.

16. K.M. Lee, D.F. Gilmore. Formulation and process modeling of biopolymer (polyhydroxyalkanoates: PHAs) production from industrial wastes by novel crossed experimental design. Process Biochem. 40 (2005) 229–246.
17. S.M. Rafigh, A.V. Yazdi, M. Vossoughi, A.A. Safekordi, M. Ardjmand. Optimization of culture medium and modeling of curdlan production from Paenibacillus polymyxa by RSM and ANN. Int. J. Biolog. Macromol. 70 (2014) 463–473.
18. W. Zhi, J. Song, F. Ouyang. Application of response surface methodology to the modeling of α-amylase purification by aqueous two-phase systems. J. Biotechnol. 118 (2005) 157–165.
19. E. Kristo, C.G. Biliaderis, N. Tzanetakis. Modelling of the acidification process and rheological properties of milk fermented with a yogurt starter culture using response surface methodology. Food Chem. 83 (2003) 437–446.
20. L. Vera-Candioti, M.M. De Zan, M.S. Cámara, H.C. Goicoechea. Experimental design and multiple response optimization. Using the desirability function in analytical methods development. Talanta 124 (2014) 123–138.
21. R. Leardi. Experimental design in chemistry: A tutorial. Anal. Chim. Acta 652 (2009)161–172.
22. B. Moras, S. Rey, G. Vilarem, P-Y. Pontalier. Pressurized water extraction of isoflavones by experimental design from soybean flour and soybean protein isolate. Food Chem. 214 (2017) 9–15.
23. J.D. Floros, A. Ekanayake, G.P. Abide, P.E. Nelson. Optimization of a diced tomato calcification process. J. Food Sci. 57 (1992) 1144–1148.
24. S. Balasubramanian, G. Paridhi, J. D. Bosco, D. M. Kadam. Optimization of process conditions for the development of tomato foam by Box-Behnken design. Food Nut. Sci. 3 (2012) 925–930.
25. A. Anastácio, I.S. Carvalho. Phenolics extraction from sweet potato peels: Key factors screening through a Placket–Burman design. Ind. Crops Prod. 43 (2013) 99–105.
26. J. Zupan, J. Gasteiger. *Neural Networks in Chemistry and Drug Design*, 2nd ed. Wiley-VCH, Weinheim, 1999.
27. S. Haykin. *Neural Networks. A Comprehensive Foundation*, 2nd ed. Prentice-Hall, Upper Saddle River, NJ, 1999.
28. O. Prakash, S. Mehrotra, A. Krishna, B.N. Mishra. A neural network approach for the prediction of in vitro culture parameters for maximum biomass yields in hairy root cultures. J. Theor. Biol. 265 (2010) 579–585.
29. G. Derringer, R. Suich. Simultaneous optimization of several response variables. J. Qual. Technol. 12 (1980) 214–219.

CHAPTER 11

Data Preprocessing

Leo M.L. Nollet
University College Ghent

CONTENTS

11.1 INTRODUCTION

Data preprocessing and reduction are essential techniques in food analysis because of increasingly large datasets. These methods are intended to reduce real word datasets.

Ramírez-Gallego et al. (1) summarized, categorized, and analyzed the contributions on data preprocessing for coping with streaming data. They present an analysis of the predictive performance, reduction rates, computational time, and memory usage of different data preprocessing techniques.

The selection of optimal preprocessing is among the main bottlenecks in chemometric data analysis (2). Preprocessing currently is a burden, since a multitude of different preprocessing methods are available for, e.g., baseline correction, smoothing, and alignment, but it is not clear beforehand which method(s) should be used for which dataset. The process of preprocessing selection is often limited to trial-and-error and is therefore considered somewhat subjective. J. Gerretzen et al. (2) present a novel, simple, and effective approach for preprocessing selection. The defining feature of this approach is a design of experiments. On the basis of design, the model performances of a few well-chosen preprocessing methods, and combinations thereof (called strategies), are evaluated. Interpretations of the main effects and interactions subsequently enable the selection of an optimal preprocessing strategy.

11.2 WHY DATA PREPROCESSING?

Three reasons exist why data are preprocessed. Data generated by analysis techniques may be incomplete and some values may be lacking. The dataset is noisy and may contain errors or outliers. Lastly, the data may be inconsistent and may contain discrepancies in codes or names (3).

11.3 TASKS IN DATA PREPROCESSING

Possible tasks in data preprocessing are:

- Data cleaning: Missing values are filled in, noisy data are smoothed, outliers are identified or removed, and inconsistencies are resolved.
- Data integration by using multiple databases, data cubes, or files.
- Data transformation: Normalization and aggregation.
- Data reduction: Here the volume of the results is reduced by producing the same or similar analytical results.
- Data discretization: This is a part of data reduction where numerical attributes are replaced with nominal ones.

11.4 DATA CLEANING

11.4.1 Fill in Missing Values (Attribute or Class Values)

- Ignoring the tuple is usually done when class label is missing.
- The attribute mean (or majority nominal value) is used to fill in the missing value.
- One is using the attribute mean (or majority nominal value) for all samples belonging to the same class.
- The missing value is predicted by using a learning algorithm. Consider the attribute with the missing value as a dependent (class) variable and run a learning algorithm (usually Bayes or decision tree) to predict the missing value.

11.4.2 Identify Outliers and Smooth Out Noisy Data

The following steps may be considered:

- Binning.
- Sorting the attribute values and partitioning them into bins.

- Then smoothing by bin means, bin median, or bin boundaries.
- Clustering: Group values in clusters and then detect and remove outliers (automatic or manual).
- Regression: Smooth by fitting the data into regression functions.

11.4.3 Correct Inconsistent Data by Using Domain Knowledge or Expert Decision

Correct inconsistent data are corrected by using domain knowledge or expert decision.

11.5 DATA TRANSFORMATION

11.5.1 Normalization

Attribute values may be scaled to fall within a specified range.

Scaling is using mean and standard deviation (useful when min and max are unknown or when there are outliers).

11.5.2 Aggregation

Aggregation is the process of moving up in the concept hierarchy on numeric attributes.

11.5.3 Generalization

Generalization is moving up in the concept hierarchy on nominal attributes.

11.5.4 Attribute Construction

Attribute construction is replacing or adding new attributes inferred by existing attributes.

11.6 DATA REDUCTION

11.6.1 The Number of Attributes May Be Reduced

Possibilities are data cube aggregation by applying roll-up, slice, or dice operations.

Irrelevant attributes may be removed by attribute selection (filtering and wrapper methods) after searching the attribute space.

Principle component analysis is searching for a lower dimensional space that can best represent the data.

11.6.2 The Number of Attribute Values May Be Reduced

Binning is reducing the number of attributes by grouping them into intervals (bins).

Clustering is grouping values in clusters.

The number of attribute values may also be reduced by aggregation or generalization.

11.6.3 The Number of Tuples May Be Reduced

The possibility exists to reduce the number of tuples.

11.7 DISCRETIZATION AND GENERATING CONCEPT HIERARCHIES

When using unsupervised discretization, class variable is not used. The processes used are equal-interval (equiwidth) binning by splitting the whole range of numbers in intervals with equal-size or equal-frequency (equidepth) binning by using intervals containing equal number of values.

When applying supervised discretization, the values of the class variable are used.

When using class boundaries there are three steps:

- Sort values.
- Place breakpoints between the values belonging to different classes.
- If there are too many intervals, merge intervals with equal or similar class distributions.

Concept hierarchies are generated by recursively applying partitioning or discretization methods.

11.8 DATA PREPROCESSING IN FOOD AND BEVERAGE AUTHENTICATION

In recent years, a wide range of techniques have emerged that, together with multivariate and multiway chemometrics, reduce the time of analysis and provide more informative results. The combination or fusion of results from different techniques increases the reliability of classification or prediction of foodstuff specifications. The combination of data from several techniques represents an important challenge for chemometricians. The review of Borràs et al. (4) provides a general overview of data fusion strategies that have been used in the field of food and beverage authentication and quality assessment.

Delabrière et al. (5) developed the *proFIA* package with innovative algorithms to preprocess flow injection analysis coupled to the high resolution mass spectrometry (FIA-HRMS) raw files. The workflow consists of three steps: (1) noise estimation, peak detection, and quantification, (2) peak grouping across samples, and (3) missing value imputation.

To prove that the experimental data are significant from the statistical standpoint has led to a steep increase in the use and development of statistical software (6). The increasing number of software tools and packages and their wide usage have created a generation of "click and do" users. They have no idea about how the statistical parameters are calculated, and about the theoretical and practical reasons of performing such tests. The authors list and discuss in detail a number of uses and misuses of some statistical methods in food science and technology using available statistical software.

Data normalization plays a critical role in metabolics. The conventional methods such as constant sum normalization (CSN) and probabilistic quotient normalization (PQN) are widely used, but both methods have their own shortcomings. Dong et al. (7) proposed a new data normalization method, called group aggregating normalization (GAN). The samples were normalized so that they aggregate close to their group centers in a principal component analysis subspace. The GAN procedure seemed to be a more robust model in the multivariate data analysis.

A series of image preprocessing techniques on TLC (thin layer chromatography) images such as compression, inversion, elimination of general baseline, removing spots shift, and concavity, and finally conversion to RGB (red, green, and blue) chromatograms were performed for the quality control of saffron (8). Subsequently, an unsupervised multivariate data analysis including principal component analysis and k-means clustering was utilized to investigate the soil salinity effect.

REFERENCES

1. S. Ramírez-Gallego, B. Krawczyk, S. García, M. Wozniak. A survey on data preprocessing for data stream mining: Current status and future directions. *Neurocomputing* 239 (2017), 39–57.
2. J. Gerretzen, E. Szymanska, J.J. Jansen, J. Bart, H.-J. van Manen, E.R. van den Heuvel, L.M.C. Buydens. Simple and effective way for data preprocessing selection based on design of experiments. *Analytiva Chemistry* 87 (24) (2015), 12096–12103.
3. www.cs.ccsu.edu/~markov/ccsu_courses/DataMining-3.html.
4. E. Borràs, J. Ferré, R. Boqué, M. Mestres, L. Aceña, O. Busto. Data fusion methodologies for food and beverage authentication and quality assessment—A review. *Analytica Chimica Acta* 891 (2015), 1–14.
5. A. Delabrière, U.M. Hohenester, B. Colsch, C. Junot, F. Fenaille, E.A. Thévenot. proFia: A data preprocessing workflow for flow injection analysis coupled to high-resolution mass spectrometry. *Bioinformatics* 33 (2017), 3767–3775.
6. C.A. Nunes, V.O. Alvarenga, A. de Souza Sant'Ana, J.S. Santos, D. Granato. The use of statistical software in food science and technology: Advantages, limitations and misuses. *Food Research International* 75 (2015), 270–280.
7. J. Dong, K.-K. Cheng, J. Xu, Z. Chen, J.L. Griffin. Group aggregating normalization method for the processing of NMR-based metabolomics data. *Chemometrics and Intelligent Laboratory Systems* 108 (2011), 123–132.
8. H. Sereshti, Z. Poursorkh, G. Aliakbarzadeh, S. Zarre, S. Ataolahi. An image analysis of TLC patterns for quality control of saffron based on soil salinity effect: A strategy for data (pre)-processing. *Food Chemistry* 239 (2018), 831–839.

CHAPTER 12

Classification and Modeling Methods

Silvana M. Azcarate, Marianela Savio, and José M. Camiña
Universidad Nacional La Pampa (UNLPam) and Consejo Nacional de Investigaciones Científicas y Técnicas (CONICET)

CONTENTS

12.1 INTRODUCTION

Ensuring food safety by means of quality and traceability assessment is a current challenge for food researchers, governments, and food industry (Azcarate et al. 2017). The development and improvement of robust, efficient, sensitive and cost-effective analytical methodologies for food analysis is continually demanded (Castro-Puyana et al. 2017, Lucci et al. 2017).

Food possess a complex matrix consisting of naturally occurring compounds such as minerals, lipids, organic acids, carbohydrates, proteins, vitamins, volatile organic compounds, etc. as well as other substances from agrochemical treatments, technological processes, or packaging materials (Lucci et al. 2017, Marini 2013). The analytical techniques involved in the qualification or quantification of these analytes include a wide range of methods which provide a huge amount of instrumental data (i.e., chromatography and spectroscopy). Food scientists all over the world are dealing with these massive amounts of data. Therefore, the use of additional disciplines and tools along with the traditional ones are needed; these tools are embedded into the discipline called chemometrics (Roberts and Cozzolino 2016, Kumar et al. 2014, Matero 2010, Munck 2007, Hopke 2003).

The word chemometrics was introduced in the 1970s by chemists and statistics researchers and according to Bruce Kowalski (U.S.) and Svande Wold (Sweden) chemometrics is "the application of mathematical and statistical tools to chemistry" (Hibbert 2016). The latest definition highlights the practical nature of chemometrics: "the science of relating measurements made on a chemical system or process to the state of the system via application of mathematical or statistical methods" (Hibbert 2016). The first uses of chemometrics were based on the tools employed in classical statistics; chemometrics was widely used after the development of specific software according to an objective: "classification". The main limitation in the development and employment of complex chemometric methods was related to the limitation in hardware technologies. Thereby, the first uses emerged in the 1980s, but their use had a great expansion since the 1990s with the improvements in new computer processors. Henceforth, the application of modern statistical concepts, experimental designs, and chemometrics turned into an integral and relevant component of the analytical process (Roberts and Cozzolino 2016).

Over the years, the advancement in the use of several chemometric methods—based on statistical multivariate models—has allowed the assessment of food traceability and fingerprinting identification. Thereby, the compositional profile (fingerprint) of foodstuffs is crucial for the characterization of food products. Fingerprinting is referred to the analysis of as many compounds as possible within a system, including their detection and the subsequent statistical treatment of the obtained results in order to look for patterns in the samples (Castro-Puyana et al. 2013, Salvatore et al. 2013). Thus, the fingerprint can be an indicator of quality, origin and authenticity. Accordingly, the aim is identifying relevant compounds (food markers) to solve the problem under study (adulteration, authentication, etc.).

Taking advantage of the information acquired with modern instrumentation (inductively coupled plasma mass spectrometry (ICP-MS), liquid chromatography MS (LC-MS), gas chromatography MS (GC-MS), Fourier-transform infrared spectroscopy (FTIR), near-infrared spectroscopy (NIR), nuclear magnetic resonance (NMR), capillary electrophoresis diode array detection (CE-DAD), flow injection MS (FI-MS), direct analysis in real time MS (DART-MS), proton transfer reaction MS (PTR-MS), etc.), the multivariate modeling has gained importance due to the possibility of using a large number of variables simultaneously, with the additional benefit of direct transfer of all the instrumental data to the software. Chemometric tools handle these large amounts of datasets and assist in extracting qualitative and quantitative relevant chemical information, aiming to assess the possible differences among samples (Azcarate et al. 2016, Tura et al. 2004, Kos et al. 2003, Yu and MacGregor 2003).

The particular statistical approach to be used will largely depend on the objectives of the study, for example, to distinguish among quality levels, authenticate product origin, or encounter variations in processing or handling (Castro-Puyana et al. 2017, Salvatore et al. 2013). In this way, several compositional descriptors can be used to chemometrically differentiate food samples that vary according to their geography, botanic origin, age, variety, quality, storage time, spoilage, fluctuations in different stages of production processing, among others (Manfredi et al. 2018, Azcarate et al. 2017, Donglu et al. 2017,

Guo et al. 2017, Hu et al. 2017, Karunathilaka et al. 2017, Longobardi et al. 2017, Savini et al. 2017, Sucupira et al. 2017, Alves Filho et al. 2016, Aru et al. 2016, Azcarate et al. 2016, Gan et al. 2016, Karunathilaka et al. 2016, Schuhfried et al. 2016, Spiteri et al. 2016, Wei et al. 2016, Azcarate et al. 2015, Tahri et al. 2015, Ammor et al. 2009). In addition, detection of adulteration is one of the most attractive purposes of food fingerprinting (Reis et al. 2017, Dall' Acqua et al. 2015, Liu et al. 2013).

After identification of the relevant variables for the problem under study, chemometric techniques assist in the model construction that characterize target samples within previously defined and validated groups.

In the last few years, due to the possibility of obtaining a considerable quantity of data the techniques of multiway data analysis are gaining a foothold in food authenticity applications (Salvatore et al. 2013). Thus, partial least squares discriminant analysis (PLSDA) and its multiway applications such as the unfolded (UPLSDA) and multiway partial least squares discriminant analysis (NPLSDA) are achieving (redundance) significance as powerful and robust tools for specific purposes (Porro-Munoz et al. 2011). In addition, soft independent modeling of class analog (SIMCA) has also been extensively employed for different classification purposes including its multiway extension, NSIMCA (Azcarate et al. 2017, Santos et al. 2016, Zhao et al. 2014, Castro-Puyana and Herrero 2013, Galtier et al. 2011).

Based on the importance of multivariate methods to evaluate traceability in foods, this chapter provides awareness on the increasing applications of two- and three-way methodologies and their advantages in the classification context. First, an explanation of data structure properties recorded from different analytical platforms with multivariate detection is provided. Then, the most explored models and algorithms for fingerprinting classification are briefly detailed along with the relevant works applying these techniques, focusing on classification results and their potential uses.

12.2 DATA STRUCTURES

The fingerprinting techniques most frequently used have shown great potential in food traceability and authentication. Depending on the analytical platform applied, the responses generated from a specific technique for a given sample can be arranged as a vector of numerical values, so-called first-order data, such as spectral data from various spectroscopic techniques, such as FTIR, NIR, NMR, ICP-MS among the most used. On the other hand, when two "hyphenated" first-order instruments are coupled, as in tandem GC-MS, LC-MS or CE-DAD, it is possible to acquire matrix data for a single sample, which is considered as second-order data (Escandar et al. 2014, Arancibia et al. 2012). Thus, multiway data structures also provide potential methodologies for food fingerprinting development.

Commonly, for fingerprinting studies a large number of samples are required in order to assess potential classifications. In these cases, when a group of samples are analyzed, the recorded data can be arranged in two- or three-dimensional arrays according to their nature, obtaining from joined samples a data matrix or a data cube, respectively (Figure 12.1). Thus, information extraction from multivariate data can be done in several ways. However, it is relevant to highlight that the choice of chemometric methods will mainly depend on the structure or arrangement of the dataset.

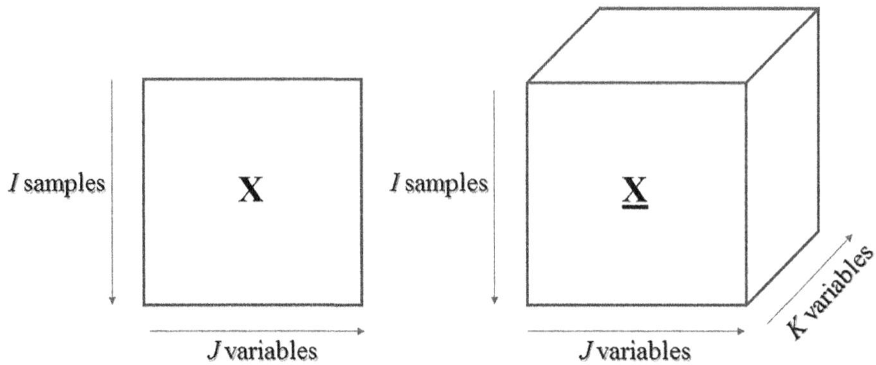

FIGURE 12.1 Schematic representation of two-way data arranged in a matrix **X** ($I \times J$) and three-way data arranged in a cube $\underline{X}$ ($I \times J \times K$).

12.3 DATA MODELING

Chemometric tools based on first- and second-order data are common and they are frequently used to classify samples according to an agreed and designated interest property. Thereby, the classification can not only provide information about quality but it can also describe a process (as degradation) in food and due to this the possibilities of information are really wide.

A classification model involves the use of 2D or 3D data matrix, which consists of objects (samples to classify) and variables (spectral data or concentration of several compounds present in the samples). Thus, a model is built from the sample pre-categorization into a known category, and the calculation algorithm is trained to classify the objects into the given categories. The aim in chemometric classification is to find the allocation rule that gives the highest percentage of correct classification. Nevertheless, this aim is not always achieved since the classification may depend on the selection and type of variable.

In case of a multiway data structure, the analysis can be done through multiway chemometric methods. However, the data structure can be simplified by rearranging the multiway data array into a data matrix that can be processed by the classical chemometric two-way classifiers. This procedure is commonly called "unfolding" and is performed by concatenating two-way arrays (matrices) extracted from the data array (Salvatore et al. 2013).

Taking into account the different possibilities of data organization, in the following section, the most frequently used methods to conduct discriminant and modeling classification are briefly described and applied on food fingerprinting studies.

12.3.1 Chemometric Methods Exploring Fingerprinting Dataset

Typically, a data analysis procedure begins with an exploratory data analysis, assayed to sum up and to visualize the data in the best possible way. These methods, better known as "unsupervised", involve those tools used as a screening phase from which the relationships between data variables are inspected, possible clusters are observed, and relevant variables are selected for classification.

According to the acquired data, there are three main methods that are used to obtain a preliminary data analysis, namely PCA (principal component analysis), PARAFAC (parallel factor analysis), and Tucker3. PCA is the most widely used method and it is applied to 2D data; whereas PARAFAC is performed on 3D data. Nonetheless, when a specific condition of a dataset for applying PARAFAC is not accomplished, the Tucker3 model may be a better choice for maintaining the data structure. Otherwise, the three-way data array can be unfolded in a two-way matrix; in this case, PCA may be suitable.

12.3.1.1 Principal Component Analysis

PCA is the most widely used method in several disciplines as an exploratory tool due to its simplicity. It assists in bilinear modeling that provides an overview of the main information contained in a 2D table. It can be named as a "projection method" because it requires information from the original variables, which are projected on a smaller number of latent variables (LVs) well known as principal components (PC). Each PC explains a fraction of the overall information enclosed in the original dataset. As an outcome, this analysis provides a visual representation of the relationships between the samples and variables. Thus, a data matrix **X** ($I \times J$) is decomposed in two matrices: score matrix (**A**) and loading matrix (**B**) (Figure 12.2), according to the mathematical expression:

$$\mathbf{X}_{IJ} = \mathbf{A}_{AI}\mathbf{B}_{AJ}^{T} + \mathbf{E}_{IJ} \tag{12.1}$$

where $\mathbf{X}_{IJ}$ is the matrix of information recorded experimentally for I samples in an arrangement of sensors with J channels, **A** is the matrix of scores, **B** is the matrix of loadings, and $\mathbf{E}_{IJ}$ is the residuals matrix. **A** indicates the relationships between samples, while **B** indicates the relationships between variables. The superscript T indicates the transpose of a vector or matrix (Bro and Smilde 2014).

Each J column of **B** defines the direction of an axis. Since the new axes are defined, each sample is projected and this information is found in the **A** matrix. Considering the correlation existing between the original variables, only a subset of them–the first A-represents the data—the remnant data contain irrelevant or random information associated with the error (**E**).

The decomposition can be carried out through algorithms such as singular value decomposition (SVD) or nonlinear iterative partial least squares (NIPALS) (Esbensen and Geladi 2010).

There are endless works reporting the employment of PCA for the assessment of food quality, as it remains a straightforward option for studying and classifying fingerprints (Pardo-Mates et al. 2017, Rohman 2017, Silva et al. 2017, Aliakbarzadeh et al. 2016, Borges et al. 2016, Azcarate et al. 2015, Huang et al. 2015, Lu et al. 2015, Basbassi et al. 2014, Camiña et al. 2014, Gao et al. 2014, López et al. 2014, Wang et al. 2014, Yudthavorasit et al. 2014, Jacobson et al. 2013, Latorre et al. 2013).

12.3.1.2 Parallel Factor Analysis

PARAFAC is the most widely used multiway decomposition method. Harshman, Caroll, and Chang (Durante et al. 2011, Castro and Manetti 2007) introduced this method in 1970 as a multivariate data analysis performed on three-way data arrays.

In general terms, the PARAFAC method achieves data compression and allows the exploratory analysis of multiway data matrices. It is based on the "parallel proportional profiles" principle in which a set of common factors can be used to explain simultaneously the variations occurring in several matrices even with disparity weighting the coefficients for each matrix. The PARAFAC decomposition aims to minimize the sum of squares of the residual e_{ijk} as indicated in Eq. (12.2) based on a least squares algorithm (Andersen and Bro 2003).

A three-way array can be decomposed by PARAFAC as a sum of triple outer product of vectors: these three sets of vectors are called loadings. Given a three-way array $\underline{\mathbf{X}}$ of dimension **I**×**J**×**K** with elements x_{ijk}, the PARAFAC model can be expressed as follows:

$$x_{ijk} = \sum_{1}^{F} a_{if} \; b_{jf} \; c_{kf} + e_{ijk} \tag{12.2}$$

where **A** ($I \times F$) with element a_{if} is the first mode score matrix, **B** ($J \times K$) with element b_{jf} and **C** ($K \times F$) with element c_{kf} are the second and the third modes loadings, respectively. **F** is the number of components used in the PARAFAC model; e_{ijk} is a residual term containing the unmodeled part of the data array $\underline{\mathbf{X}}$.

The number of factors describing the systematic variation in the data array gives the rank of the PARAFAC model. A graphical representation of the decomposition is depicted in Figure 12.2 where the three-way data array $\underline{\mathbf{X}}$ is decomposed into scores and loadings, **a**, **b**, and **c**, using two PARAFAC components, leaving the unmodeled part of $\underline{\mathbf{X}}$ in the residual array $\underline{\mathbf{E}}$, which ideally only contains unsystematic noise (Murphy et al. 2013).

The three-way data decomposition gives a mathematically unique solution for a given number of components. Therefore, if the PARAFAC model is a description of a chemically meaningful structure, then the parameters of the model will have a chemical interpretation. Thus, PARAFAC can recover the profile corresponding to each generated variance source in the analyzed data. For instance, the same profiles can be measured for the sample under study and correlated to the different defined problems (Mazina et al. 2015)

Conversely, the PARAFAC model is not always appropriate and, occasionally, for its application it is mandatory to have a low-rank trilinear structure in the dataset, otherwise the PARAFAC could provide a wrong approximation or it may even be numerically impossible to fit (Kroonenberg 2008, Smilde et al. 2004).

12.3.1.3 Tucker3

Ledyard Tucker, working with psychometric data, developed a series of methods which are known as Tucker methods (Soaresa et al. 2012). These methods are nowadays recognized as multidimensional PCA or multiway PCA (ten Berge 2004). The Tucker3 method can be considered as an extension of PCA for higherorder arrays (Knolhoff and Croley 2016, Gere et al. 2014).

Let's consider a three-way array $\underline{\mathbf{X}}$ of dimension $I \times J \times K$. It is decomposed into orthonormal triplets of loading vectors where the number of extracted components can be different for the three modes. The decomposition is illustrated in Figure 12.2 and it is mathematically expressed in Eq. (12.3).

$$x_{ijk} = \sum_{l=1}^{L} \sum_{m=1}^{M} \sum_{n=1}^{N} a_{il} b_{jm} c_{kn} g_{lmn} + e_{ijk} \tag{12.3}$$

where e_{ijk} are the elements of a three-way array **E** ($I \times J \times K$) containing unmodeled information by Tucker3 and g_{lmn} are the elements of a three-way array **G** ($L \times M \times N$) called "core". The core elements are the magnitudes of the interactions between the factors in

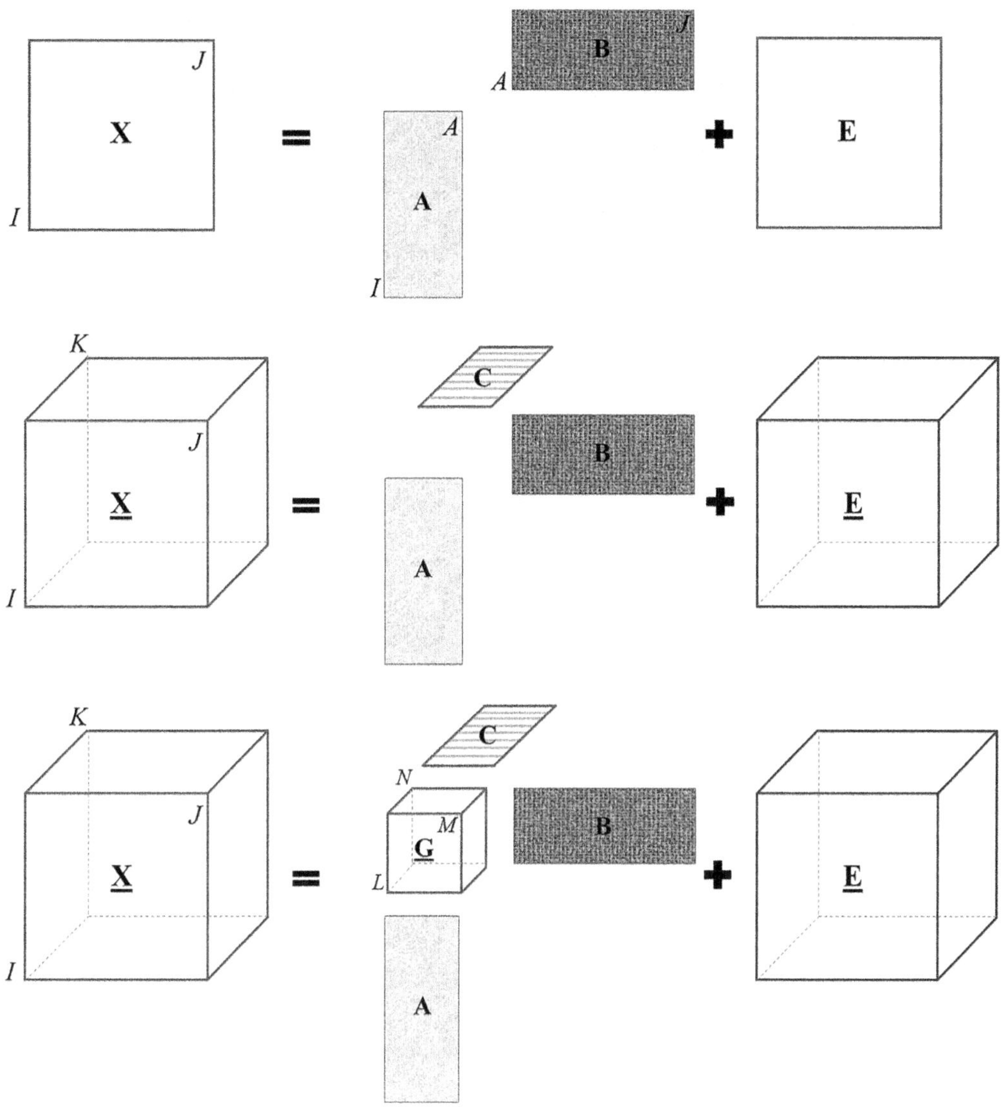

FIGURE 12.2 Schematic representation of PCA decomposition of the two-way data array **X** ($I \times J$); PARAFAC and Tucker3 decomposition of the three-way data array **X** ($I \times J \times K$).

different modes. Another striking difference of the Tucker3 method from the other multiway methods is that **G** is non-diagonal (Knolhoff and Croley 2016); this means that there is interaction between the factors.

The Tucker3 method, when applied to three-way arrays $\underline{\mathbf{X}}$ ($I \times J \times K$) with elements x_{ijk}, produces the loading matrices **A** ($I \times L$), **B** ($J \times M$), and **C** ($K \times N$), where a_{il}, b_{jm}, and c_{kn} are the elements of the loading matrices **A**, **B**, and **C**, respectively. The number of columns (L, M, and N) in the loading matrices corresponds to the number of factors in each mode. Interestingly, unlike the traditional multiway methods, in Tucker3 decomposition, different number of factors are permitted for each mode (i.e., $L \neq M \neq N$), making this attribute greatly versatile.

As an example, consider a Tucker3 model with complexity 3, 4, 3 ($L = 3$, $M = 4$, and $N = 3$), where the element g_{121} has very low value (near zero); this means that the interactions between the first factor in mode 1, second factor in mode 2, and first factor in mode 3 are not significant. On the other hand, if g_{221} has a high value, the interactions between the second factor in mode 1, second factor in mode 2, and first factor in mode 3 have a significant importance for the model. These cross-interactions between the factors found in Tucker3 models not only provide flexibility but also increase the complexity, constructing its interpretation of a nontrivial fact (ten Berge 2004).

As in other multiway methods, an important step is the factor number determination, in this case for each mode. Several approaches can be used: chemical information regarding the system, cross-validation, and the enhancement in variance explained in terms of the number of factors used (Azcarate et al. 2016, Prieto et al. 2012, Bro et al. 2008).

12.3.2 Classification by Chemometric Methods

Chemometric classification methods are an assembly of algorithm groups that are well known as supervised methods, which are fundamentally designed to find mathematical models that are able to recognize samples as members of a class based on a measurement set. For classification purposes, there are no restrictions on how the analytical platform for chemical data is obtained.

Supervised classification procedures use a common strategy regardless of the applied algorithm, consisting of the following general steps (Massart et al. 1997):

- Training the selected set consisting of objects belonging to a known class for which variables were measured.
- Variable selection involving those variables that contain information relevant to classification.
- Mathematical model construction using the relevant variables of the training set.
- Model validation using an independent sample set to evaluate the reliability of the achieved classification.

Next, brief descriptions of chemometric methods that are frequently applied in food fingerprinting classifications are provided.

12.3.2.1 Soft Independent Modeling of Class Analogy

Proposed by Wold, SIMCA is a widespread classification method, used to solve classification problems in chemistry (Massart et al. 1997). For a two-way dataset, the SIMCA method assumes that the measurement values for groups of similar samples have a uniform distribution based on PCA (Wold 1976).

Initially, a PCA model is performed for training samples of each class individually and determines the rank in each class (Q). The rank is the number of principal components (A) necessary to explain the data variance and define an A dimensional space for the Q class of a given level of significance; this A dimensional space is termed as "hyperbox" (Bro and Smilde 2014).

Once the PC decomposition is computed (Eq. (12.1)), it is possible to define the distance of the object i from the class ($D_i^{(Q)}$) according to Eq. (12.4) (Forina et al. 2008):

$$D_i^{(Q)} = \sqrt{OD_i^{(Q)^2} + SD_i^{(Q)^2}} \tag{12.4}$$

The two terms that comprise the distance of the object i from the class Q are the orthogonal distance (*OD*) and the distance of the sample i from the score space of the model (*SD*). The *OD* is the residual variance of the sample i in class Q. The *SD* term is computed when the predicted scores for a test sample are outside the hyperbox for the Q class.

Thus, it is possible to capture information reflecting the similarities of individuals within each class by class-specific PCA models using training samples from each class. Then, unknown samples are compared to the class models and assigned to different classes according to their analogy to the training samples. The number of components required by the PCA models to describe the training samples in each class is determined using cross-validation. The distance of a sample from the class model is calculated utilizing the orthogonal distance of the sample from the model space and the distance of the sample from the score space. The SIMCA classification rule is then based on the comparison of the squared distance with the class residual variance by means of a *F*-test. The unknown samples are assigned to a class if the test is passed.

In SIMCA, it might be possible that models not have discriminating power and the same sample are assigned to different classes. Another possibility is that samples that are completely different are not assigned to any of the classes. For this reason, SIMCA is practical in those situations where the variance is relevant to separate classes. However, in cases where the between-class variance is smaller than the within-class variance, variations of the classes will merge. Therefore, SIMCA will provide poor classification results (Galtier et al. 2011).

In analogy, with the two-way SIMCA, namely NSIMCA, the data of each class are modeled separately by using a multiway decomposition method like PARAFAC or Tucker3 on the training set, obtaining one model for each class. *OD* (from the class model) is based on the squared residuals from PARAFAC or Tucker3 model and the score distances (distance in model space) are based on model scores (mode 1 loadings). In extending the original SIMCA approach to the multiway case, the most critical step is the evaluation of freedom degrees that are to be used in the class variance estimation and in the *F*-test. To assess the optimal model dimensionality (number of factors for PARAFAC and number of factors in each variable mode for Tucker3), the best efficiency in cross-validation must be considered (Durante et al. 2011).

12.3.2.2 Partial Least Squares Regression Discriminant Analysis

PLSR method is the most popular regression method in chemometrics. Originally proposed for multivariate calibration, partial least square (PLS) has been extensively used in classification problems and their mathematical foundations have been described in the literature (Indahl 2014, Ballabio and Consonni 2013).

PLSDA is a linear classification method that combines the properties of PLSR with the discrimination power of a classification technique. PLSDA is based on the PLS regression algorithm (PLS1 when dealing with one dependent **Y** variable and PLS2 in the presence of several dependent **Y** variables) that searches for LVs that have a maximum covariance with the **Y** variables.

The main difference between calibration and classification application is in building of the dependent variable **Y**, which contains concentration values for the calibration model, while for discriminant analysis purposes **Y** contains coding values representing the sample class labels. When the classification problem involves more than two classes, PLS2 is employed.

PLS regression is conducted between the instrumental response in **X** block and the class label in **Y** block using training samples, and the optimal number of LVs is chosen based on the error rate by cross-validation (de Almeida et al. 2013). The final model for *A* latent variables is used to predict the class label in the test set according to Eq. (12.5):

$$\mathrm{Y_{pred}} = \mathrm{T_{pred}^{T}}\mathrm{V} \tag{12.5}$$

where $\mathbf{y}_{\mathrm{pred}}$ is the predicted label class, $\mathrm{T_{pred}^{T}}$ are the scores of test samples obtained by projection of $\mathbf{X}_{\mathrm{pred}}$ onto the training loadings, and **v** is the vector of the regression coefficients.

The main advantage of PLSDA is that the relevant sources of data variability are modeled by the so-called LVs that are linear combinations of the original variables. Consequently, it allows graphical visualization and understanding of the different data patterns and relations by LV scores and loadings. Loadings are the coefficients of variables in the linear combinations which determine the LVs and, therefore, they can be interpreted as the influence of each variable on each LV, while scores represent the coordinates of samples in the LV projection hyperspace.

Thus, the matrix **X** $(I \times J)$ is a predictor matrix and the response matrix **Y** $(I \times K)$ comprises categorical (dummy) variables that describe class memberships. *K* is equal to the number of classes. If only two classes are considered, matrix **Y** is reduced to a vector **y** $(I \times 1)$ that is set to have zero and one entry for each class (Brereton and Lloyd 2014).

PLSDA aims to improve the separation between the two groups by using the class information. When PLS regression is developed, the response value $\mathbf{Y}_{\mathrm{pred}}$ is predicted for a new sample. The decision is based on the comparison of $\mathbf{Y}_{\mathrm{pred}}$ with given categorical variables in **Y**. The sample is attributed to the class that has the minimal distance between **Y** and $\mathbf{Y}_{\mathrm{pred}}$.

PLSDA is one of the strongest classification tools in chemometrics. It is generally the reference method used in supervised classification studies for the assessment of food authenticity, food traceability, food adulteration, and food effects on human health (Gan et al. 2016, Tahri et al. 2015).

UPLSDA and NPLSDA are an extension of PLSDA for three-way data. As in the case of PLSR, they were originally proposed for multivariate calibration and, nowadays, they are also used in classification problems. In essence, both UPLS and NPLS for discriminant analysis are equally applicable for calibration purposes (Ouertani et al. 2014, Barker and Rayens 2003).

Basically, UPLSDA is a simple classification approach, used for unfolding the data array $\underline{\mathbf{X}}_{\mathrm{I,J,K}}$ to a matrix $\mathrm{X}_{(\mathrm{I,JK})}$, and then applying standard PLSDA for two-way data described above. On the other hand, NPLSDA can be applied for maintaining the three-way data structure and extracting the relevant information to model the class feature without distorting, compressing, or losing information (Salvatore et al. 2013). In this case, considering an $\underline{\mathbf{X}}$ array of dimension $I \times J \times K$, the NPLS model is obtained by modeling **X**, as in Tucker3 decomposition. Subsequently the obtained matrix $\underline{\mathrm{G}}$ is the core array of size $F \times F \times F$, where *F* is the number of NPLS components (factors). The dependent variable block in NPLSDA is a matrix **Y** holding the class information, defining for each category a *y* variable. Then, the **Y** array is modeled as in the two-way case by:

$$\mathbf{Y} = \mathbf{U}\mathbf{Q} + \mathbf{E}_Y \tag{12.6}$$

where **U** holds the **Y** scores and **Q** is the loading matrix. $\mathbf{E}_Y$ is **Y** residual arrays.

Assessment of most of the discriminant $\underline{\mathbf{X}}$ variables may be guided by inspection of the NPLS weights that—separately for each component and mode—show the most influent variables (signal features) in the NPLS model of a given *y* variable, since each *y* variable is related to one class.

A further approach of performing classification on multiway arrays is to use standard two-way classifiers on the scores obtained by applying decomposition techniques to the data arrays. This approach involves two stages: at first, multi-linear techniques such as PARAFAC or Tucker3 are used to model the three-way array. Subsequently, the loadings of the sample mode (scores), which are arranged in a two-way matrix, are used as new variables and processed by PLSDA (Gere et al. 2014, Pérez et al. 2009).

Application of the multiway classification methods, especially in the food area, is still limited. However, multiway PLSDA has been used in several food authentications. (Azcarate et al. 2017, Bernardes et al. 2016, Lu et al. 2014, Castro et al. 2007, Durante et al. 2006).

12.4 MODEL VALIDATION

In the model building step, the classification methods commonly employ cross-validation procedures to optimize the model fit (internal validation). Otherwise, the performance or predictive ability of the model for a training set or an independent test set can be evaluated through some figures of merit (external validation). Thus, for a classification model, the most common figures of merit are accuracy (*AC*), precision (*P*), sensitivity (*S*), and specificity (*SP*) (Szollosi et al. 2012).

The evaluated parameters throughout the internal and external validation are linked to the presence of errors in the results (samples assigned to wrong classes). Thus, each sample is assigned to a default class; then it can be classified as true positive (TP), true negative (TN), false positive (FP), or false negative (FN). TP is the number of positive samples that are correctly identified as positive samples, FN is the number of positive samples that are misclassified as negative samples, FP is the number of negative samples that are incorrectly identified as positive samples, and TN is the number of negative samples that are correctly identified as negative samples. Therefore, the figures of merit for a classifier using these parameters can be mathematically defined as shown in Table 12.1. Thus, *S* is the proportion of positives that are correctly identified; *SP* is the proportion of

TABLE 12.1 Figures of Merit Applied in the Classification Models Validation

Figures of Merit	Equation	Definition
Accuracy (*AC*)	$AC=(TP+TN)/(TP+TN+FP+FN)$	Percentage of samples correctly classified.
Precision (*P*)	$P=TP/(TP+FP)$	Samples accepted by the same model.
Sensitivity (*S*)	$S=TP/(TP+FN)$	Sa mples correctly classified within the class.
Specificity (*SP*)	$SP=TN/(TN+FP)$	Samples correctly classified that do not belong to the class.

TP: True positive, TN: True negative, FP: False positive, FN: False negative.

negatives that are correctly identified; *AC* characterizes the correct classification rate, i.e., the percentage of samples correctly classified; and *P* is the ratio of the number of samples correctly accepted to the total number of samples accepted by the same model (Oliveri and Downey 2012, Ballabio et al. 2009, Forina et al. 2001, González-Arjona et al. 1999).

12.5 NEW TRENDS IN THE USE OF DATA: DATA FUSION

Fingerprinting studies are normally concerned with multifactorial problems and it is adequate to explore and measure the same samples on several complementary analytical platforms (Skov et al. 2014).

For improving food authentication and traceability assessment, a further step is to join the output data from multiple instrumental sources. This strategy called data fusion is being increasingly developed to get more accurate knowledge regarding the sample and to yield superior inferences (classifications with less error rate and predictions with less uncertainty) than any one single technique with the aim of improving expected results (Borràs et al. 2015).

Data combination can be carried out at three levels. Low-level data fusion implies that the matrices describing the individual blocks are concatenated to build a single array after proper preprocessing. In the mid-level strategy, the relevant features are first extracted from each data source separately. Subsequently, the extracted features are concatenated to build a single array. In both cases, the obtained data block is then processed by the appropriate chemometric technique. Commonly, feature extraction uses PCs or LVs from PLS. Nevertheless, other methods have been applied to extract features depending on data structure and the particular problem. Finally, the high-level strategy involves building separate models for the different blocks and then the final results of the individual models, i.e., their predictions are integrated into a single final response.

Thus, measurements at multiple platforms such as NIR, NMR, CE-DAD, CE–MS, LC-MS, and GC-MS provide a larger number of data and, thus, increase the chances of obtaining a better understanding about a specific food. The challenge today is how to meaningfully combine not just single variables like in the past but blocks of them.

In many cases, the goal of data fusion is to extract patterns not only common to all sources but also specific to each source. For instance, in metabolomics, where the samples are measured by two different platforms—LC-MS and NMR—the interest may be to know whether only one of the platforms provides the metabolites explaining a biological phenomenon or if there is a synergy between the variables provided for both platforms.

In the literature, there is a wide variety of means for carrying out the mentioned steps as well as very diverse applications for the same. Spectroscopic techniques based on IR, Raman, or NMR spectroscopies are widely used for food fingerprinting, and the synergistic effects of two or more of them have been analyzed. Data from these techniques have been fused at low level by simply concatenating the data. More specifically, data fusion has been a commonly used strategy that improves authentication results (Cuevas et al. 2017, Reis et al. 2017, Biancolillo et al. 2014, Silvestri et al. 2014, Dupuy et al. 2010).

12.6 CURRENT CHALLENGES AND PERSPECTIVES

The future of chemometrics is quite difficult to predict because the advancement of multivariate methods has been continuous and they are efficiently applied. The improvements in exploratory tools and in the adaptability for the analysis of complex data, the advances in authentication, quality control of food and data assessment, especially the prediction of intrinsically complex properties that cannot be detected by a single technique—such as fraud, adulteration, origin, harvest, and raw material—indicate an increase in the range of applications of these methods. In addition, all these methods would obviously benefit with the advancement of data fusion, which still needs to be studied more thoroughly. In fact, new approaches are emerging such as applications combining low- and mid-level data fusion, fusion of second- and higher order data, i.e., data from hyphenated techniques and the combination of first-order data from evolving systems to predict the product quality. Now and in the future, the principal challenge in chemometrics will be the generalized and standardized use of multivariate methods, which are still not completely accepted by the analytical chemists. However, the use and the interest in these methods are increasing due to the need of more analyzing and data processing abilities.

REFERENCES

Aliakbarzadeh, G., Sereshti, H., and Parastar, H. 2016. Pattern recognition analysis of chromatographic fingerprints of *Crocus sativus* L. secondary metabolites towards source identification and quality control. *Anal Bioanal Chem* 408:3295–307.

Alves Filho, E.G., Almeida, F.D.L., Cavalcante, R.S., De Brito, E.S., Cullen, P.J., Frias, J.M., Bourke, P., Fernandes, F.A.N., and Rodrigues, S. 2016.^{1}H NMR spectroscopy and chemometrics evaluation of non-thermal processing of orange juice. *Food Chem* 204:102–7.

Ammor, M.S., Argyri, A., and Nychas, G.J. 2009. Rapid monitoring of the spoilage of minced beef stored under conventionally and active packaging conditions using Fourier transform infrared spectroscopy in tandem with chemometrics. *Meat Sci* 81:507–14.

Andersen, C.M., and Bro, R. 2003. Practical aspects of PARAFAC modeling of fluorescence excitation-emission data. *J Chemometr* 17:200–15.

Arancibia, J.A., Damiani, P.C., Escandar, G.M., Ibañez, G.A., and Olivieri, A.C. 2012. A review on second- and third-order multivariate calibration applied to chromatographic data. *J Chromatogr B* 910:22–30.

Aru, V., Pisano, M.B., Savorani, F., Engelsen, S.B., Cosentino, S., and Cesare Marincola, F. 2016. Metabolomics analysis of shucked mussels' freshness. *Food Chem* 205:58–65.

Azcarate, S.M., Gil, R., Smichowski, P., Savio, M., and Camiña, J.M. 2017. Chemometric application in foodomics: Nutritional quality parameters evaluation in milk-based infant formula. *Microchem J* 130:1–6.

Azcarate, S.M., Araújo Gomes, A., Vera-Candioti, L., Ugulino de Araújo, M.C., Camiña, J.M., and Goicoechea, H.C. 2016. Second-order capillary electrophoresis diode array detector data modeled with the Tucker3 algorithm: A novel strategy for Argentinean white wine discrimination respect to grape variety. *Electrophoresis* 37:1902–8.

Azcarate, S.M., Savio, M., Smichowski, P., Martinez, L.D., Camiña, J.M., and Gil, R.A. 2015. Single-step solubilization of milk samples with N,N-dimethylformamide for inductively coupled plasma-mass spectrometry analysis and classification based on their elemental composition. Talanta 143:64–70.

Ballabio, D., and Consonni, V. 2013. Classification tools in chemistry. Part 1: Linear models. PLS-DA. *Anal Methods* 5:3790–8.

Ballabio, D., and Todeschini, R. 2009. Multivariate classification for qualitative analysis. In *Infrared Spectroscopy for Food Quality Analysis and Control*, ed. S. Da-Wen. Amsterdam: Elsevier, 83–104.

Barker, M., and Rayens, W. 2003. Partial least squares for discrimination. *J Chemometr* 17:166–73.

Basbassi, M., De Luca, M., Ioele, G., Oussama, A., and Ragno, G. 2014. Prediction of the geographical origin of butters by partial least square discriminant analysis (PLS-DA) applied to infrared spectroscopy (FTIR) data. *J. Food Comp Anal* 33:210–5.

Bernardes, C.D., and Barbeira, P.J.S. 2016. Different Chemometric Methods for the discrimination of commercial aged cachaças. *Food Anal Method* 9:1053–9.

Biancolillo, A., Bucci, R, Magrì, A.L., Magrì, A.D., and Marini, F. 2014. Data-fusion for multiplatform characterization of an Italian craft beer aimed at its authentication. *Anal Chim Acta* 820:23–31.

Borges, E.M., Volmer, D.A., Brandelero, E., Lafayette, J.M., Gelinski, N., Gallimberti, M., and Barbosa, F. 2016. Monitoring the authenticity of organic grape juice via chemometric analysis of elemental data. *Food Anal Method* 9:362–9.

Borràs, E., Ferré, J., Boqué, R., Mestres, M., Aceña, L., and Busto O. 2015. Data fusion methodologies for food and beverage authentication and quality assessment. A review. *Anal Chim Acta* 891:1–14.

Brereton, R.G., and Lloyd, G.R. 2014. Partial least squares discriminant analysis: Taking the magic away. *J Chemometr* 28:213–25.

Bro, R., and Smilde, A.K. 2014. Principal component analysis. *Anal Method* 6:2812–31.

Bro, R., Kjeldahl, K., Smilde, A.K., and Kiers, H.A.L. 2008. Cross-validation of component models: A critical look at current methods. *Anal Bioanal Chem* 390:1241–51.

Camiña, J.M., Savio, M., Azcarate, S.M., Furlong, O., and Marchevsky, E.J. 2014. Chemometric methods for the classification of white wines. In: *Wines: Phenolic Composition, Classification and Health Benefits*, ed. Y. El Rayes, New York: Nova Science Publishers, 245–80.

Castro, C., and Manetti, C. 2007. A multiway approach to analyze metabonomic data: a study of maize seeds development. *Anal Bioanal Chem* 371:194–200.

Castro-Puyana, M., and Herrero, M. 2013. Metabolomics approaches based on mass spectrometry for food safety, quality and traceability. *TrAC-Trends Anal Chem* 52:74–87.

Castro-Puyana, M., Pérez-Míguez, R., Montero, L., and Herrero, M. 2017. Application of mass spectrometry-based metabolomics approaches for food safety, quality and traceability. *TrAC-Trends Anal Chem* 93:102–18.

Cuevas, F.J., Pereira-Caro, G., Moreno-Rojas, J.M., Muñoz-Redondo, J.M., and Ruiz-Moreno, M.J. 2017. Assessment of premium organic orange juices authenticity using HPLC-HR-MS and HS-SPME-GC-MS combining data fusion and chemometrics. *Food Control* 82:203–11.

Dall' Acqua, Y.G., Cunha Júnior, L.C., Nardini, V., Lopes, V.G., Pessoa, J.D.D.C., and de Almeida Teixeira, G.H. 2015. Discrimination of *Euterpe oleracea Mart* (Açaí) and *Euterpe edulis Mart* (Juçara) intact fruit using near-infrared (NIR) spectroscopy and linear discriminant analysis. *J Food Process Preserv* 39:2856–65.

de Almeida, M.R., Correa, D.N., Rocha, W.F.C., Scafi, F.J.O., and Poppi, R.J. 2013. Discrimination between authentic and counterfeit banknotes using Raman spectroscopy and PLS-DA with uncertainty estimation. *Microchem J* 109:170–7.

Donglu, F., Wenjian, Y., Kimatu, B.M., Liyan, Z., Xinxin, A., and Qiuhui, H. 2017. Comparison of flavour qualities of mushrooms (*Flammulina velutipes*) packed with different packaging materials. *Food Chem* 232:1–9.

Dupuy, N., Galtier, O., Le Dréau, Y., Pinatel, C., Kister, J., and Artaud, J. 2010. Chemometric analysis of combined NIR and MIR spectra to characterize French olives. *Eur J Lipid Sci Technol* 112:463–75.

Durante, C., Bro, R., and Cocchi, M. 2011. A classification tool for N-way array based on SIMCA methodology. *Chemometr Intell Lab* 106:73–85.

Durante, C., Cocchi, M., Grandi, M., Marchetti, A., and Bro, R. 2006. Application of N-PLS to gas chromatographic and sensory of traditional balsamic vinegars of modena. *Chemometr Intell Lab* 83:54–65.

Esbensen, K.H, and Geladi, P. 2010. Principal component analysis: Concept, geometrical interpretation, mathematical background, algorithms, history, practice. In *Comprehensive Chemometrics: Chemical and Biochemical Data Analysis*, eds. S.D. Brown, R. Tauler, and B. Walczak. Amsterdam: Elsevier, 211–26.

Escandar, G.M., Goicoechea, H.C., Muñoz de la Peña, A., and Olivieri, A.C. 2014. Second- and higher-order data generation and calibration: A tutorial. *Anal Chim Acta* 806:8–26.

Forina, M., Lanteri, S., and Rosso, S. 2001. Confidence intervals of the prediction ability and performance scores of classifications methods. *Chemometr Intell Lab* 57:121–32.

Forina, M., Oliveri, P., Lanteri, S., and Casale, M. 2008. Class-modeling techniques, classic and new, for old and new problems. *Chemometr Intell Lab* 93:132–48.

Galtier, O., Abbas, O., Le Dréau, Y., Rebufa, C., Kister, J., Artaud, J., and Dupuy, N. 2011. Comparison of PLS1-DA, PLS2-DA and SIMCA for classification by origin of crude petroleum oils by MIR and virgin olive oils by NIR for different spectral regions. *Vib Spectrosc* 55:132–40.

Gan, H.H., Yan, B., Linforth, R.S.T., and Fisk, I.D. 2016. Development and validation of an APCI-MS/GC-MS approach for the classification and prediction of Cheddar cheese maturity. *Food Chem* 190:442–7.

Gao, B., Qin, F., Ding, T., Chen, Y., Lu, W., and Yu, L. 2014. Differentiating organically and conventionally grown oregano using ultraperformance liquid chromatography mass spectrometry (UPLC-MS), headspace gas chromatography with flame ionization detection (Headspace-GC-FID), and flow injection mass spectrum (FIMS) fingerprints combined with multivariate data analysis. *J Agr Food Chem* 62:8075–84.

Gere, A., Losó, V., Györey, A., Kovács, S., Huzsvai, L., Nábrádi, A., Kókai, Z., and Sipos, L. 2014. Applying parallel factor analysis and Tucker-3 methods on sensory and instrumental data to establish preference maps: Case study on sweet corn varieties. *J Sci Food Agr* 94:3213–25.

González-Arjona, D., López-Pérez, G., and González, A. 1999. Performing procrustes discriminant analysis with Holmes. *Talanta* 49:189–97.

Guo, T., Yong, W., Jin, Y., Zhang, L., Liu, J., Wang, S., Chen, Q., Dong, Y., Su, H., and Tan, T. 2017. Applications of DART-MS for food quality and safety assurance in food supply chain. *Mass Spectrom Rev* 36:161–87.

Hibbert, D.B. 2016. Vocabulary of concepts and terms in chemometrics (IUPAC Recommendations 2016.) *Pure Appl Chem* 88:407–43.

Hopke, P.K. 2003. The evolution of chemometrics. *Anal Chim Acta* 500:363–67.

Hu, J., Ma, X., Liu, L., Wu, Y., and Ouyang, J. 2017. Rapid evaluation of the quality of chestnuts using near-infrared reflectance spectroscopy. *Food Chem* 231:141–7.

Huang, L., Liu, H., Zhang, B., and Wu, D. 2015. Application of electronic nose with multivariate analysis and sensor selection for botanical origin identification and quality determination of honey. *Food Bioprocess Tech* 8:359–70.

Indahl, U.G. 2014. The geometry of PLS1 explained properly: 10 key notes on mathematical properties of and some alternative algorithmic approaches to PLS1 modelling. *J Chemometr* 28:168–80.

Jacobson, D., Monforte, A.R., and Silva Ferreira, A.C. 2013. Untangling the chemistry of Port wine aging with the use of GC-FID, multivariate statistics and network reconstruction. *J Agr Food Chem* 61:2513–21.

Karunathilaka, S.R., Farris, S., Mossoba, M.M., Moore, J.C., and Yakes, B.J. 2016. Characterising variances of milk powder and instrumentation for the development of a non-targeted, Raman spectroscopy and chemometrics detection method for the evaluation of authenticity. *Food Addit Contam A* 33:921–32.

Karunathilaka, S.R., Farris, S., Mossoba, M.M., Moore, J.C., and Yakes, B.J. 2017. Non-targeted detection of milk powder adulteration using Raman spectroscopy and chemometrics: melamine case study. *Food Addit Contam A* 34:152–61.

Knolhoff, A.M., and Croley, T.R. 2016. Non-targeted screening approaches for contaminants and adulterants in food using liquid chromatography hyphenated to high resolution mass spectrometry. *J Chromatogr A* 1416:86–96.

Kos, G., Lohninger, H., and Krska, R. 2003. Development of a method for the determination of Fusarium fungi on corn using mid-infrared spectroscopy with attenuated total reflection and chemometrics. *Anal Chem* 75:1211–17.

Kroonenberg, P.M. 2008. *Applied Multiway Data Analysis*. New York: Wiley.

Kumar, N., Bansal, A., Sarma, G.S., and Rawal, R.K. 2014. Chemometrics tools used in analytical chemistry: an overview. *Talanta* 123:186–99.

Latorre, C., Herrero Peña Crecente, R.M., García Martín, S., and Barciela García, J. 2013. A fast chemometric procedure based on NIR data for authentication of honey with protected geographical indication. *Food Chem* 141:3559–65.

Liu, L., Yang, R., Yang, Y., and Dong, G. 2013. Application of 2D correlation infrared spectroscopy to identification of adulterated milk using NPLS-DA method. *Proceedings -3rd International Conference on Instrumentation and Measurement, Computer, Communication and Control, IMCCC 2013*, art. 6840434, 180–4.

Longobardi, F., Innamorato, V., Di Gioia, A., Ventrella, A., Lippolis, V., Logrieco, A.F., Catucci, L., and Agostiano, A. 2017. Geographical origin discrimination of lentils (Lens culinaris Medik.) using^{1}H NMR fingerprinting and multivariate statistical analyses. *Food Chem* 237:443–8.

López, M.I., Trullols, E., Callao, M.P., and Ruisánchez, I. 2014. Multivariate screening in food adulteration: untargeted versus targeted modelling. *Food Chem* 147:177–81.

Lu, W., Lu, X., Gao, B., Shi, H., and Yu, L. 2015. Differentiating milk and non-milk proteins by UPLC amino acid fingerprints combined with chemometric data analysis techniques. *J Agr Food Chem* 63:3996–4002.

Lu, Y., Gao, B., Chen, P., Charles, D., and Yu, L. 2014. Characterisation of organic and conventional sweet basil leaves using chromatographic and flow-injection mass spectrometric (FIMS) fingerprints combined with principal component analysis. *Food Chem* 154:262–8.

Lucci, P., Saurina, J., and Núñez, O. 2017. Trends in LC-MS and LC-HRMS analysis and characterization of polyphenols in food. *TrAC-Trends Anal Chem* 88:1–24.

Manfredi, M., Robotti, E., Quasso, F., Mazzucco, E., Calabrese, G., and Marengo, E. 2018. Fast classification of hazelnut cultivars through portable infrared spectroscopy and chemometrics. *Spectrochim Acta A* 189:427–35.

Marini, F. 2013. *Chemometrics in Food Chemistry*. Amsterdam: Elsevier.

Massart, D.L., Vandeginste, B.G.M., Buydens, L.M.C., De Jong, S., Lewi, P.J., and Smeyers-Verbeke, J. 1997. *Handbook of Chemometrics and Qualimetrics Part A*. Amsterdam: Elsevier.

Matero, S. 2010. *Chemometrics Methods in Pharmaceutical Tablet Development and Manufacturing Unit Operations*. Helsinki: Publications of the University of Eastern Finland.

Mazina, J., Vaher, M., Kuhtinskaja, M., Poryvkina, L., and Kaljurand, M. 2015. Fluorescence, electrophoretic and chromatographic fingerprints of herbal medicines and their comparative chemometric analysis. *Talanta* 139:233–46.

Munck, L. 2007. A new holistic exploratory approach to systems biology by near infrared spectroscopy evaluated by chemometrics and data inspection. *J Chemometr* 21:406–26.

Murphy, K.R., Stedmon, C.A., Graeber, D. and Bro, R. 2013. Fluorescence spectroscopy and multi-way techniques. PARAFAC. *Anal Method* 5:655–62.

Oliveri, P., and Downey, G. 2012. Multivariate class modeling for the verification of food-authenticity claims. *TrAC-Trends Anal Chem* 35:74–86.

Ouertani, S.S., Mazerolles, G., Boccard, J., Rudaz, S., and Hanafi, M. 2014. Multi-way PLS for discrimination: Compact form equivalent to the tri-linear PLS2 procedure and its monotony convergence. *Chemom Intell Lab* 133:25–32.

Pardo-Mates, N., Vera, A., Barbosa, S., Hidalgo-Serrano, M., Núñez, O., Saurina, J., Hernández-Cassou, S., and Puignou, L. 2017. Characterization, classification and authentication of fruit-based extracts by means of HPLC-UV chromatographic fingerprints, polyphenolic profiles and chemometric methods. *Food Chem* 221:29–38.

Pérez, N.F., Ferré, J., and Boqué, R. 2009. Calculation of the reliability of classification in discriminant partial least-squares binary classification. *Chemometr Intell Lab* 95:122–8.

Porro-Munoz, D., Duin, R.P.W., Talavera, I., and Orozco-Alzate, M. 2011. Classification of three-way data by the dissimilarity representation. *Signal Process* 91:2520–9.

Prieto, N., Rodriguez-Méndez, M.L., Leardi, R., Oliveri, P., Hernando-Esquisabel, D., Iñiguez-Crespo, M., and de Saja, J.A. 2012. Application of multi-way analysis to UV-visible spectroscopy, gas chromatography and electronic nose data for wine ageing evaluation. *Anal Chim Acta* 719:43–51.

Reis, N., Botelho, B.G., Franca, A.S., and Oliveira, L.S. 2017. Simultaneous detection of multiple adulterants in ground roasted coffee by ATR-FTIR spectroscopy and data fusion. *Food Anal Method* 10:2700–9.

Roberts, J.J., and Cozzolino, D. 2016. An overview on the application of chemometrics in food science and technology. An approach to quantitative data analysis. *Food Anal Method* 12:3470–8.

Rohman, A. 2017. Infrared spectroscopy for quantitative analysis and oil parameters of olive oil and virgin coconut oil: A review. *Int J Food Prop* 20:1447–56.

Salvatore, E., Bevilacqua, M., Bro, R., Marini, F., and Cocchi, M. 2013. Classification methods of multiway arrays as a basic tool for food PDO authentication. In *Food Protected Designation of Origin: Methodologies and Applications. Wilson and Wilson's Comprehensive Analytical Chemistry*, eds. M. de la Guardia and A. Gonzalvez Illueca. Amsterdam: Elsevier, 339–79.

Santos, P.M., Pereira-Filho, E.R., and Colnago, L.A. 2016. Detection and quantification of milk adulteration using time domain nuclear magnetic resonance (TD-NMR), *Microchem J* 124:15–19.

Savini, S., Loizzo, M.R., Tundis, R., Mozzon, M., Foligni, R., Longo, E., Morozova, K., Scampicchio, M., Martin-Vertedor, D., and Boselli, E. 2017. Fresh refrigerated Tuber melanosporum truffle: Effect of the storage conditions on the antioxidant profile, antioxidant activity and volatile profile. *Eur Food Res Technol* 243(12):2255–2263.

Schuhfried, E., Sánchez Del Pulgar, J., Bobba, M., Piro, R., Cappellin, L., Märk, T.D., and Biasioli, F. 2016. Classification of 7 monofloral honey varieties by PTR-ToF-MS direct headspace analysis and chemometrics. *Talanta* 147:213–9.

Silva, P., Freitas J., Silva C.L., Perestrelo R., Nunes F.M., and Cámara, J. 2017. Establishment of authenticity and typicality of sugarcane honey based on volatile profile and multivariate analysis. *Food Control* 73:1176–88.

Silvestri, M., Elia, A., Bertelli, D., Salvatore, E., Durante, C., Li Vigni, M., Marchetti, A., and Cocchi, M. 2014. A mid level data fusion strategy for the varietal classification of Lambrusco PDO wines. *Chemometr Intell Lab* 137:181–9.

Skov, T., Honoré, A.H., Jensen, H.M., and Naes, T. 2014. Chemometrics in foodomics: Handling data structuresfrom multiple analytical platforms. *TrAC-Trends Anal Chem* 60:71–9.

Smilde, A., Bro, R., and Geladi, P. 2004. *Multiway Analysis in the Chemical Sciences.* London: Wiley, 35–45.

Soaresa, P.K., Brunsa, R.E., and Scarminio, I.S. 2012. Principal component and Tucker3 analyses of high performance liquid chromatography with diode-array detection fingerprints of crude extracts of *Erythrina speciosa* Andrews leaves. *Anal Chim Acta* 736:36–44.

Spiteri, M., Dubin, E., Cotton, J., Poirel, M., Corman, B., Jamin, E., Lees, M., and Rutledge, D. 2016. Data fusion between high resolution 1H-NMR and mass spectrometry: a synergetic approach to honey botanical origin characterization. *Anal Bioanal Chem* 408:4389–401.

Sucupira, N.R., Alves Filho, E.G., Silva, L.M.A., de Brito, E.S., Wurlitzer, N.J., and Sousa, P.H.M. 2017. NMR spectroscopy and chemometrics to evaluate different processing of coconut water. *Food Chem* 216:217–24.

Szollosi D., Denes D.L., Firtha F., Kovacs Z., and Fekete, A. 2012. Comparison of six multiclass classifiers by the use of different classification performance indicators. *J Chemometr* 26:76–84.

Tahri, K., Bougrini, M., Saidi, T., Tiebe, C., Hassani, N.E.A.E., El Bari, N., Hubert, T., and Bouchikhi, B. 2015. Determination of safranal concentration in saffron samples by means of VE-Tongue, SPME-GC-MS, UV-Vis Spectrophotometry and multivariate analysis. 2015 *IEEE SENSORS—Proceedings, art. 7370179.*

ten Berge, J.M.F. 2004. Simplicity and typical rank of three-way arrays, with to Tucker-3 analysis with simple cores. *J Chemometr* 18:17–21.

Tura, D., Prenzler, P.D., Bedgood, D.R.Jr., Antolovich, M., and Robards, K. 2004. Varietal and processing effects on the volatile profile of Australian olive oils. *Food Chem* 84:341–39.

Wang, J., Xue, X., Du, X., Cheng, N., Chen, L., Zhao, J., Zheng J., and Cao, W. 2014. Identification of acacia honey adulteration with rape honey using liquid chromatography–electrochemical detection and chemometrics. *Food Anal Method* 7:2003–12.

Wei, Y., Chen, F., Wang, Y., Chen, L.Z., Zhang, X.W., Wang, Y.H., Wu, L.M., and Zhou, Q. 2016. Application of ICP-MS to identify the botanic source of characteristic honey in South Yunnan. *Spectroscop Spec Anal* 36:262–7.

Wold, S. 1976. Pattern recognition by means of disjoint principal component models. *Pattern Recogn* 8:127–39.

Yu, H., and MacGregor, J.F. 2003. Multivariate image analysis and regression for prediction of coating content and distribution in the production of snack foods. *Chemom Intell Lab* 67:125–44.

Yudthavorasit, S., Wongravee, K., and Leepipatpiboon, N. 2014. Characteristic fingerprint based on gingerol derivative analysis for discrimination of ginger (*Zingiber officinale*) according to geographical origin using HPLC-DAD combined with chemometrics. *Food Chem* 158:101–11.

Zhao, M., Downey G., and O'Donnell, C.P. 2014. Detection of adulteration in fresh and frozen beefburger products by beef offal using mid-infrared ATR spectroscopy and multivariate data analysis. *Meat Sci* 96:1003–11.

CHAPTER 13

Validation

Raúl Andrés Gil and Roberto Antonio Olsina
Universidad Nacional de San Luis (UNSL)

Marianela Savio
Universidad Nacional de La Pampa

CONTENTS

13.1 QUALITY CONSIDERATIONS OF MULTIVARIATE ANALYSIS IN FOOD ANALYTICAL CHEMISTRY

It is notable how statistical models/algorithms are being increasingly used in food science and technology (Riedl, Esslinger, & Fauhl-Hassek, 2015). The increasing need of analyzing large datasets has supported this trend for the last 20 years (Figure 13.1), along with the introduction of new software (and hardware) with adequate capacity of processing complex and high volumes of experimental results.

Besides, newly developed methods have been implemented in computational packages (software), and scientists have benefited from their use (Table 13.1) as manual calculations are tedious and usually imprecise for multivariate and/or multi-response processes (Granato & Calado, 2014; Passari et al., 2011; Riedl, Esslinger, & Fauhl-Hassek, 2015). This is a worrying aspect as contemporary issues demand the use of multidisciplinary approaches, which are usually based on the use of computational software and on statistical and modeling concepts. (Martens & Næs, 1989)

The food science and the food industry are becoming more concerned with quality from the consumer perspective, and this necessitates rapid actions to handle food situations. Validation systems help to minimize issues that arise as the consequences of poor quality control, thus lessening the chances of bad publicity, liability, and recalls. The up-to-date food labeling system does not guarantee food authenticity, quality, and safety. Food analytical chemistry uses multivariate methods to aid food safety and quality assurance (QA) as well as to achieve consumer confidence.

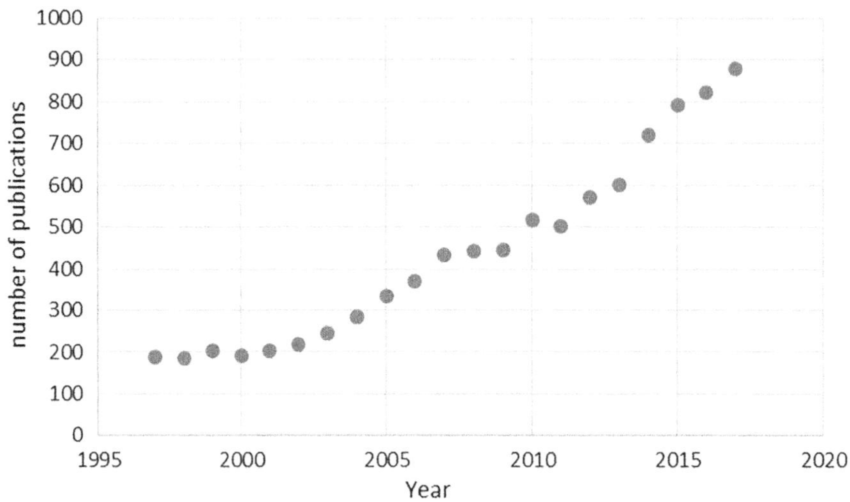

FIGURE 13.1 Increasing number of published articles from 1997–2017 using the term "statistical analysis" in the scientific search SCOPUS.

TABLE 13.1 Software Used for Multivariate Data Analysis in Food Science, Technology, and Engineering. Adapted from (Nunes et al., 2015)

Software	Website	Accessibility	Advantages and Limitations
Microsoft excel	http://products.office.com/en-us/excel	Commercial, no internet access	Advantages: Easy to conduct basic statistics (descriptive and inferential). Possible to analyze data using the add-in "Data Analysis". Correlation and linear regression analysis are included. *F*-test for two variances. Limitations: It does not include normality and homoscedasticity tests. It does not include multivariate statistical methods. Graphical representations are not of high quality. Correlation analysis cannot be evaluated by methods other than Pearson's method.
Statistica	www.statsoft.com	Commercial, no internet access	Advantages: Descriptive and inferential analysis can be easily carried out. Multivariate exploratory and classification techniques, including principal component analysis, clustering techniques, and data mining techniques (including neural networks).

(*Continued*)

TABLE 13.1 (*Continued*) Software Used for Multivariate Data Analysis in Food Science, Technology, and Engineering. Adapted from (Nunes et al., 2015)

Software	Website	Accessibility	Advantages and Limitations
			Multiple and linear regressions can be performed with the detection of outliers and residual analysis. Possibility to perform nonparametric analysis. Quality control charts can be made. Software has a user's guide manual explaining the rationale of each test. Graphical representations are of high quality. Limitations: Very expensive to acquire add-ins for all statistical analysis. User must know a priori the principle of the tests. For multivariate analysis, the software only performs auto-scaling as data preprocessing. Up to version 9, Welch test (ANOVA (analysis of variance) procedure for non-homoscedastic data) cannot be applied.
Pirouette	www.infometrix.com	Commercial, no internet access	Advantages: Graphical outputs are excellent. Multivariate exploratory, class modeling, and classification techniques can be performed. Models can be validated using leave-one-out cross-validation. Various preprocessing methods can be applied and analyses are made simultaneously. Limitations: It is not easy to perform external validation; the user must separate data randomly using other packages (Excel is an option) to perform the external validation. Software dedicated to multivariate statistical techniques.
Matlab	www.mathworks.com/products/matlab/	Commercial, no internet access	Advantages: A large number of tests (uni- and multivariate) can be implemented. Users can implement algorithms in order to improve the analyses. The software has a very elaborated and useful help. The help presents illustrative examples. Features can be improved by additional toolboxes. Users can contribute with free code/toolboxes via File Exchange (www.mathworks.com/matlabcentral/fileexchange/).

(*Continued*)

TABLE 13.1 (*Continued*) Software Used for Multivariate Data Analysis in Food Science, Technology, and Engineering. Adapted from (Nunes et al., 2015)

Software	Website	Accessibility	Advantages and Limitations
			High-quality graphics can be obtained. Limitations: It requires intermediate knowledge of programming, limiting the number of users.
Action	www.portalaction.com.br	Free, downloadable	Advantages: Both descriptive and inferential (both parametric and nonparametric) tests can be used. Homoscedasticity and normality tests are included. Software has a user's guide manual explaining the rationale of each test and how to perform the analysis. Quality control charts and statistics regarding metrology can be made. Design of experiments (response surface methodology) is included. Limitations: Only multivariate analysis of variances (MANOVA) and cluster analysis can be performed as multivariate statistics. It requires the use of Microsoft Excel to be run.
R-Project	www.r-project.org/	Free, downloadable	Advantages: It incorporates all of the standard statistical tests, models, and analyses. It is able to perform both univariate or multivariate tests. Graphical capabilities surpass that of most other statistical packages. It is open source and has been reviewed by many statisticians and computational scientists. Features can be improved by additional packages. Limitations: It requires intermediate knowledge of programming, limiting the number of users.
Chemoface	www.ufla.br/chemoface	Free, downloadable	Advantages: It has a user friendly interface. Various preprocessing methods can be easily applied. It is easy to perform cross-validation and external validation. This is a tool for data plot which is useful mainly to view continuous data, such as spectra and chromatograms.

(*Continued*)

TABLE 13.1 (*Continued*) Software Used for Multivariate Data Analysis in Food Science, Technology, and Engineering. Adapted from (Nunes et al., 2015)

Software	Website	Accessibility	Advantages and Limitations
			There are tools to detect outliers in multivariate calibration models. The software has a user guide with examples. Limitations: Software dedicated to multivariate statistical techniques. Graphic editing is limited.
Sensomaker	www.ufla.br/sensomaker	Free, downloadable	Advantages: It has a user friendly interface. There are interfaces to collect data from sensory tests. Both univariate or multivariate analysis can be performed. Parallel factor analysis (PARAFAC) can be applied to build preference maps. The software has a user guide with examples. Limitations: Software dedicated to sensory analysis. Graphic editing is limited.

Multivariate analysis handles unselective data to build a valid model capable of making predictions with certain variables concerning the analyte of interest. Multivariate methods operate the multiple data simultaneously to predict the concentration of analyte and, thus, are able to compensate for the contribution of interferences and provide quantitative information from unselective data with good results.

13.2 VALIDATION SCHEME IN CLASSIFICATION MODELS

Food analysis with multivariate techniques (either supervised or unsupervised) is a powerful tool used in authentication processes, traceability, food quality control, and others. Typically, nontargeted spectrometric or spectroscopic chemical analysis are used along with further statistical evaluation of acquired data, thereby food matrices can be investigated in terms of their geographical origin, species, variety, or adulterations (Riedl, Esslinger, & Fauhl-Hassek, 2015). The methodological steps are common for both classification and regression models (prediction through calibration). Besides, validating strategies in fingerprinting studies for the analytical method and QA are needed to guarantee the reliability of results, which is often not easy due to issues of data stability and homogeneity. Likewise, the classical qualimetric parameters (Vander Heyden, Questier, & Massart, 1998), such as precision or trueness, cannot be adapted to multivariate approaches with ease, particularly with respect to classification analysis. To date, only a few studies are known that explicitly suggest method validation and QA measures for nontargeted approaches (Dunn et al., 2012; Naz et al., 2014; Sangster et al., 2006).

Models used for classification are typically validated according to the scheme discussed in the review article of Jant Riedel et al. (Riedl, Esslinger, & Fauhl-Hassek, 2015). It presents a workflow with four well-differentiated stages: (i) data preparation, preprocessing, and pretreatment, (ii) model optimization by internal validation, (iii) model testing by external validation, and (iv) stability testing of the model by "system challenges".

At first, the raw data usually needs to be preprocessed and pretreated by different methods to obtain a data matrix accessible for further statistical analysis. For all the datasets that are processed with the same model, essentially the same data preparation procedure has to be applied. Secondly, for allowing internal and external model validation, the dataset has to be divided into a training set and a test set. At best, the test set will comprise samples derived, prepared, and measured independently from the training set. However, in practice, often the dataset of one experiment or experimental series is used for both training and test sets due to limited resources. In this case, specific algorithms for reasonable separation of the samples should be used and the model validation procedure should be repeated several times. The training dataset should be used for generating the model. An optimal model with a respective number of components (new latent variables) should be deduced by internal validation to avoid over or underfitting of the model. The resulting model performance has to be compared to a predefined study-related threshold.

If the performance requirements are not met, data should be examined for justifiable outlier samples and refined by filtering noise from the data or by selecting specific features. Based on this modified data matrix, a new model has to be generated and optimized for the number of latent variables by internal validation. Then, the performance of the so-obtained best model has to be assessed by the test dataset. The results of this external validation step describe the prediction ability of the best model generated from the training dataset. This prediction ability is expected for getting the results of new measurements on applying the model. In a final step, the stability of the model can be assessed by system challenges. The term system challenge refers to the validation of the model performance under system conditions or sample characteristics different from those under which the model was generated. At best, the system challenge will comprise all the steps of the fingerprinting approach that can influence the model performance. For instance, measurement results are subject to analytical variance of the instrument due to maintenance operations or due to biological variation depending on the environmental conditions, which may affect the long term stability of the model. A system challenge can be performed with new samples that are measured after a long time period by another operator or even by another instrument, or with "blind" samples unknown to the previously best model. The performance of the model for these new or unknown samples should compare well with the performance results of the external validation step. Study-related criteria need to be developed to decide on model sufficiency.

In case the system challenge fails, calibration of the model might become necessary. In the first step, QA measures should be taken into account to verify the analytical performance. For instance, quality control samples might be used to adjust deviations in the analytical performance resulting, e.g., from instrument drifts. Even when the analytical performance is in accordance with the validation requirements, but the new samples are still poorly predicted with the available model, then a second step related to model appropriateness should be followed. This might include a critical review of the statistical considerations such as the number and representativeness of the selected samples in the model or the appropriateness of data preprocessing and statistical modeling.

Furthermore, sample characteristics might have changed through the different harvest years, for example. As a consequence of these considerations, an update of the existing model or the rebuilding of a new model might become necessary.

13.3 VALIDATION IN MULTIVARIATE CALIBRATION: FIGURES OF MERIT

Figures of merit are quantitative parameters that assist to illustrate the performance of a technique and/or a method relative to other available methods/techniques. Its use in analytical calibration is to help in deciding the relative success of different methodologies. Indeed, research in analytical chemistry is mostly driven by the need of improving analytical figures of merit; sensitivity being the most discussed one, followed by selectivity. Whether the aim is comparing the performance of analytical procedures or optimizing experimental methods, a reliable numerical sensitivity is a requisite to decide the actual upgrade achieved (Olivieri, 2014). Analytical figures of merit are an integral part of official analysis protocols, as documented in international standards (SANCO, 2012).

Uncertainty propagation calculus is needed in order to give numerical estimations of its parameter. Error propagation is a common method of evaluation, giving insights into the relative impacts of various uncertainty sources on the prediction errors.

The best approximation to concentration variance is the well-known three-term expression (valid for the propagation of homoscedastic and uncorrelated noise):

$$\sigma_y^{\,2} = SEN_n^{\,-2}\sigma_x^{\,2} + hSEN_n^{\,-2}\sigma_x^{\,2} + h\sigma_{ycal}^{\,2} \tag{13.1}$$

where σ_x^2 is the variance in instrumental signals, h is the sample leverage, and σ_{ycal}^2 is the variance in calibration concentrations. The three terms on the right-hand side of Eq. (13.1) account for the propagation of uncertainties derived from (in the order in which they appear): (i) instrumental signals in the test sample data, (ii) instrumental signals in the calibration data, and (iii) calibration concentrations. The first and probably the most relevant of these contributions is transmitted directly via the inverse squared sensitivity, which is the most significant ingredient in Eq. (13.1). The second and third terms arise from calibration uncertainties, and are both scaled by the sample leverage h, a dimensionless parameter measuring the position of the sample relative to the calibration space.

13.3.1 Sensitivity and Selectivity

Among other important analytical parameters, sensitivity is one of the most crucial since it enables the estimation of other secondary ones including: (i) analytical sensitivity (dimensionless), (ii) selectivity, (iii) prediction uncertainty, (iv) limit of detection, and (v) limit of quantification (Currie, 1995). The International Union of Pure and Applied Chemistry (IUPAC) has set sensitivity definitions for various calibration scenarios (Danzer & Currie, 1998; Danzer, Otto, & Currie, 2004; Olivieri et al., 2006). In case of an instrumental measurement of a single signal per sample and/or single constituent (univariate calibration), the sensitivity is the ratio of the change in signal to the change in the chemical variable that is being measured (the concentration of the analyte of interest), i.e., the slope of the calibration plot obtained after linear regression (Eq. 13.2).

$$y = \frac{(x - n_0)}{m_0} \tag{13.2}$$

where m_0 and n_0 are the slope and intercept, respectively, of the zeroth-order linear calibration graph.

Apart from this particular case, analytical measurements involve several signals for one or more constituents, and in one or more samples in many other cases. When calibration is done in that way, it is known as multivariate. In addition, the experimental data can be of different types, i.e., if the data can be arranged in vector form (e.g., spectra, chromatograms, electrochemical traces, etc.), they belong to the category of first order (Olivieri, 2008, 2012).

Partial least squares (PLS) is a method of constructing predictive models when the factors are many and highly collinear. This model is based on the inverse regression model on latent variables, and it enables quantifying selected analytes in a sample without knowing the chemical identity of the interfering species. This is especially important for applications in several types of food analysis where the number and nature of interfering species are usually unknown (Burns & Ciurczak, 2008).

The definition of sensitivity in first-order multivariate calibration is rather complex (Lorber, 1986), since it does not refer to the necessary instrument response, but it depends on the analyte response in the chemical context of the sample/media analyzed. A remarkable feature of this method is that the analytes and potential interfering agents are mathematically separated by retrieving the pure constituent profiles, which can be further calibrated with multivariate signals that are the result of mathematical processing. In this field, several different sensitivity expressions have been proposed (Ho, Christian, & Davidson, 1980; Messick, Kalivas, & Lang, 1996; Olivieri & Faber, 2004).

In the simplest example, two constituents occur with the vector signal (e.g., a spectrum) for a test sample measured at a number of sensors (i.e., wavelengths) and given by

$$x = y_1 s_1 + y_2 s_2 \tag{13.3}$$

where y_1 and y_2 are analyte concentrations, and s_1 and s_2 are the pure constituent profiles at unit concentration. Following the reasoning of Olivieri (Olivieri, 2014), and aiming to assess the analyte 1 signal, the contribution of 2 can be eliminated from Eq. (13.3) by left-multiplying both sides with an orthogonal projection matrix $\left[I - s_2\left(s_2^T s_2\right) - 1 s_2^T\right]$, where I is an appropriately dimensioned unit matrix. Usually the result of $\left[\left(s_2^T s_2\right)^{-1} s_2^T\right]$ is designated as s_2^+ (with the superscript "+" implying the generalized inverse operation). Since

$$\left(I - s_2 s_2^+\right) \times s_2 = s_2 - s_2 = 0 \tag{13.4}$$

can be reordered as:

$$\left(I - s_2 s_2^+\right) x = y_1 \left(I s_2 s_2^+\right) s_1 \tag{13.5}$$

a virtual single-constituent problem is thus presented. This equation introduces the net analyte signal (NAS is defined as the portion of overall signal that can be uniquely ascribed to a given analyte (Lorber, 1986; Lorber, Fabert, & Kowalski, 1997)) for constituent 1 in the mixture (x_1*) as being proportional to its NAS at unit concentration (s_1*), the proportionality constant being the analyte concentration y_1.

$$x^* = ys^* \tag{13.6}$$

The real usefulness of the NAS lies in the fact that a plot of the length of the NAS vector as a function of analyte concentration is linear, the slope being the length of the NAS vector at unit concentration ($\|s_1^*\|$). This immediately leads to an intuitive definition of sensitivity for analyte 1 as follows:

$$SEN_1 = \frac{\|x_1^*\|}{y_1} = \|s_1^*\| = \left[s_1^T\left(I - s_2 s_2^+\right)s_1\right]^{1/2} \tag{13.7}$$

That is to say, vectorial signals at unit concentration for the pure constituents of a mixture allow estimation of the sensitivity.

The Eq. (13.8) can be generalized to the nth constituent of interest in a multi-constituent sample in different forms. One useful form is expressed as a function of the pure profiles for all the constituents, ubiquitous in classical least squares (CLS) studies (Kumar et al., 2014):

$$SEN_n = \left[\delta_n \left(S_{CLS}^T S_{CLS}\right)_n^{-1} \delta_n\right]^{1/2} \tag{13.8}$$

where δ_n is an $N \times 1$ vector selecting the analyte of interest, and the matrix S_{CLS} contains N columns, each with the pure constituent profile s_n for the nth constituent.

According to IUPAC (McNaught & Wilkinson, 2006), selectivity refers to "the extent to which a method can determine particular analytes in mixtures or matrices without interferences from other components". Analytical methods are rated qualitatively in terms of selectivity (i.e., good, excellent, extreme, and others). However, a series of numerical criteria to assess selectivity are available.

Roughly speaking, selectivity is a figure of merit evaluated through the (dimensionless) ratio of two analyte sensitivity values: the sensitivity in a mixture and the sensitivity when all other sample constituents are absent.

$$SEL_n = \frac{SEN_{n(in\,a\,mixture)}}{SEN_{n(pure)}} \tag{13.9}$$

The concept of selectivity has been generalized for multiway analysis. The defining Eq. (13.9) implies that the multiway selectivity is accessible when the pure analyte signal is adequately retrieved by the processing algorithm. This is possible in the case of multilinear (e.g., PARAFAC) analysis, for which the selectivity (SEL_n) is directly given by:

$$SEL_n = \frac{SEN_n}{m_n} \tag{13.10}$$

where m_n is the slope of the pseudo-univariate calibration graph. The degree by which SEN_n departs from m_n in Eq. (13.9) is adequately measured by the level of overlapping among the profiles for the various constituents. Since $SEN_n < m_n$, the value of SEL_n continuously varies between 0 (null selectivity) and 1 (100%, i.e., full selectivity). The adequacy of the latter approach has been revealed in the chromatographic context where a relationship between multiway selectivity and classical separation metrics has been sought. Indeed, a direct relationship has been proposed to exist between the effective peak capacity of a chromatogram and the multiway selectivity.

13.3.2 Limits of Detection and Quantification

Concerning the limits of detection (LD-LOD) and quantification (LQ-LOQ), the transition between univariate and multivariate calibration requires a special attention. When the analytical signal is univariate, it is implicit that the signal is highly selective for the specific analyte and that potential influence of the interferences is negligible; thus, the LD can be directly estimated from the univariate calibration line. Conversely, when dealing with complex data as in multivariate calibration methods, PLS regression analysis is used, and the estimation of LD may be reliant on more than one variable.

The LOD has gained significant recognition as a method quality descriptor in applied analytical chemistry, mainly due to the following factors: (i) it is stated in concentration units, allowing straightforward comparison among different methodologies and (ii) it is needed for assessing the detection capabilities, which are of central importance in specific areas including health protection, doping control, food and materials quality control, radiochemistry, monitoring traces of contaminants in environmental samples, verifying the purity of materials and products, etc. (Olivieri, 2015).

There are many terms, definitions, and calculation approaches related with detection capabilities (Belter et al., 2014; Currie et al., 1995) (Table 13.2). At present, IUPAC recognizes the definition arranged by the International Standardization Organization (document ISO 11843) (ISO, 1997) for the capability (or limit) of detection as "the lowest quantity of substance that can be distinguished from the absence of that substance (a blank value) within a stated confidence limit" (ISO 2000; McNaught & Wilkinson, 1997; Inczédy et al., 1998). ISO and IUPAC have taken this viewpoint because of the necessity for a homogenous international position, largely in response to the formal demands made by Codex for guidance on the terms due to the urgency of validating analysis methods founded on their capabilities to reliably measure essential and toxic elements in food, a problem that Codex had been attempting to elucidate since 1982 (Ortiz et al., 2003).

Accordingly, the LD expression can be achieved as proportional to the uncertainty in the predicted concentration near a blank sample (Olivieri et al., 2015).

$$LD_n = 3.3\left(SEN_n^{-2}\sigma_x^2 + h_0 SEN_n^{-2}\sigma_x^2 + h_0\sigma_{y_{cal}}^2\right) \tag{13.11}$$

where the factor 3.3 is the sum of t-coefficients accounting for type I and II errors at 95% confidence level, the subscript n distinguishes a particular analyte of interest, and h_0 is the leverage for the blank sample. The factor in front of Eq. (13.11) may be corrected for other probabilities and degrees of freedom. In the equation: (i) the LD_n is close enough to the blank so that the leverage at the LD level is equal to the blank leverage h_0, otherwise complex corrections are required and (ii) the distance from the blank to the LD is given as a sum of two confidence intervals; a more rigorous treatment suggests the use of a noncentrality parameter of a noncentral t-distribution as an alternative to the sum of classical t-coefficients. It is possible, however, that the values delivered by Eq. (13.11) and more intricate statistical approaches do not significantly differ.

Due to the complexity and diversity of multivariate regression models and analytical data that can originate, for example, from spectral analysis or hyphenated techniques, some aspects which remain outside the field of application of the ISO norm need to be taken into account carefully for multivariate LD calculation (Belter et al., 2014). There is still no generally accepted LD estimator for PLS studies. Nevertheless, there is a high interest in the topic. The main difficulty in estimating a multivariate LD is that instrumental signals are not specific for a particular analyte, so in practice it is difficult due to

TABLE 13.2 Some Universally Accepted Approaches for Estimation of the Detection and Quantification Limits. Adapted from Belter et al. 2014

Authors	Descriptions	Equations
IUPAC and ISO steps to harmonization		
IUPAC	Based on Neyman-Pearson theory of hypothesis testing, factors associated with the preparation of calibration curve (number of standards, the range of the calibration curve, number of replicates performed on the blank, standards and the unknown sample) are considered.	$x_C = \frac{t_{1-\alpha,\nu} s_0 \eta^{1/2}}{b}$ [a] $x_D = \frac{\left(t_{1-\alpha,\nu} + t_{1-\beta,\nu}\right) s_0 \eta^{1/2}}{b}$ [a] $x_Q = \frac{10\ s_Q}{b}$
	The correctness of the method was proved by Voigtman et al. (2008).	
	η - function of the OLS parameters, calculated with the equation listed below[c] ν - degrees of freedom s_Q - standard deviation of the estimated value at the quantification limit x_Q - quantification limit content value	
ISO	Method based on the noncentrality parameter of the noncentral *t*-distribution. For the assumptions: linear calibration function and homoscedastic, Gaussian measurement noise, x_C and x_D are estimated using the expressions listed alongside. Voigtman strongly advises against using this approach because of the bias.	$x_C = \frac{t_{1-\alpha,IJ} - 2S_{\underline{y}/x}}{b}\left(\frac{1}{K} + \frac{1}{IJ} + \frac{x^{-2}}{\sum_{i=1}^{I}(x_i - \bar{x})^2}\right)^{1/2}$ $x_D = \frac{\delta\left(IJ - 2_{\alpha,\beta}\right) S_{\underline{y}/x}}{b}\left(\frac{1}{K} + \frac{1}{IJ} + \frac{x^{-2}}{\sum_{i=1}^{I}(x_i - \bar{x})^2}\right)^{1/2}$
	J - number of parallel preparations for standard solutions δ - noncentrality parameter of noncentral *t*-distribution $t_{1-\alpha,IJ-2} - (1-\alpha)\%$ quantile of student's *t*-distribution with (*IJ*–2) degrees of freedom	

(*Continued*)

TABLE 13.2 (*Continued*) Some Universally Accepted Approaches for Estimation of the Detection and Quantification Limits. Adapted from Belter et al. 2014

Authors	Descriptions	Equations
Other Approaches for Computing Detection and Quantification Capabilities from 1990s to Present		
US EPA	The computation of method detection limit (MDL) is based on the one-tailed t-statistic at the 99% confidence level and s, in concentration units, for (seven or more) samples spiked at the estimated instrument detection limit (IQL).	$IDL = \frac{3S_r}{b}$ $IQL = \frac{10S_r}{b}$ $MDL = t_{1-\alpha,K-1S}$ $MQL = 3MDL$
Linnet and Kondratovich	Approach is nonparametric, i.e., no assumptions are made regarding measurement distribution. Data from blank samples are ranked by increasing values and the ninety fifth percentile is taken. $Perc_{1-\alpha}$ – $1-\alpha^{th}$ percentile of ranked blank net values n – number of blank samples	$L_C = Perc_{1-\alpha} = n\left(\frac{95}{100} + 0.5\right)$ $L_D = L_C + C_\beta \times s$ [b]
Lavagnini	Nonparametric method involving Type I and II errors similar to Hubaux-Vos, based on the one-sided simultaneous tolerance interval. It uses Theil-Sen (TS) median-based regression technique combined with Lancaster-Quade (LQ) statistics.	

(*Continued*)

TABLE 13.2 (*Continued*) Some Universally Accepted Approaches for Estimation of the Detection and Quantification Limits. Adapted from Belter et al. 2014

Authors	Descriptions	Equations
Loock and Wentzel	Method applicable with sensors, calculated with mean and standard deviation. Parametric and homoscedastic sample measurements values are assumed.	$y_D = \overline{y}_B + t_{\alpha,n} - 1s_r$ [c] $x_D = \frac{t_{\alpha,n} - 1s_r}{b}$ [b,c]
Olivieri	Limit of detection in PLS has been proposed based on the interpretation of the LD as an interval.	$LD_{min} = 3.3\left[SEN_n^{-2}\ var(x) + h_{0\,min}SEN_n^{-2}var(x) + h_{0\,min}SEN_n^{-2}var(y_{cal})\right]^{1/2}$ $LD_{max} = 3.3\left[SEN_n^{-2}\ var(x) + h_{0\,max}SEN_n^{-2}var(x) + h_{0\,max}SEN_n^{-2}var(y_{cal})\right]^{1/2}$

[a] $\eta = \left(\frac{1}{K} + \frac{1}{I} + \frac{x^{-2}}{\sum_{i=1}^{I}(x_i - \overline{x})^2}\right)^{1/2}$.

[b] $C_\beta = \frac{z_{1-\beta}}{\left[1 - \frac{1}{4 \times v}\right]}$, where $z_{1-\beta}$ denotes the standard normal deviate derived from the ninety fifth percentile of the standard Gaussian distribution.

[c] $S_r = \sqrt{\sum_{i=1}^{I}\frac{\left[\overline{y}_i - (a + bx_i)\right]^2}{1-2}}$, where $\overline{y}_i$ denotes the sample mean of replicates for standard.

two issues: the theoretical problem of obtaining the appropriate estimators and the lack of suitable calculation programs.

For first-order methodologies, SEN_n is analyte specific, but the leverage h_0 is also sample specific, meaning that diverse blank samples (samples where the analyte is absent, but which contain varying proportions of the remaining components) have different associated values of h_0. Hence, the LD_n not only turns into analyte specific but also sample specific. Further, in higher-order calibration, the value of SEN_n is analyte-, sample-, and algorithm-specific (incidentally, the leverage h_0 is not sample specific when pseudo-univariate calibration is employed). For this purpose, the detection capability of a given analyte depends on various factors beyond the instrumental signals measured for a calibration sample set.

The detection capabilities for multivariate calibration methods is presented in the IUPAC Technical Report by Olivieri where different approaches are described that were applied to diverse experimental case of studies (Olivieri et al., 2006). A recent approach for calculating the LOD in PLS has been proposed based on the interpretation of the LD as an interval, leading to a novel LD estimator composed with the corresponding results toward both simulated and experimental datasets (Allegrini & Olivieri, 2014). Although this proposal has been specifically developed for first-order PLS calibration, it can be potentially extended to other algorithms as long as they are consistent with Eq. (13.11).

The central concepts underlying this description are: (i) each test sample has a specific LD value, (ii) the universe of test samples is characterized by the calibration set of samples, (iii) the leverages for the calibration samples can be extrapolated to zero analyte concentration, and (iv) a range of LD values can be assessed for the PLS model as a whole.

The lower and upper limits of the LD interval match with the calibration samples with the lowest and largest extrapolated leverages to zero analyte concentration and can be considered using the succeeding expressions (Allegrini & Olivieri, 2014):

$$LD_{\min} = 3.3\left[SEN_n^{-2}\ var(x) + h_{0\min}SEN_n^{-2}var(x) + h_{0\min}SEN_n^{-2}var(y_{cal})\right]^{1/2} \tag{13.12}$$

$$LD_{\max} = 3.3\left[SEN_n^{-2}\ var(x) + h_{0\max}SEN_n^{-2}var(x) + h_{0\max}SEN_n^{-2}var(y_{cal})\right]^{1/2} \tag{13.13}$$

where

$$h_{0\min} = \frac{y_{cal}^{-2}}{\sum_{i=1}^{I} y_i^2} \tag{13.14}$$

$$h_{0\max} = \max\left(h_{0\max}\right) \tag{13.15}$$

The leverage in Eq. (13.11) corresponds to the value obtained in univariate calibration with a known calibration set, providing other sample constituents are absent. In Eq. (13.12), on the other hand, $h_{0\ cal}$ denotes the leverages for the projections of all the calibration samples onto h_0, which is the zero concentration hyperplane in a score space having a number of dimensions equal to the number of calibration latent variables, and $h_{0\ cal}$ can be calculated as:

$$h_{0cal} = h_{cal} + h_{\min}\left[1 - \left(\frac{y_{cal}}{y_{cal}}\right)^2\right] \tag{13.16}$$

In the concentration range LD_{min} < and < LD_{max}, a practical way to resolve if the analyte is detected or not is to estimate a specific LD value for the resultant test sample, resembling its real leverage h to the leverage h_0, which would match with its background components (i.e., in the absence of analyte). This is comparable to considering the sample as a blank, which is possible since its analyte concentration is most probably very low. The acquired LD value can then be employed to check whether the predicted concentration is below (analyte absent) or above (analyte present) the sample-specific LD.

An additional common method to estimate LOD in multivariate calibration allows some parallels with the univariate approach, and this is why it is occasionally named "pseudo-univariate LD" (Ortiz et al., 2003). In this strategy, the analyte concentrations assessed for the calibration set of samples by the PLS model are plotted alongside their reference concentrations. The outcome is a pseudo-univariate calibration chart in which the vertical scale is the assessed analyte concentration. The graph is treated as in univariate calibration, assuming that the detection limit is insensitive to linear transformations applied to the signal.

This leads to a LD_{pu} value, estimated from the classical univariate equation $(y = (x - n_0) / m_0)$:

$$LD_{pu} = 3.3\ s_{pu}^{-1} \left[\left(1 + h_{0\,min} + \frac{1}{I} \right) var_{pu} \right]^{1/2} \tag{13.17}$$

where s_{pu} is the slope of the pseudo-univariate line and var_{pu} is the variance of the regression residuals.

The parameter LD_{pu} has the benefit of being a single figure of merit characterizing the whole PLS calibration model. Conversely, one critical point about this estimator is the absence of a term accounting for calibration concentration uncertainties. In contrast, when concentration uncertainties compete with the instrumental noise in relative size, the mutual relationship among LD_{pu} and LD_{min} and LD_{max} is less clear.

This can be explained on the basis of how the errors in calibration concentrations var (y_{cal}) are incorporated into the LD definitions. In the estimation of both LD_{min} and LD_{max}, the latter contribution is scaled by the leverage, but in LD_{pu} it is directly incorporated into the first test-sample-dependent term of the LD expression. In the latter case, the "signal" is replaced by the estimated concentration. This means that although the term that takes into account the errors in calibration concentrations is not present in an explicit way, concentration errors are directly propagated to the standard error in predicted concentrations. In any case, it is important to remark that the conceptual approach to LD_{pu} is radically different from the presently proposed range of LD values, which is clearly consistent with the latest advances in error-in-variables theory and leads to a better insight into the PLS detection capabilities, because it helps to understand the effect of background and potential interfering agents on the analyte detection for complex samples (Olivieri et al., 2015)

The limit of quantitation (LQ_n), consecutively, is assessed as the critical concentration level (CL) for which the relative prediction error is 10% and is simply established at a concentration value which is ten times the related prediction uncertainty (Olivieri et al., 2006).

$$LQ_n = 10 \left(SEN_n^{-2} \sigma_x^2 + h_0 SEN_n^{-2} \sigma_x^2 + h_0 \sigma_{y_{cal}}^2 \right) \tag{13.18}$$

Analogous considerations to those for LD_n concerning the analyte and sample dependence of the LQ_n can be applied.

13.4 CONCLUSION

Multidimensional data are produced by modern analytical instrumentation in abundance, calling for new and powerful data processing techniques. Validation is a procedure that aims at providing a response to an interrogation that has to be kept in mind permanently; the nature of the question to be responded or of the hypotheses to be verified must always guide the choice of a proper validation strategy to be tracked. Research in the 21st century has developed a multitude of diverse processing algorithms, each equipped with its particular sophisticated artillery. Analysts have gradually discovered that this form of knowledge can be suitably classified, and that common aspects permeate all these apparently different ways of analyzing data. As a result, going from univariate data to multivariate data is known to provide a gain in sensitivity and selectivity, combined with analytical advantages which cannot be overestimated.

REFERENCES

Allegrini, F., and A.C. Olivieri. 2014. IUPAC-consistent approach to the limit of detection in partial least- squares calibration. *Anal. Chem.* 86:7858–7866. doi: 10.1021/ac501786u.

Belter, M., A. Sajnóg, and D. Barałkiewicz. 2014. Over a century of detection and quantification capabilities in analytical chemistry—historical overview and trends. *Talanta* 129:606–616. doi: 10.1016/j.talanta.2014.05.018.

Burns, D.A., and E.W. Ciurczak. 2008. *Handbook of Near-Infrared Analysis*. CRCPress Boca Raton Fl USA

Currie, L.A. 1995. Nomenclature in evaluation of analytical methods including detection and quantification capabilities. *Pure Appl. Chem.* 67 (10):1699–1723. doi: 10.1351/pac199567101699.

Danzer, K., and L.A. Currie. 1998. Guideline for calibration in analytical chemistry—Part 1. Fundamentals and single component calibration. *Pure Appl. Chem.* 70 (4):993–1014. doi: 10.1351/pac199870040993.

Danzer, K., M. Otto, and L.A. Currie. 2004. Guidelines for calibration in analytical chemistry Part 2. multispecies calibration: (IUPAC Technical Report). *Pure Appl. Chem.* 76 (6):1215–1225.

Dunn, W.B., I.D. Wilson, A.W. Nicholls, and D. Broadhurst. 2012. The importance of experimental design and QC samples in large-scale and MS-driven untargeted metabolomic studies of humans. *Bioanalysis* 4 (18):2249–2264. doi: 10.4155/bio.12.204.

Granato, D., and V.M.A. Calado. 2014. The use and importance of design of experiments (DOE) in process modelling in food science and technology.(Ed) D.Granato and G. Ares In. Mathematical and Statistical Methods in Food Science and Technology. Wiley Blackwell Hoboken USA.

Ho, C.N., G.D. Christian, and E.R. Davidson. 1980. Application of the method of rank annihilation to fluorescent multicomponent mixtures of polynuclear aromatic hydrocarbons. *Anal. Chem.* 52 (7):1071–1079. doi: 10.1021/ac50057a018.

Inczédy, J., T. Lengyel, A.M. Ure, A. Gelencsér, and A. Hulanicki. 1998. *IUPAC Analytical Chemistry Division, Compendium of Analytical Nomenclature*. 3rd ed. Oxford: Blackwell.

International Organization for Standardization (ISO). 1997. Capability of detection Report No. ISO 11843–1. Genève, Switzerland: ISO.

Kumar, N., A. Bansal, G.S. Sarma, and R.K. Rawal. 2014. Chemometrics tools used in analytical chemistry: An overview. *Talanta* 123 (Supplement C):186–199. doi: 10.1016/j.talanta.2014.02.003.

Lorber, A. 1986. Error propagation and figures of merit for quantification by solving matrix equations. *Anal. Chem.* 58 (6):1167–1172. doi: 10.1021/ac00297a042.

Lorber, A., K. Fabert, and B.R. Kowalski. 1997. Net analyte signal calculation in multivariate calibration. *Anal. Chem.* 69 (8):1620–1626. doi: 10.1021/ac960862b.

Martens, H., and T. Næs. 1989. *Multivariate Calibration.* Great Britain: John Wiley & Sons Ltd.

McNaught, A.D., and A. Wilkinson. 1997. *IUPAC. Compendium of Chemical Terminology.* 2nd ed. Oxford: Blackwell Scientific Publications.

McNaught, A.D., and A. Wilkinson. 2006. Compendium of Chemical Terminology. In The Gold Book. Oxford: Blackwell Scientific Publications, (1997). XML on-line corrected version: http://goldbook.iupac.org.

Messick, N.J., J.H. Kalivas, and P.M. Lang. 1996. Selectivity and related measures for nth-order data. *Anal. Chem.* 68 (9):1572–1579.

Naz, S., M. Vallejo, A. García, and C. Barbas. 2014. Method validation strategies involved in non-targeted metabolomics. *J. Chromatogr. A* 1353:99–105. doi: 10.1016/j.chroma.2014.04.071.

Nunes, C.A., V.O. Alvarenga, A. de Souza Sant'Ana, J.S. Santos, and D. Granato. 2015. The use of statistical software in food science and technology: Advantages, limitations and misuses. *Food Res. Int.* 75:270–280. doi: 10.1016/j.foodres.2015.06.011.

Olivieri, A.C. 2008. Analytical advantages of multivariate data processing. One, two, three, infinity? *Anal. Chem.* 80 (15):5713–5720. doi: 10.1021/ac800692c.

Olivieri, A.C. 2012. Recent advances in analytical calibration with multi-way data. *Anal. Methods* 4 (7):1876–1886. doi: 10.1039/c2ay25064k.

Olivieri, A.C. 2014. Analytical figures of merit: From univariate to multiway calibration. *Chem. Rev.* 114 (10):5358–5378. doi: 10.1021/cr400455s.

Olivieri, A.C., and N.M. Faber. 2004. Standard error of prediction in parallel factor analysis of three-way data. *Chem. Intell. Lab. Syst.* 70 (1):75–82. doi: 10.1016/j.chemolab.2003.10.005.

Olivieri, A.C., S. Bortolato, and F. Allegrini. 2015. Figures of merit in multiway calibration. *Data Handl. Sci. Technol.* Chapter 13. doi: 10.1016/B978-0-444-63527-3.00013-8.

Olivieri, A.C., N.M. Faber, J. Ferré, R. Boqué, J.H. Kalivas, and H. Mark. 2006. Uncertainty estimation and figures of merit for multivariate calibration: (IUPAC technical report). *Pure Appl. Chem.* 78 (3):633–661. doi: 10.1351/pac200678030633.

Olivieri C.Alejandro. 2015. Practical guidelines for reporting results in single- and multi-component analytical calibration: A tutorial. *Analytica Chimica Acta* 868:10–22

Ortiz, M.C., L.A. Sarabia, A. Herrero, M.S. Sanchez, M.B. Sanza, M.E. Rueda, D. Gimenez, and M.E. Melendez, 2003. Capability of detection of an analytical method evaluating false positive and false negative (ISO 11843) with partial least squares. *Chem. Intell. Lab. Syst.* 69:21–33. doi: 10.1016/S0169-7439(03)00110-2.

Passari, L.M.Z.G., P.K. Soares, R.E. Bruns, and I.S. Scarminio. 2011. Statistics applied to chemistry: Ten common doubts. *Quimica Nova* 34 (5):888–892. doi: 10.1590/S0100-40422011000500028.

Riedl, J., S. Esslinger, and C. Fauhl-Hassek. 2015. Review of validation and reporting of non-targeted fingerprinting approaches for food authentication. *Anal. Chim. Acta* 885:17–32. doi: 10.1016/j.aca.2015.06.003.

Sanco. 2012. Method Validation and Quality Control Procedures For Pesticide Residues Analysis in Food And Feed. Accessed 10/11/2017.

Sangster, T., H. Major, R. Plumb, A.J. Wilson, and I.D. Wilson. 2006. A pragmatic and readily implemented quality control strategy for HPLC-MS and GC-MS-based metabonomic analysis. *Analyst* 131 (10):1075–1078. doi: 10.1039/b604498k.

Vander Heyden, Y., F. Questier, and L. Massart. 1998. Ruggedness testing of chromatographic methods: Selection of factors and levels. *J. Pharm. Biomed. Anal.* 18 (1-2):43–56. doi: 10.1016/S0731-7085(98)00174-5.

Voigtman E. 2008. Limits of detection and decision. Part 1. *Spectrochim. Acta B* 63:115–128. doi: 10.1016/j.sab.2007.11.015.

SECTION 4

Applications

CHAPTER 14

Authentication and Traceability of Wine

Javed Ahamad
Mettu University

Javed Ahmad and Nehal Mohsin
Najran University

Naiyer Shahzad
Umm Al Qura University

CONTENTS

14.1 INTRODUCTION

Wine is a beverage whose value is influenced by many factors, amongst which the origin, vintage, grape variety, and growing conditions play major roles. Wine is characterized by a wide variety of sensory characteristics. The information provided on the wine label is connected to the consumer's expectations concerning the sensory and quality criteria. The label may therefore indicate significant and price determining characteristics such as brand, type, vintage, variety, and origin of the product, which represents the wine identity (Palade and Popa, 2014). Wine control is traditionally and strongly associated with the proof of authenticity. Generally, chemical falsifications that are to be dealt with are the addition of water, glycerol, alcohol, dyes, sweeteners, flavor substances, the non-authorized addition of sugars, preservatives, and acidity adjustments. However, wine authentication aims to confirm all the declarations on the label descriptions or to detect fraudulent statements by various analytical methods (Schlesier et al., 2009).

The adulteration of alcoholic beverages may not be conducted only to increase the alcoholic strength by using cheaper sources of sugar and starch besides grapes or fruit—sometimes combined with watering-down—but also to increase the presumed or apparent quality of the product such as in the case of the 1985 scandal in Austria, when diethylene glycol was added to wine to increase sweetness (Tagliabue, 1985). Another agent to sweeten wine that has been historically applied is lead acetate, which was documented in Greco-Roman times (Lessler, 1988), and chronic Pb poisoning outbreaks have been frequently recorded in history (Eisinger, 1982). The causality with wine adulteration was recognized early and the Pb addition was prohibited, for example, in 1427 in France and Spain and in 1478 in Germany (Lessler, 1988).

The analysis of wine is of great importance since wine components strongly determine its stability, organoleptic, or nutrition characteristics. In addition, wine analysis is also important to prevent fraud and to assess toxicological issues (Grindlay et al., 2011).

The great number of parameters affecting wine quality has initiated the development of different protocols for analysis. Wine constituents are strictly regulated by international organizations or government agencies to avoid fraud and health risks. Luque de Castro et al. (2005) have recently reviewed methods of analysis for the most commonly determined parameters in wine such as ethanol, sulfur dioxide, reducing sugars, polyphenols, organic acids, total and volatile acids, Fe, soluble solids, pH, and color (Grindlay et al., 2011). For the wine industry and market sector, it is particularly essential that the intended value traits created via genetics (variety), the origin of production (typicity), and the unique inputs or processing methods are preserved. In other words, it must be ensured that a product's label is accurate and not misleading, since consumers distinguish peculiar commodities from a mass of other similar ones with the belief that they bear a superior quality. However, wine is a product that can be easily adulterated, and for this reason wine authenticity is guaranteed by strict guidelines laid down by responsible national authorities and includes sensory evaluation, chemical analyses, and examination of the records kept by wine producers (Makris et al., 2006).

The EC regulations dealing with geographical indications (GI) include traceability provisions. These are built into the provisions dealing with certification schemes and the various requirements concerning the overall control of the producers' operations. For example, Article 118p of EU Regulation 1234/20073 incorporated EU Regulation 479/2008 (dealing with wine) provides for annual verification of compliance with GI product specifications by relevant bodies (e.g., product certification bodies, public authorities, etc.). In all cases, producers will need to show inspectors how and where their products were produced using reliable traceability systems (Palade and Popa, 2014).

In the certification and control processes provided for GIs under the EU regulations, traceability is a core element. Only by recording the origin of the various components and by tracking the production chain is it possible for the certification and control bodies to verify, certify, and monitor whether the final product actually has the claimed link with its geographical origin and conforms to the relevant specifications and quality requirements (Sciarra and Gellman, 2012).

Wine authenticity is very important, especially in case of quality control and consumer information. Since wine quality is dependent on the consumer demands, compliance with traceability provisions satisfies the associated economic needs. It is necessary that a traceability information system exists in the wine supply chain in order to provide a better management of all the representative events that may arise on this chain, from grape production to wine selling. Traceability documents allow the detection of certain deviations from the usual processes, which may render the final product unsafe or of a lower quality than the one expected. Therefore, identifying the safety and quality issues during the process, by means of authentication and traceability, assures compliance within the framework of international regulations with the consumers' request for certification of quality attributes (Ashurst and Dennis, 1996). Table 14.1 presents different types of fraud commonly found in the wine industry.

TABLE 14.1 The Types of Wine Fraud (Holmberg, 2010)

Type	Description	Risk for Consumer
Mislabeling Regarding Origin, Classification, or Vintage	Most common type of wine fraud. Often facilitated by a "chain of bills" when wine is sold in a chain of retailers with a stepwise "quality enhancement".	Typically, no health risk because portable, cheap wine is misrepresented as a better quality product. Deception and fraud for the consumer.
Artificial Wine	Wine based solely on additives and water. Occurs very seldom now.	Health risk such as allergenic risk dependent on the (unlabeled) additives.
Wine Enhancement	Enhancement, most often accomplished by adding sugar, glycerol, diethylene glycol, or other sweeteners, and/or other flavor compounds to the wine, or through blending with sweeter, foreign wine.	Typically, no health risk except cases of addition of toxic compounds (e.g., methanol). Deception and fraud for the consumer.

14.2 TRACEABILITY OF WINE

Traceability is defined as the ability to follow a product batch and its history through the whole, or part, of a production chain from raw materials through transport, storage, processing, distribution, and sales (called chain traceability), or internally in one of the steps of the chain, for example, the production step (called internal traceability). Simply traceability of wine is defined as a method through which anybody in the wine supply chain will be able to verify the origin and composition of each batch of wine, its conditions of storage, and all the products that came in contact with the wine after the production. Traceability in the wine industry has an important role in the quality assurance management system. It ensures the registration of specific documents for all the manipulations of raw materials, ingredients, and final products. They are created specifically to allow a rapid identification of the product history. The wine supply chain requires traceability from grape production to processing and wine distribution. Traceability of products was introduced in the 1990s and is still under investigation by scientific and industrial bodies (Bevilacqua et al., 2009). A number of traceability systems, technologies, and standards have been developed to carry out supply chain traceability and internal traceability for different business objectives (Bechini et al., 2008; Bertolini et al., 2006). Nevertheless, only large enterprises, which are characterized by a tightly aligned supply chain and supported by a considerable use of information and communication technology, employ very efficient and fully automated traceability systems. On the contrary, small enterprises only rarely implement traceability and, when they do, they add the traceability management to their normal operation, thus decreasing the efficiency and increasing the costs. Thus, today, a considerable challenge is to develop agile and automated traceability platforms for communities of small-scale enterprises.

14.2.1 Grape Growers

Grape growers are responsible for the production, harvest, and delivery of grapes. Growers should record details about the locations, types of wines, annual production records, origins, chemical contents of water used for cleaning and irrigation, and annual treatments. For each treatment product, growers should record the supplier's details, a description of the product received, as well as the applicable batch numbers. The growers provide, with each delivery, the location numbers of the plots from which the grapes came and the date of picking, so that the receiving wine producers can link the related details to the wine made from these grapes.

14.2.2 Wine Producers

Wine producers are responsible for the production, manufacture, and/or blending of wine products. Wine producers should record where, in the winery, grapes or juice was stored. They must keep accurate records of the procedures and operations performed. The wine producer is responsible for identifying each production run with a batch number. For the receipt of additives, the producers should record the supplier's details, receiving date, a description of the product received, as well as applicable batch numbers.

14.2.3 Bulk Distributor

The bulk distributor is responsible for receipt, storage, dispatch, processing, sampling, and analysis of bulk wine. The bulk distributor checks the receiving documents, records all the information including the amount of received wine, and takes samples for tasting and analysis. If the wine is rejected, the wine returns to the source; otherwise two distinct processes are performed:

i. Storage and dispatch of bulk wine without any blending or any other processing.
ii. Storage, blending of different wines, and dispatch of the new bulk blend. Identification is handled by the bulk distributor and the bulk wine container. To ensure forward tracking, it is essential to record references of the delivery items and link these to the recipients.

14.2.4 Transit Cellar

A transit cellar is responsible for the receipt, storage, dispatch, processing, sampling, and analysis of bulk wine. The transit cellar receives bulk wine from bulk distributors in different kinds of containers. Each of these containers is identified with a proper code. The transit cellar sends batches of bulk wine to the fillers/packers. Each container sent is identified with a unique number and with the associated quantity of wine (liters). In order to maintain accurate traceability throughout the chain, it is necessary that the transit cellar records the items and batch numbers, as well as the identifier of each dispatched item. To ensure forward tracking, it is necessary to record the global identifiers of the shipped items and link these to the locations of the recipients.

14.2.5 Filler/Packer

The filler/packer is responsible for the receipt, storage, processing, sampling, analysis, filling, packing, and dispatch of finished goods. The filler/packer receives containers of bulk wine from the transit cellar, and also "dry goods" in contact with wine (bottles, caps, corks, etc.). Each of the containers of bulk wine and logistic units of dry goods is identified with a proper batch number. During this stage, the wine is poured into different kinds of containers such as bottles, bags, kegs or barrels, and a lot number is allocated to them. A link between these components (bulk wine, finished product) should be maintained. The next step is packaging into cartons and pallets and the dispatch of these cartons and pallets (identified with a lot number) to the finished goods distributor. The lot number must be linked to the batch(es) of bulk wine used to fill the bottles. To ensure forward tracking, it is necessary to record the global lot numbers of the shipped items and link these to the location numbers of the recipients.

14.2.6 Finished Goods

A finished goods distributor is responsible for the receipt, storage, inventory management, and dispatch of finished goods. The finished goods distributor receives pallets and cartons from the filler/packer and dispatches them to the retailer. These trade items are identified with lot numbers. To ensure forward tracking, it is necessary to record the global lot numbers of the shipped items and link these to the location numbers of the recipients. The retailer receives pallets and cartons from the finished goods distributor and picks and dispatches goods to the retail stores. The container number of an incoming pallet is recorded and linked to the location numbers of the suppliers. The retailers keep a record of the container numbers and the lot numbers of the components of the pallets and cartons they receive. The retailers sell consumer items (bottles, cartons) to the final consumers. Each of these items is identified with a number allocated by the brand owner.

This brief description of a wine supply chain has highlighted that all the processes from the grape grower to the consumer can be traced by associating appropriate identifiers with the traceability entities managed by the single supply chain actors and by creating a record for each identifier with all the information required about the entity. Each actor of the supply chain is therefore responsible for recording traceability data corresponding to specific entities. Further, each actor has to create links between the identifiers which identify correlated entities.

14.3 AUTHENTICITY OF WINE

The authenticity of wine is guaranteed by strict guidelines laid down by the responsible national authorities that include official sensory evaluation, chemical analyses, and examination of the registers kept by the wine producers (Rapeanu et al., 2009). The authenticity of wine has been extensively investigated because wine is an easily adulterated product due to its chemical composition and its availability throughout the world (Ashurst and Dennis, 1996). Responsible and continuous controls are required to maintain the quality of wine.

Wine is a very easy product to adulterate because of its strong chemical basis (high alcoholic content, low pH) and its availability throughout the world; it is also probably one of the food products that is most often analyzed and for which legislation is the most complete. The imaginations of potential deceivers seem to be without bounds and the extent of such fraud remains unknown. Even if they are scarce, fraudulent products cause considerable damage to the wine trade and, in addition to a potential loss of prestige, the damage to the product image in terms of consumer confidence can be immense, especially in the current context of surplus where “new world” products are becoming increasingly competitive. Until recently, tasting by experts was the only way of verifying authenticity, but because experts are not readily available and because of the subjectivity of the technique, methods were developed that sometimes combined advanced mathematical tools and the latest analytical technologies to guarantee the authenticity of wine. Usually, volatile compounds are used to characterize varieties, whereas minerals are used for geographical differentiation. Amino acids as well as phenolic compounds are used for both. The development of advanced techniques for wine authentication is a challenge, which is currently being given a special attention (Ashurst and Dennis, 1996).

14.3.1 Authentication of Geographical Origin

Climatic, edaphic, and orographic factors influence the process of wine growth with direct influence on the compositional and sensory parameters of wines (Ballabio et al.,

2006). Since the area of production raises a visible mark on originality and quality characteristics of products, the determination of geographical origin is the main requirement for wine authenticity. The concept of "terroir" has been introduced in Europe, considering the specific characteristics of wine that are induced mainly by the geographical location and the characteristics of production in the concerned areas. The "denomination of origin" (DOC) regions are areas within the designated traditional wines, which have great quality features. Sensory evaluation done by specialists (tasters) was the only way to determine the geographical origin of wines. This method has a high degree of uncertainty; therefore, instrumental analysis is used to identify the compounds that are present in very small quantities (Rapeanu et al., 2009). The principal methods used for the authentication of geographical origins are discussed below.

14.3.1.1 Organic Compounds

14.3.1.1.1 Analysis of Volatile Compounds Profile Based on the content of 1-hexanol and cyclohexanone, it was possible to differentiate between Pinot noir wines of French and American origins. Using data processing and laboratory methods that generate numerous relatively accurate data (e.g., high pressure liquid chromatography (HPLC), gas chromatography (GC), atomic absorption (AA)), a pilot database can be developed (Kwan and Kowalski, 1978). Further, it was possible to identify the geographical origins of French red wines and Spanish white wines by the use of more volatile compounds (ethyl esters, isoamyl esters, aldehydes, acetals) (Rapeanu et al., 2009). More recently, by the means of purge and trap followed by capillary gas chromatography and mass spectrometry (MS), Garcia-Jares et al. (1995) performed the differentiation of some white wines of Galicean origin (Rias Baixas) using 19 highly volatile compounds.

14.3.1.1.2 Analysis of Amino Acidic Profile Amino acids are useful substances for differentiating wines. Production of wines, such as sparkling wines, where the second fermentation producing the overpressure of CO_2 is performed in the bottle leads to an increase in amino acids.

By investigating the amino acids like arginine, alanine, tyrosine, valine, and leucine that are responsible for the amino acidic profile of wines, it is possible to identify the geographical origins of wines (Flamini and De Rosso, 2006; Chambery et al., 2009).

Tusseau et al. (1994) used this technique to separate authentic Champagnes from sparkling wines; unfortunately, wines made from the Pinot Noir grape from the same place but with different vintages show wide variations that do not assist in characterization. However, Champagnes are richer than sparkling wines in all amino acids, except arginine, because of the second fermentation in the bottle and the long contact with the lees. Using a linear discriminant function and amino acid analysis on a total of 110 wines, 4% of the Champagnes and 15% of the sparkling wines were misclassified. This performance—certainly far better than any achieved by tasting—is not judged to be sufficient (Tusseau et al., 1992) and more parameters have to be introduced. Using linear discriminant analysis of 20 amino acids (Derde et al., 1983), it was possible to correctly separate wines from Bourgogne, Bordeaux, and Beaujolais.

14.3.1.2 Mineral Profile Analysis

The "fingerprinting" of the mineral profiles of wines is the most valuable method of assessing the geographical origins. Investigation of mineral elements in wine is the main procedure to authenticate the geographical origins of wines.

Minerals seem to be the primary candidates for obtaining a "fingerprint" as they are stable. The use of trace elements for the differentiation of wines closely followed the advances in analytical techniques. Strangely, 20 years ago neutron activation analysis (NAA) was capable of determining concentrations smaller than parts-per-billion (ppb), but it was costly and time-consuming. Atomic absorption, although adequate for parts-per-million (ppm) (flame) and ppb determinations (furnace), lack the ability to carry out simultaneous determinations that are mandatory in this kind of research; the ICP (inductively coupled plasma), although simultaneous, lacked adequate sensitivity. It was not until the appearance of ICP/MS that the analysis of most of the elements in wines was readily achievable (Ashurst and Dennis, 1996).

Since some of the macro and micronutrients undergo great changes during technological processes, such as Na, K, Ca, Fe, Cu, Zn, and others, attention has been focused on the elements which show very small changes during technological processes, although they are found in small quantities or as traces (Cr, Co, Sb, Cs, Sc, Eu, Hf, Ta, etc.). Li and Rb are most relevant for geographical origin authentication.

Li is found at a very low level in wine (1–200 ppb), but flame emission spectrometry using acetylene fuel is very sensitive and permits an accurate evaluation of this element. Rb is also easily determined in wines and has values in the range 0.5–5 ppm. Consistent work has been performed using these metals (Sudraud and Medina, 1993).

Investigation of rare earth elements (lanthanides) also provides valuable information in order to detect the growing areas of wines by using ICP/MS (Dutra et al., 2011).

14.3.1.2.1 Trace Elements Siegmund and Bachmann (1977; 1978) used the powerful neutron activation method for trace element analysis; today, very few authors have been able to match the level achieved by that study for 15 trace and ultra-trace elements in wines. Their method was very cumbersome and required a nuclear reactor. The analysis itself, after irradiation for more than 2 days, took more than 3 months to perform. Nevertheless, 15 elements were correctly determined using this technique.

- Elements currently determined: K, Na, Fe, Zn, and Rb at ppm and sub-ppm levels
- Trace elements: Cr, Co, Sb, Ag, Cs, and Br at ppb levels
- Ultra-trace elements: Sc, Eu, Hf, and Ta at sub-ppb levels.

The authors realized that mono-dimensional statistics, simple ratios, or graphical representations were unable to perform a good identification and, therefore, they used taxonomic analysis (cluster analysis) with Euclidian distances to group wines. With the elements Na, K, Rb, Cs, Cr, Fe, Co, Zn, and Ag they showed that clusters were formed according to the area of production as well as the grape variety. They assumed that the transfer of elements from the soil to the plant had the greatest influence on their contents in the wine. When analyzing the soil, however, the only correlation found with the content in the wine was related to the alkaline metals (Na, K, Rb, Cs) and this was not true for all the sites. The year of production also introduced some variations in the levels of the 15 elements in the same wine (place and grape). One important conclusion they drew was that if a control wine exists, then the profile of the mineral content proves the similarity in wines for a given producer. This is obviously not the case with classical determinations, such as alcoholic content, total acidity, volatile acidity, sulfur dioxide, etc. Partly because of the fact that this technique was not convenient for the normal or advanced wine laboratory, no other determinations were conducted at this level for several years. Using ARTHUR

(K nearest neighbors, discrimination analysis, Informetrix Corporation, 1981), Kowalski and his team (1979) used only two mineral elements, Ca and Ba, to classify 14 French (Bourgogne) and 26 American Pinot Noirs (there was a single misclassified sample) correctly. They used a simultaneous emission technique (dichloropropane, DCP) and 17 elements for this study. Following this pioneering work, more extensive use of minerals was applied to French and other wines. Lacasta (1982) separated wines from the French regions of Medoc and Saint-Emilion, where 64 wines (85%) were classified correctly. Medina and Van Zeller (1984) determined 14 mineral elements by atomic absorption and emission spectrometry (with flame and furnace). A correct classification of a reduced set of wines from three French regions was obtained using Li, K, Ca, Cu, and Ni and linear discriminant analysis as well as other classification techniques (soft independent modelling of class analogy, SIMCA) (Informetrix Corporation, 1981). Gonzales-Larraina et al. (1987) found that Mn, Li, K, and Fe were the most important elements for classification of the three regions of Spanish Rioja. Classification reached 93% whichever method was used (e.g., multilinear regression, K nearest neighbors, SIMCA). Using the first two elements, Mn and Li, a good distribution was obtained, the distinction occurring even between two types of productions from big wineries and a family style operation (Rioja Industrial and Rioja Cosechero) could be detected. Maarse et al. (1987) used those elements together with Na and Fe to distinguish German wines from the Rhein-Pfalz and Mosel regions. The difference in mineral elements was shown as the basis of typicity for Spanish wines from Galicia (northern Spain) and the rest of Spain (Herrero-Latorre and Medina, 1992). At the regional level in Galicia, Latorre et al. (1994) showed that using only Rb and Li, a very good differentiation can be found for Rias Baixas as compared to other white wines from two other regions: Ribeiro and Valdeorras.

Using Ca, Na, K, and Fe in the same dataset, they pointed out that these elements can be influenced by the fertilization and wine-making process and are not relevant for classification as shown by the pattern recognition techniques (Fisher weights in ARTHUR, Infometrix, 1981).

It is important to know to what extent the mineral content must reflect that of the soil and also how much mineral content is transferred from the soil to the wine. However, it is not the total content of the minerals in the soil that is important but its availability, which in turn depends on several factors such as pH, humidity, porosity, clay, humic acid complex, etc. The composition of vegetation reflects some extent that of the soil. This fact has given rise to geobotany, a geological science devoted to the search of species that are characteristic of certain ores and deposits (Brooks, 1972). It has been proven (Kim and Thornton, 1992) that vegetation (e.g., lettuce and rice) growing on soils such as uraniferous black shales that are rich in elements such as Cd, Mo, and Se accumulate these elements up to levels that may be of concern for health. Grape varieties seem to take up certain elements selectively; it must be kept in mind that grape cultivation and harvest is by no means a sterile process and that pollution by dust and soil is unavoidable. This pollution varies from year to year (because of rain and wind) and must account for a significant part of the mineral content in wine. The bunching of grapes that facilitates the retention of dust is also a factor not found in larger fruits. It is also clear that the wine-making process can alter the mineral content of wine considerably. For example, the use of technological additives or processes, such as cold treatment, can lower the content of K. In the future, more advanced technology, such as reverse osmosis, will probably introduce more factors of variation. Wine-making equipment is also of great importance for metallic elements such as Fe, Cu, Zn, and Pb (Legrand et al., 1991). When using mathematical analysis, the important thing to consider is the selection of elements that can vary in the same way in all the wines during the processing; it is not so much the absolute concentration of one element that is important but its relationship with the others. Taking all the previous arguments into consideration, the choice of the mineral elements based on experience can be summarized as follows: K, Ca, and Na are elements that can vary widely, both in the process and from wine to wine. K and Ca can combine with tartaric acid and form potassium bitartrate and calcium tartrate, respectively; these molecules can precipitate the K influenced by temperature. The deacidification of wine can be carried out with calcium carbonate or potassium bicarbonate, thus producing an increase in the content of these elements. Bentonite fining adds Na and Al to the wine (Etchats, 1986). Metallic elements that are likely to migrate into the wine from poor equipment are not the ideal choice because even though the trend toward stainless steel is making progress, traditional apparatus will be used for some time to come. That is why Fe, Cu, and Zn coming from bronze or brass are inappropriate for our purposes. High Pb content (above 150 μg/l) shows that the wine has been in contact with the material not suitable for food manufacture. This may also lead to other elements that are previously mentioned having an elevated value in the wine, sometimes going over the L'Organisation internationale de la Vigne et du Vin (OIV) recommendations: for Cu 1 mg/l; for Zn 5 mg/l; and for Pb 0.25 mg/l. Another problem arises with the oenological treatments, i.e., potassium ferrocyanide. This practice, authorized for wines by the EC regulation (JOCE, 1987) and performed only under the supervision of an oenologist, removes excess Fe and other metals, namely Zn, Mn, and Cu (Etchats, 1986). French legislation is more restrictive and this treatment is forbidden in red wines. As pointed out by Siegmund and Bachmann (1977), the only correlation found between soil and wine was shown for alkaline elements, even more so for Li, which together with Rb appears to play a key role in geographical classification. Both elements are, however, relatively easy to determine in wines using flame emission spectrometry (Lacasta, 1982). Li is found at a very low level in wine (1–200 ppb), but flame emission spectrometry using acetylene fuel is very sensitive and permits an accurate evaluation of this element. Rb is also easily determined in wines and has values in the range of 0.5–5 ppm. Consistent work has been performed using these metals.

Sudraud and Medina (1993) summarizes the elements (Table 14.2) selected for a correct classification by several authors, although it is not intended to be exhaustive. The elements most often cited by order of weight are: Rb, Li, Ba, Sr, Mg, Al, Mn. The last two elements have to be considered carefully as already indicated. These consistent results must be considered with caution because elements such as Ni, Cr, and Li are components of the glass bottle that can over time migrate into the wine (Medina and Sudraud, 1979; 1980).During aging, the relative proportion of these elements may change according to the glass composition, leading to inaccurate results in terms of geography. This fact is partially confirmed by a careful study by Siegmund and Bachmann (1977) where high Cr levels have also been found in wines (up to 2.5 ppm). Apart from these considerations, alkaline metals being easily mobilized in the soil and easily transported into the plant, are good indicators of geographical identity.

14.3.1.2.1.1 Ultra-trace Elements: The Lanthanides

Lanthanides or rare earth elements (REE) have a mass of 140–180 and their patterns give insight into the history of a particular rock or soil as pointed out by the abundant bibliography on the subject in the geological sciences. Extensive research on these elements has led to efficient determination techniques, even at the extremely low levels at which they are found in wine. ICP/MS is frequently the best technique for such determinations, largely because of the simple,

TABLE 14.2 Mineral Elements Used for Classification

Reference	Country or Region	Elements Used
Kwan et al., 1979	USA/France	Al, Ba, Ca, K, Mg, Sr
Medina and Van-Zeller, 1984	France	Ca, Cu, K, Li, Ni
Maarse et al., 1987	Germany	Fe, Na, Rb
Gonzales-Larraina et al., 1987	Spain (Rioja)	Li, Na, Rb
Etievant et al., 1988	France	Li, Ma, Rb
Herrero-Latorre and Medina, 1992	Spain	Ba, Li, Mn, Rb
Day, 1993	France	Al, Ba, Fe, K, Rb, Sr, Zn
Latorre et al., 1994	Spain (Galicia)	Li, Rb

easily interpreted spectra and the exceptionally low limits of detection possible even with standard ICP/MS instruments. This technique has evolved rapidly and the development of new instruments allows the accurate determination of ng/l or ppt in wines without pre-concentration. The authors used this technique to investigate the case of REEs in relation to the geographical determination of the wine origin. In geology, it is a widely accepted practice to normalize REE concentrations using chondrite values (Masuda et al., 1973). The chondrites are meteorites whose values represent the natural pattern of REE levels, which were unaffected by geological/geochemical processes on earth (they act as a control for the state of matter when the solar system was formed). For rocks of volcanic origin, chondrite is normally used, whereas marine sedimentary rock REE patterns are usually corrected using values from reference marine shale.

Soils are the compositional reflections of the underlying rocks that gave birth to them. Plants growing on these soils take up trace elements according to the compositions of the soils as well as their chemical properties, which the REE composition also reflects (Brooks, 1972).

14.3.1.2.2 Stable Isotopes A very appealing and sophisticated technique using "stable" isotopes, such as mass spectrometry of isotope ratios (MSIR) and nuclear magnetic resonance (NMR), has been used for geographical determination of wines. However, the results are difficult to interpret in terms of variations in the year, climate, grape varieties, fermentation process, etc. and even the reference to a physical or biochemical process remains somewhat empirical. The use of Pb isotopes is related to atmospheric pollution and seems much more practical because this phenomenon is fairly well understood and is less dependent on small, local variations.

14.3.1.2.2.1 Carbon (^{13}C), Hydrogen (^{2}H), and Oxygen (^{18}O) Isotopes Several isotopes can be used: those from the water of the wine and those from the ethanol (main organic component). The sources of variations of these products can be summarized as follows. Water absorbed by the roots is combined with carbon dioxide from the atmosphere to produce sugars by photosynthesis. During this process, there is fractionation of the different isotopes. A well-known process is the isotopic fractionation of ^{13}C during photosynthesis, which divides plants into three main groups. The evapotranspiration of water in a plant has a concentrating effect on the heavy isotopes of oxygen and hydrogen. Rain during the ripening stage can also have an influence by dilution, especially just before the harvest because, at this time, it contains fewer "heavy" isotopes. The addition of concentrated must (rectified or not) will complicate the matter further. Climate, continent, and topography can all have an influence on isotopic fractionation; even as little as a 150 m elevation is significant (Forstel, 1995). Owing to all these sources of variations and relativity of the measurements, a well-conceived database is absolutely essential for these methods to be useful and accurate.

14.3.1.2.2.2 Deuterium Deuterium is specifically distributed in the ethanol molecule produced during fermentation and is related to the deuterium content of the water and the sugar molecules. The deuterium of the methyl group (D/H) I in ethanol comes from a non-exchangeable site of the sugars. On the methylene (D/H) II site it contains a contribution from the fermentation water; and the hydroxylic group is in equilibrium with the deuterium of the wine water (Martin et al., 1988). The site-specific isotope fingerprint has been used for detecting exogenous sugars in wines (Martin and Martin, 1983). The use of these parameters permits the differentiation of wines produced in countries with substantial climatic or latitudinal differences. Using the (D/H) I, (D/H) II, and (D/H) of the water from wine and must, Martin et al. (1988) accomplished a good separation even when the wine produced from different grapes varieties gave different values for the same region, and even when the maturation effect influenced the values (non-matured grapes lead to a 0.3–0.8 ppm increase). A good distinction is made among three different types of climates:

- Cold and humid
- Cold and dry or hot and humid
- Hot and dry

Even for the same country (France), the regions (Alsace, Anjou, Bordeaux) were correctly separated for the same year of wine production.

As it is so dependent on the climate and its variation, the use of these isotopes, although useful, cannot be extrapolated from the year of production. Without this information, the sample can be attributed to several regions of origin in France (Day, 1993). These variations will be more severe for the "Atlantic" climates than for the Mediterranean ones. As discussed later in this chapter, determining the year of production is not an easy task.

14.3.1.2.2.3 Lead Pb is a wine contaminant with an average of 60 μg/l in wines (OIV, 1995). The current limit set by the OIV is 250 μg/l. This metal is a special case in the determination of authenticity. It is a very old toxicological problem known since the time of the Roman empire, whose decline was supposed to be caused partly by excess Pb intake from food, especially wine, because lead acetate was added as a sweetener. A famous theory says that the high level of Pb hampered the intellectual capacity of the ruling class at that time. Currently, wines with a Pb content under 100 μg/l are considered to be unpolluted by the material during the wine-making process or storage. This very old problem has recently found a new solution with the application of sophisticated analytical techniques, otherwise until now it was just confined to geological and environmental studies (Ashurst and Dennis, 1996).

14.3.1.2.2.4 Organo-lead Using techniques developed in the Arctic circle ice field for the determination of organo-lead compounds from gasoline additives, these compounds were found in wines. The determinations were set up at Anvers University by Lobinski and coworkers (1994), who used the special hyphenated technique (gas chromatography/

microwave induced helium plasma/atomic emission spectrometry). The authors found ethyl lead and methyl lead, but the latter was predominant. The compounds were derived from tetraethyl and tetramethyl lead, respectively, which were added as antiknock agents in gasoline. Partially burned in the motor (40–80%), they are decomposed in the atmosphere under the action of ultraviolet light, ozone etc. Extremely low levels were found—in the range of 10–500 ng/l for trimethyl lead and 0–50 ng/l for triethyl lead. The maxima were found in wines produced from grapes from vineyards close to the junction of two major highways in France in the Rhone valley. This study compared levels in the "vertical" series of wines, covering over 40 years in the same area. The concentration of trimethyl lead is no higher than 25 ng/l in rural areas, but it increases with the proximity to highways and cities. With the increased use of unleaded fuel in the EU in the last 10 years, the level of these compounds in wine has diminished sharply. For the chronological series of wines between 1950–1991, there has been an increase in trimethyl lead after the introduction of the tetramethyl compound in 1960, a maximum *circa* 1978 and then a sharp decrease from 1980 to the present. Triethyl lead is present in 1960, but the levels decrease after 1980 (Lobinski and coworkers, 1994). Until recently, there was no European wine that did not contain alkyl lead compounds. In the USA and Australia, which have used unleaded fuel for a longer period of time, there is no organic Pb in wine at present; however, before 1987, levels similar to those of European wines were found. The use of this time shift and the relative proportion of methyl and ethyl lead allows one to plot concentration curves to give an evaluation of the age of the wine from the 1960s to the present. The use of these curves permits several deductions with reference to the year of production and the origin of the wine.

- No organo-lead: The wine originates from countries that have banned these compounds many years ago (USA and Australia) or is a very recent European wine.
- According to the vintage, the relative proportion of trimethyl and triethyl lead indicate the year of production (i.e., pre-1960 there was no methyl lead present).
- High levels of organo-lead imply proximity to a heavily used road or a city (trimethyl lead > 100 ng/l and triethyl lead > 15 ng/l).

14.3.1.2.2.5 Lead Isotopes Pb has four isotopes, three of which come from the decay of U and Th. Consequently, Pb ores have different proportions of Pb isotopes according to their original content in these radioactive minerals. One major reason to use Pb isotope variation as an indicator lies in the fact that their variation is much greater than that of lighter elements such as carbon or oxygen, which are in the ppm range. The variation between two major Pb mines (Missouri, USA and Broken Hill, Australia) is close to 30%. Indeed, Pb isotope ratios have been used for 20 years to trace Pb pollutants (Chow et al., 1975). In addition, heavily polluted soil has been found in metropolitan areas and the level of Pb here is proportional to the proximity and the volume of traffic, ranging from 100–1000 ppm; these values should be compared with the level of Pb in the earth's crust, which is 16 ppm (Krauskopf, 1956).

Organo-lead is volatile and can be found in places that are as far away as the poles, but most of it is burned and remains in the same geographical area. The variation, therefore, reflects the origin of Pb that is used as an anti-knock additive: Missouri in the USA or Broken Hill in Australia. The situation in Europe is intermediate since Pb comes from different sources (Morocco, Australia, etc.). Even if the total amount of Pb emitted into the atmosphere is reduced, the current levels of pollution will remain unaffected for some time as will the Pb ratios (MacKay, 1994).

14.3.1.2.2.6 Age Using Lead Isotope Ratios The increased use of unleaded gasoline in Europe also leads to variations in specific ratios. Wine is therefore strongly marked in its composition by the environment, and this provides valuable data. The concentration trend of Pb is moving from a situation of Pb pollution toward natural earth crust ratios. Indeed, this trend was found for aerosols in a study in San Diego, California, from 1964–1974 (Chow et al., 1975) and it was confirmed by Grousset et al. (1994) who worked on European aerosols. Another factor in favor of reduced pollution is the fact that very low levels of Pb are now found in wine, regardless of the source of the constituents. It is interesting to note that Pb concentrations in aerosols are also decreasing in a similar fashion (Grousset et al., 1994).

Differences existing between the wines from Australia and the USA are mainly because of the atmospheric pollution caused by Pb present in the exhaust of automobiles that burn leaded gasoline. Organic Pb can be found in extremely low levels (0.5 μg/l), thus testifying the influence of this phenomena. Further study shows that the extent of atmospheric pollution can be traced more extensively in wine. Only a small fraction of organic Pb introduced into the gasoline exhaust remains unaffected by combustion (1%–10%); in addition, natural reactions occur that decompose the remaining compounds (ultraviolet light, etc.). That is why only very sophisticated techniques (such as GC-atomic emission detector (AED)) are able to detect alkyl lead in wine. However, isotopic ratio determination of $^{206}Pb/^{204}Pb$ shows a definite variation according to the year of production, following the increased use of unleaded fuel throughout Europe (Ashurst and Dennis, 1996).

14.3.1.3 Organic and Mineral Features

It is appealing to construct a database where minerals and organic substances are mixed in the hope of encompassing several factors of variation such as grape variety, soil, climate, and so on all in one step. Several authors have used this approach, following the pioneering work of Kowalski and coworkers (Kwan and Kowalski, 1980a; 1980b). Using the database in their work and the partial least squares (PLS) method, Frank and Kowalski (1984) showed that elemental analysis performed much better than organic analysis while differentiating French, Californian, and Pacific Northwest wines. A combination of the two chemical blocks gave the best prediction.

Medina and Van Zeller (1984) used minerals and volatile, polyphenolic, and organic acids and classical determinations (62 variables) in a study on French red wines. Wines representative of their areas of production were selected. In this feasibility study, the principal component analysis (PCA) technique was able to distinguish among wines of the three examined regions. A search for the smallest number of substances that could successfully characterize groups led to the selection of Li, propanol, procyanidin B2, and decanoic acid. Each group of substances (e.g., phenols, minerals, etc.) could be used individually for an accurate separation using group classification methods (KNN, LDA in ARTHUR, Infometrix, 1981). Following this work and having examined the most useful elements, Manterola (1985) used three mineral elements (Li, Mn, and Rb) and one all-purpose HPLC technique for 13 neutral polyphenolic substances (Salagolty-Auguste and Bertrand, 1984).

Etievant et al. (1988) used minerals and amino acids to achieve 94% accuracy when classifying 68 French red wines provided by three agronomical stations and made from pure grape varieties typical of the Loire (Angers), Southern France (Narbonne), and Bordeaux regions. Fisher's test showed that only Rb, Li, and Mg could discriminate among the regions, although six amino acids showed potential. Stepped discriminant analysis selected Rb, ethanolamine, arginine, and Mg. It is well known, however, that the amino acids are dependent on fertilization, climatic conditions, and the duration of skin contact in most cases; those indigenous to the grapes are metabolized by the yeast, and are excreted and released during the autolysis of the dead yeasts; more complex phenomena also exist. In contrast, the minerals selected do not vary even after the time needed to obtain the red wine—48 h of skin contact (Rizzon, 1985).

14.3.2 Authentication of Grape Variety

Biotic factor has a major contribution in the formation of the qualitative characteristics of wine, so the compositional and sensory parameters are largely dependent on the variety of wines.

14.3.2.1 Sensory Analysis or Tasting

The tasting was, until recently, the only way to classify wines according to vintage and origin; however, no extensive and objective research has been performed on the subject. Is tasting accurate enough for a correct classification of wine? An expert taster will answer positively and every year there is an amazing worldwide contest of "sommeliers" where even the name of the chateau is guessed. Sensory evaluation can be easily rationalized and the use of wine-tasting cards is now widespread. Frank and Kowalski (1984) showed that sensory data do not provide enough information to separate wines from France and the USA. A spin-off of this study demonstrates that elemental analysis contains more information relevant to the organoleptic characteristics than the organic composition analyzed at the time. In a very interesting study (Thurin, 1984), where the expert tasters were asked to guess the name of the pure variety (Cabernet Sauvignon, Merlot, Cabernet Franc) of grapes that the wine was made of, a surprisingly low percentage (20%) of correct answers was achieved; indeed, less than what was achieved from random guessing. However, when using all the 28 criteria of the wine-tasting card and the classification method SIMCA in ARTHUR, (Infometrix, 1981)—which performs a reduction of the variables—the classification reached 90% (using 110 tasting cards). The typicity of the grapes is characteristic enough to be inscribed in a wine score card, even when grown on different soils. The wine score cards are designed to use the tasters as sensitive machinery without a cognitive function that, in this case, interfere with the results.

Sensory evaluation is limited by the influence of several factors (ecological, agro, and technological) that modify the primary characteristics of the variety significantly. Although sensory analysis is still influenced by human subjectivity, statistical interpretation of results and the use of equipment such as "electronic nose" and "electronic tongue" improved the performance of sensory evaluation (Biernacka and Wardencki, 2012).

14.3.2.2 Mineral Profile Analysis

As the different wine varieties selectively accumulate various metals, their identification can be achieved from the content of certain elements in the wine like Li, Ni, Ca, Rb, etc. (Rapeanu et al., 2009; Dutra et al., 2011).

14.3.2.3 Amino Acid and Protein Profile Analysis

While soil, agrotechnical, and technological factors cause large variations, the content and nature of amino acids may contribute to the recognition of the wine variety. Recent research has shown that molecular weights of proteins recorded small variations in different wines depending on the variety. Proteins are stable in must and can be found in wine and even those are subject to stabilization. Protein quantification is relatively simple and can be done by conventional methods (electrophoresis) or by modern methods (mass spectroscopy, etc.) (Rapeanu et al., 2009).

14.3.2.4 Analysis of Polyphenolic Profile

Their evaluation is useful to authenticate individual wines by botanical origin. Investigation of phenolic compounds by multidimensional NMR allows the differentiation of wines by variety and even the clones of the same variety. The most representative phenolic compounds of red wines are the anthocyanins. They are in the free form—anthocyanins/anthocyanidins (malvidol, delfinidol, cyanidol, peonidol, and petunidol)—as well as in acyl and coumaryl compound forms. Cabernet Sauvignon is characterized by a higher content of malvidol and coumaryl anthocyanins, while Merlot is highlighted by a higher content of peonidol and coumaryl anthocyanins (Gonzalez-Neves et al., 2004). Because during the grape processing and wine preservation acyl and coumaryl anthocyanins are more stable, it is proposed that anthocyanin fingerprinting of red wines should be used to express the number of acyl anthocyanins and coumaryl anthocyanins and the ratio of acyl anthocyanins-to-coumaryl anthocyanins (Von Baer et al., 2008).

Since the authentication of variety by using the fingerprint anthocyanins is operational only in case of red wines, white wines, and roses, even the reds are investigated by the content of shikimic acid (Mardones et al., 2005). Shikimic acid is found in small amounts in various fruits including grapes. In wine, it is found in the concentration of 10150 mg/L. The evaluation of shikimic acid content is especially recommended for white wine authentication and as an additional indicator in red wine authentication (Makris et al., 2006).

14.3.2.5 Analysis of Volatile Compounds Profile

Quantification of the odorant substances in wines, compounds that are usually found in extremely small quantities, is achieved by their extraction with various solvents and their determination is done with modern methods such as GC coupled to MS (GC-MS, LC/ESI-MS, MALDI-TOF-MS) (Nasi et al., 2008).

14.3.2.6 Residual DNA Analysis of Grapes

Since it has been found that wines contain small amounts of DNA from grapes, DNA analysis has been done in order to identify the variety of the wine (Baleiras-Couto and Eiras-Dias, 2006). The difficulties—like low content of DNA in wine, contamination (DNA from yeast, bacteria), changes during processing, and inhibitors for the PCR reaction—were solved by using appropriate methods of DNA extraction and by improving the performance analysis using microsatellite PCR amplification.

14.3.3 Authentication of Wine Age

With time, the wines undergo a number of beneficial changes such as physical-chemical and even biochemical changes. They cause the development of a refined sensorial quality, thus old wines are more valuable. Moreover, the crop year, which benefit from favorable weather conditions, also gives great features to the wines, therefore they are appreciated by the consumers. The analytical method that is frequently used is determination of the radioactivity of the ^{14}C isotope of the ethanol molecule in wine. The method is relatively simple and consists of determination of the ^{14}C isotope radioactivity of the concentrated solution (minimum 95%) of alcohol obtained by the distillation of wine or other alcoholic beverages. For red wines, to increase certainty, the combination of these results with those obtained from the analysis of polyphenolic compounds and color indices is indicated. Determination of radioactivity by measuring the activities of wine sediment isotopes such as ^{210}Pb, ^{210}Po, ^{239}Pu, ^{240}Pu, ^{13}Cs, etc. is another way of knowing the wine age. The validity of the method was verified by appropriate dating of wine during 1850–1968 for every 6 years. Age rating wines and other alcoholic beverages by determining the radioactivity of isotopes has a high degree of reliability only for products of a considerable age (Dutra et al., 2011).

Carbon isotope ^{14}C: One of the oldest methods for determining the age of an alcohol takes advantage of nuclear fallout by determining the ^{14}C content of the ethanol (Martiniere et al., 1979). Cosmic rays colliding with nitrogen produce ^{14}C. The radioactivity of carbon in ^{14}C remained constant for two centuries (1752–1952) at a level of 13.6 dpm/g (disintegration per minute per gram). Atomic bombs and, especially the hydrogen bombs tested in the atmosphere, produced neutrons that reacted with nitrogen and produced radioactive carbon. This carbon is rapidly oxidized to form CO_2 and is incorporated in plants. Unlike deuterium (^2H) and ^{18}O, the diffusion of the radioactive $^{14}CO_2$ was worldwide and homogeneous. There is little variation (±5%) among plants or with latitude and longitude. The maximum value of 28 dpm was obtained in 1963 because most of the atmospheric nuclear testing stopped in 1964; a decrease has been observed since then but it currently remains above the natural level (Mongereau and Evin, 1993). This technique can also be applied to wines and spirits after a distillation process that must reach at least 95% by volume of ethanol (double distillation for the wines). Nevertheless, several factors can affect the results such as chaptalization, if it is carried out with sugar that is of a different age than the grape. Given the shape of the curve, it is not possible to distinguish between the age lying on each side of the curve along the same horizontal line. Given the curve's tendency to flatten out and the uncertainty associated with the technique, the method is not sufficiently accurate at present and no improvement is foreseeable. This technique is, however, useful in some circumstances such as for very old spirits or products of special vintages (±1 or 2 years). For red wines, this technique combined with the analysis of polyphenolic compounds (polymerization, tannin indicators (Glories, 1978)) or color indices (the tile color increasing with age) will narrow the margin. The addition of sugar to the wine complicates the process (because it is sugar of a different age) and widens the uncertainty to over 1 year (Mongereau and Evin, 1993).

14.3.4 Authentication Technology for Wine Production

There are a variety of techniques and processes of wine-making, authorized or used fraudulently, altering appreciably the compositional and sensorial wine parameters. This aspect of authenticity has a particular relevance to special wines like the sparkling wine, the oxidative wines, etc.

14.3.4.1 Authentication of Sparkling Wines

Sparkling wines are elite alcoholic beverages, much appreciated by consumers. There is a wide range, determined both by the diversity and quality of raw materials (wine-based) and also by developing technology. Besides the general aspects of wine authentication, sparkling wines represent a clear distinction between products which contain endogenous origin CO_2 (sparkling wine) and which have totally or partially exogenous CO_2 (Martinelli et al., 2003). The most information is provided by the component amino acid analysis, the certainty being higher when it is associated with the study of other compounds like odorant substances, polyphenol content, carbohydrates, etc. Isotopic analysis generates the most rigorous investigation results by determining the amount of isotope ratio $^{13}C/^{12}C$. In this way, it is possible to determine precisely the origin of carbon dioxide in the product, whether it comes from plants with C3 metabolism (the wine) or C4 (sugar beet, etc.) or from other sources (industrial, synthesis, etc.) (Calderone et al., 2007).

14.3.4.2 Authentication of Oxidative Type Wines

Oxidative type wines, the main representatives being Port wine, Madeira, Jerez, etc., are characterized by sensorial properties, generated mainly by biochemical oxidative processes. Authentication aims to detect the original products of imitation. In addition to sensory evaluation, content analysis of metals (Rb, Li, Mn, Fe, and Al), phenolic compounds, carbohydrates (including the ratio of glucose to fructose), presence and content of amino acids, etc. generate information that is useful in identifying the original products (Camara et al., 2006).

14.3.4.3 Authentication of Rose Wines

Rose wines lie between white and red wines, having specific chemical compounds, so it is difficult to license their origin. Authentication of rose wines has two major objectives. The first step is to identify the source of raw material, i.e., whether it comes from red grapes or a blend of white and red grapes or from white and red wines. The second step is to assess the authenticity of color, i.e., to detect possible fraud color of white wines by oenocyanin addition. Analysis of compounds and parameters such as polyphenols, sugars, volatile acidity, extract, especially the color indices, etc. and comparison of the results with reference values provide a high degree of certainty. The recommended method for defining color is the spectrophotometric method that is the reference method, which monitors the three-color indices: brightness, chromaticity, and purity (Rapeanu et al., 2009).

14.3.4.4 Authentication of Blending Wine

The blending of wine is authorized in certain well-defined circumstances. It is a problem that needs to be evaluated as it is one of the primary frauds concerning wines. According to EU regulations, Champagne rose is the only authorized blend of wines of different colors. The blending of white and red table wine is permitted in Spain, but the products may not be sold elsewhere (JOCE, 1994). In France, a distinction exists even between the "rose" wines and the "clairet' category. Clairet wines, between red and rose wines, are quite easily ascribed to their category by the analysis of their color using tri-stimulus parameters L, a, and b (Garcia-Jares et al., 1992; 1993). However, a much more complex task is to distinguish a true "rose"—made from only red grapes and a light maceration—from one produced by the blending of red and white grapes, particularly when calculating the extent of blending. Garcia-Jares and Medina (1993b) assessed this difficult problem by associating a highly sophisticated mathematical construction model such as PLS with a simple determination such as the ultraviolet-visible spectrum. In the first stage, a model was built using the spectra of red and white wines mixed in different proportions (from 1% to 20% red wine) that simulate the visual characteristics of true roses. The robustness of the model was then tested against wines of known mixed compositions and the validity of the calculation of the proportion of red and white wine was estimated. A fairly good agreement was obtained; thus, this method could be applied successfully to the detection of blending in the production of rose wines.

14.3.4.5 Authentication of Port Wine

Pessanha (1987) tackled the problem of the authenticity of Port wines in a way that mixes several components. A part of the scope of this study was to differentiate genuine Port wine, imitations (mainly from Spain but not sold under this label), and wines made with similar techniques, for example, Banyuls, a Yin Doux Naturel (VDN) from southern France. In all, 100 chemical determinations were made including:

- Classical determinations
- 14 metals
- Esters and fatty acids
- Superior alcohols and methanol
- Phenolic substances (acids and neutral)

Owing to the limited number of imitations available, all the groups of chemicals (grouped by techniques) were discriminated well. For the imitations, an unknown substance was present almost universally in the HPLC determinations of the phenolic substances. With classical determinations, because the imitations (unlike Port wines) are produced mainly with must fortified by alcohol with very little fermentation, the glucose-to-fructose ratio is above 0.9 for the imitations and below 0.8 for Port wines. The most useful metals used in discrimination were Rb, Li, Mn, Fe, and Al. Rb permits an accurate separation of the imitations (Port wines having a mean value of 1 mg/l). When halting fermentation in Port wines, wine alcohol (76%–78% vol) with very little rectification is used; the resulting product is consequently rich in higher alcohols. Isoamyl acetate and ethyl octanoate are the best indicators in the volatile series (Andrade Perez, 1961).

14.3.4.6 Authentication of Spirits

More recently, Cantagrel et al. (1993) proposed several methods, described in the following paragraphs, to distinguish the principal spirits. The distribution of common components or their ratios in spirits, for example, higher alcohols, caprylic acid, isopentanol (M3Bl/M2B1), syringaldehyde/vanillin, is the first useful step. Using only six components, it is possible to separate almost perfectly (with a 0.6% error) Cognacs, Armagnacs, and Brandies from different countries using Bayesian classification and PARVUS software (Forina et al., 1988). These components are meaningful in terms of wine-making, distillation process, and aging. Discriminant analysis using 23 variables correctly separates 120 spirits into their main classes: Brandies (French, Spanish, Italian), Cognacs, Armagnacs, Calvados, and Whiskies. Isotopic analysis of the site-specific deuterium has proven useful in the distinction of chaptalization in wines, but it requires the use of a data bank because the natural distribution of deuterium in the ethanol molecule is closely related to the geographical origin (Martin et al., 1983). Three isotopic variables, D/H1, D/H2, and R, i.e., the relative concentration of deuterium on the first carbon of ethanol ($CH_2D–CH_2–OH$), the relative concentration on the second ($CH_3–CHD–OH$), and the intramolar ratio, respectively, permit good discrimination of Cognacs from other spirits; even the Whiskies are distinguished from the pure malt (Cantagrel et al., 1993). However, if Brandies are made with wines of different geographical origins, then the distinctions are less obvious (e.g., for German Brandies produced with wines from the Cognac area).

14.3.4.7 Authentication of Botrytized Sweet Wines

Chauvet (1981) studied carefully the botrytized sweet wines of the Sauternes area. Using chemical determinations on wines and grapes, he developed an over-maturation index, which reflects the extent of growth of the so-called "noble rot"—an essential step in the wine-making process for those wines. Using his data, a PCA was performed on wines from botrytized grapes; it showed that the different stages during the development of the noble rot could be distinguished. The plots for most of the commercial wines are found between the second stage and the third when projected as supplementary parameters on this graph. While building the first axis, Mg and tartaric acids weighed the most. A two-dimensional plot of tartaric acid versus Mg gave a very useful representation of the three groups. The plane can be separated grossly into four quarters. It is well known that noble rot decomposes tartaric acid and, later, concentrates the grape. The decomposition is followed from group 1 (non-botrytized) to group 2 (fully botrytized). Concentration then occurs and the levels of Mg rise to stage 3 ("roasted" stage). The fourth quarter (upper right) would correspond to grapes concentrated by the sun (no tartaric decomposition).

14.4 CONCLUSION

The verification of wine authenticity might lie in the use of combined techniques such as those using isotopes and minerals. However, the difficult problem of chaptalization has been resolved by the use of a data bank, which may also be a possible solution to the problem of geographical origin. With the use of a database, however, problems such as represent activity, authenticity, rules of interpretation, and the transposition of results from one laboratory to another arise. Authenticity and rules of interpretation are the easiest to solve through official action. The inter-laboratory variances require the use of robust methods and, more importantly, the use of a reference material. Isotope determination requires cumbersome and expensive techniques (e.g., NMR, isotopic mass spectrometry). Isotope techniques, although enticing, require the same sophisticated mathematical tools as the easier and more affordable means, which often work equally well. Determination of the authenticity in wine calls for a pan-European effort since a database completed in one laboratory will always be limited to an informal level (pilot level or feasibility level). Nevertheless, it has been proven with consistency over many years that this problem can be solved using the mineral content of wines (stable components with time), and that the easiest cases can be solved with inexpensive determinations (e.g., Li, Rb, Mn, etc.). It has been solved for, what in our opinion is a much more difficult problem, chaptalization. The mineral profile of wine is so characteristic that it is unique to a specific wine and can be used as proof of authenticity where a control exists. It fares so much better than classical determinations. At any rate, it is possible to make accurate classifications of unknown wines and, even if for the moment, no rational explanation is completely satisfactory (e.g., transfer of the minerals from the soil to the grape), the empirical solution is sufficient for practical purposes.

The case of diethylene glycol in wine, which went undetected for some time because of it not being included in the list of tested compounds, is an excellent reminder of the limitation of targeted analysis, which is still predominant in private and governmental food-testing laboratories. These are only the first approaches into the use of nontargeted analysis, which may have the capability to detect novel or unusual compounds or may just provide a fingerprint of the beverage that can be compared against a database of authentic beverages. Various techniques may be suitable for such a nontargeted analysis, including UV/VIS, IR spectroscopy, and specifically NMR spectroscopy for which several promising approaches were published (Lachenmeier et al., 2009). The wine screening using targeted and nontargeted NMR spectroscopy have become the first method in the field of alcoholic beverages that is applied in a routine fashion (Godelmann et al., 2013). To determine the origin of a completely unknown wine, a global strategy could be conducted in three stages.

- Lead isotope ratio determination to locate the origin of wine by continent.
- 180 or site-specific deuterium for the determination of latitude, north or south.
- Metal determination such as Li, Rb, or Mn for classification according to the sub-region.

An alternative global approach would be the use of more complex fingerprints and much more site-specific elements such as the lanthanides. Other approaches such as mixing minerals, amino acids, and volatiles have already been used as a legal basis for detecting falsified products. Polyphenolic content of wine is also a good mark of authenticity (for detecting variety and region) (Ashurst and Dennis, 1996).

From a standpoint of production technology, a tighter implementation of the hazard analysis critical control point (HACCP) system was suggested to avoid adulteration. Indeed, this may detect crude forms of adulteration (e.g., methanol addition) along the supply chain, if regular sampling and analysis are implemented. However, HACCP would probably not be able to prevent intentional fraud on the last steps of the supply chain, similar to what occurred in the Czech Republic (Kourtis and Arvanitoyannis, 2001). In future, DNA sequencing and PCR techniques will probably be used as ultimate proofs of authenticity (Time Magazine, 1995).

REFERENCES

Andrade Perez, M.R. (1961–62) Doseamento dos alcools superiores no vinho do Porto e noutras biedbidas alcoolicas. Anais do Instituto do vinho do Porto.

Ashurst, P.R., Dennis, M.J. (1996). *Blackie Academic & Professional*, An Imprint of Chapman & Hall, 2–6 Boundary Row, London, SE1 8HN, 1st edition.

Baleiras-Couto, M.M., Eiras-Dias, J.E. (2006). Detection and identification of grape varieties in must and wine using nuclear and chloroplast microsatellite markers. *Analytica Chimica Acta*, 563(12); 283–291.

Ballabio, D., Mauri, A., Todeschini, R., Buratti, S. (2006). Geographical classification of wine and olive oil by means of classification and influence matrix analysis (CAIMAN). *Analytica Chimica Acta*, 570(2); 249–258.

Bechini, A., Cimino, M.G.C., Marcelloni, F., Tomasi, A. (2008). Patterns and technologies for enabling supply chain traceability through collaborative e-business. *Information and Software Technology*, 50(4); 342–359.

Bertolini, M., Bevilacqua, M., Massini, R. (2006). FMECA approach to product traceability in the food industry. *Food Control*, 17(2); 137–145.

Bevilacqua, M., Ciarapica, F.E., Giacchetta, G. (2009). Business process reengineering of a supply chain and a traceability system: A case study. *Journal of Food Engineering*, 93(1); 13–22.

Biernacka, P., Wardencki, W. (2012). Volatile composition of raw spirits of different botanical origin. *Journal of the Institute of Brewing*, 118(4), 393–400.

Brooks, R.R. (1972). *Geobotany and Biogeochemistry in Mineral Exploration*, Harper and Row, New York.

Calderone, G., Guillou, C., Reniero, F., Naulet, N. (2007). Helping to authenticate sparkling drinks with $^{13}C/^{12}C$ of CO_2 by gas chromatography-isotope ratio mass spectrometry. *Food Research International*, 40(3); 324–331.

Camara, J.S., Alves, M., Marques, J.C. (2006). Changes in volatile composition of Madeira wines during their oxidative ageing. *Analytica Chimica Acta*, 563(1–2); 188–197.

Cantagrel, R., Vidal, J.P., Snakkers, G. (1993). Le cognac: caracterisation analytique de son origine. 2nd European Symposium on Food Authenticity Isotope Analysis and Other Advanced Analytical Techniques, Nantes, France, October 1993.

Chambery, A., Monaco, G.d., Maro, A.D., Parente, A. (2009). Peptide fingerprint of high quality Campania white wines by MALDI-TOF mass spectrometry. *Food Chemistry*, 113(4); 1283–1289.

Chauvet, S. (1981). Application de l'Isotachophorese au dosage des principaux acides du yin: contribution a l'etude de la relation entre l'intensite du developpement de la pourriture noble et la composition chimique des vins blancs doux. These doctorat d'Universite de Bordeaux, France.

Chow, T.J., Snyder, C.B., Earl, J.L. (1975). Isotope ratios of lead as pollutant source indicators, United Nations FAO and International Atomic Energy Association Symposium Vienna, Austria (IAEA-SM-19114) Proceedings, pp. 95–108.

Day, M.P. (1993). Etude de la caracterisation de l'origine geographique des mouts et des vins par analyse conjointe de la composition en isotopes stables et en elements mineraux. These de doctorat Chimie Biologie Specialite Chimie Analytique, Universite de Nantes.

Derde, M.P., Massart, D.L., Ooghe, W. et al. (1983). *The Journal of Automatic Chemistry*, 5(3); 136–145.

Dutra, S.V., Adami, L., Marcon, R., Carnieli, G.J., Roani, C., Spinelli, F.R., Vanderlinde, R. (2011). Determination of the geographical origin of Brazilian wines by isotope and mineral analysis. *Analytical and Bioanalytical Chemistry*, 401(5); 1571–1576.

Eisinger, J. (1982). Lead and wine. Eberhard Gockel and the colica Pictonum. *Medical History*, 26; 279–302.

Etchats, R. (1986). Essais de dosage de quelques elements dans les vins par spectrometrie d'emission plasma DCP. Memoire DEA CEnologie Ampelologie Universite de Bordeaux II.

Etievant, P., Schlich, P., Bouvier, J.e. et al. (1988). Varietal and geographic classification of French red wines in terms of elements, amino acids and aromatic alcohols. *Journal of the Science of Food and Agriculture*, 48; 25–41.

Flamini, R., De Rosso, M. (2006). Mass spectrometry in the analysis of grape and wine proteins. *Expert Review of Proteomics*, 3(3); 321–331.

Forstel, H. (1995). Sous-Commission conventionnelle d'unification des methodes d'analyse et d'appreciation des vins. 35 erne Session, 8–10 juin PARIS. OIV, 18 rue d'Aguesseau 75008 Paris.

Forina, M, Leardi, R.1988. PARVUS—MS-DOS: a package for general pattern recognition.Elsevier

Frank, I.E., Kowalski, B.R. (1984). Prediction of wine quality and geographic origin from chemical measurements by partial least-squares regression modeling. *Analytica Chimica Acta*, 162; 241–251.

Garcia-Jares, C.M., Medina, B. (1993b). Research on white and red wine blending in the production of roses wines by means of the partial least squares method. *Journal of the Science of Food and Agriculture*, 63; 349–354.

Garcia-Jares, C.M., Medina, B., Sudraud, P. (1992). Caracterisation des vins roses et clairets par une mesure objective de leur couleuT. DIV, F. V. 913; 1915.

Garcia-Jares, C.M., Medina, B., Sudraud, P. (1993). Mise au point d'une methode pour la characterisation ct la differcntiation des vins clairets. Rev. Francaise de Encologie 140; 19–24.

Garcia-Jares, e.M., Garcia-Martin, S., Cela-Torrijos, R. (1995). Analysis of some highly volatile compounds of wine by means of purge and cols trapping injector capillary gas chromatography. Application to the differentiation of Rias Baixas Spanish white wines. *Journal of the Science of Food and Agriculture*, 43; 758–764.

Glories, Y. (1978). Recherches sur la matiere colorante des vins rouges. These Doctorat d'Etat, Universite de Bordeaux II.

Godelmann, R., Fang, F., Humpfer, E., Scheutz, B., Bansbach, M., Scheafer, H., Spraul, M. (2013). Targeted and nontargeted wine analysis by ^{1}HNMR spectroscopy combined with multivariate statistical analysis. Differentiation of important parameters: Grape variety, geographical origin, year of vintage. *Journal of Agricultural and Food Chemistry*, 61; 5610–5619.

Gonzales-Larraina, M., Gonzales, A., Medina, B. (1987). Les ions metalliques dans la differenciation des vins rouges des trois appellation d'origine Rioja. Conn. Vigne. Vins, 2; 127–140.

Gonzalez-Neves, G., Barreiro, L., Gil, G., Franco, J., Ferrer, M., Moutounet, M., Carbonneau, A. (2004). Anthocyanic composition of Tannat grapes from the south region of Uruguay. *Analytica Chimica Acta*, 513(1); 197–202.

Grindlay, G., Mora, J., Gras, L., de Loos-Vollebregt, M.T.C. (2011). Atomic spectrometry methods for wine analysis: A critical evaluation and discussion of recent applications. *Analytica Chimica Acta*, 691(1–2); 18–32.

Grousset, F.E., Quetel, C.R., Thomas, B. et al. (1994). Transient Pb isotopic signatures in western European atmosphere. *Environmental Science & Technology*, 28; 1605–1608.

Herrero-Latorre, C., Medina, B. (1992). Utilisation de quelques elements mineraux dans la differentiation de vins de Galice de ceux d'autres regions d'Espagne. Conn. Vigne Yin, 24; 147–156.

Holmberg, L. (2010). Wine fraud. *International Journal of Wine Research*, 2010; 105–113.

Infometrix (1981). ARTHUR Infometrix Corporation, Seattle, Washington.

JOCE (1987). Reglement CEE no 822/87 portant organisation commune du marche viti-vinicole.

JOCE (1994). Reglement 537/94 portant mesure transitoire en matiere de campagne des vins de table en Espagne pour l'annee 1994.

Kim, K.W., Thornton, T. (1992). Influence of Ordovician uraniferous black shales on trace element composition of soils and food crops, *Korea. Appt. Geochern*, 8(Suppl 2); 249–255.

Kourtis, L.K., Arvanitoyannis, I.S. (2001). Implementation of hazard analysis critical control point (HACCP) system to the alcoholic beverages industry. *Food Reviews International*, 17; 1–44.

Krauskopf, K.B. (1956). Sedimentary deposits of rare metals. Econ. Geot 50th anniversary volume, 411–463.

Kwan, W.O., Kowalski, B.R. (1978). Classification of wines by applying pattern recognition to chemical composition data. *Journal of Food Science*, 43(4); 1320–1323.

Kwan, W.O., Kowalski, B.R. (1980a). Pattern recognition analysis of gas chromatographic data. Geographic classification of wines of Vitis vinifera cv. Pinot Noir from France and the United States. *Journal of Agricultural and Food Chemistry*, 28; 356–359.

Kwan, W.O., Kowalski, B.R. (1980b). Correlation of objective chemical measurement and subjective sensory evaluations wines of *Vitis vinifera* variety Pinot Noir from France and the United States. *Analytica Chimica Acta*, 122; 215–222.

Kwan, W.O., Kowalski, B.R., Skogerboe, R.K. (1979). Pattern recognition analysis of elemental data; wines of *Vitis vinifera* cv. Pinot Noir from France and the United States. *Journal of Agricultural and Food Chemistry*, 27; 1321–1326.

Lacasta, F. (1982). Dosage de quelques metaux dans les vins par spectrometrie d'absorption et d'emission de flamme. Rapport de stage BTAO. DGCCRF 33405 Talence, France.

Lachenmeier, D.W., Humpfer, E., Fang, F., Sch€utz, B., Dvortsak, P., Sproll, C., Spraul, M. (2009). NMR-spectroscopy for nontargeted screening and simultaneous quantification of health-relevant compounds in foods: The example of melamine. *Journal of Agricultural and Food Chemistry*, 57; 7194–7199.

Latorre, M.J., Garcia-Jares, C., Medina, B., et al. (1994). Pattern-recognition analysis applied to classification of wines from Galicia (NW Spain) with certified brand of origin. *Journal of Agricultural and Food Chemistry*, 42; 1451–1456.

Legrand, Y., Medina, B., Grenon, J.P., et al. (1991). Les contaminants mineraux des vins. *Analusis*, 19(10); M39–M43.

Lessler, M.A. (1988). Lead and lead poisoning from antiquity to modern times. *Ohio Journal of Science*, 88; 78–84.

Lobinski, R., Witte, C., Adams, F.C., et al. (1994). Organo-lead in wine. *Nature*, 370; 24.

Luque de Castro, M.D., González-Rodríguez, J., Perez-Juan, P. (2005). Analytical methods in wineries: Is it time to change? *Food Reviews International*, 21(2); 231–265.

Maarse, H., Slump, P., Tas, A.C. Jan Schaefer et al. (1987). Classification of wines according to type and region based on their composition. *Z. Lebensm. Unters. Forsch*, 184; 198–203.

MacKay, K. (1994). International Atomic Energy Agency, Marine research station Monaco, France. Personal communication.

Makris, D.P., Kallithraka, S., and Mamalos, A. (2006). Differentiation of young red wines based on cultivar and geographical origin with application of chemometrics of principal polyphenolic constituents. *Talanta*, 70(5); 1143–1152.

Manterola, J.J. (1985). Analyse de composes phenoliques neutres de vins rouges par chromatographie liquide haute pression, d'elements mineraux par spectrometrie d'absorption atomique et d'emission, et traitement informatique des donnees, en vue de differencier les vins suivants les terroirs du Bordelais. Rapport de stage, Institut Superieur d' Agronomique de Lille. DGCCRF 33405 Talence, France

Mardones, C., Hitschfeld, A., Contreras, A., Lepe, K., Gutierrez, L., Von Baer, D. (2005). Comparison of shikimic acid determination by capillary zone electrophoresis with direct and indirect detection with liquid chromatography for varietal differentiation of red wines. *Journal of Chromatography* A, 1085(2); 285–292.

Martin, G.J. and Martin, M.L. (1983). *J. Chim. Phys*, 80; 29.

Martin, G.J., Guillou, C., Martin, M.L. et al. (1988). Natural factors of isotope fractionation and the characterization of wines. *Journal of Agricultural and Food Chemistry*. 36, 316–22.

Martinelli, L.A., Moreira, M.Z., Ometto, J.P.H.B., Alcarde, A.R., Rizzon, L.A., Stange, E., Ehleringer, J.R. (2003). Stable carbon isotopic composition of the wine and CO_2 bubbles of sparkling wines: Detecting C4 sugar additions. *Journal of Agricultural and Food Chemistry*, 51(9); 2625–2631.

Martiniere, F., Gauthier, J.M., Sudraud, P. et al. (1979). Evolution de la radioactivite par Ie carbone 14 des vins de la Gironde. Application a la recherche des millesimes. *Ann. Fals Exper. Chim*, 72(775); 263–274.

Masuda, A., Nakamura, N., Tanaka, T. (1973). Fine structures of mutually normalized rare-earth patterns of chondrites. *Geochim. Cosmochim. Acta*, 37; 239–248.

Medina, B., Sudraud, P. (1979). Teneur des vins en lithium. *Ann. Fa/s. Exp. Chim*, 72(772); 65–71.

Medina, B., Sudraud, P. (1980). Teneur des vins en chrome et nickel. Causes d'enrichissement. *Conn. Vigne. Yin.*, 14(2); 79–96.

Medina, B. and Van Zeller, A.L. (1984). Differenciation des vins de trois regions de France. *Conn. Vigne Yin*, 18; 225–235.

Mongereau, N. and Evin, J. (1993). Les applications du radiocarbone et de l'expertise. Etudes et documents. CE15, 105–110.

Nasi, A., Ferranti, P., Amato, S., Chianese, L. (2008). Identification of free and bound volatile compounds as typicalness and authenticity markers of non-aromatic grapes and wines through a combined use of mass spectrometric techniques. *Food Chemistry*, 110(3); 762–768.

OIV (1995). Le Plomb, cahiers scientifiques et techniques. OIV 75008, Paris.

Palade M., Popa M.-E. (2014). Wine traceability and authenticity - A literature review. Scientific Bulletin. Series F. Biotechnologies, Vol. XVIII, 226–233.

Pessanha, M.C.T. (1987). Caracterisation chimique et sensorielle des Porto d'origine controlee. Analyse statistique des resultats et automatisation des degustations. Memoire Dipl6me d'Etudes Approfondies <Enologie-Ampelologie, Universite de Bordeaux II.

Rapeanu, G., Vicol, C., Bichescu, C. (2009). Possibilities to asses the wines authenticity. Innovative Romanian Food Biotechnology, 5(of December)' 1–9.

Rizzon, L. (1985). Incidence de la maceration sur la composition chimique des vins. Thesis University of Bordeaux II.

Salagolty-Auguste, Bertrand, A. (1984). Wine phenolics analysis of low molecular weight components by high-performance liquid chromatography. *Journal of the Science of Food and Agriculture*, 35; 1241–1247.

Schlesier, K., Fauhl-Hassek, C., Forina, M., Cotea, V., Kocsi, E., Schoula, R., Wittkowski, R. (2009). Characterisation and determination of the geographical origin of wines. Part I: overview. *European Food Research and Technology*, 230(1); 1–13.

Sciarra, A.F., Gellman, L. (2012). Geographical indications: Why traceability systems matter and how they add to brand value. *Journal of Intellectual Property Law & Practice*, 7(4); 264–270.

Siegmund, H., Bachmann, K. (1977). Die Lagezuordnung von weinen durch bestimmung des spurenelementmusters. *Z. Lebensm. Unters.-Forsch*, 164; 1–7.

Siegmund, H., Bachmann, K. (1978). Z. Anwendung der numerischen taxonomie fiir die klassifizierung von weinen. *Lebensm. Unters.-Forsch*, 166; 298–303.

Sudraud, P., Medina, B. (1993). Determination de l'origine geographique d'un Yin par analyse des elements mineraux. 2nd European Symposium on Food Authenticity Isotope Analysis and Other Advanced Analytical Techniques, Nantes, France, October 1993.

Tagliabue, J. (1985). Scandal Over Poisoned Wine Embitters Village in Austria. NY Times. August 2.

Thurin, E. (1984). Recherche de correlations entre I'analyse physico-chimique et I'analyse sensorielle. Dipl6me National d' Enologue, Institut d' Enologie, Universite de Bordeaux II.

Time Magazine (1995). Law: blood sweat and tears. All of the above could help British police to trace criminals when a new DNA bank is completed. 15 May, 38.

Tusseau, D., Valade, M., Moncomble, D. (1994). Control of the origin and nature of wines and spirits 108th AOAC International Meeting and Exposition, Portland, Oregon 12–15 September. CIVC, BP135, 51204 Epernay, France.

Tusseau, D., Valade, M., Virion, M.e., et al. (1992). Controle de l'origine et de la nature des vins de champagne. XXeme Congrl!s Mondial de la Vigne et du Yin OIV 72 erne Assemblee Generale, Madrid, 18–26 Mai 1996. Le Vigneron Champenois No 4 Avril 1993.

Von Baer, D., Rentzsch, M., Hitschfeld, M.A., Mardones, C., Vergara, C., Winterhalter, P. (2008). Relevance of chromatographic efficiency in varietal authenticity verification of red wines based on their anthocyanin profiles: Interference of pyranoanthocyanins formed during wine ageing. *Analytica Chimica Acta*, 621(1); 52–56.

CHAPTER 15

Authentication and Traceability of Honey

Ammeduzzafar and Syed Nasir Abbas Bukhari
Al-Jouf University

Javed Ahmad and Mohammad Shabib Akhtar
Najran University

CONTENTS

15.1 INTRODUCTION

Honey is a food consumed all over the world, and it is appreciated not only as a quick source of energy but also for its antibacterial and antioxidant activity. Commercially, there are mono-floral honey (mainly made from a single botanical species) and poly-floral honey (obtained from different plant species). Honey is essentially a highly concentrated aqueous solution of two sugars, dextrose and fructose, and there may be other sugars present as well (at least 22 other more complex sugars). Many other substances also occur in honey, but the sugars are by far the major components. It has a high viscosity, "stickiness", high density, granulation tendencies, tendency to absorb moisture from the air, and it is an immunity provider with levorotatory properties. The principal physical characteristics and behavior of honey are due to their sugars, but the minor constituents–such as flavoring materials, pigments, acids, and minerals—are largely responsible for the differences among individual honey types (Graham, 1992).

Virtually every type of honey is derived from nectar, a derivative of phloem sap that is expressed from specialized cell groups called nectaries (Honeybees, 2016). Nectaries are termed floral if they occur as part of a flower or extra floral if they occur elsewhere in a plant. Nectar is an aqueous solution of sugars, amino acids, and other acids, proteins, lipids, minerals, and other components. The exact composition of nectar varies tremendously, depending on the plant species and the environmental conditions. For example, the total sugar content can range from 5% to 80% of the nectar (Graham, 1992). In many types of nectars, sucrose is the main or exclusive sugar, while in some nectars sucrose, glucose, and fructose are present in roughly equal quantities. For example, in the families *Lamiaceae* (mints) and *Ranunculaceae* (buttercups and clematis), sucrose is the dominant sugar. Some nectars have very little sucrose and varying proportions of glucose and fructose, like in the families *Brassicaceae* (mustard and cabbage) and *Asteraceae* (asters, daisies, and sunflowers) (Baker and Baker, 1983). Only rarely are other sugars found. Raffinose, galactose, and sorbitol have been found in a few plant nectars, but they are not widely found (Graham, 1992). Phloem sap contains mainly sucrose, so a chemical change must be occurring to produce glucose and fructose in the nectar. These chemical changes are mediated by transglucosidases and transfructosidases that are found in the nectarines.

Amino acids are also found in nectar, but at low quantities, typically 0.002%–4.8% of the total solids in the nectar (Graham, 1992). This is usually too low to be considered as a viable source of amino acids for the dietary needs of a honey bee; bees, typically, meet their protein needs from pollen. In a survey of 395 tropical and temperate floral nectars, Baker and Baker (1983) found a variety of amino acids in a varying proportion of samples, as listed in Table 15.1. Other components of nectar include lipids, organic acids (including ascorbic acid, vitamin C), and minerals (Bicchi et al., 1983).

The composition of honey depends on the nectar source, so it varies from nectar to nectar. The average compositions of honey samples are listed in Tables 15.1 and 15.2. The wide range of some components is due to the equally wide range of nectar sources. Single-source honey will have a lower range of values because of the greater consistency in the composition of the nectar.

Honey has various characteristics. One of the defining characteristics of honey is its low moisture content, and due to its low moisture content bacteria cannot survive in this medium that has a high osmotic pressure (Baker and Baker, 1982). However, this is not to say that honey cannot be unsafe. It may be made from nectar that is not palatable or may even be poisonous (such as that of rhododendron) or it may contain spores of mold or other organisms that can act as allergens or toxins, which is the reason honey should not be fed to babies less than one year old (Baker and Baker 1982). Honey is harvested before it becomes completely ripe; if harvesting is done after it ripens completely, it will have higher moisture content and may be vulnerable to spoilage. Also, honey is hygroscopic and will absorb water from the atmosphere if it is not kept in a well-sealed container.

The honey contains mostly 83% sugar (83 degrees Brix): this statement indicates how many grams of sucrose are dissolved in 100 g of solution; it is a common unit used with respect to honey, maple syrup, fermenting grape juice, and frozen, concentrated orange juice (White, 1992). The glucose precipitates from the solution, as many long-time

TABLE 15.1 Different Types of Amino Acids in Honeys

Amino Acid	Number of Nectars	% of Nectar
Alanin	380	96
Arginine	356	90
Serine	352	89
Proline	344	87
Glycine	332	84
Isolucine	287	73
Threonine	263	67
Valine	260	66
Leucine	255	66
Glutamic acid	245	62
Cystine	218	55
Phenylanine	216	55
Tyrosine	204	52
Tryptophan	189	48
Tysine	162	41
Glutamine	162	41
Aspartic acid	128	32
Asparagine	106	27
Methionine	80	20
Histidine	77	19

TABLE 15.2 General Composition of Honey (U.S.)

Component	Concentration (%)
Water	12.2–22.9
Fructose	30.9–44.3
Glucose	22.9–40.7
Sucrose	0.2–7.6
Other disaccharides	2.7–16
Higher sugars	0.1–3.8
Gluconic acid	0.17–1.17
Lactone	0.13–0.92
Minerals	0.02–1.03
Nitrogen	0.0–0.13

keepers of honey will confirm. So, the honey stored at cool temperatures (8 °C–16 °C) is also prone to glucose precipitation.

15.2 TRACEABILITY AND AUTHENTICATION OF HONEY

According to the definition given by Codex Alimentarius and other international honey standards, other materials, like food ingredients, added to honey make it adulterated and are not removable from honey. The identification of honey is very typical because many substances are used in honey as adulterants, like syrups (CS), high fructose corn syrups (HFCS), invert syrups (IS), or high fructose inulin syrups (HFIS). The adulteration of honey with invert sugar or syrup may not be detected readily by direct sugar analysis because its constituents are the major natural components of honey and, thus, the adulterated product would have physical properties similar to natural honey. The researchers as well as the regulatory agencies are still searching for newer, simpler, and more sensitive and economical procedures or techniques for the identification and traceability of honey (Codex Alimentarius, 2001; Cordella et al., 2003; Xiangrong et al., 2010; Ruiz-Matute et al., 2010; Kelly et al., 2006; Padovan et al., 2003).

Traceability of honey means that NSF International provides certification and testing services for honey producers along with food and personal care processors to independently verify the origin traceability of honey. This provides assurance to consumers that the product matches the claims stated on the package labels and, thus, supports the global honey industry by demonstrating compliance with U.S. and other international trade laws. True Source Honey certification traces the source of honey and verifies that it comes from an approved country. The need for traceability systems is well recognized throughout the world. In fact, traceability can protect consumers against deceptive marketing practices and/or frauds. Its objectives are to establish common definitions and to lay down overarching guiding principles and legitimate objectives for food laws in order to ensure a high level of health protection and an effective functioning of the internal market.

The possibility of tracing the origin of foodstuff is assuming an increasingly important role at the legislative level as a tool that may allow checking whether quality requirements are met. It allows establishment of the identity, history, and origin of the product. The evolution of the discipline of traceability has been accomplished in two stages: in the first step, the traceability was provided only for certain products (not-food) and for some individual food items; in the second step, it was extended to all products and food items. In the food industry, laws began to speak about traceability in relation to the organic production of agricultural products (Reg. CEE 24.06.1991 n 2092 art 9–12). The different fingerprinting techniques have been exploited for authentication and traceability of honey. The most commonly used fingerprinting techniques for honey authentication and traceability are discussed below.

Spectroscopic techniques
- Nuclear magnetic resonance (NMR)
- Near Infrared Transflectance Spectroscopy (NIR), Fourier Transform Infrared (FTIR)
- Raman spectroscopy

Chromatographic techniques
- Gas chromatography-mass spectrometry (GC-MS)
- Liquid chromatography (LC)
  - Liquid chromatography-mass spectrometry (LC-MS, or alternatively HPLC-MS)
  - Liquid chromatography coupled to isotope ratio mass spectrometry (HPLC-IRMS)
  - High performance anion exchange chromatography with pulsed amperometric detection (HPAEC-PAD)

Calorimetric methods (Application of DSC)
Fluorescence spectroscopy
Electrophoresis
DNA barcoding for honey biodiversity

The details of various fingerprinting techniques exploited for the authentication and traceability of honey are discussed below.

15.2.1 Spectroscopic Techniques

15.2.1.1 Nuclear Magnetic Resonance

NMR is a phenomenon that occurs when the nuclei of certain atoms (hydrogen) are inserted in a static magnetic field and exposed to a second oscillating magnetic field. Some nuclei experience this phenomenon, and others do not, depending upon whether they possess a property called spin. NMR has achieved general acceptance as a powerful method because of its noninvasive characteristics, high reproducibility, and sensitivity as demonstrated in a wide range of applications (Caligiani et al., 2007).

NMR is a powerful technique used to obtain structural information about compounds (Blau et al., 2008; Valente et al., 2008), which helps to understand the structures of the components in a compound along with the chemical compositions and the molecular dynamics of solids, liquids, and semisolid matrices (Cazor et al., 2006). NMR spectroscopy can also be considered as a fingerprinting technique (Bertram et al., 2005). Different types of NMR are used for detecting the purity of honey i.e., low-field NMR (^{1}H NMR) and high resolution magic angle spinning (HR MAS) NMR spectroscopy.

Low-field NMR is also known as ^{1}H NMR. It works with inexpensive permanent magnets. It is a rapid method for general application, used to investigate water mobility and other impurities in honey and other food items (Roberta et al., 2014a) because it can measure proton relaxation (Roberta et al., 2014b). The proton relaxation is described by the relaxation time constants T1 (longitudinal) and T2 (transverse), where T2 relaxation decay in food is multi-exponential; it indicates the presence of different water pools in food matrices or in honey (Bertram et al., 2001). However, we are not aware of any published reports on the use of low-field NMR (LF ^{1}H NMR) to classify honey samples of different floral origins. The relaxation time measurements are associated with the physical and chemical measurements of pH, water activity, color, water, sugar, and ash contents that might enable the classification of honey samples of different floral origins.

The water content was obtained for each type of honey, before NMR measurement came along, using the German reference refractometric method (DIN, 2012). The equivalent of 200 mg of water-free honey (about 240 mg) was weighed and mixed properly with 300 μL of NMR buffer, 700 μL of distilled water, and 100 μL of an internal standard (D_2O containing 0.1% of TSP (sodium salt of 3 (trimethylsilyl)-propionate acid-d4)). The stock solutions were prepared by mixing about 20 mg of a pure substance in 300 μL of NMR buffer, 700 μL of distilled water, and 100 μL of an internal standard. For neutralization of organic acids, 1–2 μL of 1 M NaOH was added to the solutions because the buffer capacity of our NMR buffer was not high enough to maintain a constant pH of 4.5 for these standard solutions. The calibration solution was prepared by diluting the above stock solution. The final 600 μL solution was taken in an NMR tube for direct measurement. For quantification, linear calibration curves were constructed from the standards by integrating the specific resonances for each compound against TSP as an intensity reference.

Lolli et al. (2008) analyzed 71 honey samples (robinia, chestnut, citrus, eucalyptus, poly-floral) by HR NMR using both ^{1}H NMR and heteronuclear multiple bond correlation spectroscopy (HMBC). The spectral data were analyzed by the application of unsupervised and supervised pattern recognition and multivariate statistical techniques, such as principal component analysis (PCA) and general discriminant analysis (GDA). ^{1}H-^{3}C HMBC coupled to appropriate statistical analysis seems to be an efficient technique for the classification of honey.

NMR was used for determining several saccharides with ^{13}C NMR (Mazzoni et al., 1997) and methylglyoxal and amino acids with ^{1}H NMR (Donarski et al., 2008; Donarski et al., 2010) in honey matrices. Clearly, we expanded the range of substances that can be analyzed with NMR spectroscopy without preceding separation; ^{1}H NMR is also suitable for the quantification of several aliphatic and aromatic acids as well as aldehydes.

Marc et al. (2013) investigated the purity of honey having different sugar contents like glucose, fructose, etc. using the NMR technique. This study has shown that the ^{1}H NMR spectra of honey in combination with appropriate multivariate statistics can provide qualitative information about the botanical origins and can represent a good basis for the identification of marker compounds in specific honey types. Quantitative information about a number of major components is also available from the same spectra without having to perform chromatographic separation. In combination with multivariate data analysis, NMR spectroscopy possesses the speed, simplicity, and low cost per analysis required for a suitable screening technique.

Roberta et al. (2014) used LF ^{1}H NMR for the authentication of honey collected from different botanical sources. A total of 80 honey samples from eight different botanical sources like eucalyptus, "assa-lipto", oranges, Barbados cherry, cashew tree, "assa-peixe", "cipó-uva", and poly-floral were analyzed. The biexponential fitting of the transverse relaxation (T2) data revealed two water populations in all the samples. A very close correlation ($p < 0.01$) was established between the LF ^{1}H NMR analysis of physical and chemical measurements, including water content, water activity, pH, and color. It was observed that differences in the relaxation times indicated that these are influenced by the differences in botanical origins. So, this study demonstrated that LF ^{1}H NMR can be a viable technique for the classification of honey according to their botanical origins.

Roberta et al. (2014) used LF NMR and physicochemical analytical methods for the identification of different honeys that are adulterated with different concentrations of sugar (fructose), from 0% (pure honey)–100% (pure HFCS). The physicochemical parameters like water content, water activity, pH, and color were demonstrated. The data of LF ^{1}H NMR was analyzed by using biexponential fitting and was compared with their physicochemical data. These differences were also observed by transverse relaxation (T2). Biexponential fitting of T2 showed that there are two water populations in all the samples, i.e., T21 and T22, with relaxation times in the range of 1.26–1.60 ms and 3.33–7.38 ms, respectively. It was observed that the relaxation times increased with higher percentages of high fructose syrup in adulterated honey. These linear correlations observed among the T2, T21, and T22 parameters and physicochemical data suggested that LF ^{1}H NMR can be used to discriminate pure flower honey from honey adulterated with HFCS.

Bertelli et al. (2010) evaluated an effective method for the detection of honey adulterated with different concentrations of sugar syrups using one-dimensional (1D) and two-dimensional (2D) NMR coupled to multivariate statistical analyses. A total of 63 honey samples collected from various botanical sources and seven different sugar syrups in marketed products were analyzed by NMR. The best discrimination model involved 1D spectra, and a cross-verification analysis showed a prediction capacity of 95.2% for this model. The 2D NMR analyses also gave satisfactory results (cross-verification showed

90.5% accuracy). It was concluded that the NMR method can be used for the authentication of honey.

Bertelli et al. (2010) used ^{1}H NMR relaxometry and HR NMR for the evaluation of adulteration in honey with sugar and water, along with some minor components (minor sugars, organic acids, amino acids, etc.) that can be excellent markers of adulteration. In the specific case of exogenous sugar addition, signals from some minor components of a sugar fraction (di- and oligosaccharides) are present only in the adulterated honey spectrum or, at least, are of greater intensities in the adulterated samples only (Bertelli et al., 2010).

Donarski et al. (2010) developed HR NMR spectroscopy for the verification of authenticity of honeys which are adulterated with different mono-floral honeys and commodities using different biomarkers that are developed by different researchers (Donarski et al., 2010; Schievano et al., 2012, 2013; Spiteri et al., 2015). The biomarkers like kynurenic acid are used for the authentication of sweet chestnut honey, whereas a-isophorone and 2,5-dihydroxyphenylacetic acid are used for strawberry tree honey (Donarski et al., 2010).

Schievano et al. (2012, 2013) developed an extraction and NMR process for the authentication of hydrophobic metabolites from honeys of different botanical origins, like chestnut, linden, orange, acacia, eucalyptus, honeydew, and poly-floral honeys (Schievano et al. 2012, 2013). This study is focused on retaining the aromatic compounds and those hydrophobic substances that vary the most in honeys from various botanical origins, whereas the ubiquitous and less specific carbohydrates are eliminated from the analysis.

In fact, there are many biochemical markers that are used for the identification of honey, like chrysin and pinocembrin for acacia honey and 8-hydroxylinalool and caffeine for orange honey. Manuka honey can give good qualitative and quantitative characterization with HR NMR (Le Gresley et al., 2012; Spiteri et al., 2015).

Spiteri and his associates (2015) developed an NMR method for the identification of adulteration or mislabeling in Manuka honey and other different types of honeys that are similar in pollen morphology and color like kanuka, rewarewa, or honeydew. Methylglyoxal and dihydroxyacetone are used as biomarkers for efficient quality detection and quantification of Manuka honey (Spiteri et al., 2015).

The application of chemometric methods to ^{1}H NMR spectra allowed to discriminate the eucalyptus, citrus, and wildflower honeys produced in the state of Sao Paulo, by identifying the signals from the responsible substances. Moreover, the chemometric methods for pattern recognition showed that it is possible to classify the commercial honey samples according to the nectars they are generated from. KNN (k-nearest neighbor), SIMCA (soft independent modeling of class analogs), and PLSDA (partial least square discriminant analysis) pattern recognition models had correctly classified all the samples through a validation set. However, the PLSDA method demonstrated high efficiency in NMR data analysis with the aim of increasing classification capability.

15.2.1.2 Near Infrared Transflectance Spectroscopy, Fourier Transform Infrared

NIR spectroscopy (NIRS) is a simple, quick (<30 s analysis time), and nondestructive technique that provides multi-constituent analysis for virtually any matrix with levels of accuracy and precision that are comparable to primary reference methods. NIRS is a useful technique to evaluate adulteration in honey samples in real time. It is a rapid, nondestructive, and relatively inexpensive method that may be suitable for use as a screening technique for quality control in honeys. It has the potential to detect adulterations with either beet invert or HFCS at minimum concentrations, which may be commercially useful (Xiangrong et al., 2010; Kelly et al., 2006). Spectroscopy with attenuated total reflectance (ATR) FTIR in combination with multivariate statistical techniques (chemometrics) makes it possible to obtain specific information about different parameters simultaneously, in a direct, reliable, and rapid way. In contrast to the time-consuming carbon isotope ratio analysis techniques, these FTIR spectroscopic procedures can be performed in a very short time. The use of ATR mid-infrared Fourier transform spectroscopy has made it possible to successfully quantify the content of adulterants such as CS, HFCS, and inverted sugar in honey (Gallardo-Velázquez et al., 2009).

Ruoff et al. (2006) applied the Fourier transform mid-infrared spectroscopy (FT-MIR) using an ATR for the authentication of honeys (411 samples) that were collected from different sources and the spectra were evaluated by chemometric methods. The error rates ranged from < 0.1% (poly-floral and heather honey as well as honeydew honey from metcalfa, oak, and fir)–8.3% (alpine rose honey) for both jackknife classification and validation depending on the type of honey considered. After evaluation, it was found that ATR-MIR spectroscopy is a precious tool for the authentication of honey depending on botanical origin, quality control, and geographical origin.

Almeida-Muradiana et al. (2014) used Fourier transform infrared ATR spectroscopy (FT-IR ATR) for analyzing various honey samples that were collected from northeastern Brazil. A total of nine different honey samples (six from a single origin and three mixed) were obtained from the CEARAPI Company. The samples were subjected to FT-IR ATR, palynological, and sensory analyses. The results showed that MIR spectrometry can be used as a screening method for the routine analysis of Brazilian honey because it is rapid, nondestructive, and accurate.

Woodcock et al. (2007) developed a potential NIR spectroscopy for the identification of geographical origins of various honeys. A total of 167 unfiltered honey (88 Irish, 54 Mexican, and 25 Spanish) and 125 filtered honey (25 Irish, 25 Argentine, 50 Czech, and 25 Hungarian) samples were collected. The spectra of the samples were recorded in the transflectance mode. They used various modeling methods such as PLS, regression and SIMCA. For unfiltered honey, the SIMCA model was the best, and it correctly identified about 95.5%, 94.4%, and 96% of the Irish, Mexican, and Spanish honeys, respectively. The PLS discriminant analysis correctly identified 100% of each of these classes of honeys. In case of filtered honey, the best models were produced using SIMCA; correct classifications were obtained for 91.7%, 100%, 100%, and 96% of the Argentinian, Czech, Hungarian, and Irish samples, respectively. The PLS2 discriminant analysis correctly identified 96%, 100%, 100%, and 100% of the Argentinian, Czech, Hungarian, and Irish honey samples, respectively. Overall, the SIMCA and PLS gave encouraging results, but a better classification was achieved using the PLS regression.

15.2.1.3 Raman Spectroscopy

Raman spectroscopy is a spectroscopic technique used to observe the vibrational, rotational, and other low-frequency modes in a system. Raman spectroscopy is commonly used in chemistry to obtain fingerprints for the identification of molecules as well as impurities present in the product. Raman scattering arises from the changes in the polarizability or in the shape of the electron distribution in the molecule during vibrations and, generally, symmetric vibrational modes give rise to strong Raman scattering symmetric vibrational modes generally responsible for infrared absorption. In contrast to infrared absorption, Raman and IR spectra provide complementary information about the vibrations of molecules and, in consequence, about the functional groups that constitute the product (Yang et al., 2005). The applications of this technique are in structure determination and multicomponent qualitative and quantitative analysis. FT-Raman spectroscopy is, among other recent analytical techniques, being more and more used for the assessment of the authenticity of food products, like honey, edible oils, and fats (Yang et al., 2005). The increasing use of Raman technique in the food area is due to the recent advances in instrument technology (Yang et al., 2005) like the interferometer methodology that leads to FT-Raman spectrometer type and which makes it possible to monitor many wave numbers simultaneously. Doing so, the total optical signal reaching the detector may be raised above the detector noise. An important advantage of this technique is the new sampling presentation that permits the examination of samples without any preparation in the whole range of physical states and inside classical glass vials. The major advantages of FT-Raman spectroscopy are its simplicity, rapidity, cost effectiveness, and nondestructive characteristics.

FT-Raman spectra were acquired on a Vertex 70 – RAM II Bruker FT-Raman spectrometer. This instrument is equipped with a Nd:YAG laser (Yttrium aluminum garnet crystal doped with triply ionized neodymium) with an output at 1,064 nm (9,398.5 cm^{-1}). The maximum laser power is 1.5 W. The measurement accessory is pre-aligned, only the z-axis of the scattered light is adjusted to set the sample in the appropriate position regarding the local point. The RAM II spectrometer is equipped with a liquid nitrogen cooled Ge detector. The OPUS 6.0 software was used for the spectral acquisition, manipulation, and transformation.

Samples were presented in the liquid form to the spectrometer in classical glass tubes of an internal diameter of 12 mm and a length of 75 mm (Schott Duran®). The tubes were introduced into a dedicated sample holder developed at the CRA-W and made of aluminum to assure the repeatable position of the sample in front of the laser beam. The sample holder was placed in the sample compartment. The laser power was set at 800 mW, the resolution at 4 cm^{-1}, and the number of co-added scans at 128 for each spectrum. Each spectrum was then collected in 4 mins and analyses were performed in duplicate.

Paradkar and Irudayaraj (2002) used Fourier transform Raman spectroscopy to detect added cane and beet sugar in honey. Adulterated samples were characterized by FT-Raman in the region between 200 cm^{-1}–1600 cm^{-1} and the fraction was used for quantitative and discriminant analysis. The results showed that FT-Raman spectroscopy was effective in predicting invert mixes of cane and beet sugar in all the three types of honeys. Classification of the impurities in honey by canonical variate analysis gave a minimum accuracy of 96%.

Several important and practical conclusions can be drawn from the investigation presented in this paper.

Juan et al. (2010) used the recent advances in instrument technique, i.e., FT-Raman for the authentication of natural honey. It provides complementary information about the vibrations of molecules and, in consequence, about the functional groups present in the product ingredients. Other important advantages of this technique are its simplicity, rapidity, cost effectiveness, and nondestructive characteristics. FT-Raman bands were selected and could be identified as fingerprints of the major components in honey, i.e., fructose, glucose, and sucrose. Some other bands were associated with other minor components like unknown carbohydrates and proteins. The study showed a correct classification rate of 85%–90%.

Batsoulis et al. (2005) used Raman spectroscopy with laser light in the NIR region for the detection of beet and cane sugar syrups in honey. The authors suggest that invert sugar adulterations can be detected down to the plant source from which the sugar syrup has been produced and this can be explained by the change in the $^{12}C/^{13}C$ ratio.

Goodacre et al. (2002) used Raman spectroscopy for the identification and quantitative analysis of honey because liquid chromatographic techniques showed poor repeatability of quantitative analysis for fructose and glucose. This may partly be due to the very low number of samples used for calibration. The prediction of the botanical origin using Raman spectroscopy and neuronal networks allowed 13 out of 14 honeys to be classified correctly in validation, but this study does not allow many conclusions to be drawn as only seven different honey types were studied from a total of 43 samples. From a theoretical point of view, the application of Raman spectroscopy has about the same potential as NIR.

15.2.2 Chromatographic Techniques

15.2.2.1 Gas Chromatography

Gas chromatography coupled with mass spectrometry (GC-MS) is an analytical method used to identify different substances within a test sample (low molecular weight, medium or low polarity, in ppb–ppm concentrations). The applications of GC-MS include drug detection, fire investigation, environmental analysis, explosives investigation, and identification of unknown samples as well as authentication and traceability of honey (adulteration with HFIS) or food products. Additionally, it can be used to identify trace elements. GC-MS has been widely heralded as a "gold standard" for forensic substance identification because it is used to perform a 100% specific test (Guyot et al., 1999; Mannas et al., 2007).

GC-MS consists of two major building blocks, i.e., the gas chromatograph and the mass spectrometer. In this technique, the capillary column contains the stationary phase and the separation of the components depends on the column's dimensions like length, diameter, film thickness as well as the phase properties (e.g., 5% phenyl polysiloxane). The molecules are separated in the column at different retention times and this compound allows the mass spectrometer downstream to capture, ionize, accelerate, deflect, and detect the ionized molecules separately. The fragmentation of the compound was done by breaking each molecule into ionized fragments and detecting these fragments using their mass-to-charge ratios (m/z ratios). The volatility of honey could be additionally verified on the basis of retention times and mass spectra with standards (Manyi-Loh et al., 2011; Alissandrakis et al., 2003).

The quality of honey products depends upon the flavor, aroma, color, texture, types of flowers or plants, physiology of bees and their foraging habits, climatic conditions like temperature, humidity, and post-collection processing. Naturally, bees will hunt any type of flower they can reach and, in consequence, the honey produced mostly has a mixed flavor and is commonly available in the market simply as honey or mixed-flower honey (Ampuero et al., 2004). Additionally, the honey is highly complex and contains many volatile constituents in low concentrations, whereas in a sugar matrix, polar substances are the major components (Castro-Várquez et al., 2003). So, GC coupled to MS is a good separation technique for the identification, traceability, and quality assessment of honeys that are adulterated with sugars and volatile components.

Before testing or identification of honey, an extraction process is done. Different separation techniques are available for this such as liquid-liquid extraction (LLE), simultaneous steam distillation extraction (SDE), hydrodistillation (HD), ultrasonic solvent extraction (USE), and micro-simultaneous steam distillation solvent extraction (MSDE) utilizing heat (Cuevas-Glory et al., 2007).

LLE in the continuous or batch mode is based on solubility for the extraction of liquid samples. Extraction efficiency depends upon the proper choice of solvent and its properties (polarity, boiling point, dielectric constant, hydrogen bonding capacity, availability, accessibility, cost, etc.). The extract must undergo cleanup and concentration processes.

USE is more advantageous because it reduces the extraction time as well as greatly improves the extraction efficiency. So, the mechanical effect of the ultrasound would agitate the extraction solvent, thus providing a greater penetration of the solvent into the matrix via cavitation.

The most commonly used solvents are n-hexane, acetone, chloroform, dichloromethane, diethyl ether, aqueous methanol, pentane, and a mixture of pentane and diethyl ether (1:2 v/v) (Jerkovic et al., 2007). These solvents do not require heat and help in the isolation of both low and high molecular weight compounds that are good potential markers for the determination of honey origin (Alissandrakis et al., 2005).

Jerkovic et al. (2006) developed a HD extraction method and compared it with USE for the isolation of volatile compounds from two unifloral honeys of *Robinia pseudo-acacia* and *Castanea sativa* L. They revealed that USE gave more effective results for all honey volatiles (without artifacts) and enabled the extraction of low molecular weight semi-volatile markers, especially benzoic, vanillic, and phenyl acetic acids.

Castro-Várquez et al. (2003) similarly compared three extraction techniques: SDE, LLE, and solid phase extraction (SPE) for the extraction of volatile components from rosemary honey. SDE extracts were rich in terpenes and esters, while the other two techniques embraced the formation of artifacts due to heating.

Piasenzotto et al. (2003) showed differences in the volatile profiles of chestnut honey extracted by solid phase microextraction (SPME) using polydimethylsiloxane (PDMS)-coated fiber, polyacrylate (PAC)-coated fiber, or Carboxen-coated fiber. Overall, PAC-coated fiber gave the best results, while Carboxen-coated fiber showed peak distortion due to band broadening caused by poor focalization inside the column. This could be related to slow desorption from Carboxen because of the high polarity of this phase. Conclusively, these varied isolation methods have been employed for the extraction of volatile components from the honey matrices that were analyzed by GC-MS; valuable information was obtained that enabled the determination of the botanical origins of honeys.

Manyi-Loh et al. (2010) applied a novel approach for the isolation of volatile components in the solvent extracts of honeys procured from South Africa. The honey was extracted with chloroform and n-hexane solvents having antibacterial activity. Subsequently, the solvent extracts were fractionated and purified by the silica gel column chromatographic technique with 100% n-hexane or chloroform, followed by the solvent systems n-hexane: ethyl acetate: acetic acid (7.8:3.5:1.25) and methanol:acetic acid:water (10:7:3), respectively. The isolated volatiles were analyzed by the GC-MS technique, which led to the identification of two novel compounds in honey: thiophene and N-methyl-D3-aziridine, which are of great importance.

Waldemar et al. identified the honey by using headspace SPME as a sample preparation technique and GC-MS with olfactometry (O) detection. Identification of honey volatiles was done by comparing mass spectra with the data in NIST mass spectral database. Using this technique, markers in each type of honey were indicated, like in Polish honey. Besides, the application of GC-MS technique coupled to olfactometry detection makes it possible to create aroma profiles for the investigated honeys. The employed techniques were characterized by high sensitivity and repeatability; furthermore, they are less time consuming.

Ruiz-Matute et al. (2010a) studied the sugar composition of HFCS using GC-MS. Sucrose syrups were analyzed in parallel as a control. HFCS contained fructosyl-fructose and other unknown sugars that might have been fructosyl-glucose. Honey was produced using the syrups that were used for bee feeding and that was also analyzed to detect the effect of these syrups on sugar composition. Fructosyl-fructose was detected in honey made by bees fed with HFCS. Fructosyl-fructose was also detected in honey made by free-flying bees and bees fed with sucrose syrups, but at lower levels.

Ruiz-Matute et al. (2007) determined the carbohydrate compositions of 20 honey and six syrup (HFCS and IS) samples using GC and GC-MS to detect the differences between the two groups of samples. The presence of difructose anhydride (DFAS) in these syrups was described for the first time, and its content was found to be dependent on the type of syrup. Since this compound was not detected in any of the 20 honey samples analyzed, so its presence in honey was proposed as a marker for detecting adulteration. Detection of adulteration in honey with HFCS and IS requires removal of the main sugars (monosaccharide) and the subsequent determination of the DFAS concentration. The newly developed protocol enables the detection of DFAS in adulterated honey at low concentrations.

Jun et al. (2009) developed another technique based on fingerprinting and barcoding of proteins in honey by using matrix assisted laser desorption ionization time-of-flight mass spectrometry (MALDI-TOF-MS) and MALDI Biotyper 1.1 software. The mass spectra of proteins in 16 honey samples were obtained and the peak information was extracted to generate protein fingerprints. This information was transformed into a database library of spectral barcodes that was used for differentiation of the geographical origins of honey based on pattern matching. The differentiation ability of the database library of barcodes was validated by comparing it with the results of replicate assays of five out of 16 honey samples of known Hawaiian origin obtained directly from the producers. Validation results showed that the protein fingerprints of honeys have better comparability with those honeys in the library that are from the same region than with honey samples from other regions. The protein fingerprints were used to differentiate the geographical origins of commercially purchased honey samples whose labels indicated that they were produced in different countries and in various states of the USA, including Hawaii. The results showed that the MALDI-TOF-MS Biotyper system can be a rapid, simple, and practical method for determining the geographical origin of honey sold commercially.

15.2.2.2 Liquid Chromatography

LC is an analytical chromatographic technique that is useful for separating ions or molecules that are dissolved in a solvent and it is used in the analysis of complex food mixtures and others. Currently, reversed phase LC (RPLC) followed by hydrophilic interaction chromatography (HILIC) and normal phase LC (NPLC) are the most frequently used LC-based separation techniques (Gika et al., 2014). Now, an advanced technique like ultra high performance LC (UHPLC) is used for identification because it is capable of conducting the operations at high back pressures (up to 1200–1500 bar) and it has columns with sub-2 mm particles that can be operated at high mobile phase flow rates without loss of resolution (Fekete et al., 2014).

Liquid Chromatography–Mass Spectrometry

LC-MS is a powerful technique that has a very high sensitivity, making it useful in many applications. Its applications are oriented toward the separation, general detection, and potential identification of chemicals of particular masses in the presence of other chemicals (i.e., in complex mixtures), e.g., natural products from natural product extracts and pure substances from mixtures of chemical intermediates. Preparative LC-MS systems can be used for the rapid mass-directed purification of specific substances from mixtures that are important in basic research and in pharmaceutical, agrochemical, food, and other industries. Like GC-MS, they allow the analysis and detection even for tiny amounts of a substance.

Low et al. (1995) determined the presence of sugars in honeys, i.e., HFCS sugars through isolation of the specific oligosaccharides by preparative HPLC, followed by quantitation by analytical HPLC.

Cavazza et al. (2013) studied the honey products of different botanical origins (eucalyptus, citrus, chestnut, and linden) and determined the phenolic profiles using HPLC with ultraviolet detection after SPE. A chromatogram with 58 characteristics was detected at $\lambda = 280$ nm. It was concluded that chemometric evaluation of the phenolic compound profiles yielded classification models that were able to group honey samples according to their floral sources with excellent degrees of agreement. The main advantage of this fingerprinting approach with respect to traditional methods is that identification and quantification of the analytes are not time consuming. The method proved to be effective for the assessment of honey authenticity.

Das and coworkers studied the honey for identification of phenolic compounds using the HPLC techniques. In this technique, first they separated the phenolic compounds by the given procedure. The honey was dissolved in an aqueous solution of HCl having pH = 2, filtered, and then passed through a column containing Amberlite XAD-2 as a stationary phase. Elution was first accomplished with aqueous HCl solution (pH = 2) and then with water; it allowed the separation of the phenolic fraction (retained in the column) from the polar, interfering substances like sugars. Then, the analytes were eluted with methanol and the extracted material was evaporated under reduced pressure for drying and then dissolved in water again. The cleanup phase was performed by extraction with a proper organic solvent (diethyl ether or ethyl acetate). The organic extracts were finally evaporated and dissolved again in methanol for the HPLC analysis (Das et al., 2015).

Campone and coworkers (2014) reported the quantitative and qualitative analysis of phenolic compounds from different honey samples through dispersive liquid-liquid microextraction (DLLME), Subsequent analysis with HPLC-UV method [58] for the phenolic compound proves it to be a promising tool for quality control, authentication, and characterization of honey. Different parameters were selected for high extraction efficiency, like the nature and volume of extraction and dispersive solvents, pH, and salt effect. At this optimal conditions, about 70% extraction was obtained with precision (RSD < 10%) and detection and quantification limits at ppb levels (1.4–12 and 4.7–40 ng (–1), respectively) were attained. So, it was concluded that the developed method had the simplicity of operation, the rapidity to achieve a very high sample throughput, and a low cost (Campone and coworkers, 2014).

Wabaidur et al. (2015) developed a HPLC lined with mass spectrometry method for the simultaneous separation, identification, and determination of phenolic compounds in the various honey samples collected from Yemen at the ion reaction monitoring acquisition mode. SPE was used for extraction, while multi-walled carbon nanotubes were used as a solid phase adsorbent. The chromatographic separation of all the phenolic constituents was performed on a BEH C18 column using a linear gradient elution. The 0.1% mixture of formic acid and methanol was used as a binary mobile phase. The total amounts of phenolic acids, flavonoids, and other phenols were analyzed and were found in the range of 338–3312, 122–5482 and 2.4–1342 μg/100 g of honey, respectively. 4-Hydroxybenzoic acid was found to be the major phenolic acid with chrysin, while cinnamic acid was the major component with other phenolic compounds. There was a regeneration of the solid phase adsorbent for reuse and the recovery results confirmed that the proposed method could be potentially used for the routine analysis of phenolic constituents in honey extracts (Wabaidur et al., 2015).

Badjah et al. (2014) developed a HPLC method for the simultaneous estimation of different phenolic acid and flavonoid contents in various natural honey samples collected from various regions of Yemen. The multi-walled carbon nanotubes were used as solid phase adsorbents for the extraction of 12 different samples. The total contents of phenolic acids, flavonoids, and phenolic components were 363–2658, 261–1646, and 224–1355 µg/100 g of honey, respectively, in 12 different honey samples. The 4-hydroxybenzoic acid (1410 µg/100 g), chrysin (850 µg/100 g), and cinnamic acid (1336 µg/100 g) were found to be the major phenolic acids, flavonoids, and phenolic compounds, respectively. A total of 25 compounds (ten phenolic acids, nine flavonoids, and six phenols) were analyzed, and 24 were detected, while only 23 compounds were determined quantitatively in the honey samples. So, this developed method showed potentially useful results for the polyphenols in honey extracts.

Cordella et al. (2005) described the development of HPAEC-PAD for the analysis of honey to detect adulteration and combined it with chemometric techniques to process chromatograms for a better discrimination of pure and adulterated honey. This method was developed using honey samples containing 10%–40% sugar syrups used for feeding the honey bees. Honey samples were from French beekeepers and the syrup samples were from a French supplier (Ickowicz SA). Characterization of the untampered and adulterated samples was satisfactorily completed using linear discrimination analysis (LDA) and canonical analysis (96.5%). As a control, honey bees were fed with bee-feeding syrups. This practice is particular to beekeeping, and it was shown that such feeding can affect the sugar composition of the honey produced if conducted without proper protective measures.

Liquid Chromatography Coupled to Isotope Ratio Mass Spectrometry

It is the first highly sensitive isotopic technique to resolve the individual sugars (sucrose, glucose, and fructose) using HPLC-IRMS, and it has been developed to improve the isotopic methods devoted to the study of honey authenticity. This technique has many advantages over the existing methods in terms of analysis time, sensitivity, lack of sample preparation, reduced consumption of reagents, and simplicity of the operation procedures (Cordella et al., 2003).

In the LC-IRMS method, the sample is oxidized within the aqueous solvent eluting from the HPLC. Afterwards, the generated CO_2 is separated from the liquid phase. This process is quantitative and fractionation-free. The oxidation reagent consists of two solutions, i.e., an oxidizing agent and phosphoric acid. Both are pumped separately and added to the liquid chromatographic column as the mobile phase. Within this mixture, all the individual organic compounds eluting from the HPLC column are oxidized quantitatively into CO_2 when passing through a heated reactor. In a downstream separation unit, the CO_2 is removed from the liquid phase and fixed into a stream of He. The individual CO_2 peaks in He are subsequently dried in an online gas drying unit and then admitted to the isotope ratio MS via an open split interface (Andreas, 2016).

High Performance Anion Exchange Chromatography with Pulsed Amperometric Detection

A simple and easy method based on activated charcoal treatment to fractionate honey carbohydrates followed by analysis of the oligosaccharide fraction by HPAEC-PAD has been developed in laboratories to detect the adulterations with CS and high HFCS in genuine honey samples and bee-feeding adulterations. It is also an efficient tool for the characterization of honey floral species. This method is less time consuming and less expensive than the other methods.

Pierna et al. (2011) applied the HPAEC-PAD detection for checking the authenticity of sugar purity in honey which is collected from the floral species. A total of 68 authentic honey samples (six varieties) were analyzed and interpreted by the new chemometric approach. A new algorithm was developed to automatically create the corresponding normalized data matrix, ready to be used in various chemometric procedures. This algorithm transformed the analytical profiles to produce the corresponding calibrated tables for the surfaces or intensities according to the retention times of peaks. The possibility of considering the unknown peaks (those for which no standards are available) allows most of the chemical information provided by the chromatograms to be retained. The parallel application of PCA/LDA and artificial neural networks (ANN) shows a high capability of classifying the analyzed samples (LDA: 93%; ANN: 100%) and a very good discrimination among honey groups. It was concluded that this method based on chromatographic analyses could be used for food characterization and adulteration detection.

15.2.3 Calorimetric Methods with Application of Differential Scanning Calorimeter

The differential scanning calorimeter (DSC) showed the possibility of using the glass transition temperature to distinguish between honeys and syrups and it is a powerful technique for characterizing the thermal behavior of honey and for detecting the effect of adulteration on the physicochemical and structural properties of samples.

Cordella et al. (2003) used the DSC technique to evaluate the thermal behavior of authentic honeys and sugar syrups (industrial and homemade). A total of 30 different samples of honey were analyzed (robinia, lavender, chestnut, and fir) and their glass transition temperatures (Tg) were measured. The results indicated that this parameter could be used to characterize and distinguish these honey varieties significantly . Applied for the honey samples that were artificially adulterated with different industrial syrups, DSC showed a detection level of 5%–10% depending on the type of syrup. An endothermic phenomenon occurring between 40°C –90 °C during the heating was studied by DSC and a new thermal transition similar to a glass transition was highlighted.

Zoltan et al. (1999) applied DSC using the glass transitions of pure and diluted honeys and the glass transitions of the maximally freeze-concentrated solutions of honeys for investigating the adulteration in different honeys like pure, diluted, and frozen. It was observed that the Tg of pure honey varied between –42°C and –51 °C, which indicated that the honey was unadulterated. The Tg of diluted honey (90% by wt of honey content) varied from –13°C to –20 °C. The concentration of the maximally freeze-concentrated honey solutions, expressed in terms of honey content, was approximately 102%–103%, i.e., slightly more concentrated in sugars than in honey itself.

Jolanta et al. (2015) studied the thermodynamic properties like Tg, specific heat capacity (ΔCp), and enthalpy (ΔH) of different selected honeys by using the DSC in relation to the composition, i.e., water and sugar content. The Tg of various types of honeys were determined and it was found that they significantly differed ($p < 0.05$) from each other and ranged from –49.7 °C (poly-floral) to –34.8 °C (sunflower). There was a strong correlation between the Tg values and the moisture contents in honeys ($r = -0.94$). The degree of crystallization of the honey also influenced the Tg values. It was observed that the presence or absence of sugar crystals influenced the Tg. For de-crystallized honey, the Tg values were 6°C –11°C, i.e., lower than that for crystallized honey. In conclusion, to obtain reliable DSC results, it is crucial to measure the glass transition after the complete liquefaction of honey and it is a useful technique for the identification of honey.

Sopade et al. (2006) used a modulated temperature DSC to investigate the Cp of ten kinds of Australian honeys within the processing and handling temperatures. The values obtained were found to be different from the literature values at certain temperatures, and they were not predictable by the additive model. The Cp of each honey exhibited a cubic relationship ($P < 0.001$) with the temperature (T, C). In addition, the moisture (M, %), fructose (F, %) and glucose (G, %) contents of the honey influenced its Cp. The equation: $r^2 = 0.92$ was proposed for estimating the Cp of honey and was recommended for use in the honey industry and in research.

Lenhardt et al. (2014) showed that single synchronous fluorescence spectra of different honeys differed significantly. Synchronous fluorescence spectra of three mono-floral (linden, sunflower, and acacia), poly-floral (meadow mix), and fake (fake acacia and linden) honey types (109 samples) were collected in an excitation range of 240–500 nm for synchronous wavelength intervals of 30–300 nm. Both PCA and PLSDA were performed on this dataset. A mean cross-validated classification error of 0.2% was found for a model accounting only for the mono-floral samples and of 4.8% for a model including both the mono-floral and poly-floral groups (Lenhardt et al., 2014).

According to Sergiel et al. (2014), right angle geometry three-dimensional synchronous fluorescence spectroscopy (3D-SFS) is able to differentiate and classify different honey types, including acacia, buckwheat, heather, linden, multi-flower, and rape honey. The fraction of phenolic compounds was initially isolated and preconcentrated from samples analyzed by means of polymeric reversed phase Strata X SPE.

Karoui et al. (2007) used front face fluorescence spectroscopy as a tool for the classification of honey samples according to their botanical origins. Honey fluorescence spectra from seven floral origins, namely acacia, alpine rose, chestnut, rape, honeydew, alpine poly-floral, and lowland poly-floral, were scanned after excitation at 250 nm (emission: 280–480 nm), 290 nm (emission: 305–500 nm), and 373 nm (emission: 380–600 nm) and after emission at 450 nm (excitation: 290–440 nm). Correct classification rates of 100% and 90% were observed for the calibration and the validation samples, respectively. Ruoff et al. (2006) studied the fluorescence spectra of 371 honey samples originating from Switzerland, Germany, Italy, Spain, France, Slovenia, and Denmark. The excitation spectra (220–400 nm) were recorded with emission measured at 420 nm. The samples were classified according to their geographical origins by chemometric tools with the error rates of the discriminant models ranging from 0.1%–7.5%. The samples could also be correctly classified with LDA according to their geographical origins, except for one Swiss sample that was misclassified to be of German provenance (Ruoff et al., 2006) (Table 15.3).

15.2.4 Fluorescence Spectroscopy

Fluorescence spectroscopy measures the intensity of photons emitted from a sample after it has absorbed photons. Most fluorescent molecules are aromatic. Fluorescence is an important investigational tool in many areas of analytical science due to its high sensitivity and selectivity. It can be used to investigate real time structures and dynamics, both in the solution state and under microscopes, particularly for biomolecular systems.

TABLE 15.3 Fingerprinting Techniques Used for Identification of Honeys

Fingerprinting Technique	Botanical Sources of Honey	Marker for Authenticity	Reference
SPME-GC-MS	Thyme	Ethenyl phenyl acetate	Piasenzotto et al., 2003
	Eucalyptus	Nonanoic acid and acetoin	
	Citrus	Nonanoic acid and acetoin	
USE-GC-MS	*Salvia officinalis* L.(Croatia)	Benzoic acid and phenyl acetic acid	Jerković et al., 2006
LNSDE-SE-GC-MS	*Calluna vulgaris*	Phenyl acetic acid, dehydrovomifoliol, and 3,5,5-trimethylcyclohex-2-ene derivative	Guyot et al., 1999
	Erica arborea	4-Methoxybenzaldehyde, 4-methoxybenzoic acid, and methyl vanillate	
SPME-GC-MS	*Quillajia saponaria* (Quillay)	Megastigmatrienone, 2-*p*-hydroxyphenyl alcohol, β-pineme, and linalool oxide	Montenegro et al., 2009
	Eucryphia cordifolia	Isophorone and cetosiophorone	
MSDE-GC-MS	Eucalyptus	3-caren-2-ol, p-cymene, and its derivative alcohol	Castro-Varquez et al., 2009
	Citrus	Linalool oxide, lilac alcohol, and lilac aldehydes	
	Lavender	Nerolidol oxide	

LNSDE-SE, Likens-Nickerson simultaneous distillation extraction solvent extraction; MSDE, micro-simultaneous steam distillation solvent extraction; SPME, solid phase microextraction; USE, ultrasonic solvent extraction.

Fluorescence occurs when a fluorescent capable material (fluorophore) is excited to a higher electronic state by absorbing an incident photon and it cannot return to the ground state without emitting a photon. The emission usually occurs from the ground vibrational level of the excited electronic state and it goes to an excited vibrational state of the ground electronic state. Thus, fluorescence signals occur at longer wavelengths than the absorbance signals. The energies and relative intensities of the fluorescence signals give information about the structures and environments of the fluorophores.

Chen et al. (2014) used a three-dimensional fluorescence spectroscopic technique followed by multivariate analysis for the identification of adulterated honey. The 3D spectra fluorescence spectroscopy was compressed using characteristic extraction and PCA. Then, PLS and back propagation neural network (BP-ANN) algorithms were used for modeling. The results showed that the BP-ANN model was superior to the PLS model, and the optimum prediction results were obtained as follows: root mean square error of prediction (RMSEP) = 0.0235 and R = 0.9787 in the prediction set. The study demonstrated that the 3D fluorescence spectroscopy technique combined with multivariate calibration has the potential for rapid, nondestructive, and accurate quantitative analysis of honey adulteration.

Ghosh et al. (2005) measured the front face fluorescence spectra to monitor the adulteration of honey with cane sugar syrup. Using an excitation wavelength of 340 nm, pure honey samples were characterized by a maximum located at 510 nm, which was ascribed to flavins, while cane sugar syrup adulterated samples exhibited a maximum located around 430 nm. By applying Fisher discriminant analysis (FDA) to the datasets, classification rates of 100% and 90% were observed for the calibration and validation samples, respectively.

15.2.5 Electrophoresis

Electrophoresis is the motion of dispersed particles relative to a fluid under the influence of a spatially uniform electric field. Electrophoresis of positively charged particles (cations) is called cataphoresis, while electrophoresis of negatively charged particles (anions) is called anaphoresis. Electrophoresis is a technique used in laboratories in order to separate macromolecules based on their sizes. The technique applies a negative charge, so the proteins move toward a positive charge. This is used for both DNA and RNA analysis. Polyacrylamide gel electrophoresis (PAGE) has a clearer resolution than agarose and is more suitable for quantitative analysis. In this technique, DNA footprinting can identify how proteins bind to DNA. It can be used to separate proteins by size, density, and purity. It can also be used for plasmid analysis, which develops our understanding of how bacteria become resistant to antibiotics.

Arraez-Roman et al. (2006) used capillary electrophoresis (CE) coupled to MS for the detection of phenolic compounds in rosemary honey constituents. It is an effective separation technique that allows the identification of different compounds in honey and other substances and acts as an alternative to HPLC techniques. It is a very simple and cost effective analytical method involving SPE and capillary zone electrophoresis coupled to electro spray ionization ion trap mass spectrometry (CZE-ESI-MS). The SPE, CE, and ESI-MS parameters were optimized in order to maximize the number of phenolic compounds detected and the sensitivity of their determination. All CE-ESI-MS experiments were performed with uncoated fused-silica capillaries and an alkaline volatile buffer system consisting of 100 mM NH_4OH with 10% of 2-propanol at pH 10.0. The effects of type and flow rate of the sheath liquid on the sensitivities of phenolic compounds were investigated. As result, the best sensitivity was obtained with a sheath liquid containing 2-propanol/water (60:40 (v/v)) and 0.1% (v/v) of triethylamine at 3 μL/min in the negative ion mode. So, it was concluded that electrophoresis is a good technique for the identification of honey at mg/L level.

Klaus et al. (2010) used the electrophoresis technique for the identification of filtered honey because most honeys in the market were adulterated. A method was developed that allowed the detection of mixtures of filtered and unfiltered honeys. Comparative tests showed that enzyme activities, mainly sucrose, were influenced by this process. The protein content did not decrease. Sucrose was isolated by gel chromatography and was analyzed by gel electrophoresis. One of the two dominating protein bands with 40 kDa and 65 kDa decreased significantly after filtration, which led to a shift in the natural ratio between them. The quantitative densitometry analysis of these two protein bands allowed the detection of 15% added filtered honey.

Moreno-Gonzalez et al. (2015) investigated a new analytical method based on capillary zone electrophoresis connected with tandem mass spectrometry for the identification and simultaneous estimation of nine amino glycosides in honey samples. The ion trap mass analyzer operating in the multiple reaction monitoring modes was used for detection. Different parameters were optimized in order to obtain an adequate separation along with the highest sensitivity. In order to achieve high selectivity in the sample treatment, a commercially available molecularly imprinted polymer was used for the solid phase extraction of the analytes. Under optimum conditions, recoveries for the fortified samples ranged from 88.2%–99.8%, with relative standard deviations lower than 8%. The limits of detection ranged from 0.4–28.5 μg/kg. Furthermore, the decision limit and the detection capability were evaluated, that ranged from 3.5–60.5 μg/kg and 6.0–103.1 μg/kg, respectively, demonstrating the sensitivity and applicability of this fast and simple method.

15.2.6 DNA Barcoding for Honey Biodiversity

DNA barcoding is a taxonomic method that uses a short genetic marker in an organism's DNA to identify which particular species it belongs to. It differs from molecular phylogeny in that the main goal is not to determine the patterns of relationships but to identify an unknown sample in terms of a preexisting classification (Kress et al., 2005).

Jennifer et al. (2015) used DNA barcoding to investigate the sources of honey from the different plants that were frequently used by the honey bees for better understanding the nutritional requirements of honey bees. 46 families and 25 orders were tested, but a relatively small number of plants were consistently seen across multiple honey samples. They are most commonly herbaceous species like *Rubus fruticosus*, *Filipendula ulmaria*, *Taraxacum officinale*, *Trifolium* spp., *Brassica* spp., and the non-native, invasive *Impatiens glanduliferous* as well as tree pollen was frequently seen belonging to *Castanea sativa*, *Crataegus monogyna*, and species of *Malus*, *Salix*, and *Quercus*. They concluded that although honey bees are considered to be super generalists in their foraging choices, there are certain key species or plant groups that are particularly important in the honey bee environment. It provided a starting point for investigating how we can better provide for the insects that we rely upon for pollination (Table 15.4).

TABLE 15.4 Gas Chromatographic Fingerprinting Techniques Used for Identifications of Honeys

Fingerprinting Technique	Analytes	Reference
^{1}H NMR spectroscopy	Used to identify different sugars.	Consonni et al., 2008
ATR-MIR spectroscopy	Authentication of the botanical origin and quality control test of honey.	Ruoff et al., 2006
Solid phase extraction + HPLC	Authenticates phenolic compounds from different floral sources.	Burya et al., 2007
Capillary zone electrophoresis	Authenticates phenolic compounds from different floral sources.	Andrade et al., 1997
GC and HPLC	Methyl anthranilate (MA) and hesperetin have been considered as markers for the floral origin of citrus honey.	Ferreres et al., 1994
Headspace GC-MS	Identification of aromatic compounds in honey.	Radovic et al., 2001

(Continued)

TABLE 15.4 (*Continued*) Gas Chromatographic Fingerprinting Techniques Used for Identifications of Honeys

Fingerprinting Technique	Analytes	Reference
SPME-GC/MS	Authenticates volatile compounds in honey.	Verzera et al., 2001
Fourier transform infrared micro-attenuated total reflectance spectroscopy	Quantifies the different saccharides in multiple floral honeys.	Tewari and Irudayaraj, 2004
FTIR spectroscopy	Identifies the inverted beet sugar in adulterated honeys.	Sivakesava et al., 2001b
	Identifies the inverted cane sugar in adulterated honeys.	Sivakesava et al., 2001a
Mid-infrared (MIR) spectroscopy and chemometrics	Identifies the sugar adulteration in honeys.	Kelly et al., 2004
FT-Raman spectroscopy	Identifies the beet and cane invert sugar in honeys.	Paradkar and Irudayaraj, 2002
HPLC	Identifies the carbohydrates in honeys.	Goodall et al., 1995
HPLC	Identifies the high fructose and syrups in honeys.	Lipp et al., 1998
Gas-liquid chromatography	Identifies the high fructose in adulterated honeys.	Doner et al., 1979
High performance liquid chromatography	Identifies the carbohydrates in adulterated honeys.	Goodall et al.,1995
Gas chromatography	Identifies the oligosaccharides in adulterated honeys.	Radovic et al., 2001
One-dimensional and two-dimensional high resolution nuclear magnetic resonance	Detection of honey adulteration with sugar syrups.	Bertelli et al., 2010
C-13 NMR spectroscopy	Qualitative and quantitative analysis of different carbohydrate mixtures.	Mazzoni et al., 1997

15.3 CONCLUSION

According to data available from the studied literature, it is to some extent obvious that nearly majority of the physicochemical characteristics of honey depend on the floral sources. Honey adulteration is a critical problem that can be determined by various techniques to get information about each aspect of it. Based on the type of adulterants, each applied method seems to be beneficial by itself. The PCA analysis also allowed the discrimination among honeys that showed some kind of adulteration and identification of the types of compounds involved.

Nontargeted and targeted profiling techniques based on various spectroscopic platforms have been successfully used to discriminate honey and other agro-food commodities on the basis of their geographical and botanical origins. Reliable and complete databanks are the cornerstones of such methods to unleash the full power of metabolomic approaches. Data collections containing spectra, compositional data, and metadata have to be generated using a large number of authentic samples covering as much compositional variation as possible due to geology, geography, plant species, etc. Access to such samples is the key to success. Even more important is the sharing of this information with the scientific community, perhaps by using cloud computing. Such initiatives are underway and should continue. Otherwise, research groups have to rebuild databanks in isolation, thereby wasting scarce resources. Another aspect that needs special consideration is the validation of statistical models in order to obtain robust classifications, which can be used to resolve disputes among trading partners or for official control purposes.

REFERENCES

Alissandrakis E., Daferera D., Tarantilis P.A., Polissiou M., Harizanis P.C. Ultrasound-assisted extraction of volatile compounds from citrus flowers and citrus honey. *Food Chemistry*. 2003, 82, 575–582.

Alissandrakis E., Tarantilis P.A., Harizanis P.C., Polissiou M. Evaluation of four isolation techniques for honey aroma compounds. *Journal of the Science of Food and Agriculture*. 2005, 85, 91–97.

Almeida-Muradiana L.B., Sousa R.J., Barth O.M., Gallmann P. Preliminary data on Brazilian monofloral honey from the northeast region using FT-IR ATR spectroscopic, palynological, and color analysis. *Química Nova*. 2014, 37, 716–719.

Ampuero S., Bogdanov S., Bosset J.-O. Classification of unifloral honeys with an MS-based electronic nose using different sampling modes: SHS, SPME and INDEX. *European Food Research and Technology*. 2004, 218, 198–207.

Andrade P., Ferreres F., Gil M.I., Tomás-Barberán F.A. Determination of phenolic compounds in honeys with different floral origin by capillary zone electrophoresis. *Food Chemistry*. 1997, 60, 79–84.

Andreas W.H., Michael K., Dieter J. *LC-IRMS: Authenticity Control of Honey Using the Thermo Scientific LC IsoLink LC-IRMS*, Thermo Fisher Scientific: Bremen, Germany, 2016.

Arraez-Roman D., Gomez-Caravaca A.M., Gomez-Romero M., Segura-Carretero A., Fernandez-Gutierrez A. Identification of phenolic compounds in rosemary honey using solid-phase extraction by capillary electrophoresis–electrospray ionization-mass spectrometry. *Journal of Pharmaceutical and Biomedical Analysis*. 2006, 41(5), 1648–1656.

Badjah Hadj Ahmed A.Y., Obbed M.S., Wabaidur S.M., Al Othman Z.A., Al-Shaalan N.H. High-performance liquid chromatography analysis of phenolic acid, flavonoid, and phenol contents in various natural yemeni honeys using multi-walled carbon nanotubes as a solid-phase extraction adsorbent. *Journal of Agricultural and Food Chemistry*. 2014, 62, 5443–5450.

Baker H.G., Baker I. Chemical constituents of nectar in relation to pollination mechanisms and phylogeny. In *Biochemical Aspects of Evolutionary Biology*, Nitecki M.H., Ed., University of Chicago Press: Chicago, 1982, pp. 131–71.

Baker H.G., Baker I. Chemistry of floral nectar. In *The Biology of Nectaries*, Bently B., Elias T., Eds., Columbia UniversityPress: New York, 1983.

Batsoulis A.N., Siatis N.G., Kimbaris A.C., Alissandrakis E.K., Pappas C.S., Tarantilis P.A., Harizanis P.C., Polissiou M.G. FT-Raman spectroscopic simultaneous determination of fructose and glucose in honey. *Journal of Agricultural and Food Chemistry*. 2005, 53, 207–210.

Bertelli D., Lolli M., Papotti G., Bortolotti L., Serra G., Plessi M. Detection of honey adulteration by sugar syrups using one-dimensional and two-dimensional high-resolution nuclear magnetic resonance. *Journal of Agricultural and Food Chemistry*. 2010, 58(15), 8495–8501.

Bertram H.C., Karlsson A.H., Rasmussen M., Pedersen O.D., Dontrup S., Andersen H.J. Origin of multiexponential T2 relaxation in muscle myowater. *Journal of Agricultural and Food Chemistry*. 2001, 49, 3092–3100.

Bertram H.C., Kristensen, N.B., Malmendal, A., Nielsen N.C., Bro R., Andersen, H.J., et al. A metabolomic investigation of splanchnic metabolism using ^{1}H NMR spectroscopy of bovine blood plasma. *Analytica Chimica Acta*. 2005, 536, 1–6.

Bicchi C., Belliardo F., Frattini C. *Journal of Apicultural Research*. 1983, 22, 130.

Blau L., Menegon R.F., Ferreira E.I., Ferreira A.G., Boffo E.F., Tavares L.A., et al. Synthesis and total ^{1}H- and ^{13}C-NMR assignment of cephem derivatives for use in ADEPT approaches. *Molecules*. 2008, 13, 841–854.

Burya D., Reneta G., Elke A. Analysis of phenolic acids in honeys of different floral origin by solid-phase extraction and high-performance liquid chromatography. *Phytochemical Analysis*. 2007, 18, 24–32.

Caligiani A., Acquotti D., Palla G., Bocchi V. Identification and quantification of the main organic components of vinegars by high resolution ^{1}H NMR spectroscopy. *Analytica Chimica Acta*. 2007, 585, 110–119.

Campone L., Piccinelli A.L., Pagano I., Carabetta S., Di Sanzo R., Russo M., Rastrelli L. Determination of phenolic compounds in honey using dispersive liquid-liquid microextraction. *Journal of Chromatography A*. 2014, 1334, 9–15.

Castro-Várquez L., Pérez-Coello M.S., Cabezudo M.D. Analysis of volatile compounds of Rosemary honey. Comparison of different extraction techniques. *Chromatographia* 2003, 57, 227–233.

Castro-Varquez L., Diaz-Maroto M.C., Gonzalez-Vinas M.A., Perez-Coello M.S. Differentiation of monofloral citrus, rosemary, eucalyptus, lavender, thyme and heather honeys based on volatile composition and sensory descriptive analysis. *Food Chem*. 2009, 112, 1022–1030.

Cavazza A., Corradini C., Musci M., Salvadeo P. High-performance liquid chromatographic phenolic compound fingerprint for authenticity assessment of honey. *Journal of the Science of Food and Agriculture*. 2013, 93(5), 1169–1175.

Cazor A., Deborde C., Moing A., Rolin D., This, H. Sucrose, glucose, and fructose extraction in aqueous carrot root extracts prepared at different temperatures by means of direct NMR measurements. *Journal of Agriculturaland Food Chemistry*. 2006, 54, 4681–4686.

Chen Q., Qi S., Li H., Han X., Ouyang Q., Zhao J. Determination of rice syrup adulterant concentration in honey using three-dimensional fluorescence spectra and multivariate calibrations. *Spectrochimica Acta Part A: Molecular and Biomolecular Spectroscopy*. 2014, 131, 177–182.

Codex Alimentarius. 2001. Revised codex standard for honey. (No. CODEX STAN 12–1981, Rev.1 (1987), Rev.2 (2001).

Consonni R., Cagliani L.R. Geographical characterization of polyfloral and acacia honeys by nuclear magnetic resonance and chemometrics. *Journal of Agricultural and Food Chemistry*. 2008 Aug 27, 56(16): 6873–6880.

Cordella C., Faucon J.-P., Cabrol-Bass D., Sbirrazzouli N. Application of DSC as a tool for honey floral species characterization and adulteration detection. *Journal of Thermal Analysis and Calorimetry*. 2003, 71, 279–290.

Cordella C.B., Militão J.S., Clément M.C., Cabrol-Bass D. Honey characterization and adulteration detection by pattern recognition applied on HPAEC-PAD profiles. 1. Honey floral species characterization. *Journal of Agricultural and Food Chemistry*. 2003 May 21, 51(11), 3234–3242.

Cordella C.H., Militão J.S.L.T., Clément M.C., Drajnudel P., Cabrol-Bass D. Detection and quantification of honey adulteration via direkt incorporation of sugar syrups or bee-feeding: preliminary study using highperformance anion exchange chromatography with pulsed amperometric detection (HPAEC-PAD) and chemometrics. *Anal Chim Acta*. 2005, 531, 239–248.

Cuevas-Glory L.F., Pino J.A., Santiago L.S., Sauri-Duch L.F. A review of volatile analytical methods for determining the botanical origin of honey. *Food Chemistry*. 2007, 103, 1032–1043.

Das A., Datta S., Mukherjee S., Bose S., Ghosh S., Dhar P. Evaluation of antioxidative, antibacterial and probiotic growth stimulatory activities of Sesamum indicum honey containing phenolic compounds and lignans. *Journal of Food Science and Technology*. 2015, 61, 244–250.

DIN 10752, Untersuchung von Honig; Bestimmung des Wassergehaltes, Refraktometrisches Verfahren, Berlin, Germany, 2012.

Donarski J.A., Jones, S.A., Charlton, A.J. Application of cryoprobe H-1 nuclear magnetic resonance spectroscopy and multivariate analysis for the verification of corsican honey. *Journal of Agricultural and Food Chemistry*. 2008, 56(14), 5451–5456.

Donarski J.A., Roberts D.P.T., Charlton, A.J. Quantitative NMR spectroscopy for the rapid measurement of methylglyoxal in manuka honey. *Analytical Methods*. 2010, 2(10), 1479–1483.

Donarski J.A., Jones S.A., Harrison M., Driffield M., Charlton A.J.. Identification of botanical biomarkers found in Corsican honey. *Food Chemistry*. 2010, 118, 987–994.

Doner L.W., White J.W., Phillips J.G. Gas-liquid chromatographic test for honey adulteration by high fructose corn sirup. *Journal of AOAC International*. 1979, 62(1), 186–189.

Fekete S., Schappler J., Veuthey J.L., Guillarme, D. Current and future trends in UHPLC. *Trends in Analytical Chemistry*. 2014, 63, 2–13.

Ferreres F., Giner J.M., Tomás-Barberán F.A. A comparative study of hesperetin and methyl anthranilate as markers of the floral origin of citrus honey. *Journal of the Science of Food and Agriculture*. 1994, 65, 371–372.

Gallardo-Velázquez T., Osorio-Revilla G., Zuñiga-de Loa M, Rivera-Espinoza Y. Application of FTIRHATR spectroscopy and multivariate analysis to the quantification of adulterants in Mexican honeys. *Food Research International*. 2009, 42, 313–318.

Ghosh N., Verma Y., Majudmer S.K., Gupta P.K. A fluorescence spectroscopic study of honey and cane sugar syrup. *Food Science and Technology Research*. 2005, 11(1), 59–62.

Gika H.G., Theodoridis G.A., Plumb R.S., Wilson I.D. Current practice of liquid chromatography–mass spectrometry in metabolomics and metabonomics. *Journal of Pharmaceutical and Biomedical Analysis*. 2014, 87, 12–25.

Goodacre R., Radovic B.S., Anklam E. Progress toward the rapid non-destructive assessment of the floral origin of European honey using dispersive Raman spectroscopy. *Applied Spectroscopy*. 2002, 56, 521–527.

Goodall I., Dennis M.J., Parker I., Sharman M. Contribution of high-performance liquid chromatographic analysis of carbohydrates to authenticity testing of honey. *Journal of Chromatography A*. 1995, 706, 353–359.

Graham, J.M., Ed. *The Hive and the Honeybee*, Dadant & Sons: Hamilton, IL, 1992.

Guyot C., Scheirman V., Collin S. Floral origin markers of Heather honeys: *Calluna vulgaris* and *Erica arborea*. *Food Chem*. 1999, 64, 3–11.

Honeybees. http://en.wikipedia.org/wiki/Honeybees (accessed June 2016).

Jerković J., Mastelić J., Marijanović Z. A variety of volatile compounds as markers in unifloral honey from Dalmatian sage (*Salvia officinalis* L.). *Chemistry & Biodiversity*. 2006, 3, 1307–1316.

Jerkovic I., Mastelic J., Marijanović Z., Klein Z., Jelic M. Comparison of hydrodistillation and ultrasonic solvent extraction for the isolation of volatile compounds from two unifloral honeys of *Robinia pseudoacacia* L. and *Castanea sativa* L. *Ultrasonics Sonochemistry*. 2007, 14, 750–756.

Jolanta T.-G., Sławomir B., Kamila G., Krzysztof M. Differential scanning calorimetry for determining the thermodynamic properties of selected honeys. *Journal of Apicultural Science*. 2015, 59(1), 109–118.

Juan A., Fernández P., Ouissam A., Pierre D., Vincent B. Discrimination of Corsican honey by FT-Raman spectroscopy and chemometrics. BASE (Biotechnologie, Agronomie, Société et Environnement/Biotechnology, Agronomy, Society and Environment) 2010.

Jun W., Michael M.K., Weiyue Q., Soojin J., Gongyi S., Qing X.L. Rapid determination of the geographical origin of honey based on protein fingerprinting and barcoding using MALDI TOF MS. *Journal of Agricultural and Food Chemistry*. 2009, 57(21), 10081–10088.

Karoui R., Dufour E., Bosset J.O., De Baerdemaeker J. The use of front face fluorescence spectroscopy to classify the botanical origin of honey samples produced in Switzerland. *Food Chemistry*. 2007, 101(1), 314–323.

Kelly J.D., Petisco C., Downey G. Potential of near infrared transflectance spectroscopy to detect adulteration of Irish honey by beet invert syrup and high fructose corn syrup. *Journal of Near Infrared Spectroscopy*. 2006, 14, 139–146.

Lenhardt L., Zeković I., Dramićanin T., Dramićanin M.D., Bro, R. Determination of the botanical origin of honey by front-face synchronous fluorescence spectroscopy. *Applied Spectroscopy*. 2014. 68(5), 557–563.

Lipp J., Ziegler H., Conrady E. Detection of high fructose- and other syrups in honey using high-pressure liquid chromatography. *Z. Lebensm. Unters. Forsch*. 1988, 187, 334–338.

Jennifer H., Natasha d.V., Adelaide G., Col R.F., Joel A., Matthew J.H., Les B., Beverley A.-G. Using DNA metabarcoding to identify the floral composition of honey: A new tool for investigating honey bee foraging preferences. *PLoS One*. 2015, 10(8), e0134735.

Kelly J.F.D., Downey G., Fouratier V. Initial study of honey adulteration by sugar solutions using mid-infrared (MIR) spectroscopy and chemometrics. *Journal of Agricultural and Food Chemistry*. 2004, 52, 33–39

Kress W.J., Wurdack K.J., Zimmer E.A., Weigt L.A., Janzen D.H. Use of DNA barcodes to identify flowering plants. *Proceedings of the National Academy of Sciences of the United States of America*. June 2005, 102(23), 8369–8374.

Klaus B., Gudrun B., Cord L., Karl S.. Characterization of filtered honey by electrophoresis of enzyme fractions. *Apidologie*. 2010. doi:10.1051/apido/2010036.

Le Gresley A., Kenny J., Cassar C., Kelly A., Sinclair A., Fielder M.D.. The application of high resolution diffusion NMR to the analysis of manuka honey. *Food Chemistry*. 2012, 135, 2879–2886.

Lolli M., Bertelli, D., Plessi, M., Sabatini, A.G., Restani, C. Classification of Italian honeys by 2D HR-NMR. *Journal of Agricultural and Food Chemistry*. 2008, 56(4), 1298–1304.

Low N., South W. Determination of honey authenticity by capillary gas chromatography. *Journal - Association of Official Analytical Chemists*. 1995, 78, 1106–1113.

Mannas D., Altug T. SPME/GC/MS and sensory flavour profile analysis for estimation of authenticity of thyme honey. *Journal of Food Science and Technology*. 2007, 42, 133–138.

Manyi-Loh C.E., Clarke A.M., Munzhelele T., Green E., Mkwetshana N.F., Ndip R.N. Selected South African honeys and their extracts possess *in vitro* anti-*Helicobacter pylori* activity. *Archives of Medical Research*. 2010, 41, 324–331.

Manyi-Loh E., Roland N.N., Anna M.C. Volatile compounds in honey: A review on their involvement in Aroma, botanical origin determination and potential biomedical activities. *International Journal of Molecular Sciences*. 2011, 12, 9514–9532.

Marc O., Yulia B.M., Thomas K., Dirk W.L. Qualitative and Quantitative Control of Honeys Using NMR Spectroscopy and Chemometrics. ISRN Analytical Chemistry 2013, Article ID 825318, doi:10.1155/2013/825318.

Mazzoni V., Bradesi P., Tomi, F., Casanova, J. Directqualitative and quantitative analysis of carbohydrate mixtures using C-13 NMR spectroscopy: Application to honey. *Magnetic Resonance in Chemistry*. 1997, 35, S81–S90.

Montenegro G., Gómez M., Casaubon G., Belancic A., Mujica A.M., Peña R.C. Analysis of volatile compounds in three unifloral native Chilean honeys. *International Journal of Experimental Botany*. 2009, *78*, 61–65.

Moreno-Gonzalez D., Lara F.J., Jurgovska N., Gamiz-Gracia L., Garcia-Campana A.M. Determination of aminoglycosides in honey by capillary electrophoresis tandem mass spectrometry and extraction with molecularly imprinted polymers. *Anal Chim Acta*. 2015, 3(891), 321–328.

Padovan G.J., De Jong D., Rodrigues L.P., Marchini J.S. Detection of adulteration of commercial honey samples by the 13C/12C isotopic ratio. *Food Chemistry*. 2003, 82, 633–636.

Paradkar M.M., Irudayaraj J. Discrimination and classification of beet and cane inverts in honey by FT-Raman spectroscopy. *Food Chemistry* . 2002, 76, 231–239.

Piasenzotto L., Gracco L., Conte L. Solid phase micro extraction (SPME) applied to honey quality control. *Journal of the Science of Food and Agriculture*. 2003, 83, 1037–1044.

Pierna J.A.F., Abbas O., Dardenne P., Baeten V.. Discrimination of Corsican honey by FT-Raman spectroscopy and chemometrics. *Biotechnology, Agronomy and Society and Environment*. 2011, 15(1), 75–84.

Press release from the National Honey Board, Longmont, CO, Feb 2005. www.honey.com/media/presskit/industry.asp (accessed Feb 2017).

Radovic B.S., Careri M., Mangia A., Musci M., Gerboles M., Anklam E. Contribution of dynamic headspace GC-MS analysis of aroma compounds to authenticity testing of honey. *Food Chemistry*. 2001, 72, 511–520.

Radovic B., Parker I., Sharman M., Geiss H., Anklam A. Contribution of high temperature gas chromatographic analysis of oligosaccharides and ion chromatographic cations and anions to authenticity testing of honey. *Dtsch. Lebensm. Rundsch.* 2001, 97, 380–384.

Roberta d.O.R.R., Eliane M., Carla d.S.C., Edgar d.J. Detection of honey adulteration of high fructose corn syrup by low field nuclear magnetic resonance (LF ^{1}H NMR). *Journal of Food Engineering*. 2014, 135, 39–43.

Roberta d.O.R.R., Eliane T.M., Carla d.S.C., Maria L.G.M., Carlos A.C.J., Sérgio M., Edgar F.O.d.J. Classification of Brazilian honeys by physical and chemical analytical methods and low field nuclear magnetic resonance (LF ^{1}H NMR). *LWT - Food Science and Technology*. 2014, 55, 90–95.

Ruiz-Matute A.I., Rodríguez-Sánchez S., Sanz M.L, Martínez-Castro I. Detection of adulterations of honey with high fructose syrups from inulin by GC analysis. *Journal of Food Composition and Analysis*. 2010, 23, 273–276.

Ruiz-Matute A.I., Soria A.C., Martínez-Castro I., Sanz M.L. A new methodology based on GC-MS to detect honey adulteration with commercial syrups. *Journal of Agricultural and Food Chemistry*. 2007, 55, 7264–7269.

Ruiz-Matute A.I., Weiss M., Sammataro D., Finely J., Sanz M.L. Carbohydrate composition of high-fructose corn syrups (HFCS) used for bee feeding: Effect on honey composition. *Journal of Agricultural and Food Chemistry* 2010a, 58, 7317–7322.

Ruoff K., Luginbühl W., Künzli R., Iglesias M.T., Bogdanov S., Bosset J.O., von der Ohe K., von der Ohe W., Amado R. Authentication of the botanical and geographical origin of honey by mid-infrared spectroscopy. *Journal of Agricultural and Food Chemistry*. 2006 Sep 6, 54(18), 6873–6880.

Schievano, E., Morelato, E., Facchin, C., Mammi, S.. Characterization of markers of botanical origin and other compounds extracted from unifloral honeys. *Journal of Agricultural and Food Chemistry*. 2013, 61, 1747–1755.

Schievano E., Stocchero M., Morelato E., Facchin C., Mammi S. An NMR-based metabolomic approach to identify the botanical origin of honey. *Metabolomics*. 2012, 8, 679–690.

Sergiel I., Pohl P., Biesaga M., Mironczyk A. Suitability of three-dimensional synchronous fluorescence spectroscopy for fingerprint analysis of honey samples with reference to their phenolic profiles. *Food Chemistry*. 2014, 145, 319–326.

Sivakesava S., Irudayaraj J. Prediction of inverted cane sugar adulteration of honey by Fourier transform infrared spectroscopy. *Journal of Food Science*. 2001a, 66, 972–978.

Sivakesava S., Irudayaraj J. Detection of inverted beet sugar adulteration of honey by FTIR spectroscopy. *Journal of the Science of Food and Agriculture*. 2001b, 81, 683–690.

Sopade P., Ahalley P.J., Darcy B.R. Specific heat capacity of Australian honeys from 35 to 165c as a function of composition using differential scanning calorimetry. *Journal of Food Processing and Preservation*. 2006, 30(2), 99–109.

Spiteri M., Jamin E., Thomas F., Rebours A., Lees M., Rogers K.M., Rutledge D.N. Fast and global authenticity screening of honey using ^{1}H-NMR profiling. *Food Chemistry*. 2015, 189, 60–66.

Tewari J., Irudayaraj J. Quantification of saccharides in multiple floral honeys using Fourier transform infrared micro-attenuated total reflectance spectroscopy. *Journal of Agricultural and Food Chemistry*. 2004, 52, 3237–3243.

Valente A.M.M.P., Boffo E.F., Costa F.G., Melo F.M.P., Melo I.S., Ferreira A.G. Complete NMR assignments of a new prodigiosin isolated from Streptomyces violaceusniger violaceusniger. *Annals of Magnetic Resonance*. 2008, 7, 1–11.

Verzera A., Campisi S., Zappala M., Bonaccorsi I. SPME-GC/MS analysis of honey volatile components for the characterization of different floral origin. *American Laboratory.* 2001, 18–21.

Wabaidur S.M., Ahmed Y.B.H., Alothman Z.A., Obbed M.S., Al-Harbi N.M., Al-Turki T.M. Ultra high performance liquid-chromatography with mass spectrometry method for the simultaneous determination of phenolic constituents in honey from various floral sources using multiwalled carbon nanotubes as extraction sorbents. *Journal of Separation Science.* 2015, 38, 2597–2606.

Waldemar W., Tomasz C., Tomasz D., Paulina B. and Beata P. Application of gas chromatography, mass spectrometry and olfactometry for quality assessment of selected food products. *Ecological Chemistry and Engineering.*

White Jr, J.W. Honey. In The Hive and the Honeybee, Graham J.M., Ed., Dadant & Sons: Hamilton, IL, 1992, pp. 869–925.

Woodcock T., Downey G., Kelly J.D., O'Donnell C. Geographical classification of honey samples by nearinfrared spectroscopy: A feasibility study. *Journal of Agricultural and Food Chemistry.* 2007, 55, 9128–9134.

Xiangrong Z., Shuifang L., Yang S., Zhuoyong Z., Gaoyang L., Donglin S., Feng L. Detection of adulterants such as sweeteners materials in honey using near-infrared spectroscopy and chemometrics. *Journal of Food Engineering.* 2010, 101, 92–97.

Yang H., Irudayaraj J., Paradkar M.M. Discriminant analysis of edible oils and fats by FTIR, FT-NIR and FT-Raman spectroscopy. *Food Chemistry.* 2005, 93, 25–32.

Zoltan K., Guido P., Jan T. Glass transition temperature of honey as a function of water content as determined by differential scanning calorimetry *Journal of Agricultural and Food Chemistry.* 1999, 47(6), 2327–2330.

CHAPTER 16

Application of Fingerprinting Techniques in Authentication and Traceability of Meat

Sadaf Jamal Gilani
Al-Jouf University

Syed Sarim Imam
Glocal University

Md. Rizwanullah
Jamia Hamdard

CONTENTS

16.1 INTRODUCTION

Many consumers are concerned about the meat they eat, and accurate labeling is important to inform consumer choice. The problem of food adulteration is by no means a contemporary phenomenon and is likely to be as old as food processing and food production (Ballin, 2010 Cordain et al., 2005). The term adulteration is defined as "to debase, to make impure by addition of a foreign or inferior substance" and "to prepare for sale by replacing more valuable with less valuable or inert ingredients". It can also be defined as food which is not "of the nature or substance or quality demanded by the consumer". Nonauthenticity can take different forms:

1. Complete or partial removal of valuable constituents.
2. Complete or partial replacement of food components with an undeclared alternative.
3. Concealment of damage or inferior foodstuffs.
4. Addition of undeclared materials to increase bulk or make the product appear better (Hargin, 1996; Nakyinsige et al., 2012).

The application of fingerprinting in food detection has been important due to the large requirement of food materials across the world. Different fingerprinting methods have been explored for the detection of food, material sources, available content, added substances, objectionable substances, and presence of microorganisms and pesticides (Valeria, Caterina & Antonio, 2005; Rezzi et al., 2007; Zarkadas, Gagnon & Poysa, 2007; Santonico, Pittia & Pennazza, 2008; Casale, Oliveri & Armanino, 2010). Further, they are also involved in the detection of substandard substances and in ensuring that the quality remains above the normal or accepted limit. The German analytical chemist, Frederick, first published in the year 1820 about the food adulteration issue and completed a treatise on adulterations of food and culinary poisons (Accum, 1820). This was the first serious attempt to expose both the extent and dangers of food adulteration (Ellis et al., 2012).

The term meat is defined as "the flesh of animals or animal tissues used as food". In 1984, UK Meat Products and Spreadable Fish Products Regulations defined meat as "the flesh, including fat, and the skin, rind, gristle and sinew in amounts naturally associated with the flesh used, of any animal or bird which is normally used for human consumption". In case of meat adulteration, the high-quality meat species may be partially or completely replaced with lower quality meat species. The meat product may be piled with fat to get an excess amount with the flesh and, further, may be extended by the addition of water, non-meat proteins, and other substances.

The identification of meat species after butchering, dicing, or mincing has been a difficult task for the analysts for many years and the different problems in meat authenticity

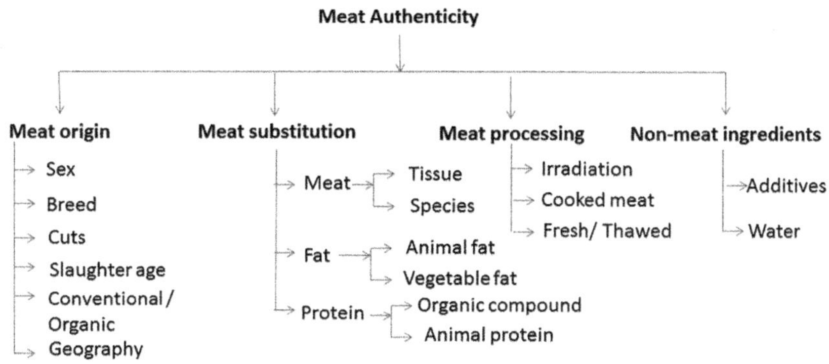

FIGURE 16.1 Different meat authenticity problems.

are depicted in Figure 16.1. The species identification for smoked and cooked sausages as well as for canned and cooked meats is an even more tedious process. Traders or customs and health authorities want to confirm that the consignments of meats and meat products that are being purchased, imported, and exported belong to the declared species. There have been various examples in the last decade of the substitution of an expensive meat species with meat from a cheaper species. Gross substitution can occur where the analyst would be looking for high percentages (e.g., > 10%) of undeclared meat species, while adventitious contamination may mean that the analyst is looking for low percentages (e.g., 1%–10% m/m) of undeclared species; this could be in raw or cooked products.

16.2 COMPONENTS OF MEAT PRODUCTS

16.2.1 Meat Content

The meat content can be calculated to determine the exact amounts of ingredients present. The analysts calculate the meat content of a sample with little or no knowledge of the ingredients present, composition, and processing history. In the year 1919, Stubbs and More developed a method with slight modifications and this method is still in use to estimate the meat content of meat products and is prescribed in a range of texts (e.g., Kirk & Sawyer, 1991; Commission Regulation (EEC), 1989). This method of estimation involves the determination of fat-free meat content by calculating the total nitrogen content in the meat and, further, calculation of the meat content by the use of an appropriate "nitrogen factor". This microscopic method has been used to detect the presence of soya proteins and milk proteins in meat. The obtained fat-free raw meat can be used by the analyst to calculate the "lean meat" content of a product. In the UK, the lean meat is defined as "meat free, when raw, of visible fat" (The Meat Products and Spreadable Fish Products Regulations, 1984). An available range has been given for intramuscular fat along with the nitrogen contribution from the "visible fat". There is a formula to calculate the total lean meat content that is used by different researchers with appropriate corrections (Pearson, 1968, 1970; Cottam, 1975).

16.2.2 Excess Gelatin

The presence of excess gelatin in meat also constitutes the same problem as the presence of excess connective tissues in canned ham products. It is important for the analysts to calculate the content of the gelatin that can contribute nitrogen to the meat content. Richardson and Tarassuk (1934) reported an alternative procedure to determine the gelatin content in the meat.

16.2.3 Excess Connective Tissue

The presence of excess connective tissues is assessed by the analyst to check whether it is present in the product in an amount above that expected to be naturally associated with the flesh used. The presence of "excessive" connective tissues will contribute nitrogen and, therefore, give false high meat content. Lord and Swan (1984) detected the problem of connective tissues in meat.

16.2.4 Added Water

The presence of added water is calculated in a fat-free meat product and it is determined using the following formula:

$$\%\text{added water} = 100 - \left[\%\text{fat free meat} + \%\text{fat} + \%\text{NAA}\left(\text{non} - \text{aqueous additions}\right)\right]$$

Authenticity of meat products can be effectively judged by the conclusions drawn.

16.3 EXAMINATION PROCEDURES

16.3.1 Microscopy

Microscopy has an important role to play in the examination and analysis of meats and meat products. It will give a rough estimation of the relative proportions of constituents and it assesses their forms. It often allows a judgment to be made on the quality and authenticity of a product. The major advantage of this method is that the analyst requires a relatively short period of time to identify the major constituents in a meat product. The microscopic examination of meat can reveal the muscle cell structure (intact, natural form, or disruptions in structure) and the nature of meat used in a product (present as cuts of muscle, as pieces, mechanically recovered), which can give an idea about whether the product has been prepared from restructured or reformed meat. The tissues and organs can also be identified and detected using the microscopic technique, for example, lung, kidney, intestine, pancreas, uterus, etc. Further, the screening and identification of the presence of non-meat proteins, cereal additives, wheat rusk, and the form of soya can also be done by the microscopic technique. The cryosections of meat required are approximately 10–12 μm in thickness and are best prepared using a freezing microtome. Chemical stains can be applied to enhance the appearance of particular features during polarization microscopy as many ingredients are viewed between crossed polarizing filters. (Flint & Firth, 1981). The unique structural features and ingredients such as muscle cells, collagen, elastin, and added proteins (soya, milk, and cereal) can all be located and identified by using the stain. There are many other reagents that are available for the detection of specific parts or adulterants (Bayliss-High, 1984). Picrosirius Red F3BA and polarization microscopy are used to distinguish raw and heat processed collagen from each other (Flint & Pickering, 1984). The detection, identification, and characterization of the finely comminuted product by microscopy may be very difficult. So, the antibodies labeled with enzyme detection systems or immunohistological staining can be applied to the cryosections of such meat products. Microscopy has an important role to play in the examination and characterization of meats and meat products. It is an excellent complementary technique to chemical and immunological analysis.

16.4 DIFFERENT ANALYTICAL METHODS AVAILABLE FOR DETECTION

Different analytical methods (Table 16.1) are available for the species identification. Initially, lipid analyses, protein analyses and anatomical, histological, organoleptic, chemical, biochemical, spectrophotometric, chromatographic, and electrophoretic techniques were widely used for speciation studies. With the advancement of analytical techniques, authenticity and accuracy of the detection methods are also increased. New techniques based on molecular approach have raised the hopes of developing better identification methods, which can overcome the limitations of the existing methods. Newer methods are used for meat species identification, such as DNA analysis, electrophoretic, immunological, immunoelectrophoretic assays, etc.

TABLE 16.1 Examples of Analytical Techniques Applicable for the Authentication of Meat and Meat Products

S. No	Authenticity problem	Technique Involved	Reference
1	Meat species	Real-time PCR	Laube et al. (2007a; 2007b); Martín et al. (2009)
		Traditional PCR	Rodriguez et al. (2003); Pascoal et al. (2005)
		ELISA	Koppelman et al. (2004)
		LC	Ashoor and Osman (1988); Chou et al. (2007)
2	Meat tissue	ELISA	Kim et al. (2005); Muldoon et al. (2004)
		Mid-infrared spectroscopy	Al Jowder et al. (1999)
3	Animal protein	LC	Ashoor et al. (1988); Chou et al. (2007)
		ELISA	Muldoon et al. (2004); Kim et al. (2005)
4	Geographical origin	MS	Franke et al. (2008); Franke et al. (2008)
5	Organic vs. conventional meat	Fluorescence microscopy	Kelly et al. (2006)
		GC	Angood et al. (2008); Husak et al. (2008); Kim et al. (2009)

16.4.1 Lipid Analysis

Capillary gas chromatography, gas chromatography (GC), and/or HPLC (high performance liquid chromatography) with mass detection have been widely used to examine the fatty acid isomer profile by the determination of fatty acids at the 2-position in triglycerides after enzymatic hydrolysis. The examination of the triglyceride profile can provide a tentative identification of some meat species (e.g., horse, pork, and kangaroo). The interpretation is very difficult for highly processed, cooked food and for blends of more than one meat species. The composition of animal lipid varies with season and diet. The beef and horse lipids can be detected and differentiated by looking at the relative ratios of the fatty acid isomers: 18:0, 18.3Q3, 18:1Q9, and 18:2Q6, but differentiating fat from other species is much more difficult. Saeed et al. (1986) used fatty acid 20:2Q6 as an indicator to detect the presence of pork fat in beef and lamb fats. Lipid analysis is sometimes useful to provide complementary evidence in cases of gross substitution, but the detection limits are generally too high for the detection of adulteration at the 0%–15% level.

16.4.2 Chromatographic Techniques

Different types of chromatographic techniques like HPLC and cation exchange chromatography are utilized for the identification of meat species. These fingerprinting techniques provide rapid methods for the detection of meat adulteration. The fingerprinting region of the test sample is compared with standards and the purity is identified (Toorop et al., 1997). Cation exchange chromatography was applied for the separation of hemoglobin, followed by filtration with cellulose acetate paper. Diode array detection was done at 416 nm. The characteristic peak patterns of meat species could be specified using this technique (Ashoor et al., 1998).

16.4.3 Electrophoresis

The use of electrophoresis in the identification of meat species has been done since the 1970s. Isoelectric focusing is probably the most widely used technique of electrophoresis for meat species identification and this technique can identify a wide range of adulterating species. This relatively simple approach of identification can be applied to binary mixtures of meats, but in such cases the electrophoretograms become complex and interpretation may be difficult. Electrophoretic methods are very useful for screening purposes and they are also capable of detecting binary or ternary meat species mixtures and may be applied to "cooked" products. Kaiser et al. (1980) used aqueous meat extracts (e.g., beef, pork, venison, roe, impala, chicken, turkey), followed by the collection of Coomassie Brilliant Blue stained proteins to identify the species in raw binary mixtures. A quantitative analysis was performed by densitometric evaluation of the electrophoretograms, which allowed a minimum detection limit of 2%–5% (m/m). Sinclair and Slattery (1982) published a similar procedure for the differentiation of raw beef, pig, buffalo, horse, and kangaroo; however, sheep and goat could not be distinguished using this method. Different researchers have used different markers like esterase patterns to identify meat species (Heinert & Klinger, 1980; Prasad & Mishra, 1981).While Slattery and Sinclair (1983) used muscle lactate dehydrogenase and esterase isoenzymes for the detection, King and Kurth (1982) stained isoelectric gels for phosphoglucomutase to detect pig, horse, and buffalo in beef with detection levels of 1% m/m. King (1984) reported a method to extract the proteins from cooked meats with 6 M guanidine hydrochloride. The extract was dialyzed with triton X-100 (1%), and adenylate kinase (AK) and creatinine kinase (CK) activities could be recovered sufficiently to allow the isoelectric focusing, enzyme staining, and subsequent species identification. The AK and CK activities could not be recovered when meat samples were heated above 120°C and 105°C, respectively. The detection limits were found to be very high: 10% (m/m) after a concentration step and 25% (m/m) without concentration.

Kim and Shelef (1986) identified binary mixtures of beef, pork, chicken, and turkey meats with detection limits above 5% (clear differentiation at 25% m/m) using agarose gel electrophoresis with the identification of isoenzyme MM and myoglobin bands. The staining of myoglobins used for the detection of chicken meat mixed with beef and pork was done by isoelectric focusing with peroxidase (Kauffer et al., 1990). Hofmann and Bluchel (1986) identified the original species of raw meat by examining the myoglobin patterns after isoelectric focusing, and the sensitivity of the method was later improved by taking advantage of the pseudo-peroxidase activity of myoglobin and by applying peroxidase staining (Bauer & Hofmann, 1986). Janssen et al. (1990) developed a method for myoglobin staining that was applicable to meats that were heated to 120°C for 5 mins. The extracted meat proteins were treated with 8 M urea and 0.1% dithioerythritol. The proteins were subjected to isoelectric focusing in ultrathin gels containing urea and ampholytes. Detection of less than 10% (m/m) pork, horse, kangaroo, and sheep meats in beef was possible when the beef was heated to an internal temperature of 120°C for 5 mins. This method of identification is interesting for the selective recognition of species myoglobin bands. McCormick et al. (1992) investigated the identification of meat from different domestic species using enzyme staining for CK and AK. Pork and bison could be differentiated from all the examined species, but beef, elk, caribou, red deer, and goat could not be differentiated from each other. Skrokki and Hormi (1994) used a polyacrylamide gel (PAGE) method for the detection of adulteration in minced meat to identify the pork content.

16.4.4 DNA Methods

DNA method is used for the identification of specific recognition of unique DNA sequences in samples. It can form the basis of a highly specific assay for species identification. These are far more user friendly for applications in food laboratories. DNA-based methods for meat species identification provide a way to identify the species of origin in raw and cooked meat products. Total genomic probes can be used to identify distantly related species, but tailored oligonucleotide probes need to be designed for the differentiation of closely related species. Identifying meat species and detecting its content in highly processed and heated meat products are difficult. The assay specificity is a problem while differentiating between closely related species, such as sheep, goat, beef, and buffalo. The principle involved in DNA-based methods is relatively simple and straightforward. DNA is extracted from the meat product, partially purified and, after denaturation, immobilized on a solid surface. Radiolabeled DNA probes can be used and now colorimetric, fluorescent, or chemiluminescent labels may also be used in the process. The potential of DNA methods have been explored by Bauer et al. (1987), Wintero et al. (1990), Chikuni et al. (1990), and Ebbehoj and Thomsen (1991a; 1991b). These researchers used total genomic species DNA as the probe and highly repetitive base sequences were attached to the probes. These DNA probes were found to be homologous between closely related species.

Ebbehoj and Thomsen (1991b) used unlabeled DNA from cross-hybridizing species to discriminate between some ruminants and to differentiate between some sheep and goats. DNA amplification method (Meyer et al., 1993) included a procedure with polymerase chain reaction (PCR) to detect raw and cooked meat speciation using growth hormone gene-specific primers. Further, a modified technique used sheep satellite I DNA primers for PCR amplification, followed by Apa II restriction digestion to differentiate the meats of sheep and goat (Chikuni et al., 1994). The use of forensically informative nucleotide sequencing (FINS) for identifying the animal origins of biological specimens of gourmet meats were reported by two research groups (Bartlett & Davidson, 1992; Forrest & Carnegie, 1994). The matching of the most closely related species was done by determining the nucleotide sequence of the amplified DNA sequence and by phylogenetic analysis. These PCR and sequence analysis-based techniques are relatively expensive but show extreme sensitivity to the reaction. The DNA-based procedure is a simple and robust method that can be used by food analysts to identify the meat species present in raw and highly processed products. UK Laboratory of the Government Chemist developed a simple slot-blot assay method for the specific identification of pork, beef, rabbit, sheep, and goat in admixture (Lumley, 1996; Hunt et al., 1997). Meat samples are washed with buffer to remove blood and plasma, the remaining muscle cells are lysed, and the total genomic DNA is extracted either from authentic tissues or from cooked and processed products with a buffer-SDSEDT A mixture. After incubation with ribonuclease A and proteinase K, DNA is extracted with phenol-chloroform and precipitated with ethanol. Sample DNAs are added to digoxigenin labeled species-specific probes for the hybridization stage. The DNA detected from the lysed muscle cells indicates the presence of meat. The goat and sheep meats can be differentiated in highly processed meat products using the assay procedure.

16.4.5 Immunodiffusion Method

This method has been found to be ideal for routine species identification of raw meats. This method is simple to use, inexpensive, robust, gives rapid results, and is easily available as some commercial kits containing all the reagents. A number of papers have been published that explain the production of species-specific antisera by the immunization of different animals with unidentified antigens extracted from meats. Different commercial AGID (agar gel immunodiffusion) kits for meat speciation have been produced, but they didn't survive for long periods in the market. Rhone-Poulenc Diagnostics Ltd now sells the ORBIT (overnight rapid bovine identification test), PROFIT (poultry rapid overnight field identification test), PRIME (porcine rapid identification method), and SOFT (serological ovine field test) test kits, which can be conveniently used to identify meat species present in raw meat products. Binding Site (Birmingham, UK) and Sigma-Aldrich (Saint Louis, Missouri) market preprepared AGID plates and anti-species affinity-purified antisera for meat species testing. The species specificity of each batch purchased is evaluated against all the possible animal species that may be present. An alternative to AGID meat speciation is the FAST Immunostick Test Kit produced by Cortecs Diagnostics (UK), which can be used for beef, horse, pork, sheep, goat, poultry, rabbit, and kangaroo detection. The identification of meat species can be done via a color change in less than 1 h with detection limits of about 1% (m/m) for the species in admixture in lean meat samples. The AGID method was used by Hayden et al. (1978) to detect the serum proteins in pig, horse, and rabbit flesh extracts. An 1%–3% level of detection and also the qualitative detection of each species in beef was possible by this method. In another research by Hayden (1979), the production of antisera for ovine, porcine, and equine myoglobins and the collected antigens were found to be relatively thermostable. The use of antisera in AGID helps in the detection of raw flesh of the corresponding species and mildly heated meat. The reliability of the assay detection result decreased with increased sample heating. In another study, Hayden (1981) developed an AGID procedure to detect rabbit antisera raised against thermostable antigens of horse, pig, and lamb. The procedure could be performed at a temperature of 71°C to detect the flesh of the target species in sausages cooked with detection levels between 5% and 10% (m/m). Swart and Wilks (1982) used commercially available anti-species antisera in an AGID assay method (anti-kangaroo antiserum prepared by immunizing rabbits with kangaroo blood) with detection limits of 5%–20% for sheep, horse, and kangaroo meats in beef. The effects of cross-reacting antibodies were nullified by the addition of relevant species' proteins. The commercial antisera showed a detection limit of 3% (m/m) that was obtained in the authors' laboratory using AGID. Patterson and Spencer (1983) reported a rapid dipstick assay method to detect meat speciation in which the solid phase was coated with species-specific antisera. A detection limit of 1% (m/m) was reported for this method and the positive reaction was indicated by a color change of meat in admixture. Schweiger et al. (1983) reported an AGID procedure for the quantification of chicken and turkey muscles in beef and pork products. They used turkey muscle protein, troponin, as the antisera and the results indicated that the detection of chicken or turkey could be done at a temperature of 70°C in cooked products. Kangethe et al. (1986) also worked on the development of AGID methods to detect the meat species in heated products. Thermostable muscle antigens (TMAs) cattle antisera were used to detect the presence of beef in meat products. Many research groups (Mageau et al., 1984; Cutrufelli et al., 1986; 1987; 1988; 1989) used AGID techniques for the detection of bovine, avian, porcine, and ovine meats. The method used Bovine serum albumin as the immunogen. They used reagent-impregnated filter paper discs to overcome the need for reagent handling in the procedure. The ORBIT method used AGID plates and paper discs impregnated with rabbit, sheep, or goat anti-bovine antiserum. The PROFIT method used goat anti-poultry antisera raised against chicken and turkey serum albumin; the PRIME method used goat anti-porcine antisera; and the SOFT method used calf anti-ovine antisera for detection. Cutrufelli et al. (1987) analyzed 30 raw meat samples using the detection methods mentioned earlier and reported that the techniques were reliable and capable of detecting adulteration at the 10% (m/m) level. The raw meats are identified via blood or plasma proteins rather than via muscle proteins, and a few methods can also be applied to cooked samples. The analysts preferred to use commercially available kits to detect the contents in fresh and cooked meats. The results showed good quality, good species specificity; and these methods are relatively cheap, easy to use, and are suitable for screening large numbers of samples. Commercial kits are used, but some in-house AGID methods have also been validated and these employ commercially available anti-species antisera. This allows simple and cost effective routine screening of raw meat samples using a suite of immunological reagents with no dependence on one particular commercial supplier. Commercially available AGID kits can only be applied to raw meats and they only detect blood or plasma proteins. There is a need for methods that can be applied to cooked samples for detecting muscle proteins.

16.4.5.1 Ring Precipitation Test

This detection involves an antigen-antibody reaction. At the point of interaction between antigens and antibodies if a ring is formed, the test will be considered as positive for a particular meat. It is a qualitative evaluation test, but it also has some disadvantages. It is not suitable for detecting meat in heat-treated meat species and, sometimes, it gives false positive results by forming a diffused ring.

16.4.5.2 Double Immunodiffusion Test

This test is suitable for both qualitative and quantitative assessment of meat adulteration. This test works on the same principles as the ring precipitation test. Here also an antigen and antibody reaction takes place. The point of detection is the band that is formed at the interaction point. The basic differences between these two tests are that the band is held for a longer period of time and the visibility of the band is also enhanced. The antiserum is used to test a mixture of meats or meat samples and is suitable for the detection of meat adulteration up to 5%. This test is helpful for the identification of meat cooked at 80°C for 10 mins. The total time required for performing the test is 2–3 days.

16.4.5.3 Isoelectric Focusing

In this method of meat fingerprinting, the principle is the migration of proteins in a pH gradient by the formation of species-specific bands. The identification of origin can be done on the basis of location, density, and area of bands. This technique can be helpful in the identification of fresh as well as cooked meat, but it is not suitable for closely related species and frozen meat (Skarpeid et al., 1998). The whole muscle visualization can be done by using coomassie blue, but the identification of low levels of buffalo, pig, or horse meats in beef can be done by the utilization of phosphoglucomutase. King (1984) reported the use of AK and phosphor gluconate dehydrogenase (PGD) for the identification of low levels of kangaroo or horse meat in beef and for the differentiation of mutton with chevon.

16.4.6 Enzyme-Linked Immunosorbent Assay

Since the early 80s, a wide range of ELISA (enzyme-linked immunosorbent assay) techniques have been used for meat speciation. They are able to perform a rapid screening of raw meat samples, and they are further designed to detect blood plasma proteins. The detection and estimation of meat contents are difficult in meat mixtures. This technique cannot be applied to cooked products. There is a complex process involved in identifying the species of origin of meats in cooked products. The process uses purified antisera, and the extraction of thermostable antigens is done to achieve specificity for the identification. A prolonged cooking or heating is not ideal for the assay response as the specificity diminishes. These methods cannot be used quantitatively. There are several ELISA techniques to detect muscle proteins and they may be capable of detecting and identifying meat species contents in a mixture of meats.

16.4.6.1 Raw Meats

Hitchcock et al. (1981) first introduced the application of ELISA methods for the estimation of soya proteins in meat products. The advantage of this method is the requirement of fewer antisera. The results can be obtained in several hours; a large number of samples can be processed and, thereafter, "quantitative" results may be obtained. Whittaker et al. (1982) and Kangethe et al. (1982) published the first reports on the application of ELISA methods for meat specification. They used anti-species antisera raised in rabbits against serum albumin from different target species (e.g., beef, sheep, horse, kangaroo, pig, and camel). They applied an indirect ELISA method for the estimation and the antigens were extracted from the samples with buffer or saline. They reported detection limits of 3% (m/m). Whittaker et al. (1983) published a more extensive observation of the ELISA procedure. In 1984, Patterson et al. reported about using a double sandwich ELISA procedure to detect less than 1% (m/m) of meat species in admixture. They concluded that the ELISA technique was capable of differentiating sheep, goat, beef, and buffalo. The same research group (Patterson & Spencer, 1985) used monospecific antisera produced by the immunization in target-related species to differentiate phylogenically related species. The cattle were immunized with serums of buffalo, sheep, and goat, and the horses with the donkey serum. The detection level was 1% (m/m) or less for each target meat species using ELISA. Jones and Patterson (1985) reported an ELISA method to detect pig meat in meat mixtures by using porcine serum albumin (PSA) antisera from rabbits and sheep. They reported detection levels as low as 0.5% (m/m) of pork in beef sausages. In 1986, Jones and Patterson reported another ELISA method to identify meat speciation. They used commercially available anti-species antisera of different animals (e.g., anti-cow, anti-horse, anti-pig, anti-sheep/goat) and performed an integral assay inhibition of heterologous cross-reactivity by the addition of heterologous serum albumin prior to ELISA. Ayob et al. (1989) published similar methods for the detection of meat. Another research group applied a competitive ELISA method by using the commercially available anti-species antisera. They developed a procedure by using beef serum proteins for the estimation of beef content in beef joints (Griffiths & Billington, 1984). The results from the ELISA method matched very well with the well-established beef contents of model mixtures (but calibration was from the same meats). A wide variation in the contents of beef joints was observed for the fat-free beefs. A difference in the amount of residual blood in the various joints was observed. Martin et al. (1988a) reported in a study that most anti-species antisera are raised against serum albumins or whole blood serum. These antisera cannot be used to detect the amounts of undeclared meat species due to the variable contents of residual blood in meats. They extracted sarcoplasmic antigens from lean pig, horse, chicken, and beef meats for adding the antisera to muscle proteins. They used isotonic saline and raised antibodies against the pig antigens in rabbits. The antisera showed affinity toward the sarcoplasmic extracts from other meat species. The authors showed evidence of the specificity for the antisera of anti-porcine sarcoplasmic protein (anti-PSP). The results from the study stated a 1%–50% (m/m) detection rates of pig substitution in raw lean beefs. The preparation of anti-chicken sarcoplasmic protein (anti-CHSP) antisera was reported by Martin et al. (1988b). They explained in their study that the sarcoplasmic extracts contained hemoglobin and chicken serum albumin. The study disclosed that the method was used to detect muscle not the effect of blood and plasma. Neither data has shown natural variations of sarcoplasmic proteins extracted from various muscles within and between the animals. The preparation of monoclonal antibodies for CHSP was done by Martin et al. (1991). They presented quantification calibration lines from 1% to 100% for chicken.

16.4.6.2 Cooked Meats

The antisera used in the ELISA method are thermolabile proteins for the detection of raw meat products. For the identification of target antigens in cooked and canned products, Berger et al. (1988) explained a complex and lengthy procedure to extract antigens from the muscles of pork and chicken that produced potent species-specific antisera. The immunizing antigens are heat resistant and species specific. Neither heat resistance nor specificity is absolute and the method cannot be used for quantitative estimation.

Patterson and Jones (1989) reported an ELISA method for heated pig meat. They used two ELISA methods to detect heated lean meats (pork) in a variety of meat products. They used an indirect ELISA method to give a linear calibration plot of optical density versus percentage of pig meat in lean meat mixtures with beef, lamb, and chicken. The assay gives a permissible limit of detection of 5% pork lean. The assay could not be used to evaluate absolutely the lean meat contents in products. Sawaya et al. (1990a; 1990b) explained the applications of different analytical methods like ELISA, HPLC of triglycerides and ophidine dipeptides, and GC of fatty acids for the detection of pork in processed meats. In the ELISA method, the used antigens were extracted from pork by using saline, followed by the autoclaving process at 120°C for 30 mins and the filtered extract was then used to immunize sheep. To perform a competitive ELISA procedure, the sheep anti-porcine antisera were mixed with extracts from autoclaved beef, sheep, and horse meats to block cross-reactions with heterologous species. The detection level was 2% in pork model meat mixtures. The meat had been heated up to 1200°C for 30 mins. The identification of heated meats was done with thermostable antigens prepared from adrenal glands and muscles by grinding, sonicating, and heating at 98°C for 15 mins, followed by centrifugation and autoclaving in normal saline solution (Sherikar et al., 1993). Commercially available kits were used to perform the ELISA procedure. The identification of meat species in raw and cooked meats was done with a kit of Cortecs Diagnostics (UK). The cooked meat ELISA kit is capable of detecting beef, pork, poultry, and sheep, whereas the ELISA kit for the raw meat is capable of detecting cow, horse, pig, sheep, poultry, and kangaroo meats in admixture. These kits are suitable for screening meat products and about 1%–3% of each meat species in admixture can be detected. All the reagents required for the assay of fresh meats or packed meats are included in each kit. These kits cannot be used to quantify the contents of a particular meat species in a mixture.

16.4.7 Polymerase Chain Reaction

Mullis et al. (1986) invented this test and it is found to be a rapid in vitro technique in which multiple copies of a specific piece of DNA sequence can be obtained in a few hours. Due to its lesser complexity and speed, it is reliable, sensitive, cost effective, authentic, and is potentially applicable for the authentication of meats and meat products. It can detect even a single copy of a sequence from a single cell sample. It is also useful in solving the issues of traceability for live animals and derived products (Cunningham & Meghen, 2001). This technique was also capable of differentiation of the meats from male and female animals. The benefits of PCR over other conventional methods include detection of a wide variety of meat samples. This technique is also helpful in the detection of fresh or processed meats for very small amounts of adulteration (up to 1%). A genetic marker has been used for meat speciation and it may be a nuclear gene or mitochondrial gene markers. The mitochondrial gene is more convenient and applicable because mt-DNA isolation is easier. With this method, very small samples can be tested due to the presence of multiple copies in a cell; mt-DNA copies range from 100–10,000 per cell. These markers are also capable of detecting very old biological samples because of the higher stability of mt-DNA and its strength in comparison to nuclear DNA. Walker et al. (2004) developed a SYBR Green-based PCR assay for the quantification of DNA from equine, canine, feline, rat, hamster, guinea pig, and rabbit meats. The minimum effective detection levels ranged from 0.1 ng–0.1 pg of starting DNA template for estimation. In another study, the SYBR Green real-time PCR assay was involved using ruminant-specific Bov-B SINE primers for the detection of bovine meat and bone meal. In this study, the melting curve analysis was used to detect 0.1% contamination in bovine, sheep, and bone meals heated at 133°C for 20 mins (Aarts et al., 2006).

16.4.7.1 Polymerase Chain Reaction-Restriction Fragment Length Polymorphism

This technique involves PCR amplification of a gene followed by digestion with restriction enzymes. It is a convenient, rapid, sensitive and, versatile assay for meat species identification. It uses different types of enzymes and nuclear and mitochondrial gene markers. Meat species can be detected by the PCR amplification of DNA followed by the species-specific cleavage with a restriction enzyme (Verma et al., 2013).

16.4.7.2 Real-Time PCR

This is an advanced version of the PCR technique. Reactions can be monitored in an early stage. Fajardo et al. (2008) developed for the first time a rapid, real-time PCR using the SYBR Green detection system. This test was used for the quantification of red deer, fallow deer, and roe deer DNAs in meat mixtures. The method combined the use of cervid-specific primers that amplified 134, 169, and 120 bp of the 12S rRNA gene fragment in red deer, fallow deer, and roe deer, respectively and universal primers that amplified a 140 bp fragment of the nuclear 18S rRNA gene from eukaryotic DNA. Frezza et al. (2008) evaluated four different species-specific probes for the quantification of bovine, ovine, swine, and chicken mitochondrial DNAs by real-time PCR. They detected short, species-specific sequences amplifiable from the heat-treated materials with a detection limit of 0.5 ng for swine, 0.5 ng for chicken, 0.01 ng for ovine, and 0.01 ng for bovine genomic DNA.

16.4.7.3 Multiplex PCR

This PCR assay was used for a simultaneous, precise, and rapid identification of multiple meat species in a single PCR reaction. This method uses several primer pairs targeting genomic and mitochondrial DNA. Multiple species identification can be done by setting different primer sets for each species and species-specific reverse or vice versa (Kumar et al., 2015). The identification of cattle, pig, and chicken meats was done by Fei et al. (1996) using multiplex PCR primers based on mt D-loop sequences. The use of the multiplex PCR assay method was reported by another research group, Behrens et al. (1999), for the identification of chicken, turkey, beef, pork, goat, mutton, donkey, and horse meats. They reported in their study that the heated and processed meat products were detected at a detection limit of 1%. Matsunaga et al. (1999b) used a mixture targeting 18S rDNA gene to get specific PCR products of 293, 254, and 267 bp fragments to differentiate mammalian, poultry, and fish species in meat. Matsunaga et al. (1999a; 1999b) targeted mitochondrial cytochrome b gene variability using a common forward primer and six reverse primers. They identified goat, chicken, cattle, sheep, pig, and horse meats. The identification was performed for fresh and cooked meats, including autoclaved meat and heated meat at various temperatures. Rodriguez et al. (2003) clearly identified the adulteration in goose, mule duck, chicken, turkey, and swine using the common forward primer designed on mitochondrial 12S rRNA gene. They detected approximately 1% of adulteration using the species-specific reverse primers. A PCR assay method was used for the specific and quantitative detection of pork, beef, sheep, and goat meats using common forward primers designed on mitochondrial 12S rRNA (Rodriguez et al., 2004). They detected raw and heat-treated meat mixtures using species-specific reverse primers with detection limits of 1% (wt/wt). Different researchers reported different assays with detection limits of the multiplex assays between 0.002% and 3.6% (Behrens et al., 1999; Matsunaga et al., 1999b; Dalmasso et al., 2004).

An evaluation of the presence of accidently added meats in food compounds was done by the multiplex PCR assay method for the identification of different species. Ten samples from each species were collected and the DNA was extracted successfully. The results reported that none of the samples were contaminated with porcine residuals, but the samples of sausages and cold cuts were contaminated with poultry residuals (Ghovvati et al., 2009). In another study, a multiplex PCR-based method was proposed for the rapid identification of meat from yak and cattle using three primers (Yin et al., 2009). They reported possible adulteration of yak meat with cattle meat, based on the sequence of mitochondrial 12S rRNA gene. The assay described satisfactorily that the amplification was accomplished in the analysis of raw and heat-treated binary meat mixtures of yak/cattle with a detection limit of 0.1% for cattle meat.

16.4.7.4 Other PCR Techniques

A number of different PCR assays were used for meat species identification. Sasazaki et al. (2004) used the amplified fragment length polymorphism (AFLP-PCR) technique to differentiate between the Japanese Black and F1 (Japanese Black x Holstein) breeds for the elimination of fake breed labeling of meat. Guha and Kashyap (2005) designed the heminested PCR assay methods by two consecutive amplifications of the mitochondrial 16S rRNA gene for the identification of nilgai, hog deer, blackbuck, goral, chital, sambar, and thamin deer. Fajardo et al. (2007) used extracted DNA from pasteurized (72°C for 30 min) and sterilized (121°C for 20 min) meats and the results showed adequate specificity and reproducibility for a number of domestic meats. Martin et al. (2007) reported the qualitative detection of four duck species in meat mixtures using primer pairs based on targeting mitochondrial 12S rRNA gene by PCR assay method. The result showed that the detection limit ranged from 0.1%–1.0% (w/w) and the technique was applied for the detection of raw as well as sterilized muscular binary mixtures.

16.4.8 TaqMan Assays

Dooley et al. (2004) developed TaqMan assays for the first time for meat species identification in beef, pork, lamb, chicken, and turkey. The meat speciation was achieved using species-specific primers, but the absolute level of detection, for each species, was not determined. The experimental limits for beef, lamb, and turkey were below 0.1% (Kesmen et al., 2009). The assays work with small regions of the mitochondrial cytochrome b (cytb) gene. There are two TaqMan probes developed for meat species speciation:

- Type I probe is specific for the mammalian species (beef, lamb, and pork).
- Type II probe is specific for the poultry species (chicken and turkey).

The normal end points for the TaqMan PCR conditions were limited to 30 cycles. Each species can be detected when it is spiked in any other species at a detection level of 0.5%. The absolute level of detection for beef, lamb, and turkey was below 0.1% level (Kesmen et al., 2009; Singh et al., 2014). Laube et al. (2007a) developed a method to detect trace amounts of beef, pork, lamb, goat, chicken, turkey, and duck in processed food using TaqMan TM®-PCR systems. Laube et al. (2007b) used TaqMan TM®-PCR method to detect minute amounts of animal species (concentration of 0.1%) in relation to the total proportion of meats in food products. TaqMan real-time PCR method was used by Lopez-Andreo et al. (2005). The method used minor groove binding (MGB) probes for the detection/quantitation of bovine, porcine, lamb, chicken, turkey, and ostrich DNA in complex samples. The detection limit ranged from 0.03 to 0.80 pg of template DNA. The experimental mixtures contained 2–4 different species. The method was suitable for detecting more than 1% of pork, chicken or turkey and more than 5% of cattle or lamb.

16.5 DETECTION AND DETERMINATION OF NON-MEAT PROTEINS

The non-meat proteins are useful ingredients for the meat products because of their emulsifying and binding properties. The adulteration of meat is an issue when non-meat proteins are added to extend the apparent meat content of the product in an attempt to overcome the minimum meat content criteria. A wide range of ELISA procedures have been proposed for the detection and determination of soya proteins in meat products.

16.5.1 Soya Protein

The detection and determination of soya proteins in meat products have been creating problems for food analysts for many years. The presence of soya can be detected and the type can be identified by examining the characteristic features associated with each form. Soya proteins in meat products can be detected using stereological point counting techniques (Flint & Meech, 1978). This approach has been essentially overtaken by the application of immunological methods. Glynn (1939) used an immunoprecipitation test to detect soya flour in sausages. This procedure was modified by Fergusson et al. (1942). Commercially available anti-soya antisera are used to perform simple immunodiffusion methods for screening raw meat products to detect the presence of soya proteins with detection limits of 1% (m/m). Since 1980, ELISA methods have been developed to detect and determine soya proteins in cooked products. Hitchcock et al. (1981) described an ELISA procedure based on antibodies raised to denatured-renatured soya proteins. Initially, soya proteins were solubilized using urea and mercaptoethanol. The mixture was dialyzed, which allowed proteins to renature, and the solubilized proteins were used to immunize rabbits. The result of the assay reported that the 7S soya protein fraction (f3-conglycinin) was most antigenic after renaturation and the response for whey and 11S fraction (glycinin) was 30% and 18%, respectively. Hitchcock et al. (1981) used a modified method to use commercially available antisera. They used commercially available antisera raised against native and heat-denatured (but not renatured) soya proteins. The model meat products contained soya protein isolate and they were analyzed after heating at 100°C for 30 min and at 121°C for 30 min. The extraction procedure for samples involved a maceration process with a number of organic solvents to remove fat prior to protein extraction using urea and mercaptoethanol. Griffiths et al. (1984) used commercial anti-soya antisera to react with the experimental antisera. The reaction showed two different responses: strong binding to the 7S protein and relatively weak binding to the 11S and whey proteins. The results of this work were found significantly promising for a collaborative study and the further findings were published by Crimes et al. (1984). Rittenburg et al. (1987) published an improved indirect competitive ELISA method in which a commercial assay kit was used along with standardized and stabilized reagents.

The sample extraction procedure was modified to allow the use of an aqueous buffer. In most of the soya product examinations, the percentage of soya protein measured by the immunoassay and compared with the total protein content was between 80% and 100%. Ravestein and Driedonks (1986) developed an experimental procedure for the denaturation and solubilization of soya protein. They isolated soya from glycinin subunits (11S protein), boiled it in sodium dodecyl sulphate (SDS), added buffer and used it for immunization of rabbits to produce anti-soya antisera. The meat products were extracted with buffer containing SDS and mercaptoethanol for 10 min. The supernatants collected after centrifugation and dilution were used for the ELISA test. The quantitative analysis was done with a detection limit of 0.5% (m/m). This method can be applied to sterilized products (120°C, 30 min). LC-MS/MS (liquid chromatography–mass spectrometry/mass spectrometry) can be used to detect bovine (Grundy et al., 2007) and porcine (Grundy et al., 2008) fibrinopeptides A and B in concentrations down to 5%. As the fibrinopeptides A and B are species-specific, the methods are capable of discriminating between the blood-based binding agents of bovine and porcine origins. The presence of proteins and phosphates in meat products impart increased water binding capacities. In this case, detection of foreign proteins or phosphates might provide proof of fraudulent practice.

16.5.2 Non-meat Proteins Other Than Soya

16.5.2.1 Electrophoresis

Electrophoresis coupled with immunostaining is applicable for cooked products. PAGE has been used by many researchers to detect non-meat proteins in meat products. However, problems can be experienced when examining heated products, and the effectiveness of a procedure is largely dependent upon the type of meat product being examined and the non-meat protein present. Dysseler et al. (1986) explained a PAGE procedure for the identification and detection of wheat gluten in cooked foods. Further, Kim et al. (1987) and Woychick et al. (1987) used an SDS-PAGE method to detect soya and milk proteins in chicken products and to detect soya proteins in frankfurters. Janssen et al. (1986) used SDS electrophoresis coupled with nitrocellulose blotting and an immunoperoxidase staining system to detect soya proteins in heated meat products. The detection of wheat gluten, whey protein, casein, ovalbumin, and soya proteins have been done in heated meats using a similar procedure by Janssen et al. (1987).

16.5.2.2 Immunoassay

Soya-derived products are the most common sources of non-meat proteins added to meat products and there are a few methods available for the detection and determination of other non-meat proteins. The immunoassay method was used to detect and determine wheat as undeclared meat extenders. Turin and Bonomi (1994) published a research report on the detection of egg white proteins (ovalbumin) in dairy products and the approach may be applicable to meat products. The detection was performed by blotting onto nitrocellulose and by immunochemical staining with commercially available rabbit anti-ovalbumin serum. Immunoassay kits for the detection and semiquantitative determination of bovine casein in foodstuffs are commercially available and can be applied to meat products. A selection of commercially available antisera exists for recognizing non-meat proteins, such as soya, wheat gluten, bovine whey, bovine casein, and chicken ovalbumin. Immunoassays are commercially available for the detection and determination of gluten and casein, and some antisera are available for recognizing non-meat proteins.

16.6 MECHANICALLY RECOVERED MEAT

Traditional manual butchery procedures produce hand-deboned meat (HDM), but this practice results in residual meat remaining on the bones of a carcass. The mechanically recovered meat (MRM) is the type of meat that machines can remove from animal bones and poultry carcasses. The UK Food Standards Committee (FSC/REP/72,1980) declares that MRM will not be considered as meat if present in an amount more than 5% of the declared lean meat content of any meat product. Since MRM is disrupted muscle and is less expensive than whole muscle meat, it may be prepared from whole muscles or pieces of meats without an appropriate label declaration. Meech and Kirk (1986) discussed the chemical compositions of MRMs and assessed the data to check the compositional indices that could be used to indicate the presence of MRM in a product. The various parameters for detecting MRM are moisture, fat, protein, ash, bone, fatty acids, sterols, pH, hydroxyproline, and trace elements. Crosland et al. (1995) checked different types of meats like beef, lamb, pork, chicken, turkey MRMs and also HDMs to detect the presence of purine, nitrogen, fat, collagen, moisture, ash, calcium, and iron. Several differences between MRM and HDM were found. Another research group, Savage et al. (1995), used gel electrophoresis to examine MRMs from different machines and animal species. High pressures were applied to the bones for mechanical recovery and for releasing the bone fluids into the MRM, which could then be used as markers to differentiate MRM from HDM. A comparison of myoglobin, hemoglobin, and other protein bands was done in raw beef, raw lamb, raw pork HDMs and MRMs by using electrophoretograms. The identification of MRM and its percentage incorporation in minced meat samples were worked out. Pickering et al. (1995a; 1995b) differentiated MRM and HDM using the application of light microscopy. Samples from different species and MRMs produced by different machines and processing conditions were evaluated by Savage et al. (1995). The test sample was cut into cryosections, stained with toluidine blue, and observed by light microscopy to detect the presence of hyaline cartilage. All the MRM samples showed the presence of hyaline cartilage but not the HDM. The presence of hyaline in a sample could be used as an indicator of the presence of MRM. The detection level given for beef MRM is 10% (m/m) and for chicken it is 25% (m/m), but the detection was difficult for pork MRM and HDM. Pickering et al. (1995b) explained a procedure to detect chicken bone marrow antigens using an immunological approach. Chicken MRM is the most commonly used MRM in meat products in the UK.

REFERENCES

Aarts, H.J., Bouwel, M., Bumjer, J.B., Lenstra, J.A., Van Raamsdonk, L.W. 2006. Detection of bovine meat and bone meal in animal feed at a level of 0.1%. *J AOAC Int* 89(6):1443–1446.

Accum, F. 1820. *A Treatise on Adulterations of Food and Culinary Poisons, and Methods of Detecting Them*. Evergreen Review, Inc., New York.

Al Jowder, O., Defernez, M., Kemsley, E.K., Wilson, R.H. 1999. Mid-Infrared spectroscopy and chemometrics for the authentication of meat products. *J Agric Food Chem* 47(8):3210–3218.

Angood, K.M., Wood, J.D., Nute, G.R., et al. 2008. A comparison of organic and conventionally-produced lamb purchased from three major UK supermarkets: Price, eating quality and fatty acid composition. *Meat Sci* 78(3):176–184.

Ashoor, S.H., Monte, W.C., Stiles, P.G. 1998. Liquid chromatographic identification of meats. *J Assoc Off Anal Chem* 71(2):397–403.

Ashoor, S.H., Osman, M.A. 1988. Liquid chromatographic quantitation of chicken and turkey in unheated chicken–turkey mixtures. *J Assoc Off Anal Chem* 71(2):403–405.

Ayob, M.K., Ragab, A.A., Allen, J.E. 1989. An improved rapid ELISA technique for detection of pork in meat products. *J Sci Food Agric* 49(1):103–116.

Ballin, N.Z. 2010. Authentication of meat and meat products. *Meat Sci* 86(3):577–587.

Bartlett, S.E., Davidson, W.S. 1992. FINS (Forensically Informative Nucleotide Sequencing): A procedure for identifying the animal origin of biological specimens. *Bio Tech* 17(1):408–411.

Bauer, F., Hofmann, K. 1986. Increased sensitivity by peroxidase staining of myoglobins. *Meat Ind* 67:861–867.

Bauer, V.C., Teifel-Greding, J., Liebhardt, E. 1987. Species identification of heat denaturized meat samples by DNA analysis. *Arch Food Hyg* 38:149–176.

Bayliss-High, O. 1984. *Royal Microscopical Society Microscopy Handbooks: Lipid Histochemistry*, Oxford University Press, Oxford.

Behrens, M., Unthan, M., Bronkmann, Y., Buchholz, R., Lotus, N. 1999. Identification of animal species in heated and complex meat products using specific PCR reactions. *Fleschwirtschaft* 79(5):97–100.

Berger, R.G., Mageau, R.P., Schwab, B. 1988. Detection of poultry and pork in cooked and canned meat foods by enzyme-linked immunosorbent assays. *J Assoc Off Anal Chem* 71(2):406–409.

Casale, M., Oliveri, P., Armanino, C. 2010. NIR and UV-vis spectroscopy, artificial nose and tongue: Comparison of four fingerprinting techniques for the characterization of Italian red wines. *Anal Chim Acta* 668(2):143–148.

Chikuni, K., Ozutsumi, K., Koishikawa, T., Kato, S. 1990. Species identification of cooked meats by DNA hybridisation assay. *Meat Sci* 27(2):119–128.

Chikuni, K., Tabata, T., Kosugiyama, M., Monma, M., Saito, M. 1994. Polymerase chain reaction assay for detection of sheep and goat meats. *Meat Sci* 37(3):337–345.

Chou, C.C., Lin, S.P., Lee, K.M., et al. 2007. Fast differentiation of meats from fifteen animal species by liquid chromatography with electrochemical detection using copper nanoparticle plated electrodes. *J Chromatogr B Analyt Technol Biomed Life Sci* 846(1–2):230–239.

Commission Regulation (EEC). 1989. Commission Regulation (EEC) No 226/89. *Official Journal of the European Communities 3111189*, NOL 2YIII-12.

Cordain, L., Eaton, S.B., Sebastian, A., et al. 2005. Origins and evolution of the Western diet: health implications for the 21st century. *Am J Clin Nutr* 81(2):341–354.

Cottam, J. 1975. A note on the calculation of lean meat content of sausages. *J Assoc Public Analysts* 13:133–134.

Crimes, A.A., Hitchcock, C.H.S., Wood, R. 1984. Determination of soya protein in meat products by an enzyme-linked immunosorbent assay procedure: Collaborative study. *J Assoc Public Analysts* 22:59–78.

Crosland, A.R., Patterson, R.L., Higman, R.C., Stewart, C.A., Hargin, K.D. 1995. Investigation of methods to detect mechanically recovered meat in meat products—I: Chemical composition. *Meat Sci* 40(3):289–302.

Crutrufelli, M.E., Mageau, R.P., Schwab, B., et al. 1988. Development of porcine rapid identification method (PRIME) by modified agar-gel immunodiffusion. *J Assoc Off Anal Chem* 71(2):444–445.

Cunningham, E.P., Meghen, C.M. 2001. Biological identification systems: Genetic markers. *Rev Sci Tech* 20(2):491–499.

Cutrufelli, M.E., Mageau, R.P., Schwab, B., Johnston, R.W. 1986. Development of poultry rapid overnight field identification test (PROFIT). *J Assoc Off Anal Chem* 69(3):483–487.

Cutrufelli, M.E., Mageau, R.P., Schwab, B., Johnston, R.W. 1987. Detection of beef and poultry by serological field screening tests (ORBIT and PROFIT): Collaborative study. *J Assoc Off Anal Chem* 70(2):230–233.

Cutrufelli, M.E., Mageau, R.P., Schwab, B., Johnston, R.W. 1989. Development of serological ovine field test (SOFT) by modified agar gel immunodiffusion. *J Assoc Off Anal Chem* 72(1):60–61.

Dalmasso, A., Fontanella, E., Piatti, P., et al. 2004. A multiplex PCR assay for the identification of animal species in feedstuffs. *Mol Cell Probes* 18(2):81–87.

Dooley, J.J., Paine, K.E., Garrett, S.D., Brown, H.M. 2004. Detection of meat species using Taq Man real-time PCR assays. *Meat Sci* 68(3):431–438.

Dysseler, P., Jacqmain, D. 1986. Electrophoretic determination of gliadins in cooked foods. *Acta Gastroenterol Belg* 49(4):409–414.

Ebbehoj, K.F., Thomsen, P.D. 1991a. Species differentiation of heated meat products by DNA hybridisation. *Meat Sci* 30(3):221–234.

Ebbehoj, K.F., Thomsen, P.D. 1991b. Differentiation of closely related species by DNA hybridisation. *Meat Sci* 30(4):359–366.

Ellis, D.I., Brewster, V.L., Dunn, W.B., Allwood, W., et al. 2012. Fingerprinting food: current technologies for the detection of food adulteration and contamination. *Chem Soc Rev* 41:5706–5727.

Fajardo, V., González, I., López-Calleja, I., et al. 2007. PCR identification of meats from chamois (Rupicaprarupicapra), pyrenean ibex (Capra pyrenaica), and mouflon(Ovisammon) targeting specific sequences from the mitochondrial D loop region. *Meat Sci* 76(4):644–652.

Fajardo, V., Gonzalez, I., Martın, I., Rojas, M., Hernandez, P.E., Garcıa, T. 2008. Real-time PCR for quantitative detection of chamois (Rupicapra rupicapra) and pyreneanibex (Caprapyrenaica) in meat mixtures. *J Assoc Off Anal Chem*, 91(1): 103–111.

Fei, S., Okayama, T., Yamanoue, M., et al. 1996. Species identification of meat and meat products by PCR. *Anim Sci Tech (Jpn)* 67:900–905.

Fergusson, C.S., Racicot, P.A., Rane, L. 1942. Study of use of precipitin test for determination of soy flour in sausages. *J Assoc Off Agric Chem* 25(4):533–537.

Flint, F.O., Firth, B.M. 1981. A toluidine blue stain mountant for the microscopy of comminuted meat products. *Analyst* 106:1242–1243.

Flint, F.O., Meech, M.V. 1978. Quantitative determination of texturised soya protein by a stereo logical technique. *Analyst* 103:252–258.

Flint, F.O., Pickering, K. 1984. Demonstration of collagen in meat products by an improved picro-sirius red polarisation method. *Analyst* 109(11):1505–1506.

Forrest, A.R., Carnegie, P.R. 1994. Identification of gourmet meat using FINS (forensically informative nucleotide sequencing). *Biotechniques* 17(1):24–26.

Franke, B., Haldimann, M., Gremaud, G., et al. 2008. Element signature analysis: its validation as a tool for geographic authentication of the origin of dried beef and poultry meat. *Eur Food Res Tech* 227(3):701–708.

Franke, B., Koslitz, S., Micaux, F., et al. 2008. Tracing the geographic origin of poultry meat and dried beef with oxygen and strontium isotope ratios. *Eur Food Res Tech* 226(4):761–769.

Frezza, D., Giambra, V., Chegdani, F., et al. 2008. Standard and Light-Cycler PCR methods for animal DNA species detection in animal feedstuffs. *Innov Food Sci Emerg Technol* 9(1):18–23.

Ghovvati, S., Nassiri, M.R., Mirhoseini, S.Z., Moussavi, A.H., Javadmanesh, A. 2009. A fraud identification in industrial meat products by multiplex PCR assay. *Food Control* 20(8):696–699.

Glynn, J.H. 1939. Quantitative determination of soy bean protein in sausage and other protein mixtures. *Science* 89(2315):444.

Griffiths, N.M., Billington, M.J. 1984. Evaluation of an enzyme-linked immunosorbent assay for beef blood serum to determine indirectly the apparent beef content of beef joints and model mixtures. *J Sci Food Agric* 35(8):909–914.

Griffiths, N.M., Billington, M.J., Crimes, A.A. 1984. An assessment of commercially available reagents for an enzyme-linked immunosorbent assay of soya protein in meat products. *J Sci Food Agric* 35(11):1255–1260.

Grundy, H.H., Reece, P., Sykes, M.D., et al. 2007. Screening method for the addition of bovine blood-based binding agents to food using liquid chromatography triple quadrupole mass spectrometry. *Rapid Commun Mass Spectrom* 21(18):2919–2925.

Grundy, H.H., Reece, P., Sykes, et al. 2008. Method to screen for the addition of porcine blood-based binding products to foods using liquid chromatography/triple quadrupole mass spectrometry. *Rapid Commun Mass Spectrom* 22(12):2006–2008.

Guha, S., Kashyap, V.K. 2005. Development of novel heminested PCR assays on mitochondrial 16S rRNA gene for identification of seven pecora species. *BMC Genet* 6:42.

Hargin, K.D. 1996. Authenticity issues in meat and meat products. *Meat Sci* 43(1):277–289.

Hayden, A.R. 1978. Determination of residual species serum albumin in adulterated ground beef. *J Food Sci* 43(2): 476–478, 492.

Hayden, A.R. 1979. Immunochemical detection of ovine, porcine and equine flesh in beef products with antisera to species myoglobin. *J Food Sci* 44(2):494–500.

Hayden, A.R. 1981. Use of antisera to heat-stable antigens of adrenals for species identification in thoroughly cooked beef sausages. *J Food Sci* 46(6):1810–1813.

Heinert, H.H., Klinger, A. 1980. Animal-specific protein excretion: protein and enzyme pattern Dream and deer. *Fleischwirtsch* 60:1682–1685.

Hitchcock, C.H.S., Bailey, F.J., Crimes, A.A., Dean, D.A.G., Davis, P.J. 1981. Determination of soya proteins in food using an enzyme-linked immunosorbent assay procedure. *J Sci Food Agric* 32(2):57–65.

Hofmann, K., Bluchel, E. 1986. Determining the species of raw muscle meat on the basis of myoglobin pattern in the pH gradient gel. *Meat Econ* 66(1):916–921.

Hunt, D.J., Parkes, H.C., Lumley, I.D. 1997. Identification of the species of origin of raw and cooked meat products using oligonucleotide probes. *Food Chem* 60:437–442.

Husak, R.L., Sebranek, J.G., Bregendahl, K. 2008. A survey of commercially available broilers marketed as organic, free-range, and conventional broilers for cooked meat yields, meat composition, and relative value. *Poult Sci* 87(11):2367–2376.

Janssen, F.W., Hogele, G.H., Voorpostel, A.M.B., de Baaij, J.A. 1990. Myoglobin analysis for determination of beef, pork, horse, sheep and kangaroo meat in blended cooked products. *J Food Sci* 55(6):1528–1530.

Janssen, F.W., Voortman, G., Baaij, J.A. 1987. Detection of wheat gluten, whey protein, casein, ovalbumin and soy protein in heated meat products by electrophoresis, blotting and immunoperoxidase staining. *J Agricol Food Chem* 35, 563–567.

Janssen, F.W., Voortman, G., de Baaij, J.A. 1986. Detection of soya proteins in heated meat products by blotting and dot-blot. *Z Lebensm Unters Forsch* 182(6):479–483.

Jones, S.J., Patterson, R.L.S. 1985. Double-antibody ELISA for detection of trace amounts of pig meat in raw meat mixtures. *Meat Sci* 15(1)1–13.

Jones, S.R., Patterson, R.L.S. 1986. A modified indirect ELISA procedure for raw meat speciation using crude anti-species antisera and stabilized immunoreagents. *J Sci Food Agric* 37(8):767–775.

Kaiser, K.P., Matheis, G., Knita-Durrmann, C., Belitz, H.D. 1980. Qualitative and quantitative analysis of raw binary meat mixtures by means of isoelectric focussing. *Z Letensm Unters Forch* 171:415–419.

Kangethe, E.K., Gathuma, J.M., Lindqvist, K.J. 1986. Identification of the species of origin of fresh, cooked and canned meat and meat products using antisera to thermostable muscle antigens by Ouchterlony's double diffusion test. *J Sci Food Agric* 37, 157–164.

Kangethe, E.K., Jones. S.J., Patterson, R.L.S. 1982. Identification of the species of origin of fresh meat using an enzyme-linked immunosorbent assay procedure. *Meat Sci* 7(3):229–240.

Kauffer, K., Beneke, B., Bentler, W. 1990. Detection of comminuted chicken meat in native and heated meat mixtures by means of isoelectric focussing in polyamide gels. *Fleischwirtsch* 70(10):1188–1194.

Kelly, M., Tarbin, J.A., Ashwin, H., Sharman, M. 2006. Verification of compliance with organic meat production standards by detection of permitted and nonpermitted uses of veterinary medicines (tetracycline antibiotics). *J Agric Food Chem* 54(4):1523–1529.

Kesmen, Z., Gulluce, A., Sahin, F., Yetin, H. 2009. Identification of meat species by Taq Man based real-time PCR assay. *Meat Sci* 82(4):444–449.

Kim, H., Shelef, L.A. 1986. Characterisation and identification of raw beef, pork, chicken and turkey meats by electrophoretic patterns of their sarcoplasmic proteins. *Food Sci* 51, 731–41.

Kim, S.H., Huang, T.S., Seymour, T.A., et al. 2005. Development of immunoassay for detection of meat and bone meal in animal feed. *J Food Prot* 68(9):1860–1865.

Kim, Y.H., Park, G.B., Lee, M.H., et al. 1987. Studies on the determination of non-meat proteins in chicken products. *Korean J Animal Sci* 29(9):416–419.

Kim, D.H., Seong, P.N., Cho, S.H., et al. 2009. Fatty acid composition and meat quality traits of organically reared Korean native black pigs. *Livestock Sci* 120(1–2):96–102.

King, N.L. 1984. Species identification of cooked meats by enzyme staining of isoelectrofocusing gels. *Meat Sci* 11(1):59–72.

King, N.L., Kurth, L. 1982. Analysis of raw beef samples for adulterant meat species by enzyme-staining of isoelectricfocussing gels. *J Food Sci* 47(5):1608–1612.

Kirk, R.S., Sawyer, R. 1991. *Pearson's Composition and Analysis of Foods*, (9th edn). Longman, Harlow, UK.

Koppelman, S.J., Lakemond, C.M., Vlooswijk, R., Hefle, S.L. 2004. Detection of soy proteins in processed foods: Literature overview and new experimental work. *J Assoc Off Anal Chem* 87(6):1398–1407.

Kumar, A., Kumar, R.R., Sharma, B.D., et al. 2015. Identification of species origin of meat and meat products on the DNA basis: A review. *Crit Rev Food Sci Nutr* 55(10):1340–1351.

Laube, I., Zagon, J., Broll, H. 2007a. Quantitative determination of commercially relevant species in foods by real-time PCR. *Int J Food Sci Technol* 42(3):336–341.

Laube, I., Zagon, J., Spiegelberg, A., et al. 2007b. Development and design of a 'ready-to-use' reaction plate for a PCR-based simultaneous detection of animal species used in foods. *Int J Food Sci Technol* 42(1):9–17.

Lopez-Andreo, M., Lugo, L., Garrido-Pertierra, A., Prieto, M.I., Puyet, A. 2005. Identification and quantification of species in complex DNA mixtures by real-time polymerase chain reaction. *Anal Biochem* 339(1): 73–82.

Lord, D.W., Swan, K.J. 1984, Connective tissue in Beef burgers. *J. Assoc. Publ. Analysts* 22:131–140.

Lumley I.D. 1996. Authenticity of meat and meat products. In: Ashurst P.R., Dennis M.J. (eds) *Food Authentication*. Springer, Boston, MA. pp. 108–139.

Mageau, R.P., Cutrufelli, M.E., Schwab, B., Johnston, R.W. 1984. Development of an overnight rapid bovine identification test (ORBIT) for field use. *J Assoc Off Anal Chem* 67(5):949–954.

Martin, I., Garcia, T., Fajardo, V., Lopez-Calleja, I., Rojas, M., Hernandez, Y.E., Gonzalez, I., Martin, R. 2007. Mitochondrial markers for the detection of four duck species and the specific identification of Muscovy duck in meat mixtures using the polymerase chain reaction. *Meat Sci* 76(4): 721–729.

Martin, I., García, T., Fajardo, V., et al. 2009. SYBR-Green real-time PCR approach for the detection and quantification of pig DNA in feedstuffs. *Meat Sci* 82(2):252–259.

Martin, R., Azcona, J. I., Casas, L. 1988a. Sandwich ELISA for detection of pig meat in raw beef using antisera to muscle soluble proteins. *J Food Prot* 51(10):790–794.

Martin, R., Azcona, J.I., Tormo, J., Hernandez, P.E., Sanz, B. 1988b. Detection of chicken meat in raw meat mixtures by a sandwich enzyme immunoassay. *Int J Food Sci Technol* 23(3):303–310.

Martín, R., Wardale, R.J., Jones, S.J., Hernández, P.E., Patterson, R.L. 1991. Monoclonal antibody sandwich ELISA for the potential detection of chicken meat in mixtures of raw beef and pork. *Meat Sci* 30(1):23–31.

Matsunaga, T., Chikuni, K., Tanabe, R., et al. 1999a. A quick and simple method for the identification of meat species and meat products by PCR assay. *Meat Sci* 51(2):143–148.

Matsunaga, T., Shibata, K., Yamada, J., Shinmura, Y. 1999b. Effects of processing condition on species identification of meat products. *J Jap Soc Food Sci Tech* 46(3):187–194.

Matsunaga, T., Shibata, K., Yamada, J., Shinmura, Y., Chikuni, K. 1998. Identification of meat species based on the difference of 18S ribosomal RNA-genes. *J Jap Soc Food Sci Tech* 45(12):719–723.

McCormick, R.J., Collins, D.A., Field, R.A., Moore, T.D. 1992. Identification of meat from game and domestic species. *J Food Sci* 57(2):516–517.

The Meat Products and Spreadable Fish Products Regulations (1984). Statutory Instrument Food Composition and Labelling: No. 1566, pp. 4687–4707.

Meech, M.V., Kirk, R.S. 1986. Chemical characterisation of mechanically recovered meats. *J Assoc Public Analysts* 24:13–26.

Meyer, R., Candrian, U., Luthyl, J. 1993. Species differentiation and detection of soya in heated meat products by the polymerase chain reaction (PCR). *Mitt Gebiete Lebensm Hyg* 84:112–121.

Muldoon, M.T., Onisk, D.V., Brown, M.C., Stave, J.W. 2004. Targets and methods for the detection of processed animal proteins in animal feedstuffs. *Int J Food Sci Tech* 39(8):851–861.

Mullis, K., Faloona, F., Scharf, S., et al. 1986. Specific enzymatic amplification of DNA in vitro: The polymerase chain reaction. *Cold Spring Harb Symp Quant Biol* 51(1):63–73.

Nakyinsige, K., Man, Y.B., Sazili, A.Q. 2012. Halal authenticity issues in meat and meat products. *Meat Sci* 91(3):207–214.

Pascoal, A., Prado, M., Calo, P., Cepeda, A., Barros-Velazquez, J. 2005. Detection of bovine DNA in raw and heat-processed foodstuffs, commercial foods and specific risk materials by a novel specific polymerase chain reaction method. *Eur Food Res Tech* 220(3–4):444–450.

Patterson, R.L.S., Jones, S.J. 1989. Species identification in heat processed meat products. *Proceedings of 35th International (Copenhagen) Congress on Meat Sci Technol* 2:529–536.

Patterson, R.M., Spencer, T.L. 1983. A rapid on-site test for speciation of meat. *Aust Vet J* 60(12):381–382.

Patterson, R.M., Spencer, T.L. 1985. Differentiation of raw meat from phylogenically related species by enzyme-linked immunosorbent assay. *Meat Sci* 15(3):119–123.

Patterson, R.M., Whittaker, R.G., Spencer T.L. 1984. Improved species identification of raw meat by double sandwich enzyme-linked immunosorbent assay. *J Sci Food Agric* 35(9):1018–1023.

Pearson, D. 1968. The calculation of the lean meat content of pork sausages. *J Assoc Public Analysts* 6:129–134.

Pearson, D. 1970. A general formula for the calculation of the lean meat content. *J Assoc Public Analysts* 8:22–24.

Pickering, K., Evans, C.L., Hargin, K.D., Stewart C.A. 1995a. Investigation of methods to detect mechanically recovered meat in meat products-III: Microscopy. *Meat Sci* 40(3):319–326.

Pickering, K., Griffin, M., Smethurst, P., Hargin, K.D., Stewart, C.A. 1995b. Investigation of methods to detect mechanically recovered meat in meat products-IV: Immunology. *Meat Sci* 40(3):327–336.

Prasad, V., Mishra, D. 1981. Differentiation of meats of different species of animals by muscle esterase pattern in different age and sex groups. *Indian J Animal Sci* 51:211–214.

Ravestein, P., Driedonks, R.A. 1986. Quantitative immunoassay for soya protein in raw and sterilised meat products. *J Food Technol* 21(1):19–32.

Rezzi, S., Giani, I., Héberger, K., et al. 2007. Classification of gilthead sea bream (Sparusaurata) from 1H NMR lipid profiling combined with principal component and linear discriminant analysis. *J Agric Food Chem* 55(24):9963–9968.

Richardson, G.A., Tarassuk, N.P. 1934. Detection of gelatin in cultured buttermilk and cottage cheese. *J Assoc Off Anal Chern* 17:314–319.

Rittenburg, J.H., Adams, A., Palmer, J., Allen, J.C. 1987. Improved enzyme-linked immunosorbent assay for determination of soy protein in meat products. *J Assoc Off Anal Chem* 70(3):582–587.

Rodriguez, M.A., Garcia, T., Gonzalez, I., et al. 2003. Identification of goose, mule duck, chicken, turkey, and swine in foiegras by species-specific polymerase chain reaction. *J Agric Food Chem* 51(6):1524–1529.

Rodriguez, M.A., Garcia, T., Gonzalez, I., et al. 2004. PCR identification of beef, sheep, goat and pork in raw and heat-treated meat mixtures. *J Food Prot* 67(1):172–177.

Saeed, T., Abu-Dagga, F., Rahman, H.A. 1986. Detection of pork and lard as adulterants in beef and mutton mixtures. *J Assoc Off Anal Chem* 69(6):999–1002.

Santonico, M., Pittia, P., Pennazza, G., et al. 2008. Study of the aroma of artificially flavoured custards by chemical sensor array fingerprinting. *Sens Actuators B Chem* 133(1):345–351.

Sasazaki, S., Itoh, K., Arimitsu, S., et al. 2004. Development of breed identification markers derived from AFLP in beef cattle. *Meat Sci* 68(2):275–280.

Savage, A.W., Ian Richardson, R., Jolley, P.D., Hargin, K.D., Stewart, C.A. 1995. Investigation of methods to detect mechanically recovered meat in meat products-II: Gel electrophoresis. *Meat Sci* 40(3):303–317.

Sawaya, W.N., Saeed, T., Mameesh, M., et al. 1990a. Detection of pork in processed meat: experimental comparison of methodology. *Food Chem* 37(3):201–219.

Sawaya, W.N., Mameesh, M.S., El-Rayes, E., Usain, A., Dashti, B. 1990b. Detection of pork in processed meat by an enzyme-linked immunosorbent assay using antiswine antisera. *J Food Sci* 55(2):293–297.

Schweiger, A., Baudner, S., Gunther, H.O. 1983. Isolation by free-flow electrophoresis and immunological detection of troponin T from turkey muscle: An application in food chemistry. *Electrophoresis* 4(2):158–163.

Sherikar, A.T., Karkare, U.D., Khot, J.B., Jayarao, B.M., Bhilegaonkar, K.N. 1993. Studies on thermostable antigens, production of species-specific antiadrenal sera and comparison of immunological techniques in meat speciation. *Meat Sci* 33(1):121–136.

Sinclair, A.J., Slattery, W.J. 1982. Identification of meat according to species by isoelectric focussing. *Aust Vet J* 58(2):79–80.

Singh, V.P., Pathak, V., Nayak, N.K., Verma, A.K., Umaraw, P. 2014. Recent developments in meat species speciation - a review. *J Livestock Sci* 5:49–64.

Skarpeid, H.J., Kvaal, K., Hildrum, K.I. 1998. Identification of animal species in ground meat mixtures by multivariate analysis of isoelectric focusing protein profiles. *Electrophoresis* 19(18):3103–3109.

Skrokki, A., Hormi, O. 1994b. Composition of minced meat: survey of commercial ground meat. *Meat Sci* 38(3):503–509.

Skrokki, A., Hormi, O. 1994b. Composition of minced meats: Part A. Methods. *Meat Sci* 38(3):497–501.

Slattery, W.J., Sinclair, A.J. 1983. Differentiation of meat according to species by electrophoretic separation of muscle lactate dehydrogenase, esterase isoenzymes and isoelectric focusing of soluble muscle proteins. *Aust Vet J* 60(2):47–51.

Stubbs, G., More, A. 1919. The estimation of the approximate quantity of meat in sausages and meat pastes. *Analyst* 44:125–127.

Swart, K.S., Wilks, C.R. 1982. An immunodiffusion method for the identification of the species of origin of meat samples. *Aust Vet J* 59(1):21–22.

Toorop, R.M., Murch, S.J., Ball, R.O. 1997. Development of rapid and accurate method for separation and quantification of myofibrillar proteins in meat. *Food Res Int* 30(8):619–627.

Turin, L., Bonomi, F. 1994. Immunochemical detection of ovalbumin in dairy-based foods. *J Sci Food Agric* 64(1):39–45.

Valeria, T., Caterina, M., Antonio, G., et al. 2005. DNA-based methods for identification and quantification of small grain cereal mixtures and fingerprinting of varieties. *J Cereal Sci* 41(3):213–220.

Verma, R., Saluja, B., Singh, R.R. 2013. Differentiation of Adulterated Meat Products through Molecular Technique: PCR-RFLP. *Octa J Biosci* 1(1):17–23.

Walker, J.A., Hughes, D.A., Hedges, D.J., et al. 2004. Quantitative PCR for DNA identification based on genome-specific interspersed repetitive elements. *Genomics* 83(3):518–527.

Whittaker, R.G., Spencer, T.L., Copland, J.W. 1982. Enzyme-linked immunosorbent assay for meat species testing. *Aust Vet J* 59(4):125.

Whittaker, R.G., Spencer, T.L., Copland, J.W. 1983. An enzyme-linked immunosorbent assay for species identification of raw meat. *J Sci Food Agric* 34(10):1143–1148.

Wintero, A.K., Thomson, P.D., Davies, W. 1990. A comparison of DNA-hybrid is at ion, immunodiffusion, countercurrent immunoelectrophoresis and isoelectric focusing for detecting the admixture of pork and beef. *Meat Sci* 27(1):75–85.

Woychick, J.H., Happich, M., Trinh, H., Scilcrs, R. 1987. Quantification of soya protein in frankfurters by gel electrophoresis. *J Food Sci* 52(6):1532–1534.

Yin, R.H., Bai, W.L., Wang, J.M., et al. 2009. Development of an assay for rapid identification of meat from yak and cattle using polymerase chain reaction technique. *Meat Sci* 83(1):38–44.

Zarkadas, C.G., Gagnon, C., Poysa, V., et al. 2007. Protein quality and identification of the storage protein subunits of tofu and null soybean genotypes, using amino acid analysis, one and two-dimensional gel electrophoresis, and tandem mass spectrometry. *Food Res Int* 40(1):111–128.

CHAPTER 17

Application of Fingerprinting Techniques in Authentication and Traceability of Fruits and Vegetables

Sadaf Jamal Gilani and Ammeduzzafar
Al-Jouf University

Syed Sarim Imam
Glocal University

Javed Ahmad
Najran University

CONTENTS

17.1 INTRODUCTION

17.1.1 Food Authenticity

Traditional analytical methods for food authentication are very time-consuming processes. There are a number of disadvantages such as slow speed of detection, the necessity for sample pretreatments, the requirement of highly skilled personnel, and sample destruction. Nowadays, many new fast and nondestructive instrumental methods have been introduced to overcome these disadvantages. Among them, UV-visible, infrared spectroscopic, and chromatographic techniques have proven their applications to be successful analytical methods for food analysis (Wang et al., 2006; Woodcock et al., 2008). These methods are reported to be fast, accurate, nondestructive, and less time consuming and they also require minimal or no sample preparation. These methods do not require expensive reagents and thus no wastes are produced. They give a characteristic profile reflecting the chemical compositions for the evaluation and quality control of the analyzed sample (Tistaert et al., 2011).

Fingerprinting techniques are described as a combination of analytical techniques which can inform about the compositions of some food items in a nonselective way, with the main aim of characterizing or authenticating the food (Cuadros-Rodriguez et al., 2016). Fingerprinting is the combined use of a variety of analytical techniques or methods. The term "profiling" comes from the metabolomic terminology. Horning and Horning, in the year 1971, first proposed the concept of "metabolite profiling" in relation to the determination of metabolites using mass spectrometry (MS) (Horning et al., 1971; Wolfendera et al., 2015). At the end of the 20th century, Fiehn introduced the term "metabolomic fingerprinting" from the idea of infrared fingerprint (Fiehn et al., 2001). The counterparts of all the low molecular weight metabolites in a biological system (typically < 1500 Da) are known as the metabolomes. The metabolome or its fraction can be quantitatively analyzed by the use of advanced analytical techniques. The analysis also involves multiple steps like sampling, sample preparation, chemical analysis, and data analysis. The analysis of a given set of known metabolites, for instance, a set of amino and organic acids is known as "metabolite profiling", whereas the "metabolic fingerprinting" is an unspecific analysis of a test material, e.g., a set of mass peaks obtained with mass spectrometry. The "metabolomics" provide direct functional information and the data can be integrated into metabolic models, whereas the "metabolomics fingerprinting" provides a pattern that is used to classify samples based on the origin of either their biological relevance or origin. The main focus of current metabolomics is on the field of pharmacy and toxicology, but the use of these techniques is becoming more and more important in the analysis of food. The concepts of food fingerprinting and food profiling were made in accordance with the corresponding definitions in metabolomics (Koek

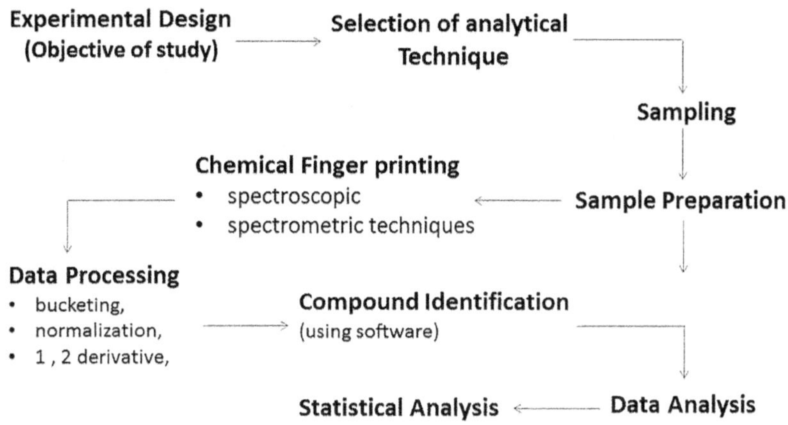

FIGURE 17.1 Schematic representation of steps involved in different analytical techniques of food fingerprinting.

et al., 2011). The food fingerprinting techniques are not applied for the identification of all the metabolites but for the recognition of fragments patterns; thus they are called matrix fingerprints (Antignac et al., 2011).

Detection of adulterations provides a great challenge to the analytical chemists because of increasing product diversity and the continuous development of new production technologies. In general, the fingerprinting technique gives a chemical picture by the direct analysis of crude sample extracts, typically by pyrolysis mass spectrometry (Py-MS) (Dettmer et al., 2007), direct infusion mass spectrometry (DIMS) (Kirwan et al., 2015), ambient mass spectrometry (Ambient MS) (also designated as direct ionization mass spectrometry (Klampfl and Himmelsbach, 2015)), or nuclear magnetic resonance spectrometry (NMR) (Krishnan et al., 2015). Both techniques, MS and NMR, generate metabolite profiles in which the signals are assigned to metabolite fingerprints. The analysis of the metabolome mainly depends on the distribution of intensities in the corresponding spectra rather than on the assignment of the signals. Prior to the detection of compounds by different spectrometric techniques such as fluorescence spectrometry, MS, UV spectrometry, or NMR, the separation of the compounds must be done by gas chromatography (GC), liquid chromatography (LC), or by capillary electrophoresis (CE) (Cavaliere et al., 2011; Wolfendera et al., 2015). The fingerprinting approaches are typically based on a high-throughput screening of samples with the purpose of differentiation of samples. After identification of the sample patterns in food matrices, the aim is to differentiate among the various food fingerprints in terms of, e.g., their botanical or geographical origins to detect possible adulterations. Analytical techniques that can be used for the investigations are depicted in Figure. 17.1.

17.1.2 Strengths and Limitations of the Fingerprints

The different fingerprinting techniques have been classified as per the physical and chemical principles they follow. Each technique has its own intrinsic strengths and limitations as depicted in Tables 17.1 and 17.2.

TABLE 17.1 Principles, Advantages, and Disadvantages of Different Analytical Techniques Used in Fingerprinting of Food Materials

					Spectroscopy Technique				Separation Technology	
PE	IE	RFLP	RAPD	AFLP	NMR	UV	MS	IR	GC	LC
Working Principle										
Differences in protein structure and electric charge	Differences in enzyme structure and electric charge	Differences at DNA restriction enzyme cutting sites	Differences in DNA sequence	Differences in DNA sequence	Differences in element com-position	Differences of molecular structure	Differences in element composition	Differences of molecular structure	Differences of molecular structure	Differences in molecular structure
Advantages										
Accurate distinguish of different species	Simple, stable, good repeatability and applicability	Mature technology, low detection limit, good repeatability	Fast, easy to operate, little sample	convenient; reprod-ucible; little DNA template	Non-destructive testing, high accuracy and rapid detection	Broad spectrum peak, low detection limits	High sensitivity, little sample consumption, rapid analysis	Each compound has its specific infrared spectroscopy	Short separating time, high resolution, high sensitivity	High separation efficiency, good sensitivity, wide application
Disadvantages										
Difficult to detect different genetic relationship	Small number of available enzyme, poor polymorphism	Vulnerable to individuals mutations and false negative	Poor stability Of method	High require-ments of DNA and enzyme purity	Expensive equipment, Complex testing data	Less spectrum.	Expensive, Require higher content of component.	Complex spectrum, Poor accuracy	Complex sample preparation;	High solvent consumption.

TABLE 17.2 Overview of the Analytical Techniques and their Characteristics Used for Food Authentication

S. No.	Techniques	Sensitivity	Simplicity	Application	Analysis Time
1	Nuclear magnetic resonance spectroscopy	C	B	Identification and Profiling	B
2	Infrared spectroscopy	B	A	Profiling	A
3	Fluorescence spectroscopy	A	A	Profiling	A
4	Atomic spectroscopy	B	B	Identification and Profiling	B
5	High performance liquid chromatography	B	A	Profiling	B
6	Gas Chromatography	A	A	Profiling	B
7	Capillary electrophoresis	C	A	Profiling	B
8	Isotope ratio mass spectrometry	A	B	Identification and Profiling	B
9	Inductively coupled plasma mass spectrometry	A	B	Identification and Profiling	A
10	Proton transfer reaction mass spectrometry	A	A	Profiling	A
11	Gas chromatography mass spectrometry	A	A	Identification and Profiling	B
12	Sensor technology	C	A	Profiling	A
13	Sensor Analysis	B	B	Profiling	C
14	Polymerase chain reaction	A	B	Identification and Profiling	B

a. For the qualitative and quantitative purpose:
   - The electrophoretic fingerprinting technique is based on the principle of differences in protein/DNA structures and electric charges and is used for the qualitative analysis technology.
   - The spectral fingerprinting and chromatographic fingerprinting techniques are used for the quantitative analysis. These are not only able to confirm the existence of the component but are also able to measure the quantity of the component.

b. For the nondestructive/destructive purpose:
   - The working of spectral fingerprinting is based on the principle of differences in element compositions and it identifies the sample by on-line inspection.
   - Both the above techniques, electrophoretic fingerprinting and chromatographic fingerprinting, require several hours for identification.

c. With the cost and accuracy aim:
   - The chromatographic fingerprinting is the costliest method. It requires and implies the most expensive equipment.
   - The most accurate results will be obtained from chromatographic fingerprinting for quantifying the compounds.

d. With the aim of developing image processing technologies:
   - The images of electrophoretic fingerprinting can be visualized with the naked eye using ultraviolet light.
   - The images from spectral fingerprinting and chromatographic fingerprinting are visualized by a computer using the corresponding software.

17.2 APPLICATION OF DIFFERENT TECHNIQUES USED FOR FINGERPRINTING

In this chapter, different analytical approaches for the detection of fruits and vegetables based on their geographical origins are discussed (Table 17.3). These analytical techniques have been subdivided into four different types based on their principles: mass spectrometry techniques, spectroscopic techniques, separation techniques, and other techniques.

17.2.1 Nuclear Magnetic Resonance Spectroscopy

The principle of NMR is based on the measurement of the absorption of radiofrequency radiation by atomic nuclei with non-zero spins in a strong magnetic field (Ibanez and Cifuentes, 2001). The absorption by atomic nuclei leads to small variations in the magnetic field of the region. The most commonly used nuclei in NMR with non-zero spins are the isotopes of ^{1}H (spin = 1/2) and ^{13}C (spin = 1/2). The frequently used isotopes are ^{15}N (spin = 1/2), ^{17}O (spin = 5/2), ^{19}F (spin = 1/2), or ^{31}P (spin = 1/2). The various advantages of NMR technique are high precision, accuracy, good repeatability, and maintenance of sample integrity, sample size, and shape. There are two types of NMR techniques that are used in food analysis: (i) low resolution NMR (LR-NMR; frequencies of 10–40 MHz) (ii) high resolution NMR (HR-NMR; frequencies above 100 MHz) (Ibanez and Cifuentes, 2001). The LR-NMR instruments are small, easy to handle, inexpensive, and suitable for performing fast and reproducible measurements. The use of HR-NMR has been reported more widely in many food authenticity studies than the use of LR-NMR (Reid et al., 2006). The main advantage of HR-NMR over LR-NMR is that the former gives much more detailed information regarding the molecular structure of a food sample. ^{1}H NMR or ^{13}C NMR fingerprinting for food analysis are combined with advanced chemometric methods. The determination of the geographical origin of food is done on the basis of the isotopic ratio of a given nucleus found in a constituent of the analyzed food. It is used in the nondestructive testing of fruit, microbial contamination analysis, and pesticide residues detection. Applications of NMR in food fingerprinting are: analysis of food components, determination of water content, and identification of lipids, amino acids, and carbohydrates. Further, determination of adulteration in fruit juices (Belton et al., 1998), wines (Ogrinc et al., 2003), beer (Lachenmeier et al., 2004), and spirits (Petrakis et al., 2005) have also been reported.

The stable isotope ratios are particularly important for the authenticity analysis of food and food ingredients as demonstrated by Rossmann (2001). Different plants use different metabolic pathways to generate sugars: the C3 pathway is used by the sugar beet plants to fix CO_2 or the C4 pathway is used by sugar cane (Lawlor, 1989). Combined studies (^{1}H NMR and LC-MS) have been used for the direct analysis of aromatic components in the identification of adulterants in different beverages (like wines, beer, grape juice) (Gil et al., 2003). For determination of the oil content in soybean, pulsed NMR technique has been used by Song et al. (2006). They found a good correlation coefficient between NMR and the national standard Soxhlet extraction method by testing 408 soybean samples. Wang et al. (2007) studied stored oranges and determined the water content. They got a satisfactory result while determining the change of water content in oranges. Similarly, Jin et al. (2007) studied the changes of water dynamics in strawberries from maturation to the corruption of strawberries. Their study reported that the

TABLE 17.3 Application of Different Fingerprinting Techniques for the Detection of Adulterants in Different Vegetables and Fruits

S. No.	Fruit/Veg	Technique	Run Time (min)	Detection	Inference	Reference
1	Apple	Gas chromatography coupled to mass spectrometry with a time-of-flight analyser	46	Volatile	Classification of apple varieties	Aprea et al. (2011)
2	Apple juice	Gas chromatography coupled to mass spectrometry	40	Volatile	Discrimination of juices by apple varieties and geographical origin	Guo et al. (2012)
3	Citrus fruits peels	Gas chromatography coupled to mass spectrometry	91	Secondary metabolites	Classification by fruit types	Parastar et al. (2012)
4	Raspberry	High performance liquid chromatographic coupled to a diode array detector	30	Flavonoids	Identitation of raspberry	Chen et al. (2009)
5	Kiwi and pomelo	High performance liquid chromatographic coupled to a electrospray ionization mass spectrometry	30	Catechins, phenols and flavonoids	Classification according to the fruit species and subspecies	Sarbu et al. (2012)
6	Pomegranate peel	High performance liquid chromatographic coupled to an ultraviolet detector	80	Phenols	Identitation of Chinese pomegranate from different orchards	Li et al. (2015)
7	Tofu-type soybean	High performance liquid chromatographic coupled to an ultraviolet detector	40	Isoflavones	Differentiation of Tofu-type soybeans	Shen et al. (2012)
8	Sugarcane	High performance liquid chromatographic coupled to a diode array detector	60	Flavonoids	Discrimination of genotypes	Leme et al. (2014)
9	Red wine	Liquid chromatography- electrospray ionization- Quadrupole time of flight analyser	45	Non-volatiles	Discrimination between grape varieties	Arbulu et al. (2015)

(Continued)

TABLE 17.3 (*Continued*) Application of Different Fingerprinting Techniques for the Detection of Adulterants in Different Vegetables and Fruits

S. No.	Fruit/Veg	Technique	Run Time (min)	Detection	Inference	Reference
10	Red wine	High performance liquid chromatographic coupled to an ultraviolet detector	20	Phenols	Discrimination of Spanish wines from PDO, grape varieties, ageing periods and vintage	Serrano et al. (2012)
11	Cherry wines	Gas chromatography coupled to mass spectrometry	49	Volatile	Discrimination of Sherry wines	Xiao et al. (2014)
12	White apricot almond	Gas chromatography coupled to mass spectrometry	31	Fatty acid	Quality control and detection of frauds	Tian et al. (2014)
13	White wine	Gas chromatography coupled to mass spectrometry	48	Volatile	Authentication of grape varieties	Springer et al. (2014)
14	Mushrooms	Gas chromatography coupled to mass spectrometry with a trap ionic ioniser	121	Volatile	Discrimination mushrooms species	Malheiro et al. (2013)
15	Pomegranate	FTIR principal component analysis partial least squares regression analysis	NA	NA	Pomegranate concentrate authentication	Vardin et al. (2008)
16	Cabbage	NMR- principal component analysis	NA	NA	Geographical Origin of cabbage	Kim et al. (2013)
17	Potato	Gas chromatography coupled to mass spectrometry	NA	NA	Cultivars discrimination of potato	Dobson et al. (2008)
18	Tomato	Nuclear magnetic resonance gas chromatography mass spectrometry	NA	NA	Geographical Origin of tomatoes	Consonni et al. (2009)
19	Grapefruits	Flow injection electrospray ionisation time-of-flight mass spectrometry	NA	NA	Farming practice of grapefruits	Chen et al. (2010)
20	Tomatoes and pepper	Direct analysis in real time mass spectrometry	NA	NA	Farming practice	Novotná et al. (2012)

corruption had a strong relationship with water migration. Zhao et al. (2008) used ^1H NMR fingerprinting technique to identify beer. For the identification of adulterants, they have used 19 types of commercial canned fruit beers. With pattern recognition analysis, the technique successfully identified the geographical origins of all kinds of beers.

17.2.2 Infrared Spectroscopy

Fourier-transform infrared (FT-IR) spectroscopy is a widely used technique in the determination of adulterants in food material. It is an extremely rapid, high-throughput, and nondestructive analysis of a wide range of samples (Peres et al., 2007). This technique works on the principle that when a sample is treated with an infrared radiation (4,000–400 cm^{-1} mid-infrared region or 14,000–4,000 cm^{-1} near-infrared region), the functional groups of the sample absorb the radiation and give rise to stretching or bending vibrations (Ellis et al., 2012). The mid-IR spectrum is further divided into specific regions, such as C–Hx stretching vibrations of 3,050–2,800 cm^{-1} for fatty acids; 1,750–1,500 cm^{-1} for proteins and peptides (C=O, N–H and C–N); 1,200–900 cm^{-1} for polysaccharides (C–O and C–O–C); and 1,090–1,085 cm^{-1} for nucleic acids (P–O vibrations: 1,245–1,220 cm^{-1}). These vibration/absorption spectra can be interpreted directly with reference species. The resultant infrared absorbance spectra can be referred to as chemical "fingerprints". Every specific chemical or biochemical substance has its own unique infrared profile that is characteristic of the particular sample under analysis (Dunn and Ellis, 2005; Ellis et al. 2007). Due to simple working and minimum sample preparation, it is a valuable analytical technique for the metabolic fingerprinting of food materials (Ellis and Goodacre, 2006; Ellis et al., 2007). It can analyze different biochemical parameters, such as carbohydrates, proteins, lipids, amino acids, fatty acids, and polysaccharides of vegetables and fruits, rapidly and simultaneously. The segregation and analysis were performed for different wines (Picque et al., 2005) and honey (Ruoff et al., 2006) on the basis of geographical origin using mid-IR techniques. Soy sauce (Iizuka and Aishima, 1997), grapes (Arana et al., 2005), and wines (Liu et al., 2006) have been analyzed to detect the geographical origins by near-IR method. All the results were further interpreted by chemometric methods. Bing et al. (2005) have used near-IR spectroscopy to study genetically modified corn and its derivatives. Another research group, Fu et al. (2006), used diffuse reflectance FT-IR spectroscopy for the study of fruit firmness. They used a partial least squares (PLS) method to study a mathematical model between solid degree and diffuse reflectance spectra. The total sugar and acid content and the available vitamins in citrus fruits were determined by PLS (Yuan et al., 2005). The method was found to be suitable for detecting citrus quality, and the related coefficient between the predicted value and the chemical determination reached 95%. Another study carried out by Chen et al. (2006) reported that the above method is feasible for identifying the authenticity of tea. They used 138 samples of Biluochun tea and they got significant regression and differential rates of 93.48% and 84.44%, respectively, in the prediction of 90 independent samples. Two research groups, Zhao et al. (2006) and Wu et al. (2009), have used this technique to study the tenderness of beef. The model of the study was established to explain beef tenderness using an IR technique, and the result showed an 84.21% correct prediction rate. Zhang et al. (2008) used a near-IR spectroscopic method combined with a cluster type of soft independent modeling method to detect lamb origin. The result of the study confirmed that the near-IR spectroscopic method was a valid method for lamb origin tracing.

17.2.3 Ultraviolet Spectroscopy

UV fingerprinting techniques work on the principle of UV absorption spectra and they reflect the transitions of valence electron energy levels of elements. It is basically used in the analysis and detection of the main ingredients in food. Hu et al. (2005) used the UV method to detect nitrite concentration by adding safranine T to the sample in acidic condition. The sample reacts with nitrite and gives a maximum absorption peak at 351 nm after nitrification. The method was found to be highly sensitive in the detection of nitrite in food material. The UV method was also reported to be suitable for the detection of vitamin C content in fruit and vegetables. Zhao et al. (2006) showed good recovery rates of 97.9%–99.0% with a detection limit of 0.014 mg mL^{-1}. Chen et al. (2006) determined lecithin in health care products. They reported a rapid detection of lecithin with good reproducibility and stability. The total sugar content in honey was determined with the UV method by Yang et al. (2007). The average sugar content was 41.97% in samples tested at 286 nm. Fei and Mei (2008) used ion chromatography UV to detect paraquat residues in vegetables. The result was found to be satisfactory with a detection limit of 5.1 mg L^{-1}. This method, used for cabbage samples, was reported to be very effective. Han et al. (2009) developed a robust higher order derivative spectrometric method with Sudan red to test the presence of lycopene.

17.2.4 Mass Spectrometry

MS fingerprinting is a powerful analytical technique that works on the principle of formation and measurement of negatively or positively charged ions (Aebersold and Mann, 2003). The samples must be ionized, and the ions are separated when they pass through the electric and magnetic field. Finally, the detection of the mass-to-charge (m/z) ratios of these ions takes place, thus giving the mass spectra of samples. MS technique is usually used in combination with tandem GC or LC for food analysis. Quantification of analytes is typically done with LC coupled to triple quadrupole mass spectrometers (LC-MS/MS), which provides sufficiently high levels of sensitivity and specificity (Li et al., 2010; Lutter et al., 2011). Toxic secondary metabolic products (mycotoxins) are synthesized by fungi that grow on food products and cause irritation in a weakened immune system. The different types of mycotoxins are aflatoxins, ochratoxin A, B, and C, and fumonisins for which the growth has been studied in dried fruits (Karaca et al., 2010). These secondary metabolic products have been widely found in Brazil nuts (aflatoxins) (Pacheco et al., 2009), Belgian beers, barley grains (ochratoxins) (Olsson et al., 2002; Anselme et al., 2006), and also in Hong Kong foodstuffs. About 32% of samples contained detectable levels of these secondary metabolites.

17.2.4.1 Gas Chromatography Mass Spectrometry

GC-MS comprises a gas chromatograph coupled to a mass spectrometer and it is used as an instrumental analytical technique in the detection of impurities (Dion et al., 2008). GC separates the different components present in the mixture and MS characterizes each of the components individually. Complex mixtures can be analyzed qualitatively and quantitatively using both these techniques in combination. For analysis with GC-MS, the analyte must be volatile and thermally stable (Cubero-Leon et al., 2014). Xu et al. (2011) developed a GC-MS method for the metabolic profiling of *Salmonella typhimurium* in meat samples. This was not a particularly rapid method. Research by the same group led to the development of a rapid protocol for the collection of volatile organic compounds (VOCs) and the analysis by thermal desorption GC-MS (Xu et al., 2010). The authentication of many food materials have been reported using GC-MS and Py-MS methods, especially orange juice (Goodacre, 1997; Goodacre et al., 1997) and oysters (Ratel et al., 2008). Recently in food analysis, direct analysis in real time (DART)-MS is a rapidly developing technique used for the target and nontarget analysis of complex food matrices. It is a simple and high-throughput technique used for the qualitative identification of chemical species and for metabolomic fingerprinting/profiling (Zhou et al., 2010). It is also used for the quantification of low molecular weight food components, such as trace organic contaminants (Hajslova et al., 2011). In another research, fumonsins have been detected in cereal products (Li et al., 2010) and dried vine fruits (Varga et al., 2010) using this technique. The capability of DART-MS was compared to the traditional HPLC (high performance liquid chromatography) method while detecting curcumins directly from the turmeric plant (*Curcuma longa*) and turmeric containing beverages and curry powders. DART-MS could be used for the semiquantitative determinations of curcumin in the range of 5–100 mg mL^{-1}, and it could also be applied for the chemical profiling and quality control of a range of medicinal plants as well as food ingredients (Kim et al., 2009).

17.2.4.2 Isotope Ratio Mass Spectrometry

Isotope ratio mass spectrometry (IRMS) works on the principle of identification of chemically identical compounds on the basis of their isotopic contents (Dion et al., 2008). The isotopic contents of different constituents like carbohydrates, fats, proteins, and minerals depend on various factors. Some of these factors can be expected to be indicative of the geographical origins. This technique alone or in combination with other techniques has been used to detect the geographical origins of a variety of food products, ranging from vegetables (Georgi et al., 2005) and fruit juices (Rossmann, 2001; Perez et al., 2006) to natural flavors, honey, and wines (Ogrinc et al., 2001; Rossmann, 2001; Martinelli et al., 2003). The adulteration of royal jelly was also checked by IRMS using changes in the carbon and nitrogen ($^{15}N/^{14}N$) isotope ratios, related to the adulteration with high fructose corn sugar (using carbon isotope ratio) and yeast powder as a protein source (using nitrogen isotope ratio) (Stocker et al., 2006).

17.2.4.3 Proton Transfer Reaction Mass Spectrometry

Proton transfer reaction mass spectrometry (PTR-MS) is a quantitative on-line monitoring method for the identification of VOCs (Hansel and Mark, 2004; Dion et al., 2008). This method is a soft ionization method used to ionize organic molecules, and after the mass range scanning, the fingerprints of the volatile compounds are obtained. This technique enables the monitoring of a variety of organic species in complex matrices within seconds with low detection limits. The typical aromas of volatile compounds can be measured in regional products. This method has been used for its capabilities in the geographical authentication of olive oils and in detecting the origins of truffles, Grana cheeses, and wines (Aprea et al., 2007; Spitaler et al., 2007; Dion et al., 2008) because of its capabilities. PTR-MS has the potential to become a more suitable technique for determining the geographical origins of food products.

17.2.4.4 Inductively Coupled Plasma Mass Spectrometry

Inductively coupled plasma mass spectrometry (ICP-MS) technique has been used for the quantitative determination of a wide range of metals and nonmetals in a wide variety of food materials. It is a powerful tool and can detect the metals or nonmetals in trace and ultra-trace concentrations (Lachas et al., 2000; Dion et al., 2008). This technique can be used in combination with IRMS and atomic absorption/emission spectrometry (AAS or AES) for simultaneous elemental analysis (Gomez-Ariza et al., 2006). Both solid and liquid samples can be injected for identification. It can be used to detect the different geographical origins of food products by the analysis of numerous inorganic elements (Ariyama and Yasui, 2006). Many edible food materials like onions (Ariyama et al., 2006), nuts (Gomez-Ariza et al., 2006), vegetables (Ariyama and Yasui, 2006; Ariyama et al., 2006; Ariyama et al., 2007), teas (Moreda-Pineiro et al., 2003), and wines (Coetzee et al., 2005) have been successfully analyzed for the identification of their geographical origins. In these studies, chemometric techniques were used for the statistical analyses of the obtained ICP-MS data (Ariyama et al., 2004).

17.2.5 Raman Spectroscopy

Raman spectroscopy is a nondestructive analytical technique that offers distinctive fingerprint spectra of different analytes. The small instrumentation facility makes this a promising technique for the fingerprinting of food materials. This technique is widely used for the identification of foodstuffs due to the variety of analytes that can be used—lipids, proteins (Herrero, 2008) carbohydrates, pigments, and flavorings (He et al., 2008). Its application is greatly increased due to the possibility of coupling it to 1 mm resolution microscopes that are ideal for trace analyses (Gu et al., 2010; Huang et al., 2010). This technique is able to identify analytes in solutions with minimal interference from water. A major flavonoid glycoside (naringin) is found naturally in grapefruits (metabolized to flavonone naringenin in humans). A detection pattern was developed for the detection of orange juice adulteration with grapefruit juice (Shi et al., 2007; Ellis et al., 2011). A wide range of applications for Raman spectroscopy coupled with other analytical techniques have been reported, e.g., the detection of yeast in apple juices (Mizrach et al., 2007), Sudan I in chilli powder (Cheung et al., 2010), production of clarified apple and apricot juices (Camerlingo et al., 2007), and quality of tomato juice (Malekfar et al., 2010). Further, this technique has been used for analyzing the presence of pesticides in juices and on fruit surfaces (Shende et al., 2004), and also to monitor the tertiary changes in protein isolates from kidney beans (Tang and Ma, 2009). Similar approaches have previously been used to confirm the botanical and geographical origins of European honey (Ellis et al., 2012).

17.2.6 Atomic Spectroscopy

Atomic spectroscopy is an advanced analytical method used to detect vaporized atoms of metals and nonmetals in a variety of food samples. The AAS technique works on the principle of determining the concentration of gas phase atoms by measuring the absorption of light (Hou et al., 2001; Dion et al., 2008). The analyte concentration is calculated from the amount of light absorbed. AAS is performed in combination with chemometrics for inorganic elemental analysis in determining the geographical origins of various food products. The mineral elements (selenium) can be detected in honey (Paramas et al., 2000) and wines (Frias et al., 2001) to identify the corresponding geographical regions. The combination of AAS and AES techniques is used to identify the different geographical origins of orange juice, nuts (Schwartz and Hecking, 1991), and potatoes (Rivero et al., 2003). Further, AAS techniques combined with other analytical techniques were used for the chemical profiling of teas (Moreda-Pineiro et al., 2003), fruits (Perez et al., 2006), and onions (Ariyama et al., 2004; Ariyama et al., 2007) to determine the geographical growing regions.

17.2.7 Application of Chromatographic Techniques in Fingerprinting

Chromatography works on the principle of separation mechanism. Nowadays, it is being widely used in the fingerprinting analyses of food compositions due to its high separation efficiency. It can easily detect trace amounts of the adulterants present in food materials.

17.2.7.1 Gas Chromatography Fingerprinting

GC techniques have been widely applied for the authentication and identification of a wide range of food for quality prediction. It has short a separation time, high resolution, high sensitivity, and is also suitable for detecting volatile liquids or solids and gas mixtures. There are many sensitive detectors that have been coupled to GC and their detecting capacities are increased to a large extent as compared to HPLC detectors. GC-MS fingerprinting techniques are improved by multivariate curve resolution (MCR) to detect secondary metabolites in citrus fruit peels (Parastar et al., 2012). The assessments of the geographical origins of extra virgin olive oils and table wines were done with GC-MS using a three-way data collection (Salvatore et al., 2013). This method provides important information for analyzing the mixtures of vegetable oil and olive oil and for determining the percentages of oils used. Solid phase dispersion GC was applied to detect 31 types of pesticide residues (including pyrethroids and organochlorines) in fruits and vegetables (Feng et al., 2005). The method was found to be accurate with a relative standard deviation of less than 10% and a recovery rate of 80.4%–112.7%. Zhong et al. (2005) introduced a method to detect acrylamide in fried starchy food (potato chips and French fries) with the GC-MS technique. They found an acrylamide content between 278 and 4,518 mg kg^{-1} with a detection limit of 5 mg kg^{-1}. He et al. (2005) applied the GC-MS technique to detect the adulteration in edible oil. Zhan et al. (2008) used this technique for the detection of commonly used additives in food materials (methylparaben, propylparaben, ethylparaben, butylparaben, BHA, BHT, and TBHQ). The result of this study showed that the recovery rate was between 94.4% and 105.1% with a detection limit of 0.01–0.5 mg kg^{-1}. Similarly, Jiang et al..(2009) developed a GC-MS method for the simultaneous determination of 36 pesticide residues in tea samples. The method was simple, fast, sensitive, and accurate. Wang et al. (2008) determined carbofuran residues in grass carp and *Penaeus vannamei* with GC with a detection limit of 0.02 mg kg^{-1}. Again, this technology had been employed by Tian et al. (2006) for the detection of volatile flavor compounds in Jinhua ham. They demonstrated that the main components of the volatile flavor compounds in the Jinhua Ham were aldehyde, acid, and ketone compounds combined with some sulfur and heterocyclic compounds. The profile of coffee flavor was qualitatively and quantitatively analyzed with the help of SFA (saturated fatty acids), which was used as a supporting tool to resolve the overlapped peaks of the different volatile constituents (Huanga et al., 2007).

17.2.7.2 Liquid Chromatography Fingerprinting

LC is a robust analytical technique used for the detection of adulterants in food materials. It is a widely used technique due to its high sensitivity, accuracy, separation efficiency, rapidity, and reproducible results. This technique is mainly applied for the detection of pesticides, food additives, and toxic substances. The selection of a detector depends upon the nature of the compound to be analyzed. Further, the combination of MS detector with

HPLC has led to a greater applicability and a higher detection efficiency. Higher efficiency and better permeability were achieved with the applications of new monolithic HPLC columns as compared to the conventional HPLC columns. It also provides a reduction in analysis time with a higher flow rate (Cabrera, 2004). The analysis and quantification of amino acids, biogenic amines, phenolic compounds (Rodriquez-Delgado et al., 2002; Herberger et al., 2003), and contaminant ochratoxin A (Bellı et al., 2002) were done with HPLC. HPLC detection has also been applied to geographically differentiate among honey, cheeses, and nuts (Tomas-Barberan et al., 1993; Di Cagno et al., 2003; Gomez-Ariza et al., 2006) on the basis of detection profiles of flavonoids, metal-binding proteins, triglycerides, and peptides. Zou et al. (2006) developed a robust HPLC method for the determination of sulfite in garlic by testing five samples with a detection limit of 10 mg L^{-1} and a recovery rate of 98.0%–100.0%. Hao et al. (2007) detected the benzopyrene content in meat products using HPLC with fluorescence detector by the external standard quantification method. They reported that the developed method was found to be simple, robust, and faster in comparison to the international standard test method. Another research group introduced the ultra performance liquid chromatography quadrupole mass spectrometry (UPLC-MS/MS) technique to quantify the presence of 24 types of sulfa drug residues in milk (She et al., 2008). This method was found to be sensitive enough to provide a high recovery rate and it analyzed all the samples within 10 mins.

In another study, the HPLC-MS method was developed by Huang et al. (2008) to detect the melamine content in food and food additives. The method showed higher recovery rates (80%–90%) with a detection limit of 0.01 mg L^{-1}. Li et al. (2009) used the integrated liquid chromatography with electrospray ionization tandem mass spectrometry method (LC-ESI-MS/MS) to determine different kinds (156 types) of pesticide residues in five concentrated vegetable and fruit juices. The method showed a good recovery rate between 70% and 120% with a detection range of 0.10–56.77 mg kg^{-1} and a quantification range of 0.33–189.23 mg kg^{-1}. Li et al. (2009) developed a rapid and robust HPLC method to detect and analyze the residues of malachite green (MG) and its metabolite, leucomalachite green (LMG), in eel. Recovery rates of 85.7%–89.7% and 87.6%–91.2% were found, respectively. In another study, a satisfactory result was reported by Huang et al. (2009) using the reversed phase liquid chromatography mass spectrometry method for the estimation of melamine in milk and milk powder. Li et al. (2009) used a LC method for the determination of 11 kinds of synthetic food colorants with a detection limit of 0.5 mg kg^{-1}. Liao et al. (2009) developed a suitable method using high fat and high protein food like "Peking Duck" for the extraction and detection of polycyclic aromatic hydrocarbons (PAH). Three main detectable amounts of PAH were found between 0.56 and 3.19 mg kg^{-1}, lower than the required standard limit of 5 mg kg^{-1}. HPLC fingerprinting techniques were used to discriminate the different qualities of soybean, raspberry, and pomegranate. The calculation was done by applying the similarity analysis of some parameters [correlation (r) and congruence (c) coefficients] and by distance measurements. The comparison of dissimilarity indices of the chromatographic fingerprints is done only in a visual way by observing the overlapped chromatograms (Dumarey et al., 2010).

17.2.7.3 Capillary Electrophoresis

CE can be used to analyze and characterize a wide variety of analytes ranging from food materials to proteins and nucleic acids (Kvasnicka, 2005). It works on the principle of electrokinetic separation based upon the differences in electrophoretic mobilities. This mobility can be controlled by the parameters like isoelectric point, charge-to-mass ratio, molecular size, and separation conditions. The most commonly employed methods of CE include capillary isoelectric focusing, capillary gel electrophoresis, capillary electrochromatography, and capillary zone electrophoresis. Delgado et al. (1994) first introduced CE as a technique for discrimination of the geographical origins of various food products. They used this method to detect the presence of flavonoids in different proportions in honey depending on their geographical origins. Another research group used the same technique to geographically discriminate Chinese fruit extracts (Peng et al., 2006). This technique has also been used to detect the contents of six metals in a variety of Spanish red wines in a single analysis (Nunez et al., 2000). These results showed that the classification of the wines was done according to their geographical origins.

17.2.8 Other Techniques

17.2.8.1 DNA Fingerprinting

DNA fingerprinting is a biological technique used to identify components on the basis of differences in the DNA patterns. It has a number of advantages, i.e., high accuracy, sensitivity, and speed, and a large number of samples are automatically processed by it (De La Fuente and Juarez, 2005). The identification is done by the polymerase chain reaction (PCR) using molecular markers. The determination of sequences is done via amplification of the DNA strains. It is used to detect very low amounts of nucleic acid probes (Ibanez and Cifuentes, 2001). It can not only differentiate the adulterated species in food materials and identify the species of origin in food but can also identify the subspecies. PCR was found to be a suitable approach for the geographical discrimination of grapes and for the determination of growth and of the number of yeasts and yeast species in grapes (Raspor et al., 2006). In a similar way, PCR was also used for the identification of the geographical origin and the microbial diversity in natural whey cultures (Mauriello et al., 2003) and to discriminate cheeses (Bonizzi et al., 2006) and olive oils (Woolfe and Primrose, 2004). It is a widely accepted method in food identification. Other variants of DNA fingerprinting are restriction fragment length polymorphism (RFLP), random amplified polymorphic DNA (RAPD), and amplified fragment length polymorphism (AFLP).

17.2.8.1.1 Restriction Fragment Length Polymorphism This was the first DNA fingerprinting technique invented for food detection and it has high accuracy. It works on the principle of comparing the protein electrophoretic fingerprints with close genetically related species. Liu et al. (2006) reported a model to check the substandard mung bean components using this technique in soybean, cowpea, and other legumes. The result was found to be accurate and helpful in identification. RFLP technique has been used by Wang et al. (2006) to identify four major farming scallops in China's coastal areas. A PCR-RFLP method was used by Shao et al. (2008) to detect aflatoxin. The results showed that the enzyme-cutting pattern of PCR products was in line with expectation. A PCR-RFLP method was used for the detection of raw materials in Ejiao (donkey hide glue) (Wang et al., 2006). In this method, DNA was extracted from the dry skins of horses, cattle, and donkeys. The result showed that this method was highly accurate in identifying the skin species (100%). Further, this technique has also been adopted in the detection of fish adulteration (Hung-Sheng et al., 2007; Wen and Deng-Fwu, 2007).

17.2.8.1.2 Random Amplified Polymorphic DNA The RAPD technique is a simple and fast method which uses genetic markers. More information can be acquired with this method than with the RFLP fingerprinting technique. This technique does not require hybridization and cloning preparation steps and can be carried out on the whole genome. Zhou et al. (2006) used RAPD and cluster analysis to study the differences between *Pelteobagrus eupogon* and *Pelteobagrus fulvidraco*. Xu et al. (2008) developed a new method to identify the milk origins from eight different farms on different days. The results of the study showed that the amplified fragments of milk were same in the samples coming from the same farms on different days. The milk collected from different farms had different fragments, which differed mainly in the number and brightness of electrophoretic bands. Further, Zheng et al. (2007) developed a rapid identification method using RAPD for the detection of bacterial contamination in beer. Initially, a standard pollution bacteria database was prepared and then bacteria from the test samples were identified by fingerprint cluster analysis. The study reported that the bacteria from the same sources were bunched together. In a similar way, another research group used RAPD technique to detect foodborne *Shigella* in different sources (Guo et al., 2008). The results showed that the classification of *Shigella* using the primer was completely in line with the results of traditional serotype typing.

17.2.8.1.3 Amplified Fragment Length Polymorphism AFLP technique works via the combination of RAPD and RFLP techniques, so it has a higher reliability and repeatability than RFLP. This technique also has a greater convenience, safety, and efficiency than RAPD. Li et al. (2008) used this method to detect the fingerprints of ten major varieties of brassica in Hangzhou. They selected five pairs of primers for the study that helped them to differentiate among ten varieties of brassica with an accuracy of 100%. In another study by Lin et al. (2008), the fingerprints of antelope, black bear, and family cattle were checked with two enzymes and two primers.

17.2.8.2 Enzyme-Linked Immunosorbent Assay

In the last decade, the identification of food products has been done primarily with genetic (Bottero et al., 2002; Terzy et al., 2005) and immunological techniques (Carrera et al., 1997; Liu et al., 2006). The genetic methods require an expert person and expensive laboratory equipment. They are the most specific and sensitive methods reported for food authentication. The use of immunological assays can reduce the experimental costs and testing times. In quantitative estimation by ELISA (enzyme-linked immunosorbent assay), the optical density of the sample is interpolated into a standard curve of the standard by serial dilution of the target (Goldsby et al., 2003).

There are two variants of ELISA: indirect and sandwich ELISA. In case of indirect ELISA, two different antibodies are used, one antibody is specific to the antigen and the other is attached to an enzyme. This second antibody gives the assay its name: "enzyme-linked", and it causes a chromogenic substrate to produce a signal during the detection. Whereas, in case of the sandwich ELISA technique, the antigen is bound to the detection antibody and the capture antibody. The detection antibody can be coupled to an enzyme (enzyme-linked antibody) that will produce a biochemical reaction for detection (Goldsby et al., 2003). An increasing number of research papers are coming out related to the adulteration of fruit juice beverages, highlighting the significance of this important problem (Jezek and Suhaj, 2001; Tzouros and Arvanitoyannis, 2001). The adulteration can be done by a simple dilution with water or by substitution with cheaper artificial ingredients (sugars, acids, and colorants). In most of these methods, citrus juice is replaced by the pulp and peel extracts of some cheaper fruits (Mears and Shenton, 1973). The research group, Sass-Kiss and Sass (2000; 2002), developed polyclonal antibodies against the characteristic grapefruit and orange juice and peel peptides. These developed antibodies were used in an immunoassay and they allowed the detection of adulterated fruit juice products. In another study, Lopez-Calleja et al. (2006)

applied an indirect ELISA technique using a monoclonal antibody (MAb) against bovine b-casein (Anguita et al., 1995) to detect the adulteration in undeclared cow cheeses. Hurley et al. (2004; 2006) developed an indirect competitive and sandwich ELISA for the detection of low percentages of undeclared bovine milk in goat and buffalo milk.

17.2.8.3 Sensor Technology

Sensor technology fingerprinting technique is used to detect the presence of volatile compounds in the headspace of the food sample. This study is performed by an array of semi-selective gas sensors of the volatile compounds (Strike et al., 1999). This technique performs global fingerprint analysis and provides results and representations that can be easily interpreted. There are many advantages that are reported for the sensor technology. These include speed, simplicity of detection, less amount of sample required, and low cost of analysis (Franke et al., 2005). There are a large number food materials that are reported on the basis of differentiation in the geographical origins of olive oils, orange juice, and wines (Guadarrama et al., 2001; Steine et al., 2001; Penza and Cassano, 2004).

17.2.8.4 Sensory Analysis

Sensory analysis (or sensory evaluation) is found to be an important analytical technique for detecting the product quality of food materials. It works on the principle of sensory analysis panels of human assessors. The recorded responses may be assessed by applying statistical techniques to the results. It comprises a set of techniques that are used for accurate measurements of food quality (appearance, odor, flavor, and texture) (Perez Elortondo et al., 2007). Sensory analyses have also been applied to geographically discriminate spirit drinks (absinthe) (Lachenmeier, 2007) and for the differentiation between wines (Vilanova and Vilarino, 2006). The detection is performed in different phases using multiple descriptors to be identified by the wine tasters. In the first and second phases, the wine tasters identify; further, the rate of intensity of the remaining descriptors are evaluated for each wine. Finally, characteristic descriptors may be found in each wine depending on the corresponding geographical areas. Further, the grape cultivar and wine production area are linked together to get accurate results.

17.2.8.5 Stable Carbon Isotope Ratio Analysis

SCIRA (stable carbon isotope ratio analysis) technique is a useful analytical technique for the fingerprinting of food materials. Its working principle is based on the ^{13}C:^{12}C ratio of plant materials involved in the photosynthetic pathways used by individual plants (Winkler, 1984; Firestone and Reina, 1996). Winkler (1984) reviewed the application of stable isotope mass spectrometry in food control. Most of the land plants use the Calvin cycle to incorporate CO_2 by the carboxylation of ribulose diphosphate to phosphoglycerate. Whereas, few plants such as corn, millet, sorghum, sugarcane, and some tropical grasses use the alternative Hatch-Slack pathway for the carboxylation of phosphoenolpyruvate to oxalacetate. A much smaller group of plants (pineapple and cacti) use the crassulacean acid metabolic (CAM) pathway, which is capable of C_3 or C_4 CO_2 fixation (Winkler, 1984). The naturally occurring stable carbon isotopes, ^{12}C and ^{13}C, are useful for measuring the authenticity of apple juice (Doner et al., 1980; Doner and Phillips, 1981; Brause and Raterman, 1982), orange juice (Brause et al., 1984), and other food items (Gaffney et al., 1979).

17.2.8.6 Biochemical Fingerprinting

There are two types of biochemical fingerprinting that are used for species identification and purity analysis of food materials.

17.2.8.6.1 Protein Electrophoresis This method works on the principle of producing characteristic protein markers and it enables the discrimination of different species. The qualities of clinically infected mastitis cow milk and healthy cow milk were tested via the application of two-dimensional gel electrophoresis to distinguish the whey proteins (He et al., 2007). Tang et al. (2006) employed this method to detect the adulteration of dairy products. In a similar way, another research group, Li et al. (2006), used this technique to identify different types of edible birds' nests using the adulterated products.

17.2.8.6.2 Isoenzyme Electrophoresis This technique involves the specificity of isoenzymes. Zhihua et al. (2009) used six isoenzymes from two different organizations to make distinctions among the protein tags of different groups of clams. Meng Yan et al. (2009) used two isoenzymes for the identification of six tissues of *Channa argus* and *Channa asiatica*. There were marked differences between the tissue-specific and species-specific in the systems of isoenzymes in the two fish. Further, this method was also found to be suitable for reporting the valid identification of loach groups from different regions (Wu and Chang, 2009).

17.3 CONCLUSION

The interest of consumers in high-quality food materials along with a clear knowledge of geographical identities has grown rapidly. So, suitable techniques for the identification of geographical origins of food products (fruits and vegetables) are highly desirable. Fingerprinting techniques have become important technologies for the identification of food (fruits and vegetables). In the last two decades, a number of analytical methods have been applied for identification and profiling, which are becoming more sensitive, reliable, and faster now. These techniques are also used in combination to get more accurate results. The analytical equipment are also operated in combination with computers to store the results in databases. Among these analytical methods, the spectrum fingerprinting is the sole method that can track the food origin. The testing of additives, objectionable constituents, pesticides, or drug residues have been performed by the techniques like UV, MS, GC, LC. DNA and electrophoretic fingerprinting methods have shown the ability to detect microorganisms as well as to identify species in food materials. These analytical methods are able to detect the specific characteristics of food materials which are influenced by geographically specific factors. It is clear that all the analytical techniques have individual pros and cons. So, an accurate determination of the geographical origin of a food product seems feasible with the combination of these techniques.

REFERENCES

Aebersold, R., Mann, M. 2003. Mass spectrometry-based proteomics. *Nature*, 422, 198–207.

Anguita, G., Martın, R., Garcıa, T., Morales, P., Haza, A.I., Gonzalez, I. 1995. Indirect ELISA for detection of cows'milk in ewes' and goats' milks using a monoclonal antibody against bovine b-casein. *Journal of Dairy Research*, 62, 655–659.

Anselme, M., Tangni, E.K., Pussemier, L., Motte, J.C., Van Hove, F., Schneider, Y.J., Van Peteghem, C., Larondelle, Y. 2006. Comparison of ochratoxin A and deoxynivalenol in organically and conventionally produced beers sold on the Belgian market. *Food Additives and Contaminants*, 23(9), 910–918.

Aprea, E., Carlin, S., Versini, G., Mark, T.D., Gasperi, F. 2007. Assessing truffle origin by PTR-MS fingerprinting and comparison with GC-MS for the headspace analysis. In *3rd International Conference on Proton Transfer Reaction Mass Spectrometry and its Applications*, A. Hansel and T.D. Mark (Eds.). Innsbruck University Press, Austria, pp. 132–135.

Aprea, E., Gika, H., Carlin, S., Theodoridis, G., Vrhovsek, U., Mattivi, F. 2011. Metabolite profiling on apple volatile content based on solid phase microextraction and gas-chromatography time of flight mass spectrometry. *Journal of Chromatography A*, 1218, 4517–4524.

Arana, I., Jaren, C., Arazuri, S. 2005. Maturity, variety and origin determination in white grapes (Vitis Vinifera L.) using near infrared reflectance technology. *Journal of Near Infrared Spectroscopy*, 13, 349–357.

Arbulu, M., Sampedro, M.C., Gomez-Caballero, A., Goicolea, M.A., Barrio, R.J. 2015. Untargeted metabolomic analysis using liquid chromatography quadrupole time-of-flight mass spectrometry for non-volatile profiling of wines. *Analytica Chimica Acta*, 858, 32–41.

Ariyama, K., Aoyama, Y., Mochizuki, A., Homura, Y., Kadokura, M., Yasui, A. 2007. Determination of the geographic origin of onions between three main production areas in Japan and other countries by mineral composition. *Journal of Agricultural and Food Chemistry*, 55, 347–354.

Ariyama, K., Horita, H., Yasui, A. 2004. Chemometric techniques on inorganic elements composition for the determination of the geographic origin of Welsh onions. *Analytical Sciences*, 20, 871–877.

Ariyama, K., Nishida, T., Noda, T., Kadokura, M., Yasui, A. 2006. Effects of fertilization, crop year, variety, and provenance factors on mineral concentrations in onions. *Journal of Agricultural and Food Chemistry*, 54, 3341–3350.

Ariyama, K., Yasui, A. 2006. Review—The determination technique of the geographic origin of Welsh onions by mineral composition and perspectives for the future. *Japan Agricultural Research Quarterly*, 40(4), 333–339.

Belli, N., Marın, S., Sanchis, V., Ramos, A.J. 2002. Review: Ochratoxin A (OTA) in wines, musts and grape juices: Occurrence, regulations and methods of analysis. *Food Science and Technology International*, 8(6), 325–335.

Belton, P.S., Colquhoun, I.J., Kemsley, E.K., Delgadillo, I., Roma, P., Dennis, M.J., Sharman, M., Holmes, E., Nicholson, J.K., Spraul, M. 1998. Application of chemometrics to the 1 HNMR spectra of apple juices: discrimination between apple varieties *Food Chemistry*, 61, 207–213.

Bing, Y., Luo, Y.-B., Huang, K.-L. 2005. Application of near-infrared diffuse reflectance spectroscopy to the detection and identification of transgenic corn. *Spectroscopy and Spectral Analysis*, 25(10), 1580–1583.

Bonizzi, I., Feligini, M., Aleandri, R., Enne, G. 2006. Genetic traceability of the geographical origin of typical Italian water buffalo Mozzarella cheese: A preliminary approach. *Journal of Applied Microbiology*, 102(3), 667–673.

Bottero, M.T., Civera, T., Anastasio, A., Turi, R.M., Rosati, S. 2002. Identification of cows' milk in buffalo cheese by duplex polymerase chain reaction. *Journal of Food Protection*, 65, 362–366.

Brause, A.R., Ratterman, J.M. 1982. Verfication of authenticity of apple juice. *Journal—Association of Official Analytical Chemists*, 64(4), 846–849.

Brause, A.R., Ratterman, J.M., Petrus, D.R. 1984. Verification of authenticity of orange juice. *Journal—Association of Official Analytical Chemists*, 67(3), 535–539.

Cabrera, K. 2004. Applications of silica-based monolithic HPLC columns. *Journal of Separation Science*, 27(10–11), 843–852.

Camerlingo, C., Zenone, F., Delfino, I., Diano, N., Mita, D.G., Lepore, M. 2007. Investigation on clarified fruit juice composition by using visible light micro-Raman spectroscopy. *Sensors*, 7, 2049–2061.

Carrera, E., Garcıa, T., Cespedes, A., Gonzalez, I., Sanz, B., Hernandez, P.E. 1997. Immunostick colorimetric ELISA assay for the identification of smoked salmon (Salmo Salar), trout (Oncorhynchus mykiss) and bream (Brama raii). *Journal of the Science of Food and Agriculture*, 74, 547–550.

Chen, J., Lu, Y.H., Wei, D.Z., Zhou, X.L. 2009. Establishment of a fingerprint of raspberries by LC. *Chromatographia*, 70, 981–985.

Chen, P., Harnly, J.M., Lester, G.E. 2010. Flow injection mass spectral fingerprints demonstrate chemical differences in Rio red grapefruit with respect to year, harvest time, and conventional versus organic farming. *Journal of Agricultural and Food Chemistry*, 58(8), 4545–4553.

Chen, Q.S., Zhao, J.W., Zhang, H.D. 2006. Identification of authenticity of tea with near infrared spectroscopy based on support vector machine. *Acta Optica Sinica*, 26(6), 933–937.

Chen, W., Zhang, D., Zhang, B. 2006. Determination of lecithin in functional food by UV spectrophotometry. *Journal of the Chinese Cereals and Oils Association*, 21(3), 189–191.

Cheung, W., Shadi, I.T., Xu, Y., Goodacre, R. 2010. Quantitative analysis of the banned food dye Sudan-1 using surface enhanced Raman scattering with multivariate chemometrics. *Journal of Physical Chemistry* C, 114, 7285–7290.

Coetzee, P.P., Steffens, F.E., Eiselen, R.J., Augustyn, O.P., Balcaen, L., Vanhaecke, F. 2005. Multi-element analysis of South African wines by ICP-MS and their classification according to geographical origin. *Journal of Agricultural and Food Chemistry*, 53, 5060–5066.

Consonni, R., Cagliani, L.R., Stocchero, M., Porretta, S. 2009. Triple concentrated tomato paste: Discrimination between Italian and Chinese products. *Journal of Agricultural and Food Chemistry*, 57(11), 4506–4513.

Cuadros-Rodriguez, L., Ruiz-Samblas, C., Valverde-Som, L., Estefanía Perez-Castano, E., Antonio Gonzalez-Casado, A. 2016. Chromatographic fingerprinting: An innovative approach for food 'identitation' and food authentication—A tutorial. *Analytica Chimica Acta*, 909, 9–23.

Cubero-Leon E, Penalver R, Maquet A, 2014. Review on metabolomics for food authentication. *Food Research International*, 60, 95–107.

De La Fuente, M.A., Juarez, M. 2005. Authenticity assessment of dairy products. *Critical Reviews in Food Science and Nutrition*, 45, 563–585.

Delgado, C., Tomas-Barberan, F.A., Talou, T., Gaset, A. 1994. Capillary electrophoresis as an alternative to HPLC for determination of honey flavonoids. *Chromatographia*, 38(1/2), 71–78.

Di Cagno, R., Banks, J., Sheehan, L., Fox, P.F., Brechany, E.Y., Corsetti, A. 2003. Comparison of the microbiological, compositional, biochemical, volatile profile and sensory characteristics of three Italian PDO ewes' milk cheeses. *International Dairy Journal*, 13, 961–972.

Dobson, G., Shepherd, T., Verrall, S.R., Conner, S., McNicol, J.W., Ramsay, G. 2008. Phytochemical diversity in tubers of potato cultivars and landraces using a GC–MS metabolomics approach. *Journal of Agricultural and Food Chemistry*, 56(21), 10280–10291.

Doner, L.W., Krueger, H.W. and Reesman, R.H. 1980. Isotope composition of carbon in apple juice. *Journal of Agricultural and Food Chemistry*, 28(2), 362–364.

Doner, L.W., Phillips, J.G. 1981. Detection of high-fructose corn syrup in apple juice by mass spectrometric 13C/12C analysis: A collaborative study. *Journal—Association of Official Analytical Chemists*, 64(1), 85–90.

Dumarey, M., Smets, I., Vander Heyden, Y. 2010. Prediction and interpretation of the antioxidant capacity of green tea from dissimilar chromatographic fingerprints. *Journal of Chromatography B*, 878, 2733–2740.

Dunn, W.B., Ellis, D.I. 2005. Metabolomics: Current analytical platforms and methodologies. *Trends in Analytical Chemistry*, 24, 285–294.

Ellis, D.I., Dunn, W.B., Griffin, J.L., Allwood, J.W., Goodacre, R. 2007. Metabolic fingerprinting as a diagnostic tool. *Pharmacogenomics*, 8, 1243–1266.

Ellis, D.I., Goodacre, R. 2006. Metabolic fingerprinting in disease diagnosis: Biomedical applications of infrared and Raman spectroscopy. *Analyst*, 131, 875–885.

Ellis, D.I., Victoria, L.B., Warwick, B.D., Allwood, J.D., Alexander, P.G., Goodacre, R. 2012. Fingerprinting food: Current technologies for the detection of food adulteration and contamination. *Chemical Society Reviews*, 41, 5706–5727.

Fei, D., Mei, L. 2008. Determination of paraquat in vegetable by ion chromoatography with UV detection. *Environmental Pollution and Control*, 30(6), 31–33.

Feng, X., Chen, W., Wang, Z.-Y. 2005. Determination of multipesticide residues in fruits and vegetables by matrix solid-phase disperse and gas chromatography. *Food Science*, 26(7), 194–197.

Fiehn, O. 2001. Combining genomics, metabolome analysis, and biochemical modelling to understand metabolic networks. *Comparative and Functional Genomics*, 2, 155–168.

Firestone, D., Reina, R.J. 1996. Authenticity of vegetable oils. In *Food Authentication*, P.R. Ashurst and M.J. Dennis (Eds.). Blackie Academic & Professional, an imprint of Chapman & Hall, London, pp. 232–234.

Frıas, S., Trujillo, J.P.P., Pena, E.M., Conde, J.E. 2001. Classification and differentiation of bottled sweet wines of Canary Islands (Spain) by their metallic content. *European Food Research and Technology*, 213, 145–149.

Fu, X., Ying, Y., Liu, Y. 2006. Detection of pear firmness using near infrared diffuse reflectance spectroscopy. *Spectroscopy and Spectral Analysis*, 26(6), 1038–1041.

Georgi, M., Voerkelius, S., Rossmann, A., Grassmann, J., Schnitzler, W.H. 2005. Multielement isotope ratios of vegetables from integrated and organic production. *Plant and Soil*, 275, 93–100.

Gil, M., Le Duarte, I.F., Godejohann, M., Braumann, U., Maraschin, M., Spraul, M. 2003. Characterization of the aromatic composition of some liquid foods by nuclear magnetic resonance spectrometry and liquid chromatography with nuclear magnetic resonance and mass spectrometric detection *Analytica Chimica Acta*, 488, 35–51.

Goldsby, R.A., Kindt, T.J., Osborne, B.A., Kuby, J. 2003. *Enzyme linked immunosorbent assay. In Immunology* (pp. 148–150). W.H. Freeman, & Company, New York.

Gomez-Ariza, J.L., Arias-Borrego, A., Garcıa-Barrera, T. 2006. Multielemental fractionation in pine nuts (pinus pinea) from different geographic origins by size-exclusion chromatography with UV and inductively coupled mass spectrometry detection. *Journal of Chromatography A*, 1121, 191–199.

Goodacre, R. 1997. Use of pyrolysis mass spectrometry with supervised learning for the assessment of the adulteration of milk of different species. *Applied Spectroscopy*, 51, 1144–1153.

Goodacre, R., Hammond, D., Kell, D.B. 1997. Quantitative analysis of the adulteration of orange juice with sucrose using pyrolysis mass spectrometry and chemometrics. *Journal of Analytical and Applied Pyrolysis*, 40–41, 135–158.

Gu, C., Yang, X.A., Zhang, J., Newhouse, R., Cao, L.C. 2010. *Information Optics and Optical Data Storage*, F. Song, S. Tao, F.T.S. Yu, S. Jutamulia, K.A.S. Immink and K. Shono (Eds.). Spie-Int Soc Optical Engineering, Bellingham, vol. 7851.

Guadarrama, A., Rodriguez-Mendez, M.L., Sanz, C., Rios, J.L., de Saja, J.A. 2001. Electronic nose based on conducting polymers for the quality control of the olive oil aroma—Discrimination of quality, variety of olive and geographic origin. *Analytica Chimica Acta*, 432, 283–292.

Guo, A., Hou, J.K., Wan, P. 2008. Establishment of RAPD identification method for shigella. *Journal of Anhui Agricultural Sciences*, 36(11), 4455–4456, 4469.

Guo, J., Yue, T., Yuan, Y. 2012. Feature selection and recognition from nonspecific volatile profiles for discrimination of apple juices according to variety and geographical origin. *Journal of Food Science*, 77, 1090–1096.

Hajslova, J., Cajka, T., Vaclavik, L. 2011. Challenging applications offered by direct analysis in real time (DART) in food-quality and safety analysis. *Trends in Analytical Chemistry*, 30, 204–218.

Han, B., Jin, J., Wu, Q.Y. 2009. High order derivative spectrophotograph determination of Sudan red in the samples containing lycopene. *Chinese Journal of Pharmaceutical Analysis*, 29(5), 710–713.

Hansel, A., Mark, T.D. 2004. Foreword. *International Journal of Mass Spectrometry*, 239, vii–viii.

Hao, X., Hu, J.Z., Zhong, H. 2007. Detection of Benzo (a) Pyrene in meat food by high pres sure liquid chromatography (HPLC). *Food Science and Technology*, 7, 219–221.

He, L.L., Kim, N.J., Li, H., Hu, Z.Q., Lin, M.S. 2008. Use of a fractal-like gold nanostructure in surface-enhanced raman spectroscopy for detection of selected food contaminants. *Journal of Agricultural and Food Chemistry*, 56, 9843–9847.

He, X.Q., Xu, D., Luo, M. 2005. A microwave-assisted derivatization and GC-MS method coupled with calibration transformation matrix for determination of fatty acid in edible oils. *Journal of Instrumental Analysis*, 24(1), 25–28.

He, Y., Yang, Y.X., Zhao, X. 2007. Identification of differential whey protein from healthy and clinic mastitis cows by two-dimensional gel electrophoresis. *Journal of Gansu Agricultural University*, 6, 41–44.

Herberger, K., Csomos, E., Simon-Sarkadi, L. 2003. Principal component and linear discriminant analyses of free amino acids and biogenic amines in Hungarian wines. *Journal of Agricultural and Food Chemistry*, 51, 8055–8060.

Herrero, A.M. 2008. Raman spectroscopy a promising technique for quality assessment of meat and fish: A review. *Food Chemistry*, 107, 1642–1651.

Horning, E.C., Horning, M.G. 1971. Human metabolic profiles obtained by GC and GC/MS. *Journal of Chromatographic Science*, 9, 129–140.

Hou, X., Levine, K.E., Salido, A., Jones, B.T., Ezer, M., Elwood, S. 2001. Tungsten coil devices in atomic spectrometry: Absorption, fluorescence and emission. *Analytical Sciences*, 17(1), 175–180.

Hu, Y.F. 2005. Study on the photometry of nitrites in meat products. *Physical Testing and Chemical Analysis Part B: Chemical Analysis*, 41(3), 188–189.

Huang, F., Huang, X., Wu, H. 2008. Determination of melamine in feeds and food additives by LC-MS. *Journal of Instrumental Analysis.*, 27(3), 313–315.

Huang, W.E., Li, M.Q., Jarvis, R.M., Goodacre, R., Banwart, S.A. 2010. *Advances in Applied Microbiology*. Elsevier Academic Press Inc, San Diego, 70, 153–186.

Huanga, L.F., Wu, M.J., Zhongb, K.J., Sunb, X.J., Liang, Y.Z., Dai, Y.H., Huanga, K.L., Guoa, F.Q. 2007. Fingerprint developing of coffee flavor by gas chromatography mass spectrometry and combined chemometrics methods. *Analytica Chimica Acta*, 588, 216–223.

Hung-Sheng, H., Tuu-Jyi, C., Deng-Fwu, H. 2007. Using the PCR-RFLP method to identify the species of different processed products of billfish meats. *Food Control*, 18, 369–374.

Hurley, I.P., Coleman, R.C., Ireland, H.E., Williams, J.H.H. 2004. Measurement of bovine IgG by indirect competitive ELISA as a means of detecting milk adulteration. *Journal of Dairy Science*, 87, 215–221.

Hurley, I.P., Coleman, R.C., Ireland, H.E., Williams, J.H.H. 2006. Use of sandwich IgG ELISA for the detection and quantification of adulteration of milk and soft cheese. *International Dairy Journal*, 16, 805–812.

Ibanez, E., Cifuentes, A. 2001. New analytical techniques in food science. *Critical Reviews in Food Science and Nutrition*, 41(6), 413–450.

Iizuka, K., Aishima, T. 1997. Soy sauce classification by geographic origin based on NIR spectra and chemometrics pattern recognition. *Journal of Food Science*, 62, 101–104.

Jezek, J., Suhaj, M. 2001. Application of capillary isotachophoresis for fruit juice authentication. *Journal of Chromatography A*, 916, 185–189.

Jin, Z.Q., Zhang, J.S., Lin, X.Y. 2007. Study on water loss rate and decay of strawberry by NMR and MRI. *Food Science*, 28(8), 108–111.

Jiang, J.-S., Zhao, B., Zhou, L. 2009. Simultaneous determination of 36 pesticide residues in tea by GC-MS. *Food Science*, 30(14), 276e330.

Karaca, H., Velioglu, Y.S., Nas, S. 2010. Mycotoxins: Contamination of dried fruits and degradation by ozone. *Toxin Reviews*, 29, 51–59.

Kim, H.J., Jang, Y.P. 2009. Direct analysis of curcumin in turmeric by DART-MS. *Phytochemical Analysis*, 20, 372–377.

Kim, J., Jung, Y., Song, B., Bong, Y.S., Ryu, D.H., Lee, K.S. 2013. Discrimination of cabbage (Brassica rapa ssp. pekinensis) cultivars grown in different geographical areas using 1H NMR-based metabolomics. *Food Chemistry*, 137(137), 68–75.

Kvasnicka, F. 2005. Capillary electrophoresis in food authenticity. *Journal of Separation Science*, 28, 813–825.

Lachas, H., Richaud, R., Herod, A.A., Dugwell, D.R., Kandiyoti, R. 2000. Determination of trace elements by ICP-MS of biomass and fuel oil reference materials using milligram sample sizes. *Rapid Communications in Mass Spectrometry*, 14(5), 335–343.

Lachenmeier, D.W. 2007. Assessing the authenticity of absinthe using sensory evaluation and HPTLC analysis of the bitter principle absinthin. *Food Research International*, 40, 167–175.

Lachenmeier, D.W., Frank, W., Humpfer, E., Schafer, H., Keller, S., Mortter, M., Spraul, M. 2004. *European Food Research and Technology*, 220, 215–221.

Lawlor, D.W. 1989. *Photosynthesis: Metabolism, Control and Physiology*. Longman Scientific and Technical, Harlow.

Leme, G.M., Coutinho, I.D., Creste, S., Hojo, O., Carneiro, R.L., Bolzani, V. 2014. HPLC-DAD method to metabolic fingerprinting to the phenotyping of sugarcane genotypes. *Analytical Methods*, 6, 7781–7788.

Li, H.B., Wang, S., Wang, H. 2008. Application of AFLP marker technology in the identification of greengrocery variety "Youdonger". *Journal of Anhui Agricultural Sciences*, 36(1), 72–73.

Li, J., Dong, Y., Zhou, H. 2006. Identification of edible bird's nest with electrophoresis. *World Science Technology/Modernization of Traditional Chinese Medicine*, 8(3), 30–32.

Li, J., He, X., Li, M., Zhao, W., Liu, L., Kong, X. 2015. Chemical fingerprint and quantitative analysis for quality control of polyphenols extracted from pomegranate peel by HPLC. *Food Chemistry*, 176, 7–11.

Li, N., Li, S.H., An, J. 2009. Determination of eleven kinds of synthetical pigments in food by rapid resolution liquid chromatography. *Science and Technology of Food Industry*, 30(1), 313–315.

Li, W., Herrman, T.J., Dai, S.Y. 2010. Rapid determination of fumonisins in corn-based products by liquid chromatography-tandem mass spectrometry. *Journal of AOAC International*, 93, 1472–1481.

Li, Y., Gao, H.P., Yao, A. 2009. Detection on the residues of malachite green and its metabolite in eel by high performance liquid chromatography. *Journal of Anhui Agricultural Sciences*, 37(1), 31–33.

Li, Y., Zheng, F., Wang, M. 2009. Rapid screening and confirmation of 156 pesticide residues in concentrated fruit and vegetable juices using liquid chromatography-tandem mass spectrometry. *Chinese Journal of Chromatography*, 27(2), 127–137.

Liao, Q., Gao, Z.J., Zhang, S.X. 2009. Determination of polycyclic aromatic hydrocarbons in skin of "Peking Roast Duck" by high performance chromatography- fluorescence detection. *Food Science*, 30(10), 149–152.

Lin, J.X., Chen, Z.H., Zhang, M. 2008. Application of AFLP fingerprinting in animal species identification. *Journal of Nanjing Forestry University (Natural Sciences Edition)*, 32(2), 142–144.

Liu, J., Wei, C., Ma, L.H. 2006. Detect mungbean ingredients by use specific RFLP fragment amplification. *Plant Quarantine*, 6, 336–338.

Liu, L., Cozzolino, D., Cynkar, W.U., Gishen, M., Colby, C.B. 2006. Geographical classification of Spanish and Australian Tempranillo red wines by visible and near-infrared spectroscopy combined with multivariate analysis. *Journal of Agricultural and Food Chemistry*, 54, 6754–6759.

Liu, L.H., Chen, F.C., Dorsey, J.L., Hsieh, Y.H.P. 2006. Sensitive monoclonal antibody-based sandwich ELISA for the detection of porcine skeletal muscle in meat and feed products. *Journal of Food Science*, 71, M1–M6.

Lopez-Calleja, I.M., Gonzalez, I., Fajardo, V., Hernandez, P.E., Garcıa, T., Martın, R. 2006. Application of an indirect ELISA and a PCR technique for detection of cows' milk in sheep's and goats' milk cheeses. *International Dairy Journal*, 17, 87–93.

Lutter, P., Savoy-Perroud, M.C., Campos-Gimenez, E., Meyer, L., Goldmann, T., Bertholet, M.C., Mottier, P., Desmarchelier, A., Monard, F., Perrin, C., Robert, F., Delatour, T. 2011. Screening and confirmatory methods for the determination of melamine in cow's milk and milk-based powdered infant formula: Validation and proficiency-tests of ELISA, HPLC-UV, GC-MS and LC-MS/MS *Food Control*, 22, 903–913.

Luykx, D.M.A.M., van Ruth, S.M. 2008. An overview of analytical methods for determining the geographical origin of food products. *Food Chemistry*, 107, 897–911.

Malekfar, R., Nikbakht, A.M., Abbasian, S., Sadeghi, F., Mozaffari, M. 2010. Evaluation of tomato juice quality using surface enhanced Raman spectroscopy. *Acta Physica Polonica A*, 117, 971–973.

Malheiro, R., Guedes de Pinho, P., Soares, S., Cesar da Silva Ferreira, A., Baptista, P. 2013. Volatile biomarkers for wild mushrooms species discrimination. *Food Research International*, 54, 186–194.

Martinelli, L.A., Moreira, M.Z., Ometto, J.P.H.B., Alcarde, A.R., Rizzon, L.A., Stange, E. 2003. Stable carbon isotopic composition of the wine and CO2 bubbles of sparkling wines: Detecting C-4 sugar additions. *Journal of Agricultural and Food Chemistry*, 51, 2625–2631.

Mauriello, G., Moio, L., Genovese, A., Ercolini, D. 2003. Relationships between flavoring capabilities, bacterial composition, and geographical origin of natural whey cultures used for traditional water buffalo mozzarella cheese manufacture. *Journal of Dairy Science*, 86, 486–497.

Mears, R.G., Shenton, A.J. 1973. Adulteration and characterization of orange and grapefruit juices. *Journal of Food Technology*, 8, 357–389.

Mizrach, A., Schmilovitch, Z., Korotic, R., Irudayaraj, J., Shapira, R. 2007. Yeast detection in apple juice using raman spectroscopy and chemometric methods. *Transactions of the ASABE*, 50, 2143–2149.

Moreda-Pineiro, A., Fisher, A., Hill, S.J. 2003. The classification of tea according to region of origin using pattern recognition techniques and trace metal data. *Journal of Food Composition and Analysis*, 16(2), 195–211.

Novotna, H., Kmiecik, O., Gałązka, M., Krtkova, V., Hurajova, A., Schulzova, V. 2012. Metabolomic fingerprinting employing DART-TOFMS for authentication of tomatoes and peppers from organic and conventional farming. *Food Additives & Contaminants: Part A*, 29(9), 1335–1346.

Nunez, M., Pena, R.M., Herrero, C., Garcia-Martin, S. 2000. Analysis of some metals in wine by means of capillary electrophoresis. Application to the differentiation of Ribeira Sacra Spanish red wines. *Analysis*, 28, 432–437.

Ogrinc, N., Kosir, I.J., Kocjancic, M., Kidric, J. 2001. Determination of authenticity, regional origin, and vintage of Slovenian wines using a combination of IRMS and SNIF-NMR analyses. *Journal of Agricultural and Food Chemistry*, 49, 1432–1440.

Ogrinc, N., Kosir, I.J., Spangenberg, J.E., Kidric, J. 2003. The application of NMR and MS methods for detection of adulteration of wine, fruit juices, and olive oil. *A Review Analytical and Bioanalytical Chemistry*, 376, 424–430.

Olsson, J., Borjesson, T., Lundstedt, T., Schnurer, J. 2002. Detection and quantification of ochratoxin A and deoxynivalenol in barley grains by GC-MS and electronic nose. *International Journal of Food Microbiology*, 72, 203–214.

Paramas, A.M.G., Barez, J.A.G., Garcia-Villanova, R.J., Pala, T.R., Albajar, R.A., Sanchez, J.S. 2000. Geographical discrimination of honeys by using mineral composition and common chemical quality parameters. *Journal of the Science of Food and Agriculture*, 80, 157–165.

Parastar, H., Jalali, H.M., Sereshti, H., Varnosfaderani, A.M. 2012. Chromatographic fingerprint analysis of secondary metabolites in citrus fruits peels using gas chromatography-mass spectrometry combined with advanced chemometric methods. *Journal of Chromatography A*, 1251, 176–187.

Peng, Y., Liu, F., Ye, J. 2006. Quantitative and qualitative analysis of flavonoid markers in Frucus aurantii of different geographical origin by capillary electrophoresis with electrochemical detection. *Journal of Chromatography B*, 830, 224–230.

Penza, M., Cassano, G. 2004. Chemometric characterization of Italian wines by thin-film multisensors array and artificial neural networks. *Food Chemistry*, 86, 283–296.

Peres, B., Barlet, N., Loiseau, G., Montet, D. 2007. Review of the current methods of analytical traceability allowing determination of the origin of foodstuffs. *Food Control*, 18, 228–235

Perez, A.L., Smith, B.W., Anderson, K.A. 2006. Stable isotope and trace element profiling combined with classification models to differentiate geographic growing origin for three fruits: Effects of subregion and variety. *Journal of Agricultural and Food Chemistry*, 54, 4506–4516.

Perez Elortondo, F.J., Ojeda, M., Albisu, M., Salmeron, J., Etayo, I., Molina, M. 2007. Food quality certification: An approach for the development of accredited sensory evaluation methods. *Food Quality and Preference*, 18, 425–439.

Petrakis, P., Touris, I., Liouni, I., Zervou, M., Kyrikou, I., Kokkinofta, I., Theocharis, C.R., Mavromoustakos, T.M. 2005. Authenticity of the Traditional Cypriot Spirit "Zivania" on the Basis of 1HNMR Spectroscopy Diagnostic Parameters and Statistical Analysis *Journal of Agricultural and Food Chemistry*, 53, 5293–5303.

Picque, D., Cattenoz, T., Corrieu, G., Berger, J.L. 2005. Discrimination of red wines according to their geographical origin and vintage year by the use of mid-infrared spectroscopy. *Sciences des Aliments*, 25, 207–220.

Raspor, P., Milek, D.M., Polanc, J., Mozina, S.S., Cadez, N. 2006. Yeasts isolated from three varieties of grapes cultivated in different locations of the Dolenjska vine-growing region, Slovenia. *International Journal of Food Microbiology*, 109, 97–102.

Ratel, J., Berge, P., Berdague, J.L., Cardinal, M., Engel, E. 2008. Mass spectrometry based sensor strategies for the authentication of oysters according to geographical origin. *Journal of Agricultural and Food Chemistry*, 56, 321–327.

Reid, L., O'Donnel, C.P., Downey, G. 2006. Recent technological advances for the determination of food authenticity. *Trends in Food Science & Technology*, 17, 344–353.

Rivero, R.C., Hernandez, P.S., Rodrıguez, E.M.R., Martın, J.D., Romero, C.D. 2003. Mineral concentrations in cultivars of potatoes. *Food Chemistry*, 83(2), 247–253.

Rodriquez-Delgado, M.A., Gonzalez-Hernandez, G., Conde-Gonzales, J.E., Perez-Trujillo, J.P. 2002. Principal component analysis of the polyphenol content in young red wines. *Food Chemistry*, 78, 523–532.

Rossmann, A. 2001. Determination of stable isotope ratios in food analysis. *Food Reviews International*, 17, 347–381.

Ruoff, K., Luginbuhl, W., Kunzli, R., Iglesias, M.T., Bogdanov, S., Bosset, J.O. 2006. Authentication of the botanical and geographical origin of honey by mid-infrared spectroscopy. *Journal of Agricultural and Food Chemistry*, 54, 6873–6880.

Salvatore, E., Bevilacqua, M., Bro, R., Marini, F., Cocchi, M. 2013. Classification methods of multiway arrays as a basic tool for food PDO authentication. In *Food Protected Designation of Origin: Methodologies and Applications*. M. de la Guardia and A. Gonzalvez (Eds.). Elsevier, Amsterdam, pp. 339–382.

Sarbu, C., Nacu-Briciu, R.D., Kot-Wasik, A., Gorinstein, S., Wasik, A. Namiesnik, J. 2012. Classification and fingerprinting of kiwi and pomelo fruits by multivariate analysis of chromatographic and spectroscopic data. *Food Chemistry*, 130, 994–1002.

Sass-Kiss, A., Sass, M. 2000. Immuno analytical method for quality control of orange juice products. *Journal of Agricultural and Food Chemistry*, 48, 4027–4031.

Sass-Kiss, A., Sass, M. 2002. Distribution of various peptides in citrus fruits (grapefruit, lemon, and orange). *Journal of Agricultural and Food Chemistry*, 50, 2117–2120.

Schwartz, R.S., Hecking, L.T. 1991. Determination of geographic origin of agricultural products by multivariate analysis of trace element composition. *Journal of Analytical Atomic Spectrometry*, 6, 637–642.

Serrano, D.L., Saurina J., Hernandez C.A.S., Checa, A. 2012. Classification and characterisation of Spanish red wines according to their appellation of origin based on chromatographic profiles and chemometric data analysis. *Food Chemistry*, 135, 1425–1431.

Shao, B.Y., Wang, C., Zheng, J. 2008. Study on DNA extraction and PCR-RFLP detection method of Aspergillus flavus. *Food Science*, 29, 393–396.

She, Y.X., Liu, J.J., Wang, J. 2008. Determination of 24 sulfonamides in milk by ultra performance liquid chromatography combined with Quadrupole mass spectrometry. *Journal of Instrumental Analysis,* 27(12), 1313–1317.

Shen, D., Wu, Q., Sciarappa, W.J., Simon, J.E. 2012. Chromatographic fingerprints and quantitative analysis of isoflavones in Tofu-type soybeans. *Food Chemistry*, 130, 1003–1009.

Shende, C., Inscore, F., Gift, A., Maksymiuk, P., Farquharson, S. 2004. Analysis of pesticides on or in fruit by surface-enhanced Raman spectroscopy. *Proceedings of the SPIE*, 5587, 170–176.

Shi, P.Y., He, Q., Song, Y., Qu, H.B., Cheng, Y.Y. 2007. Characterization and identification of isomeric flavonoid O-diglycosides from genus Citrus in negative electrospray ionization by ion trap mass spectrometry and time-of-flight mass spectrometry. *Analytica Chimica Acta*, 598, 110–118.

Song, D.Y., Li, P., Zhao, Y. 2006. Application of pulsed NMR for intact measurement of oil content in soybean seed. *Chinese Journal of Oil Crop Sciences*, 28(2), 199–202.

Spitaler, R., Araghipour, N., Mikoviny, T., Wisthaler, A., Via, J.D., Mark, T.D. 2007. Discrimination of different wine varieties through direct headspace analyses by PTR-MS. In *3rd International Conference on Proton Transfer Reaction Mass Spectrometry and its Applications*, A. Hansel and T.D. Mark (Eds.) Innsbruck university press . Austria, pp. 136–139.

Springer, A.E., Riedl, J., Esslinger, S., Roth, T., Glomb, M.A., Fauhl-Hassek, C. 2014. Validated modeling for German white wine varietal authentication based on HS-SPME-GC-MS fingerprinting. *Journal of Agricultural and Food Chemistry*, 62, 6844–6851.

Stocker, A., Rossmann, A., Kettrup, A., Bengsch, E. 2006. Detection of royal jelly adulteration using carbon and nitrogen stable isotope ratio analysis. *Rapid Communications in Mass Spectrometry*, 20, 181–184.

Strike, D.J., Meijerink, M.G.H., Koudelka-Hep, M. 1999. Electronic noses—A mini-review. Fresenius. *Journal of Analytical Chemistry*, 364(6), 499–505.

Tang, C.H., Ma, C.Y. 2009. Heat Induced modification in the functional and structural properties of vicilin rich protein isolate from kidney (Phaseous vulgaris L) bean. *Food Chemistry*, 115, 859–866.

Tang, P., Tian, J., She, Z.B. 2006. Determination of protein in dairy products by capillary electrophoresis studies. *Journal of Analytical Science*, 22(1), 5–8.

Terzy, V., Morcia, C., Gorrini, A., Stanca, A.M., Cegrı, P.R., Faccioli, P. 2005. DNA-based methods for identification and quantification of small grain cereal mixtures and fingerprinting of varieties. *Journal of Cereal Science*, 41, 213–220.

Tian, H., Zhan, P., Zhang, H. 2014. Development of a fatty acid fingerprint of white apricot almond oil by gas chromatography and gas chromatography-mass spectrometry. *European Journal of Lipid Science and Technology*, 116, 126–133.

Tian, H.X., Wang, Z., Xu, S.Y. 2006. Separation and identification of volatile flavors of Jinhua Ham by gas chromatography-mass spectrometry coupled with head space solid phase microextraction. *Chinese Journal of Chromatography*, 24(2), 177e180.

Tistaert, C., Dejaegher, B., Vander Heyden, Y., 2011. Chromatographic separation techniques and data handling methods for herbal fingerprints: a review. *Analytica Chimica Acta*, 690(2), 148–161.

Tomas-Barberan, F.A., Ferreres, F., Garcıa-Viguera, C., Tomas- Lorente, F. 1993. Flavonoids in honey of different geographical origin. *Zeitschrift fur Lebensmittel-Untersuchung und –Forschung*, 196, 38–44.

Vardin, H., Tay, A., Ozen, B., Mauer, L. 2008. Authentication of pomegranate concentrate using FTIR spectroscopy and chemometrics. *Food Chemistry*, 108(2), 742–748.

Varga, J., Kocsube, S., Suri, K., Szigeti, G., Szekeres, A., Varga, M., Toth, B., Bartok, T. 2010. Fumonisin contamination and fumonisin producing black Aspergilli in dried vine fruits of different origin. *International Journal of Food Microbiology*, 143, 143–149.

Vilanova, M., Vilarino, F. 2006. Influence of geographic origin on aromatic descriptors of Spanish Albarino wine. *Flavour and Fragrance Journal*, 21, 373–378.

Wang, M., Li, R., Zou, S. 2008. Determination of carbofuran residue in aquatic products by gas chromatography. *Chinese Journal of Chromatography*, 26(6), 775–777.

Wang, N., Lin, X.-Y., Yuan, R.-S. 2007. Study on navel orange storage processing with NMR. *Food Science*, 28(7), 83–87.

Wang, S., Bao, Z., Zhang, L.L. 2006. Species identification of four scallop species mainly cultured in China with application of PCR-RFLP technique. *Food Science*, 27(7), 201–214.

Wang, X.L., Pan, J., Wang, S. 2006. Identification of the raw material of Ejiao using the CytB gene PCR-RFLP method. *Periodical of Ocean University of China*, 36(4), 645–648.

Winkler, F.l. 1984. Application of natural abundance stable isotope mass spectrometry in food control. In *Chromatography and Mass Spectroscopy in Nutrition Science and Food Safety*, A. Frigerio and H. Milon (Eds.). Elsevier, Amsterdam, pp. 173–190.

Wolfendera, J.L., Martia, G., Thomas, A., Bertrand, S. 2015. Current approaches and challenges for the metabolite profiling of complex natural extracts. *Journal of Chromatography A*, 1382, 136–164.

Woodcock, T., Downey, G., O'Donnell, C.P. 2008. Review: Better quality food and beverages: The role of near infrared spectroscopy. *Journal of Near Infrared Spectroscopy*, 16(1), 1–29.

Woolfe, M., Primrose, S. 2004. Food forensics: Using DNA technology to combat misdescription and fraud. *Trends in Biotechnology*, 22, 222–226.

Wu, J.H., Peng, Y.K., Jiang, F. 2009. Hyperspectral scattering profiles for prediction of beef tenderness. *Transactions of the Chinese Society for Agricultural Machinery*, 40(12), 135–138, 150.

Wu, Y., Chang, C. 2009. Biochemical genetic analysis of isozymes in three populations of Misgurnus anguillicaudatus. *Journal of Northwest A&F University (Natural Science Edition)*, 37(11), 37–42.

Xiao, Z., Liu, S., Gu, Y., Xu, N., Shang, Y., Zhu, J. 2014. Discrimination of cherry wines based on their sensory properties and aromatic fingerprinting using HSSPME-GC-MS and multivariate analysis. *Journal of Food Science*, 79, C284–C294.

Xu, X., Chen, Q., Yong, K. 2008. Application of PCR in analyzing farm sources. *Journal of Food Science and Technology*, 33(12), 258–261.

Xu, Y., Cheung, W., Winder, C.L., Goodacre, R. 2010. VOC-based metabolic profiling for food spoilage detection with the application to detecting Salmonella typhimurium-contaminated pork. *Analytical and Bioanalytical Chemistry*, 397(6), 2439–2449.

Xu, Y., Cheung, W., Winder, W.L., Dunn, W.B., Goodacre, R. 2011. Metabolic profiling of meat: assessment of pork hygiene and contamination with Salmonella typhimurium. *Analyst*, 136, 508–514.

Yang, W., Tian, S., Dong, W. 2007. Determination of total sugar in honey by ultraviolet spectrophotometry. *Chinese Journal of Spectroscopy Laboratory*, 24(4), 739–742.

Yuan, H., Fu, H., Rao, P.F. 2005. Analysis of trans fatty acids by infrared spectroscopy and gas chromatography. *Journal of Cereals & Oils*, 24(1), 25–28.

Zhan, J., Yu, X.J., Huang, W. 2008. A simple method for the simultaneous determination of 7 food additives in foods with complex matrices by GC/MS. *Journal of Food and Drug Analysis*, 24(6), 729–731.

Zhang, N., Zhang, D., Li, S. 2008. Preliminary study on origin traceability of mutton by near infrared reflectance spectroscopy coupled with SIMCA method. *Transactions of the Chinese Society of Agricultural Engineering*, 24(12), 309–312.

Zhao, J., Zhai, J., Liu, M. 2006. The determination of beef tenderness using near infrared spectroscopy. *Spectroscopy and Spectral Analysis* 26(4), 640–642.

Zhao, R., Zhang, W., Ouyang, J. 2008. 1H NMR fingerprinting spectra and pattern recognition of beer. *Food Science*, 29(12), 564–567.

Zheng, Y., Zhu, J., Li, Y. 2007. Study on application of RAPD fingerprint on rapid identification for beer spoilage. *Food Science*, 28(11), 331–337.

Zhihua, L., Shouju, J., Yinghui, D., Xueliang, C., Rong-mao, L., Yongpu, Z. 2009. The isozymes in different populations of clam Meretrix meretrix. *Journal of Dalian Fisheries University*, 6, 525–530.

Zhong, W., Chen, D., Yong, W. 2005. Determination of acrylamide in fried starchy foods by gas chromatography-mass spectrometry. *Chinese Journal of Chromatography*, 23(3), 312–314.

Zhou, M., McDonald, J.F., Fernandez, F.M. 2010. Optimization of a direct analysis in real time/time-of-flight mass spectrometry method for rapid serum metabolomic fingerprinting. *Journal of the American Society for Mass Spectrometry*, 21, 68–75.

Zhou, Q.B., Li, F.B., Zhou, L. 2006. RAPD markers between yellow catfish pelteobagrus fulvidraco and long yellow catfish pelteobagrus eupogon. *Acta Hydrobiologica Sinica*, 30(4), 482–485.

Zou, Y.L. 2006. HPLC determination of sulfite in pickle garlic using post-column fluorescence detection. *The Food Industry*, 4, 52–53.

CHAPTER 18

Application of Authentication and Traceability in Chocolate

Saima Amin and Showkat R. Mir
School of Pharmaceutical Education and Research

CONTENTS

18.1 INTRODUCTION

Chocolate is a sweet food product obtained from *Theobroma cacao* L. seeds belonging to the *Malvaceae* family (Aprotosoaie et al., 2016), which contains not less than 35% dry cocoa solids in total, including not less than 18% cocoa butter and not less than 14% dry, non-fat cocoa solids such as alkaloids (theobromine, phenylethylamine, and caffeine) and flavonoids.

After grinding the roasted seeds, cocoa powder is obtained to form liquid, paste, and bars. Since the cocoa seeds are bitter in taste they are fermented to develop flavor. The overall processing of cocoa beans is shown in Figure 18.1. Cocoa is a popular crop that serves as the raw material for chocolate manufacturing industries. It is also used as a flavor enhancer in confectioneries, baking, and dairy products. Cocoa trees are very delicate and sensitive and, therefore, producers protect the crop from the wind and sun by planting them between other trees, for example, between coconut and nutmeg trees. Ivory Coast is the major producer of cocoa beans followed by Ghana, Indonesia, Nigeria, and Cameroon (Cidell and Alberts, 2006; Sunjoj et al., 2016b).

The chocolate products need to be labeled for their authenticity. The labeling and composition of chocolate products are controlled by the Cocoa and Chocolate Products Regulations 2003, which implement the EC Directive 2000/36/EC2 of European Parliament. The quality of cocoa beans wholly depends on the fermentation process during which flavors are produced. Fermentation improves the flavor of chocolate by reducing astringency and bitterness. Due to the high demand of cocoa beans, there are chances of adulterations by traders. One possibility of adulteration is to mix the under-fermented beans with the fermented ones. Also, all food containing chocolate and cocoa products have nutritional values that are related to the amount and kind of carbohydrates, fats, and proteins that they contain. The cocoa, milk, and sugar are the main ingredients of chocolate. Some allergens such as peanut antigens and egg antigens have also been found as adulterants in chocolate (Khuda et al., 2015). Therefore, under the Nutrition Labeling and Education Act (NLEA) regulations (www.fda.gov/downloads/food/guidanceregulation/ucm265446.pdf), most food products are required to display nutritional information on their packages. This information is considered necessary by the government for providing nutritional guidance to the consumers. The main fat in cocoa and chocolate is the cocoa butter, but they can also contain other edible fats. Despite the presence of a variety of amino acids in cocoa protein, an approximately pure protein content of 25% in cocoa solids, cocoa, and plain chocolate is not an important source of protein as might be expected. Some manufacturers add collagen for health-conscious consumers to prevent aging and wrinkles, and for this halal authentication is required in some countries (Shariff and Abdlah, 2014). Carbohydrates in the form of sugars are present in high quantities in chocolate products, therefore chocolates are made from many different recipes and contain extra ingredients in addition to cocoa products. Thus, the nutritional value will vary with the ingredients. For example, dark chocolate has a high proportion of cocoa solids and, therefore, its nutritional value is higher than that of white milk chocolate that has a lower proportion of cocoa. In milk chocolate, the protein value is higher than that in dark chocolate (Moros et al., 2007). Since the chocolate products display nutritional value on the packing, it can be adulterated by using cheap ingredients to beat the brand driven market competition. Therefore, chocolate manufacturers are required to label the composition of fats also to satisfy legislation requirements (Food Information for Consumers Regulation, 2014).

18.2 PROCESSING OF COCOA BEANS

Traceability, as defined by the European community, is the ability to trace and follow a food, feed, and food producing animal or substance, which is intended or expected to be incorporated into food or feed, through all the stages of production and distribution (European Community, 2011). Since the cocoa supply chain consists of growers, collectors, local traders, exporters, multinational companies (MNC), local processors, and local manufactures, we can divide the traceability into two parts, tracking and tracing, as proposed by Schwagele (2005), thus enabling the supply chain to recognize the flow of the product from the initial stages. Therefore, in case of any recalls due to uncontrolled use of pesticides, farmers are held responsible or the suppliers would bear the cost of recalls. Thus, the efficiency of a traceability system depends on the ability to identify each unit that is produced and distributed in a way that enables continuous tracking, from primary production to the retail point of sale (Manikas et al., 2008). For easy traceability, three

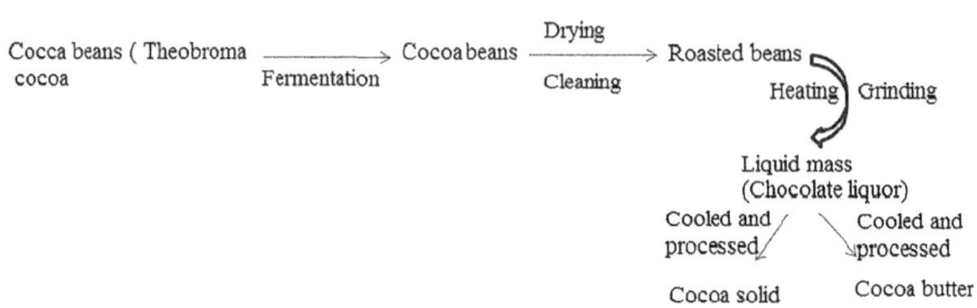

FIGURE 18.1 Processing of cocoa beans.

varieties of chocolates have been identified by the U.S. Food and Drug Administration (FDA, Part 163), namely:

1. Sweet chocolate: It is a widely used cocoa product containing cocoa solids, cocoa butter, added oil, and sugar.
2. Milk chocolate: It contains milk in addition to sweet chocolate.
3. White chocolate: It contains cocoa butter, sugar, and milk, but cocoa solid is not added to it.

The practice of combining chocolate with foreign ingredients or of replacing one or more natural or synthetic ingredients to give the desired properties is known as chocolate adulteration. The adulteration of food items with physical or chemical substances is a serious health threat for the consumers (Lakshmi, 2012). There has been a rising number of cases related to food adulteration, such as adulteration of wheat gluten with melamine to increase the protein content; milk adulteration with detergent, fat, and urea; contamination of chili powder with red brick powder, etc. In such scenarios, chocolate branding is the only option available to the consumers to avoid adulteration. For example, Cadbury, established in 1866, was the first to brand their chocolate as fat-free as they squeezed out the cocoa butter from the cocoa beans using a cocoa press (https://chocolateclass.wordpress.com). This process eliminated the need to add starch to mask the taste and texture of cocoa butter. Thus, Cadbury was the first brand to announce the authenticity of their chocolates. Their product, Cadbury cocoa essence, was labeled as the pure chocolate product. Later, some companies, such as van Houten and Walter Baker & Co., also branded their chocolate products as pure. The breakfast cocoa product by Walter Baker (W. Baker & Co. Advertisement) was claimed to be free of alkaline chemicals that were used by others to darken the appearance of chocolate or to mask the unpleasant color. Similarly, another way of adulterating the chocolate originated from cocoa beans is to substitute the cocoa butter with animal fats, such as butter, lard, tallow, etc. These fats are cheaper, thereby the production of chocolate is possible at a lower cost. Such substitutions affect the melting properties of the original chocolate bars and they do not melt in the mouth easily. Therefore, mouth melting is an indication of high purity and quality of chocolates (Grivetti and Shapiro, 2009a,b; Bahri and Che Man, 2016).

Adulteration of chocolate was traced in the early 1700s in Europe when they had low quality chocolates that contained cocoa shells, sweet almonds, beef marrow, sugar, and flavor. In 1850, a British medical journal published a study on the adulteration of chocolate with red bricks, as a consequence of which the British Food and Drug Act, 1860 and Adulteration of Food Act, 1972 (https://chocolateclass.wordpress.com) were framed to implement regulations related to the purity and quality of chocolates. In the 19th century, issues related to the safety of such tempting products were highlighted (Muller and Steinhart, 2007). For example, the U.S. Food and Drug Administration highlighted the presence of insect filth (more than 75 fragments/50 g cocoa powder) and rodent filth (2 rodent hair/50 g cocoa powder) as a compromise to the authenticity of chocolate powder (US-FDA, 03/20/2015).

Authentication is important for both—the official bodies in charge of labeling and the industries where incoming batches of raw materials and finished products are tested for compliance with specifications. The authenticity of chocolate must also be verified to maintain quality and consumer satisfaction and, possibly, to prevent economic fraud. With increasing trends of adulteration, fast and reliable analytical methods are needed to tackle the authentication challenges and ensure product quality. A range of physical and chemical methods have been used for the detection of adulteration and authenticity in food and agricultural commodities, but no single method could provide the means to allow the unequivocal identification of cocoa and cocoa mixtures. However, it seems most probable that the combined application of several methods and the subsequent chemometric data evaluation could provide the means for a reliable quantification of mixtures of cocoa butter and other vegetable fats (Winkler, 2014).

The following major adulterants have been identified in chocolates (Gabriela, 2012; Anonymous, 1995):

1. Fats and oils
    a. Cocoa butter substitutes
    b. Cocoa butter equivalents
    c. Cocoa butter replacer

Vegetable fat is added to chocolate depending on its functional compatibility with cocoa butter, and it is classified as follows: (i) Cocoa butter equivalent (CBE)- They are non-lauric plant fats, which are similar in their physical and chemical properties to cocoa butter and are miscible with it in any amount without altering the properties of cocoa butter. (ii) Cocoa butter replacer (CBR)- They are non-lauric fats in which the distributions of fatty acids are similar to that in cocoa butter, but where the structures of triglycerides are completely different; they are compatible with cocoa butter only in small proportions. (iii) Cocoa butter substitutes (CBS)-They are lauric plant fats, chemically totally different from cocoa butter with only some physical similarities; they are suitable only for 100% substitution of cocoa butter (Bohacenko et al., 2004). Triacylglycerol (TAG) compositional analyses have been extensively explored as a means of detecting adulterations with oils and fats. Determination of the TAG compositions for lipid products helps to detect the adulterations in milk fat and cocoa butter. As genuine cocoa butter comprises only three major TAG species, namely 1,3-dipalmitoyl-2-oleyl-glycerol (POP), 1-palmitoyl-2-oleoyl-3-stearoyl-glycerol (POS), and 1,3-distearoyl-2-oleoyl-glycerol (SOS), a straight line relationship is found between C_{50} and C_{54} contents based on their natural variability. Since the TAG compositions of CBEs and cocoa butter (CB) are not identical, the differences between them are used to detect the presence of CBEs in confectionery products (Lipp and Anklam, 1999; Simoneau et al., 1999). A specific TAG of CB would help to identify deviations produced by adulteration practices (Marikkar et al., 2016). Therefore, the addition of vegetable fats such as palm oil or coconut oil is permissible up to 5% of the finished product and must be declared on the label (AOAC, 1995).

1. Starch such as soybean flour (Altug and Gonul, 1987).
2. Bacterial contamination: *Salmonella* and *E. coli* have also been found as significant contaminants in chocolate. In a case study, Brazilian chocolates were found to be contaminated due to dirty equipment, utensils, and food handlers (Nascimento et al., 2015). The risks associated with such contaminated products had been recognized by the European industry organizations for chocolate, confectionery, and biscuits. In the 1990s, the International Office of Cocoa, Chocolate, and Confectionary (IOCCC) (IOCCC, 1991) published two codes of practices: one based on the hazard analysis and critical control point (HACCP) and one for specific GMPs (good manufacturing practices) for cocoa in the chocolate, biscuit, and confectionery industries (caobisco/ accessed January 2, 2017).The microbiological safety of chocolate products can be ensured by the

consequent application of the HACCP concept and by adherence to prerequisite programs to ensure good manufacturing and agricultural practices throughout the whole processing chain. This includes not only the final processing steps of making chocolate, but it starts at a point where raw agricultural materials, like cocoa and nuts, are sourced for chocolate making. Microbial data can play an important role in the verification of implemented controls, but their validity and limitations need to be understood (Kvenberg et al., 2000; Swanson et al., 2000; Kornacki, 2006).

3. Heavy metal contamination: The Food Safety and Standard Authority of India (FSSAI) has set upper levels for two major heavy metals—lead (5 ppm by weight of dry fat-free substances) and copper (70 ppm by weight of fat-free substances)—which can cause damage to the central nervous system leading to lower intelligence quotients (FSSAI regulations, 2011). Cadmium is carcinogenic and can damage to kidneys, bone, and liver. Copper causes general weakness, depression, and low blood pressure (Rankin et al., 2005). Some essential elements have also been reported in chocolates, which might significantly contribute to the nutritional values of products, but aluminum, cadmium, and nickel have to be taken into consideration (Sager, 2012). Due to health concerns, the manufacturers of chocolate are adopting technical reformulation strategies where antioxidants, in a concentration of 40%–50%, and sugar, at low contents, are added. Therefore, categorization of chocolates into premium quality chocolates and main stream chocolates has been adopted. The premium chocolate is less vulnerable to health-related issues. Authentication and/or traceability studies must therefore be developed to safeguard this product against fraud, which can occur even when the product is partially or totally substituted with hazelnuts of lower quality (Oddone et al., 2009).

18.3 QUALITY ASPECTS OF CHOCOLATE

High-quality cocoa beans are essential for chocolate production with adequate nutritional values and desired taste. Quality controls are done with various laboratory tests, which represent the mandatory starting point for a proper traceability system. Apart from the bean size, color, and general appearance that serve as the physical attributes for estimating the quality of cocoa beans, another important physical evaluation includes grading of beans as moldy beans, slaty beans, and insect damaged beans. Quartering tool, secateur, moisture analyzer, and cut test board are some of the instruments used for bean counting, segregation of defective beans, and identification of fermentation levels. The validation of chocolate authenticity relies mostly on the analysis of proteins and/or DNA sequences (Ellis et al., 2016). DNA barcoding is a new emerging technique to authenticate the chocolate quality. It is a vital tool to trace the adulteration in chocolate as the gene sequence of a genuine cocoa bean is different from that of its adulterants. The barcode of life database (BOLD) coordinated by the International Barcode of Life Project (IBOL) is a repository, which supports the collection of DNA barcodes, with the aim of creating a reference library for all living species (Ratnasingham and Hebert, 2007; Galimberti et al., 2013). However, the development of DNA marking of cocoa beans could open up increased opportunities for the branding and marketing of single-origin fine cocoa beans; strict quality checks and control over the chocolate manufacturing processes are equally important. Apart from the screening of TAG, some of the simple tests for analyzing the quality of chocolates are as follows (Afoakwa et al., 2015; Sunoj et al., 2016a,b):

1. **Fermentation index**
   Fermentation index is estimated by an established method (Gourieva and Tserevitinov, 1979). According to this method, cocoa nibs are powdered and extracted in a mixture of methanol:hydrochloric acid. It is filtered after keeping the mixture for some time in a refrigerator at 8°C. Absorbance of the filtrate is measured at 460 and 530 nm using a UV-Visible spectrophotometer. A fermentation index value greater than one is indicative of well-fermented cocoa, while less than one refers to under-fermented cocoa.
2. **Measurement of pH**
   The pH of cocoa beans is determined according to the method described in AOAC (1995). Generally, the ground beans are blended with hot, distilled water and then the homogenate is filtered and cooled. The pH is measured with a digital pH meter. A sharp decrease in pH is observed for pods stored for nearly 7 days. This is due to the decrease in pulp volume per seed that occurs because of the water evaporation and inversion of sucrose, which diminishes the total sugar content leading to a reduced acid production during fermentation. After the fourth day onward, the adhering pulp is released in the form of sweat, which releases acid and alcohol during fermentation (Afoakwa et al., 2013). However, the acidic beans have a direct impact on the flavor of chocolate, imparting an astringent taste and it seems to cause problems while manufacturing chocolate. Therefore, pod storage treatments result in beans with slightly higher pH (i.e., less acidity).
3. **Total polyphenols**
   The total polyphenols in cocoa beans are estimated using the Folin-Ciocalteu reagent (Singleton and Rossi, 1965). Total polyphenol is expressed in terms of mg g^{-1} of the ground beans. In general, the total polyphenol content decreases with increasing storage time at any level of fermentation. The decrease in polyphenol could be due to the diffusion of polyphenols into the fermentation sweat. Also, the reduction in pulp volume during pod storage might facilitate the oxidation and polymerization of polyphenols, which will reduce the astringency and bitterness in cocoa and cocoa products made from the cocoa beans.

18.4 CHOCOLATE AUTHENTICATION THROUGH TRACEABILITY MEASUREMENT

Food safety management for cocoa and chocolate focuses mainly on the incoming hazards and their controls at different stages of processing as well as the prevention of recontamination during further processing. Adulteration with oils and fats is an important commercial issue as it affects the authenticity of chocolate. Instrumental techniques based on chromatography and infrared (IR) spectroscopy have received particular attention from researchers worldwide since they are fast and efficient (Awad et al., 2012). Majority of the past studies have suggested the use of assays based on fatty acids, triacylglycerol components, minor constituents, and spectral characteristics as they are really useful for determining the adulteration in chocolate lipids. Physical methods usually look for physical parameters, such as density, texture, color, and solubility. Chemical techniques, such as chromatography techniques and mass spectroscopy (MS), are based on the chemical properties of the samples and have shown good robustness in component identification and adulteration detection in such commodities. For regulatory compliance, monitoring of chocolate quality at various stages of production, processing, and distribution is very important. Some of the analytical techniques available for preventing frauds during manufacturing are described here.

1. **Spectroscopic techniques**
   Ultraviolet and visible (UV-Vis) absorption spectroscopy is the measurement of a beam of light after it passes through a sample or after reflection from a sample surface. Absorption measurements can be done at a single wavelength or over an extended spectral range. UV-Visible methods were first introduced by the American chemist, Arnold Beckman, in 1940. He invented a spectrophotometer that had the ability to split the beam of light into sample and reference beams. When a substance was placed in the spectrophotometer, it absorbed UV-Visible light, resulting in a signature called spectrum. The UV instrument was equipped with hydrogen and deuterium as the source of ultraviolet light, while a tungsten lamp was used for the visible measurements. Those light sources emitted radiations of a specific wavelength range, for example, the deuterium lamps emitted light in the range of 190–380 nm, while the tungsten halogen lamp generated light in the range of 380–800 nm. The sample to be screened for analysis was placed in the instrument and the absorbance of light by that placed sample is taken as the concentration of the analyte present. Similarly, the IR spectra are recorded on the FTIR instruments, such as Jasco FT/IR-410, Japan by using the potassium bromide pressed disc method. The spectrum is scanned over a frequency range of 4,000–400 cm^{-1}. Such a spectroscopy provides signals based on the absorption of a single wavelength that allows the identification of oils and fats or the identification of individual fatty acids (Lipp and Anklam, 1998; Di Anibal et al., 2009).
2. **Gas chromatography (GC)**
   GC involves a sample being vaporized and injected onto the head of the chromatographic column. The sample is transported through the column by the flow of inert, gaseous mobile phase. The column itself contains a liquid stationary phase, which is adsorbed onto the surface of an inert solid. Triglyceride profile in cocoa butter has been identified with high temperature GC using a short column (0.5 m) packed with a nonpolar stationary phase or a nonpolar capillary column. The triglyceride fractions are separated according to their acyl-carbon number, i.e., the total number of carbon atoms in the fatty acid chains (Bohacenko et al., 2004). The volatile flavor components of unroasted and roasted cocoa butters can also be investigated by GC. Pyrazines are found to play a very important role in flavor. Most of them are present in roasted cocoa butter, while some are present in unroasted cocoa butter also. The total concentration of pyrazines in roasted cocoa butter is about twice as high as in unroasted cocoa butter (Hashim and Chaveron, 1994). Similarly, estimation of the linalool content by simultaneous steam distillation is found to be suitable for the quality classification of roasted cocoa butters (Pino and Roncal, 1992). The presence of illipe fat in cocoa butter can be detected by the determination of fatty acids in mixtures containing 10% and more of illipe fat. Analysis of the sterols, especially the quantification of a-spinsterol, allows the detection of 5% and more illipe fat in cocoa butter, while the analysis of tri-terpenes leads to a detection of 1% illipe fat in cocoa butter. The three major triglycerides in cocoa butter and chocolate, namely POP, POS, and SOS, have been extensively investigated by GC (Ulberth et al., 2003).

Detailed analyses of several fractions of the unsaponifiables (sterols, methylsterols, triterpenes, etc.) have also been carried out (García Regueiro et al., 1994).

3. **High performance liquid chromatography (HPLC)**
HPLC makes use of a high pressure pump to deliver a mobile phase solvent at a uniform rate at pressures that typically range from 500–5,000 psi. The most obvious advantage of HPLC is that samples can be separated much more quickly. Instruments like Shimadzu model HPLC equipped with quaternary LC-10AVP pumps, a variable wavelength programmable UV detector, a SPD-10AVP column oven (Shimadzu), a SCL-10AVP system controller (Shimadzu), a Rheodyne injector fitted with a 20 µL loop, and the Class-VP software are used for the analysis of samples. Chromatography is performed with a reverse phase column at ambient conditions. The mobile phase is pumped through the column at a flow rate of 1.0 mL min^{-1} and an injection volume of 20 µL is usually taken and the elute is analyzed at a wavelength that represents the wavelength for maximum response. The amount of cocoa present in a chocolate product is generally regarded as a guide to quality and its accurate quantification has been attempted for over 100 years (Lipp and Anklam, 1998; Richards and Wailes, 2012). HPLC is a powerful analytical technique, but it is undesirable when the work flow is relatively fast and the samples are expensive (Stefanoa et al., 2012). Total phenolic and flavonoid contents have been determined in commercial chocolates with HPLC (Meng et al., 2009). It takes a higher elution time (Cserhati et al., 2005; Fanali et al., 2013) when a UV detector equipped with an HPLC system is used for the separation of components in chocolate. Bromination leads to a better separation of the monounsaturated triglycerides from the di- and polyunsaturated triglycerides. The brominated monounsaturated triglycerides give four times stronger absorption and the analysis time is also considerably reduced. The application of isocratic mobile phase conditions in combination with an evaporative light scattering detector offers the possibility of a fast separation of triglycerides in vegetable fats and oils. The time needed for analysis is only about (Caboni et al., 1991) 20 mins, thus providing the possibility of separating triglycerides (Rombaut et al, 2009). The impregnation of the solid phase with silver ions followed by a stepwise elution, give pure trisaturated and disaturated monoene species together with a mixed fraction of saturated dimonoene and disaturated diene and a fraction of highly unsaturated species. Silver ion HPLC is used to determine the amount of *cis*- and *trans* triglycerides in vegetable fats. Triglycerides containing trans-fatty acids are formed during the hardening of fats and oils, which is based on the hydrogenation of double bonds. This technique can be applied for the detection of hardened vegetable fats in chocolates. An HPLC system with fluorescence detector has been used along with a-tocotrienol biomarker to detect palm mid fraction in cocoa butter. The minimum detection limit in such a system was 5% (w/w) (Elham et al., 2008a,b)

4. **Thin layer chromatography (TLC)**
This technique is used to separate nonvolatile mixtures. It is performed on a sheet of glass, plastic, or aluminum foil that is coated with a thin layer of silica gel or cellulose. It is predominantly used as an analytical technique that offers speed and versatility, due to the range of stationary phases available. The spots related to procyanidins and/or anthocyanins from cocoa beans may be visualized using a UV lamp (Jan and Elke, 2000). A TLC method for the detection of at least 5% kokum butter in cocoa butter has been developed using a mobile phase of benzene, ethyl acetate, and acetic acid. A bluish green spot indicates the

presence of kokum butter. This phenomenon could not be observed with other fats, such as fats from mango kernel or mahua (Deotale et al., 1990).

5. **Differential scanning calorimetry (DSC)**
The physical state of a substance is characterized by the DSC thermogram analysis using instruments such as Pyris 6, PerkinElmer, USA. The samples are placed in standard aluminum pans and then scanned at a constant heating rate of 10°C min^{-1} at temperatures between 20°C and 450°C under a nitrogen gas flow of 20 mL min^{-1}. DSC helps in characterization of the component mixtures in chocolate by calculating the partial melting enthalpy. The pure triglycerides, POP, POS, and SOS have been investigated for their crystallization and melting behavior (Lipp and Anklam, 1998). DSC is also described as a very convenient tool for classifying the different types of confectionery fats. It can be used to identify and quantify the progressive changes during the formulation of a typical cocoa butter equivalent and the effects caused by the presence of trisaturated triglycerides and diglycerides. Soft cocoa butters are characterized by a low percentage area of the polymorph II endotherm at 21°C. Hard cocoa butter shows the opposite phenomenon, i.e., low content of 1-palmitoyl-2, 3-dioleoyl-glycerol (POO) and 1-stearoyl-2, 3-dioleoyl-glycerol (SOO), low iodine value, and a high percentage area of the polymorph II endotherm (Cebula and Smith, 1992).
6. **Mass spectroscopy (MS)**
MS is an analytical technique that ionizes chemical species and sorts the ions based on their mass-to-charge ratios. In simpler terms, a mass spectrum measures the masses within a sample. MS is used in many different fields and is applied to pure samples as well as complex mixtures. The equipment is capable of sensitive determinations at the picogram level, but sample pretreatment is essential for the removal of endogenous substances and for facilitating the ionization of the drug substances. Appropriate sample pretreatment helps in avoiding a decline in sensitivity and improves reproducibility. The samples must not contain any detergents or salts and are dissolved in deionized water alone or in necessary amounts of organic solvents (acetonitrile, methanol). It is important to always blow with clean air to remove dust from the sample tube. Pyrolysis MS is used for the classification of different fats and oils comprising cocoa butter. The analysis time is less than 3 mins and is based on the thermal degradation of the compound followed by a mass spectrometric analysis of the fragments. When principal component analysis (PCA) is applied to the data, a good discrimination of cocoa butter from all the other vegetable fats can be achieved (Anklam et al., 1997).
7. **X-ray microanalysis**
The characteristic X-rays are formed by the ionization of inner shell electrons. The inner shell electrons are ejected and the outer shell electrons replace them. The energy difference is released as X-rays. It is a vital tool to know the atomic (chemical) configuration of the compound. The diffraction pattern is considered as a scanning electron microscopic (SEM) image. Milk powder, cocoa solids, and chocolates are investigated for their metal contents using X-ray microanalysis (Brooker, 1990). Cocoa solid particles are rich in potassium but contain very small amounts of calcium. This difference in elemental composition allows two different particles to be distinguished in digital X-ray maps. The time needed for this procedure and the remaining inaccuracy, however, seem to make the application of X-ray microanalysis less suitable for routine control of the milk fat content in chocolate. X-ray diffraction has been used for the characterization of a

mixture of 50% cocoa butter and 50% hydrogenated vegetable fat. The addition of vegetable fats clearly affects both the molecular structure and the morphology of the fat crystals (Brooker, 1990).

8. **Nuclear magnetic resonance (NMR)**
NMR is based on the properties of magnetically active nuclei, which respond to a perturbation induced in a sample by applying a radio wave pulse. The nuclei, under intense magnetic field, respond to the pulse by oscillating at a particular frequency, thus generating a signal which is recorded and transformed into a spectrum, reporting the intensity of the response as a function of oscillation frequency. Deuterated solvents in which ^1H atoms are replaced with ^2H atoms are used in NMR. Common solvents include chloroform-D, acetone-D6, benzene-D6, deuterium oxide (D_2O), DMSO-D6, ethanol-D6, and methanol-D4. The sample is dissolved in one of these solvents and the spectrum is recorded.

    The frequency of each nucleus is characteristic of its position in the molecule which depends on the physicochemical states of the substances. It depends not only on the structure of the molecule but also on the chemical environment around it. It can be used to investigate the polymorphic forms of solid fats. Polymorphic forms of the fats include unstable and the more stable forms and their modifications. NMR analysis of chocolate also offers spatial detection of the polymorphic states of cocoa butter in chocolate and clearly resolves the various components present. NMR is used for analyzing the fluidification of cocoa butter at 27.5°C. The parameters of the resulting bimodal curve are used for the separation of cocoa according to the processes and the types of roasters used (Ellis et al., 2012; Diomande et al., 2015).

9. **Enzyme-linked immunosorbent assay (ELISA)**
Although genetic methods are specific and sensitive for the various food components, they require expensive laboratory equipment and a certain degree of expertise. As an alternative, ELISA is widely used for regulatory purposes in detecting food authenticity because of its specificity, simplicity, and sensitivity. It is a highly efficient assay to identify eight food allergens simultaneously on the surfaces of optical thin-film biosensor chips (i.e., soybean, wheat, peanut, cashew, shrimp, fish, beef, and chicken). Here, the specific allergen DNA is isolated and amplified by two tetraplex PCR systems developed and validated in the same study, followed by an enzyme-labeled indirect detection. The enzymatic reaction products are precipitated on the thin-film surface and the modified interference pattern of light on the biosensor produces a significant color change on the surface, indicating the presence of allergens. The main advantage of this approach is the color change that can be visualized with naked eyes. It avoids the use of sophisticated equipment and is suitable for the routine analysis of a large number of samples (Asensio et al., 2008; Pilolli et al., 2013).

    ELISA can allow, together with other analytical methods, consumer protection and confidence, and an accurate implementation of traceability for successful regulatory food controls. On the whole, ELISA is an immunological technique that involves an enzyme (a protein that catalyzes a biochemical reaction) to detect the presence of an antibody or an antigen in a sample. The two commonly used variants of ELISA for food authentication are the indirect and the sandwich ELISA. The indirect ELISA utilizes two antibodies, one of which is specific to the antigen and the other is coupled to an enzyme. This second antibody gives the assay its name: "enzyme-linked", and it causes a chromogenic or fluorogenic substrate to produce a signal. Sometimes, this second antibody may be linked to a protein. In the sandwich ELISA, the antigen is bound between two antibodies: the capture antibody and the detection antibody. The detection antibody can be coupled to an enzyme or can bind the conjugate (enzyme-linked antibody) to produce a biochemical reaction. ELISA tests may be run in qualitative or quantitative formats. Qualitative results provide simple positive or negative results for the samples. The cut-off between positive and negative is determined by the analyst and may be statistical. In quantitative ELISA, the optical density or fluorescent unit of the sample is interpolated into a standard curve which is typically obtained by a serial dilution of the target (Goldsby et al., 2003). Polyclonal antibodies offer a number of benefits such as recognition of a mixture of different epitopes of the antigens and more tolerance to small changes in the natures of antigens, like polymerization or slight denaturation. They are the preferred choice for detecting denatured proteins. However, they present limitations such as variable affinity, limited production, and extensive purification procedures to eliminate cross-reactivity for particular species identification. In contrast, Monoclonal antibodies (MAbs) are a homogeneous population of antibodies produced by the hybridoma technology that have a defined biological activity, consistent specificity, and whose production is not limited. Both polyclonal antibodies and MAbs are used in the ELISA variants for food authentication. However, the great limitation of ELISA methodology is the denaturation of target proteins during food processing and, therefore, the target protein, epitope, may not be present in the conditions detectable by the antibodies. This limitation has been solved by the development of antibodies against thermostable proteins (Ansfield et al., 2000). The advances in ELISA technology have led to the rapid development of different commercial immunoassay kits for use in the food industry. The most commonly used immunoassay kits for food component identification are ELISA and lateral flow tests. ELISA test kits may be used in microtiter plates and in immunostick formats. Lateral flow tests or "dipsticks" employ the same immunoassay principles as the ELISA tests. The antibodies and other reagents are coated on a nitrocellulose membrane rather than on the inside of test wells or paddles. They use colloidal gold, dye, or latex bead conjugates to generate signals rather than enzyme bead conjugates. The simplicity of tests and the short time required for the analysis make them suitable for the screening of a large number of food samples. Also, trace levels of peanuts, hazelnuts, almonds, etc. can elicit adverse reactions, and an unintentional exposure to the allergens may have devastating consequences for sensitive individuals. Some researchers have successfully developed ELISA techniques (using polyclonal antibodies and MAbs) to detect the ingredients in ice creams, chocolates, or sauces (Kiening et al., 2005).

10. **Surface plasmon resonance-based detection**

Biosensors are the potential alternatives to ELISA as they are fast, reproducible, and cheap modes of multi-analyte detection. It is a device that converts biologically recognized events into physical and chemical signals to represent the target concentration. Optical biosensors are (Pilolli et al., 2013) used to monitor food quality, particularly for detecting allergens such as ovalbumin added to whole milk or the *Arachis hypogaea* allergen (Ara h2) epitomes in chocolates. It measures the changes in the surface properties of the analytes when they bind to respective targets.

11. **Vibrational spectroscopic techniques**
IR radiation was discovered by F.W. Herschel in 1800. This is an electromagnetic radiation extending from 780–1 mm. The IR range is divided into the following three bands: near-infrared (NIR: 780–5 μm), mid-infrared (MIR: 5–30 μm), and far-infrared (FIR: 30–1,000 μm). IR spectroscopy is a technique based on the vibrations of the atoms in a molecule. An IR spectrum is commonly obtained by passing IR radiation through a sample and then determining what fraction of the incident radiation is absorbed at a particular energy (Nawrocka and Lamorska, 2013). Four vibrational spectroscopic techniques, namely near-infrared (NIR), mid-infrared (MIR), Raman spectroscopy, and hyperspectral imaging (HSI) are used for the determination of food authenticity and adulteration or contamination. These techniques have the advantage of being nondestructive and have a relatively low analysis cost. In addition, these spectroscopic techniques can be adopted for both qualitative and quantitative analyses of food products and confectionaries. They provide an alternative to wet-chemical and time-consuming techniques. NIR and fourier transform infrared (FTIR) spectroscopies are very useful for the identification of functional groups in molecules. NIR spectrometers can be categorized into three groups depending on the instrumental devices: (i) sequential instruments where the absorbance is sequentially collected with time and the instrument is equipped with a monochromator or filters; (ii) Fourier transform or multiplex instruments where several frequencies are detected simultaneously in the form of an interferogram; and (iii) multichannel instruments where several detectors separately detect the absorbance at several wavelengths (Lohumi et al., 2015).

NIR has been proved to be a reliable method for the determination of moisture, fat, and sucrose in powdered cocoa products. It has emerged as a major analytical technique for a wide range of food analyses. It has become an attractive option because of its high speed analysis and its ease of operation. Considerable efforts have been made so far to use this technique for the detection of adulterants in oils and fats. NIR signals are associated with molecular vibrations, specifically the overtones and combinations of fundamental vibrations (Ellis et al., 2012). Chemical bonds between light atoms, such as C-H, O-H, and N-H, generally have high vibrational frequencies, which result in overtone and combination bands that are detectable in the NIR region of 780–2,500 nm. Use of NIR spectroscopy for the prediction of nutritional parameters (like sucrose, lactose, fat, and moisture) and rheological properties (viscosity and yield) in chocolates has been reported (Moros et al., 2007; Vesela et al., 2007).

FTIR spectroscopy has been employed to characterize pure vegetable oils, butter, fats, and margarines and to distinguish between plant oils and lard as well as lard and other animal fats (Bahri and Che Man, 2016). FTIR is a fast and reliable method for detecting the amount of trans-fatty acids in vegetable fats and oils. Its reproducibility is stated to be high with good accuracy as compared to the analysis of samples by chemical methods. The observed differences in the spectral characteristics of such fats is due to the compositional and structural differences of the molecules, such as degrees of unsaturation and chain lengths, monounsaturated-to-polyunsaturated acyl group ratios, and variations in trans-fatty acid contents among the oils. These differences are exploited to check the adulteration practices in oils and fats (Marikkar et al., 2016). FT-NIR spectroscopy was developed to overcome the limitations encountered with dispersive NIR instrumentation. The use of a prism or a moving grating separates the frequencies emitted by the NIR source. A detector measures the amount of energy that passes through the sample at each frequency. The use of an interferometer reduces the analysis time per sample by producing a unique type of signal that has all the infrared frequencies encoded in it.

Cocoa butter can be distinguished from CBEs on the basis of FTIR frequencies ranging from 4,000 to 600 cm^{-1} (Goodacre and Anklam, 2001). To quantify the lard contamination in chocolate lipids, FTIR spectroscopy with attenuated total reflectance (ATR) element and PLS (partial least squares) regression analysis has been used with a minimum detection level of 3% (w/w) (Che Man et al., 2005). The manufacturers are bound to disclose the contents legally. Chocolate is usually defined either as containing no non-cocoa vegetable fat (NCVF), except milk fat, or as containing up to 5% of the chocolate as an SOS-type confectionery fat. SOS-type confectionery fats lower the quality of the chocolate. For this, FT-NIR spectroscopy is used to predict the quality parameters of cocoa beans, such as fermentation index, pH, and total polyphenol content (Afoakwa, 2011). Fermentation index is the degree of fermentation of the beans and is a measure of "brownness". During the initial stages of fermentation, the cocoa beans are purple but when anthocyanin pigments are decomposed to form other condensation products, it turns brown. The pH values during and at the end of fermentation have a major role in flavor development. Beans of higher pH (5.5–5.8) are considered as unfermented with low fermentation index, while those of lower pH (4.75–5.19) are considered as well-fermented. Chocolates made from the cocoa beans of high pH (5.5–5.8) are astringent due to the presence of strong flavor (Jinap et al., 1995). Cocoa beans are a rich source of polyphenols, which serve as antioxidants. The presence of polyphenols imparts astringency and bitterness to the cocoa beans. Application of FT-NIR spectroscopy in the food industry for rapid and quality measurements was well researched. Quantification of fat, nitrogen, pH, fermentation index, moisture (Krahmer et al., 2015), and the identification of cocoa bean adulteration can be done with this method. FT-NIR spectroscopy is also used to predict the total polyphenol contents in cocoa beans nondestructively. Here, a tungsten halogen lamp is used as the light source (Sunoj et al., 2016a) A Michelson interferometer equipped with a lead sulfide (PbS) detector detects the interference of a light beam passing through the sample. The interferogram is converted into a spectrum by Fourier transformation of the interferogram. In FT-NIR, the water absorption regions of a chocolate show prominent peaks at around 4,000, 5,200, 6,800, and 8,200 cm^{-1}. The spectral regions between 6,033–5,493 cm^{-1} and 4,953–3,849 cm^{-1} are found responsible for predicting the fermentation index of cocoa beans. The total polyphenol content predicted by FT-NIR spectroscopy is in the range of 7.02–16.23 mg g^{-1} (Sunoj et al., 2016b), which is closely related to the chemical analysis range of 6.48–15.58 mg g^{-1}. It is well known that cocoa butter is the main ingredient in chocolates, besides sugar and cocoa mass. These ingredients contribute to the spectral signals because of the peptide and carbohydrate bonds. This sense, it is possible to distinguish peaks at 1,728 nm (5,786 cm^{-1}), 1,761 nm (5,678 cm^{-1}), and 1,213 nm (8,247 cm^{-1}) from the NIR spectra, which match the C-H bonds in $-CH_2$ groups (Moros et al., 2007). Besides the common peaks, cocoa butter also has absorbance peaks corresponding to the C-H bonds in both -CH and $-CH_2$ groups. The spectra also show amide II and amide III bands at 2,010 nm (4,975 cm^{-1}) and 2,050 nm (4,878 cm^{-1}), which means that cocoa butter and, consequently chocolate, contain small amounts of proteins. Moreover, chocolate spectra show some peaks, which indicate that they contain sugars, e.g., O-H absorbance in sucrose at 1,440 nm (6,944 cm^{-1}) and 2,080 nm (4,808 cm^{-1}). The NIR spectra of fatty acids and esters contain bands at 1,180, 2,143, and 2,190 nm. Other specific bands of fats are located at 2,304 and 2,343 nm due to the presence of milk in chocolate (Lipp and Anklam, 1998; Che Man et al., 2005).

12. **Diffuse reflectance infrared Fourier transform (DRIFT) analysis**
Determination of a series of compounds in chocolate, like theobromine, caffeine (methylxanthines), serotonin, histamine, phenethylamine, spermidine (biogenic amines), lead, cadmium, and nickel (minerals), has been reported using traditional labor-intensive and/or time-consuming methods. In DRIFT analysis, samples are directly deposited, covering the sample compartment cell and then the triplicate DRIFT spectrum is recorded for each sample. Usually, the spectrum of a chocolate sample is scanned from 714–2,857 nm (14,000 to 3,500 cm^{-1}) by averaging 36 scans per spectrum using a nominal resolution of 8 cm^{-1}, which takes 16.8 s per spectrum (Moros et al., 2007). Spectra are recorded in Kubelka-Munk units that are more linear in concentration than the absorbance units. For the measurement of each sample replicate, the sample is rotated to achieve a higher sample volume for measurement taking into account the heterogeneity of the solid samples.

13. **Multivariate techniques**
Multivariate analysis involves many variables instead of one (univariate) or two (bivariate). Therefore, it summarizes the strengths of the linear relationships between each pair of response variables using the correlations table, it identifies dependencies and clusters using the scatter plot, and examines multiple variables such as principal components (Lopez et al., 2014). It is based on the discriminating power such as the K-nearest neighbor algorithm (Di Anibal et al., 2009), linear discriminant analysis, and partial least squares discriminant analysis (Vitale et al., 2013). All these methods require at least two classes to be defined, which implies a prior knowledge of the possible adulterants. Other multivariate techniques are based on their modeling abilities. Soft independent modeling of class analogy (SIMCA) and unequal dispersed classes (UNEQ) are the most well known of these. They are used to test whether a sample is adulterated or not regardless of which adulterant might be present. In some cases of food adulteration, one might have information about the possible adulterants based on experience or literature information. Therefore, targeted modeling is proposed in which, in addition to the unadulterated class, a class for the known possible adulterants is modeled. Therefore, the feasibility of NIR spectroscopy coupled to SIMCA is assessed to detect adulteration. Two approaches are used for this: nontargeted and targeted modeling. If the main interest is to examine if a sample is adulterated or not, untargeted modeling is a good approach. One of its advantages is that the classification model can be established by analyzing the unadulterated samples only. The model is easy to implement and, once it has been established, it can be used to predict any sample, even when there is a lack of knowledge about the potential adulterants. Targeted modeling is a suitable approach in situations where testing is required for a common or known adulterant. One of the main advantages of this approach is that it can provide detailed information about the problem. Comparing both the approaches, it can be said that untargeted modeling is more restrictive as it offers a single response; a sample either fits or does not fit the model. On the other hand, with targeted modeling, a sample can have several responses: a response that fits the model

of just one class; fits the model of more than one class (double classification); or does not fit any model. Therefore, the targeted model provides more information about the tested samples than the untargeted model does. To guarantee food safety and quality, most analytical strategies are based on the knowledge of the contaminants (Stefenoa et al., 2012). Spectroscopic techniques are fast and easy to use, but their combinations with multivariate chemometric techniques turn them into powerful tools for adulteration testing.

14. **Chemometric techniques**

The chemometric techniques are used in research to improve the understanding of chemical information and to correlate quality parameters or physical properties to analytical instrument data through mathematical and statistical methods (Teye et al., 2014). The only requirements are appropriate instruments and software to interpret the data. Usually, standard normal variate (SNV), direct orthogonal signal correction (DOSC), principal component analysis (PCA), support vector machine (SVM), and partial least squares (PLS) models are used in such analyses. They are used to gain efficiency in assessing product quality. PCA of the spectra is performed after preprocessing with SNV and DOSC. SNV is used to remove slope variation and to correct scatter effects, while DOSC method developed by Westerhuis and coworkers is done to remove spectral variance sources (Westerhuis et al., 2001; Luypaert et al., 2007). PCA is an exploratory method that reduces the dimension of the data matrix and compresses the information into a few interpretable variables called principal components (PCs), which are linear combinations of the original variables (Luna et al., 2013). The first three PCs (PC1, PC2, and PC3) give useful information about the score plot so that similar samples are clustered closer to each other. SVM is performed to solve the classification problem after PCA to get the cluster trends of the samples. These models are used for developing the predictive model, which measures the unfermented cocoa beans in fermented ones (Chen et al., 2008). The dataset used in such experiments is divided into two subsets, called the calibration set and prediction set. The calibration set is used to develop the model, while the prediction set is used for evaluating the actual predictive abilities of the developed models. The individual samples in each set are selected in order to come to a 2/1 division of calibration set/prediction set. To avoid a bias in subset division, a substantial number of samples are picked; three spectra for each sample are randomly selected as the calibration set, while the remaining samples are used for the prediction set. In order to classify cocoa beans as fermented, unfermented, and adulterated, a model for predicting the amount of unfermented cocoa in fermented cocoa (% weight basis) has been developed where algorithms are carried out using Matlab (Mathworks Inc., USA).

REFERENCES

Afoakwa EO 2010. Chapter 2: Cocoa cultivation, bean composition and chocolate flavour precursor formation and character. Afoakwa EO. (ed.). In *Chocolate Science and Technology*. John Wiley & Sons, Chichester, 22.

Afoakwa EO 2011. Cocoa cultivation, bean composition and chocolate flavour precursor formation and character. Afoakwa EO.(ed.). In *Chocolate Science and Technology*. John Wiley & Sons, Chichester, 12–34.

Afoakwa EO, Kongor JE, Budu AS, Mensah-Brown H, Takrama JF 2015. Changes in some biochemical qualities during drying of pulp pre-conditioned and Fermented Cocoa (theobroma cacao) Beans. *African Journal of Food, Agriculture, Nutrition and Development* 15(1):9651–9670.

Afoakwa EO, Kongor JE, Takrama JF, Budu AS 2013. Changes in nib acidification and biochemical composition during fermentation of pulp preconditioned cocoa (Theobroma cacao) beans. *International Food Research Journal* 20(4):1843–1853.

Altug T, Gonul M 1987. The detection of soyabean flour in cocoa powder. *Journal of Food Quality* 10(5):295–298.

Anklam E, Bassani MR, Eiberger T, Kriebel S, Lipp M, Matissek M 1997. Characterisation of cocoa butters and other vegetable fats by pyrolysis mass spectrometry. Fresenius Journal of Analytical Chemistry 357:981–984.

Anonymous, Official Methods of Analysis of AOAC International 1995. AOAC Official Method 963.15 Fat in Cacoa Products.

Ansfield M, Reaney SD, Jackman R 2000. Performance assessment and validation of a sensitive immunoassay for detection of ruminant and porcine heat stable proteins in compound animal feedstuffs. *Food and Agricultural Immunology* 12:285–297.

Aprotosoaie AC, Luca SV, Miron A 2016. Flavor chemistry of cocoa and cocoa products—An Overview. *Comprehensive Reviews in Food Science and Food Safety* 15(1):73–91.

Asensio L, Gonzalez I, Garcıa T, Martın R 2008. Determination of food authenticity by enzyme-linked immunosorbent assays (ELISA). *Food Control* 19:1–8.

Awad TS, Moharram HA, Shaltout OE, Asker D, Youssef MM 2012. Applications of ultrasound in analysis, processing and quality control of food: A review. *Food Research International* 48:410–427.

Bahri S, Che Man YB 2016. Rapid detection of lard in chocolate and chocolate - based food products using fourier transform infrared spectroscopy. *Journal of Tropical Agriculture and Food Science* 44(2):253–263.

Bohacenko I, Kopicova Z, Pinkrov J 2004. Chocolate authenticity control concerning compliance with the conditions for adding cocoa butter equivalents as laid down by directive 2000/36 EC. *Czech Journal of Food Sciences* 23(1):27–35.

Brooker, BE 1990. Identification and characterization of cocoa solids and milk proteins in chocolate using X-ray microanalysis. *Food Structure* 9(1):9–21.

Caboni MF, Conte LS, Lercker G 1991. HPLC analysis of lipids in food by light scattering detection in quality control in strategies for food quality control and analytical methods in Europe. In *Proceedings of the 6th European Conference on Food Chemistry*, Hamburg, Germany, September 22–26, 468473.

Cebula DJ, Smith KW 1992. Differential scanning calorimetry of Confectionery Fats: Part II, effects of blend and minor components. *Journal of the American Oil Chemists Society* 69:992–998.

Che Man YB, Syahariza ZA, Mirghani MES, Jinap S, Bakar J 2005. Analysis of potential lard adulteration in chocolate and chocolate products using Fourier transforms infrared spectroscopy. *Food Chemistry* 90(4):815–819.

Chen Q, Zhao J, Liu M, Cai J, Liu J 2008. Determination of total polyphenols content in green tea using FT-NIR spectroscopy and different PLS algorithms. *Journal of Pharmaceutical and Biomedical Analysis* 46(3):568–573.

Cidell JL, Alberts HC 2006. Constructing quality: The multinational histories of chocolate. *Geoforum* 37:999–1007.

Cserhati T, Forgacs E, Deyl Z, Miksik I 2005. Chromatography in authenticity and traceability tests of vegetable oils and dairy products: A review. *Biomedical Chromatography* 19:183–190.

Deotale, MY, Patil, MN, Adinarayaniah, CL 1990. Thin layer chromatographic detection of kokum butter in cocoa butter. *Journal of Food Science and Technology* 27:230–236.

Di Anibal CV, Odena M, Ruisanchez I, Callao MP 2009. Determining the adulteration of spices with Sudan I–IV dyes by UV–visible spectroscopy and multivariate classification techniques. *Talanta* 79(3):887–892.

Diomande D, Antheaume I, Leroux M, Lalande J, Balayssac S, Remaud GS, Tea I. 2015. Multi-element, multi-compound isotope profiling as a means to distinguish the geographical and varietal origin of fermented cocoa (Theobroma cacao L.) beans. *Food Chemistry* 188:576–582.

Elham MF, Jinap S, Che Man YB, Noraini I 2008a. Application of alpha-tocotrienol for detection of palm mid-fraction in dark chocolate formulation. *European Food Research and Technology* 228(2):163–168. doi:10.1007/s00217-008-0919-6.

Elham MF, Jinap S, Che Man YB, Noraini I 2008b. Detection and quantification of palm mid fraction in chocolate model system. *International Journal of Food Science & Technology* 43(6):1083–1087. doi:10.1111/j.1365-2621.2007.01570.

Ellis DI, Brewster VL, Dunn WB, Allwood, JW, Alexander P, Golovanovc, Goodacre R 2012. Fingerprinting food: Current technologies for the detection of food adulteration and contamination. *Chemical Society Reviews* 41:5706–5727.

Ellis DI, Muhamadali H, Allen DP, Elliott CT, Goodacre R 2016. A flavour of omics approaches for the detection of food fraud. *Current Opinion in Food Science* 10:7–15.

European Community. http://ec.europa.eu/food/food/foodlaw/traceability. Accessed on April 15, 2011.

Fanali C, Dugo L, Dugo P, Mondello L 2013. Capillary-liquid chromatography (CLC) and nano-LC in food analysis. *Trends in Analytical Chemistry* 52:226–238.

FDA PART 163—Cacao products Subpart B—Requirements for Specific Standardized Cacao Products Sec. 163.124 White chocolate.

FDA PART 163—Cacao products Subpart B—Requirements for Specific Standardized Cacao Products Sec. 163.130 Milk chocolate.

Food Safety and Standards (Contaminants, Toxins and Residues) Regulations, 2011. www.fssai.gov.in/. Accessed on April 20, 2017.

Gabriela BM 2012. Tracking down the chocolate adulteration by replacing the cocoa butter. *Annals: Food Science and Technology* 13(1):15–18.

Galimberti A, Mattia FD, Losa A, Bruni I, Federici S, Casiraghi M, Martellos S, Labra M 2013. DNA barcoding as a new tool for Food traceability. *Food Research International* 50:55–63.

García Regueiro JA, Gibert J, Díaz I. 1994. Determination of neutral lipids from subcutaneous fat of cured ham by capillary gas chromatography and liquid chromatography. *Journal of Chromatography A* 667(1–2):225–233.

Goldsby RA, Kindt TK, Osborne BA, Kuby J 2003. *Immunology*, 5th ed. W.H. Freeman and Company, New York.

Goodacre R, Anklam E 2001. Fourier transform infrared spectroscopy and chemometrics as a tool for rapid detection of other vegetable fats mixed in cocoa butter. Journal of the American Oil Chemists' Society 78(10):993–1000. doi:10.1007/s11746-001-0377-x.

Gourieva KB, Tserevitinov OB 1979. Method of evaluating the degree of fermentation of cocoa beans. USSR patent 646.

Grivetti LE, Shapiro H-Y 2009a. *Adulteration: The Dark World of Dirty Chocolate: History, Culture, and Heritage*. John Wiley and Sons, Hoboken, NJ, 625–633.

Grivetti LE, Shapiro H-Y. 2009b. *Adulteration: The Dark World of "Dirty" Chocolate. Chocolate: History, Culture, and Heritage*. John Wiley and Sons, Hoboken, NJ, 684–695. http://content.yudu.com/Library/A1nrdq/CookingChocolateHist/resources/686

Hashim L, Chaveron H 1994. Extraction and determination of methyl pyrazines in cocoa beans using coupled steam distillation-microdistillator. *Food Research International* 27:537–544.

https://chocolateclass.wordpress.com/2016/03/11/dirty-chocolate-a-case-study-of-food-adulterion. Accessed on April 19, 2017.

https://en.wikipedia.org/wiki/Massspectrometry. Accessed on April 19, 2017.

New Legislation Food Information for Consumers Regulation (FIR) 2014. http://www.food.gov.uk/enforcement/regulation/fir/labelingproposals

www.caobisco.com/page.asp?p=220. Accessed on June 2, 2011.

www.fda.gov/downloads/food/guidanceregulation/ucm265446.pdf. Accessed on April 19, 2017.

International Office of Cocoa, Chocolate and Confectionary, 1991. The IOCCC code of hygienic practice based on HACCP for the prevention of salmonella contamination in coca, chocolate and confectionary products. IOCCC, Brussels.

Jan W, Elke A 2000. Review on polyphenols in Theobroma cacao: Changes in composition during the manufacture of chocolate and methodology for identification and quantification. *Food Research International* 33:423–447.

Jinap S, Dimick PS, Hollender, R 1995. Flavour evaluation of chocolate formulated from cocoa beans from different countries. *Journal of Food Control* 6:105–110.

Khuda E, Sefat E, Lauren J, Fu T, Williams Kristina M 2015. Effects of processing on the recovery of food allergens from a model dark chocolate matrix. *Food Chemistry* 168:580–587.

Kiening M, Niessner R E, Baumgartner S, Krska R, Bremer M, Tomkies V, et al. 2005. Sandwich immunoassays for the determination of peanut and hazelnut 514 traces in foods. *Journal of Agricultural and Food Chemistry* 53:3321–3327.

Kornacki JL 2006. Microbiological sampling in the dry foods processing environment. Food Safety Magazine, 66.

Krahmer A, Engel A, Kadow D, Ali N, Umaharan P, Kroh LW, Schulz H 2015. Fast and neat—Determination of biochemical quality parameters in cocoa using near infrared spectroscopy. *Food Chemistry* 181:152–159.

Kvenberg JE, Schwalm DJ 2000. Use of microbial data for hazard analysis and critical control point verification—Food and Drug Administration perspective. *Journal of Food Protection* 63(6):810–814.

Lakshmi V 2012. Food adulteration. *International Journal of Science Inventions Today* 1(2):101–113.

Lipp M, Anklam E 1998. Review of cocoa butter and alternative fats for use in chocolate—Part B. analytical approaches for identification and determination. *Food Chemistry* 62(1):99–108. doi:10.1016/S0308-8146(97)00161-1.

Lohumi S, Sangdae L, Hoonsoo L and Cho BK 2015. A review of vibrational spectroscopic techniques for the detection of food authenticity and adulteration. *Trends in Food Science & Technology* 46:85–98.

Lopez MI, Esther Trullols E, Callao MP, Ruisanchez I 2014. Multivariate screening in food adulteration: Untargeted versus targeted modelling. *Food Chemistry* 147:177–181.

Luna AS, da Silva AP, Pinho JSA, Ferre J, Boque R 2013. Rapid characterization of transgenic and non-transgenic soybean oils by chemometric methods using NIR spectroscopy. *Spectrochimica Acta, Part A: Molecular and Biomolecular Spectroscopy* 100:115–119. doi:10.1016/j.saa.2012.02.085.

Luypaert J, Heuerding S, Massart DL, Heyden YV 2007. Direct orthogonal signal correction as data pretreatment in the classification of clinical lots of creams from near infrared spectroscopy data. *Analytica Chimica Acta* 582(1):181–189.

Manikas I, Manos B 2008. Design of an integrated supply chain model for supporting traceability of dairy products. *International Journal of Dairy Technology* 62:126–138.

Marikkar JMN, Mirghani MES and Jaswir I 2016. Application of chromatographic and infra-red spectroscopic techniques for detection of adulteration in food lipids: A review. *Journal of Food Chemistry and Nanotechnology* 2(1):32–41.

Meng CC, Mhd Jalil AM, Ismail A 2009. Phenolic and theobromine contents of commercial dark, milk and white chocolates on the malaysian market. *Molecules* 14:200–209.

Moros J, Inon FA, Garrigues S, de la Guardia M 2007. Near-infrared diffuse reflectance spectroscopy and neural networks for measuring nutritional parameters in chocolate samples. *Analytica Chimica* Acta 584(1):215–222.

Muller A, Steinhart H 2007. Recent developments in instrumental analysis for food quality. *Food Chemistry* 101:1136–1144.

Nascimento MS, Reolon EM, Santos ARB, Moreira VE, Silva N 2015. Enterobacteriaceae contamination in chocolate processing. *Food Control* 47:291–297.

Nawrocka A, Lamorska J 2013. Chapter 14: Determination of food quality by using spectroscopic methods. In *Advances in Agrophysical Research*, 347–367. http://dx.doi.org/10.5772/52722.

Oddone M, Aceto M, Baldizzone M, Musso D, Osella D 2009. Authentication and traceability study of hazelnuts from piedmont, Italy. Journal of Agricultural and Food Chemistry 57(9):3404–3408.

Pilolli R, Monaci L, Angelo V 2013. Advances in biosensor development based on integrating nanotechnology and applied to food-allergen management. *Trends in Analytical Chemistry* 47:12–26.

Pino J, Roncal E 1992. Linalool content in roasted cocoa butter as a characteristic of several flavor grade cocoas. *Die Nahrung* 36(3):299–301.

Rankin CW, Nriagu JO, Aggarwal JK, Arowolo TA, Adebayo K, Flegal AR 2005. Lead contamination in cocoa and cocoa products: Isotopic evidence of global contamination. *Environmental Health Perspectives* 113(10):1344–1348.

Ratnasingham S, Hebert PDN 2007. BOLD: The barcode of life data system. *Molecular Ecology Notes* 7:355–364. www.barcodinglife.org

Richards A, Wailes B 2012. Estimation of fat-free cocoa solids in chocolate and cocoa products—Global survey of typical concentrations of theobromine and caffeine determined by HPLC. *Journal of Association Public Analysts* 40:1–12.

Rombaut R, De Clercq, Foubert I, Dewettinck K 2009. Triacylglycerol analysis of fats and oils by evaporative light scattering detection. *Journal of the American Oil Chemists' Society* 86:19–25.

Sager M 2012. Chocolate and cocoa products as a source of essential elements in nutrition. *Journal of Nutrition & Food Science* 2:123. doi:10.4172/2155-9600.1000123.

Schwagele F 2005. Traceability for a European perspective. *Meat Science* 71:164–173.

Shariff SM, Abdlah NA 2014. Halal certification on chocolate products: A case study. *Procedia—Social and Behavioral Sciences* 121:104–112.

Simoneau C, Hannaert P, Anklam E 1999. Detection and quantification of cocoa butter equivalents in chocolate model system: Analysis of triglyceride profiles by high resolution GC. *Food Chemistry* 65(1):111–116. doi:10.1016/S0308-8146(98)00106-X.

Singleton VL, Rossi JA 1965. Colorimetry of total phenolics with phosphomolybdic-phosphotungstic acid reagents. *American Journal of Enology and Viticulture* 16:144–158.

Stefanoa VD, Avellonea G, Bongiornoa D, Cunsolob V, Muccillib V, Sforzac S, Dossenac A, Drahosd L, Vekeyd K 2012. Applications of liquid chromatography–mass spectrometry for food analysis. *Journal of Chromatography A* 1259:74–85.

Sunoj S, Igathinathane C, Visvanathan R 2016a. Quantification of fermentation levels in cocoa beans using FT-NIR spectroscopy. ASABE Annual International Meeting, 1–9. doi:10.13031/aim.20162461813.

Sunoj S, Igathinathane C, Visvanathan R 2016b. Nondestructive determination of cocoa bean quality using FT-NIR Spectroscopy. *Computers and Electronics in Agriculture* 124:234–242.

Swanson K MJ, Anderson JE 2000. Industry perspective on the use of microbial data for hazard analysis and critical control point validation and verification. *Journal of Food Protection* 63(6):815–818.

Teye E, Huang X, Lei W, Dai H 2014. Feasibility Study on the use of Fourier transform near-infrared spectroscopy together with chemometrics to discriminate and quantify adulteration in cocoa beans. *Food Research International* 55:288–293.

Ulberth F, Buchgraber M, Anklam E 2003. EU chocolate analysis project—Current status and future work. LPS 132/2003, 1–5. www.soci.org.

USFDA 03/20/2015. CPG Sec. 515.700 Chocolate and chocolate liquor—adulteration with insect and rodent filth. www.usfda.gov. Accessed on January 30, 2017.

Vesela A, Barros AS, Synytsya A, Delgadillo I, Copikova J, Coimbra MA 2007. Infrared spectroscopy and outer product analysis for quantification of fat, nitrogen, and moisture of cocoa powder. *Analytica Chimica* Acta 601(1):77–86.

Vitale R, Bevilacqua M, Bucci R, Magrì AD, Magrì AL, Marini F 2013. A rapid and non-invasive method for authenticating the origin of Pistachio samples by NIR spectroscopy and chemometrics. *Chemometrics and Intelligent Laboratory Systems* 121:90–99.

W. Baker & Co. Advertisement. Digital image. N.p., n.d. Web.

Westerhuis JA, de Jong S, Smilde AK. 2001. Direct orthogonal signal correction. *Chemometrics and Intelligent Laboratory Systems* 56(1):13–25.

Winkler A 2014. Chapter 10—Coffee, cocoa and derived products food safety management. In *Food Safety Management: A Practical Guide for the Food Industry*, eds. Y Motarjemi, H Lelieveld. Academic Press, Cambridge. 251–282.

CHAPTER 19

Authentication and Traceability of Rice

Abu Tariq
Universiti Sains Malaysia

Showkat Ahmad Bhawani and Ahmad Husaini
UNIMAS

Abdul Moheman
M. J. P. Rohilkhand University

CONTENTS

19.1 INTRODUCTION

Rice is one of the most important staple food crops in the world with a rich genetic diversity. The increased growth of the world population has enhanced the demand for rice. In Asia, 90% of the population considers rice as its main source of calories. The price of rice in the world market is not homogenous because of the differences in quality, variety, and processing (Nagaraju et al. 2002). Unscrupulous traders are involved in adulteration of rice with low grade, low cost, and low nutritious grains to gain more profits with less efforts. It has been reported that after the green revolution in 1960 thousands of varieties of rice have been introduced worldwide (Siddiq et al. 2012). The term adulteration means addition of low standard additional substances to a food item in order to increase the quantity. It is a widespread practice for almost all agricultural food products, and distinguishing the authentic products from the look-alike food products is difficult with naked eye/visual observation. Many of the high commercial value food products or the largely produced products are targeted for adulteration. Adulterations of food items result from the following practices: (1) addition of look-alike materials of low standards and costs, (2) addition of substances with low qualities, (3) dilution of the original products, and (4) wrong information on the age and origin of the materials. These four major practices are being followed by most of the unscrupulous cultivators and traders for illegal adulteration of food items to obtain huge profits. Similar conditions are prevalent in case of rice, as it is sold either as brown rice or as polished rice for export or for domestic purposes. Hence, high-quality rice is being adulterated with low quality and low price varieties by the traders and cultivators to obtain large profits.

Adulteration of food products is a major crime and it is a matter of great concern not only for preventing economic fraud but also for safety reasons. The addition of undeclared food ingredients, such as allergens or toxins, is harmful for the consumer's health, which in turn represents a potential public health risk. Moreover, provision of wrong information on the food product labels may have adverse effects on the eating habits of certain groups of consumers, who make food choices on religious grounds or due to specific lifestyle choices. Many regulations have been enforced to fight food adulteration and the misbranding of food products (European Parliament 2002, 2003, 2012, 2013 2014; European Commission, 2012; U.S. Food and Drug Administration 2014), enforcing the cultivars and traders to outline the correct and complete information about the food products that are being traded.

A comprehensive definition for traceability was given by Olsen and Borit (2013): Traceability is the ability to access any or all recorded identification throughout its entire life cycle. Tracing and tracking are two parts of traceability. Tracing is used to identify the origin by means of history or records in supply chains and is said to be a backward process, while tracking is used to identify the end users and trading parties by location and it is said to be a forward process (Petersen and Green 2005). Traceability has addressed various issues related to the management of food crisis, bulk product traceability, preservation of quality and identity, prevention of fraud, and anti-counterfeiting (Dabbene et al. 2014), and it also minimizes adulteration in food (Spink et al. 2016).

Adulteration is a threat to health and life, so authentication and traceability of food labeling claims must be guaranteed (Spink and Moyer 2011). Therefore, authentic, accurate, and reliable analytical tools and methods are needed to check and verify the components used in a food product. Thus, a compatible relationship between the purchaser and the seller must be achieved. Several advanced and sophisticated methods or protocols based on proteins or metabolites, chemical composition, and the material or DNA have been introduced over the past decades for univocal detection and precise quantification of the adulterants. However, none of the methods were able to detect the adulterants in every kind of agricultural product, resulting in the application of combined methods/protocols for overcoming this setback effectively. Depending upon the source material used, methods employed for the detection and quantification of adulterants were broadly classified as non-DNA based and DNA based.

The authenticity of the agricultural food products can be established with methods that are based on morphological and physicochemical parameters, proteins, DNA, and metabolites (Vaingankar and Kulkarni, 1989, Findlay et al. 1997, Li and Rutger 2000, Thind and Sogi 2005 Arvanitoyannis 2008), but there is an uncertainty in choosing one of the valuable methods for detection and quantification of adulteration on a commercial scale. However, Vemireddy et al. (2007) showed that the capillary electrophoresis method can reveal the threat of adulteration. The protein-based method rely on immunological (Asensio et al. 2008) or electrophoretic and chromatographic assays (Berrini et al. 2006), whereas the metabolite analysis is based on instrumentation techniques, such as HPLC (high performance liquid chromatography) (Moco et al. 2006, Cavaliere et al. 2007), NMR (nuclear magnetic resonance) (Le Gall et al. 2001; Cuny et al. 2008), and MS (mass spectrometry) (Ogrinc et al. 2003, Castro-Puyana and Herrero 2013). The farming systems and processing methods are affected by the environmental conditions and industrial procedures, which unfortunately lead to inconsistent results when HPLC, NMR, and MS techniques are applied for the detection and quantification of adulterants (Martinez et al. 2003, Woolfe and Primrose 2004). On the contrary, DNA-based methods have shown more reliability due to the stability of DNA during the production and processing techniques applied along the food chain. Such techniques were proved to be robust and handy in the detection and quantification of adulteration in agricultural food products (Woolfe and Primrose 2004, Primrose et al. 2010). The availability of complete genome sequences of several species including rice varieties such as *indica* (9311) and *japonica* (Nipponbare) cultivars has resulted in the discovery of molecular markers and has opened up unprecedented avenues in the quality control of food products. Therefore, DNA markers offer a powerful tool to address the validation of food authenticity and traceability of primary products entering the food chain both in fresh and processed food.

However, these methods, which are result oriented in case of quality, are unsuitable for detecting the geographical areas of cultivation. Kelly and coworkers (2002) explained the detection of geographical areas of cultivations using isotopic and multi-element analyses of rice varietal samples.

There are many modern techniques for food authenticity and traceability, such as restriction fragment length polymorphism (RFLP), polymerase chain reaction (PCR), real-time PCR (RTi-PCR), DNA sequencing, radio-frequency identification (RFID), near field communication (NFC), isotope analysis, fingerprinting techniques, and DNA barcoding. In the early 1970s, isotope ratio mass spectrometry (IRMS) was used to detect adulteration in food production. In recent years, stable isotope analysis has been gaining increasing attention in the control of food quality and provenance due to the high efficiency and precision of this method (McLeod et al. 2013, Zhao et al. 2014). Generally, isotope analysis focuses on three aspects of food authenticity and traceability: (1) discrimination of food adulteration, (2) geographical origin of food, and (3) verification of organic food.

19.2 METHODS FOR AUTHENTICATION AND TRACEABILITY

19.2.1 Non-DNA-based Methods

In the recent past, methods based on morphological parameters have been used for differentiating the varieties of rice (Vainganka r and Kulkarni 1989, Thind and Sogi 2005). Techniques such as electron microscopy (Beerh and Srinivas 1991) and image analysis (Kim et al. 1997, Carter et al. 2006) were used to examine the morphologies of kernel traits. These methods were unsuccessful in differentiating the cultivars due to significant variations in prices within similar kinds of kernel morphologies. Also, for the detection and identification of adulterants on a commercial scale, these morphological or histological-based techniques were too expensive and inefficient. Moreover, for cultivar identification and quantification of adulteration, several physicochemical properties of starch such as water uptake, loss of solids in cooking water (Vaingankar and Kulkarni 1989), protein characteristics (Thind and Sogi 2005), and grain length after cooking (Siddiq 1982) were used. Rice cultivars were also differentiated on the basis of water solubility and sugar compositions of the rice cell wall materials (CWMs) (Lai et al. 2007). The physicochemical techniques were efficient, reliable, and cost effective in the identification and quantification of adulteration on a smaller scale, but none of the techniques were found suitable at a commercial level due to a highly influenced environment and the genotype of the plant. This condition compelled the scientific community to develop different approaches for obtaining stable results.

The protein-based methods are especially used for grouping the cultivars and they play a very important role in the detection of adulterants. Some of the important protein-based methods used for the detection of adulterants are: isozyme-based gel electrophoresis, immunoassays (Vlachos and Arvanitoyannis 2008), sodium dodecyl sulfate polyacrylamide gel electrophoresis (SDS-PAGE) (Montalvan et al. 1998), Western blotting (Singh et al. 2004, Li et al. 2011), Kjeldahl method, and the use of Soxhlet apparatus (Storck et al. 2005). Siddiq et al. (1972) used acrylamide gel electrophoresis and reported variations in the soluble proteins for three subspecies of *Oryza sativa* (indica, japonica, and javanica). Subsequently, eight isozyme markers were used for six rice varietal groups: two major groups [Group I (indica) and Group VI (japonica)], two minor groups [Group II (aus) and Group V (Basmati)], and two satellite groups [Group III and Group IV (deepwater rices)] (Glaszman, 1987). Montalvan et al. (1998) reported the separation of rice grain proteins for the Brazilian and Japanese upland rice cultivars electrophoretically by using SDS-PAGE. The electrophoretic profiles were used for estimating the concentrations of the protein fractions by using densitometric scanning. The expressions of the target proteins representing different rice tissues at different developmental stages were detected by the method of western blotting (Singh et al. 2004, Li et al. 2011).

19.2.2 DNA-Based Methods

DNA-based techniques were found to have wider acceptability in the identification, detection, and quantification of adulterants due to their simplicity, accessibility, repeatability, and rapidness. The introduction, development, and availability of DNA markers and the cost-effective sequencing techniques made DNA-based methods more popular (Table 19.1). These markers have offered unprecedented applications to resolve the authentication crisis in the food industry in general and the rice market in particular. For varietal identification, there are several DNA markers available, such as RLFP, random amplified polymorphic DNA (RAPD), amplified fragment length polymorphism (AFLP), simple sequence repeat (SSR)—also known as microsatellite, and single nucleotide polymorphism (SNP). Among these markers, the microsatellite or SSR markers are the most informative in terms of reproducibility, genetic informativeness, and applicability (McCouch et al. 2002, Vemireddy et al. 2007, Ashfaq and Khan 2012).

Microsatellites are short repeated regions of less than 100 bp, extensively used in mapping and tagging of genes, phylogeny and population genetics, disease diagnosis, forensic investigations, and study of diversity. The microsatellite DNA markers have many advantages over other markers like abundance and a codominant nature. In rice alone, more than 50,000 microsatellite markers are available. Recent studies showed that evenly-distributed hypervariable microsatellite (hvRM) markers were more polymorphic than the non-hypervariable SSRs (Narshimulu et al. 2011) and also that estimations of the temporal trends in genetic diversity over decadal periods (Choudhury et al. 2013)

TABLE 19.1 DNA Markers for Rice Cultivars

Rice Variety	DNA Markers/(Marker Type)	References
Indica and Japonica varieties	43 RFLP probes	Zhang et al. (1992)
Basmati varieties	OPA 13, OPA 16, OPA 17, OPA 19, OPB 8, OPB 11, OPB 13, OPB 18, OPN 1, and OPN 17 /(RAPD)	Choudhury et al. (2001)
	RM1, RM19, RM208, RM55, RM201, RM219, RM225, RM229, RM202, M2, and M16/(Microsatellites)	Bligh (2000)
	RM163, RM171, RM72, RM238, RM102, RM16, RM302, RM330, and RM16/(Microsatellites)	Nagaraju et al. (2002)
	RM1, RM21, RM38, RM170, RM210, RM226, and RM229/(Microsatellites)	Siwach et al. (2004)
	RM1, RM5, RM103, RM135, RM171, RM174, RM222, and RM252/(Microsatellites)	Jain et al. (2004)
	RM1, RM5, RM44, RM110, RM229, RM234, RM171, RM242, and RM255/(Microsatellites)	Pal et al. (2004)
	RM1, RM171, RM202, RM44, RM55, RM348, RM72, and RM241/(Microsatellites)	Archak et al. (2007)
	RM1, RM171, RM202, RM44, RM55, RM348, RM72, and RM241/(Microsatellites)	Vemireddy et al. (2007)
	R4M17, R4M43, R4M50, R5M13, R5M30, R6M14, R10M30, R11M23, and R12M27/(Indels)	Steele et al. (2008)
BR-11	RM11/147/(Microsatellites)	Rahman et al. (2007)
Badshabhog	RM151/289/(Microsatellites)	
BR-19	RM153/178/(Microsatellites)	
Parental lines and hybrids and some popular varieties	RM70, RM334, RM475, RM219, RM206, RM336, RM547, RM164, RM335, and RM276/(Microsatellites)	Sundaram et al. (2008)
Hybrid purity	RM 206, RM 216, RM 258, and RM263/ (Microsatellites)	Nandakumar et al. (2004)
Koshihikari in processed rice products	B1, B7, B43, B43a, E30, F6, G22, M2CG, M11, S13, and Wka9/(SCAR)	Ohtsubo and Nakamura (2007)
43 Rice cultivars	8 SNPs/(SNP)	Shirasawa et al. (2006)
Cultivars of rice wine	A7, B43, M11, and G22/ (STS)	Ohtsubo et al. (2008)
Rice specific primer	PRR/(STS)	Nakamura and Ohtsubo (2010)

were achieved through these markers. The fluorescent-labeled microsatellite markers in combination with either slab-gel or capillary electrophoresis provide more accurate allele sizes. Recently, padlock probe ligation and microarray detection (PPLMD), a DNA-based multiplex detection tool, was developed that extended to a 15-plex traceability with a focus on the products of wheat and Basmati rice (Voorhuijzen et al. 2012).

Quantification of the adulterants in food products is as important as detection of the adulterants. Authenticity can be achieved only when both accurate detection and quantification can be done. Vlachos and Arvanitoyannis (2008) established this fact for the authentication of rice. Farm products such as wheat, maize, or rice during the postharvesting period of cultivation get adulterated with certain amounts of inadvertent mixtures. For the accurate quantification of adulterants, very few methods are available, unlike the methods of detecting adulteration where a large number of techniques are available such as DNA markers and fragment detection systems. DNA-based microsatellite markers coupled to capillary electrophoresis or RTi-PCR method were found suitable for the accurate quantification of adulterants.

19.2.2.1 PCR Techniques

PCR is an in vitro genetic technique which allows the enzymatic amplification (synthesis of large quantities) of a targeted region of DNA. It is undoubtedly the most powerful and result-oriented laboratory technique ever invented. It is easy to apply, cost effective, and its unique combination of specificity and sensitivity coupled with great flexibility has led to a revolution in the field of genetics. The concept of PCR was conceived and operationalized by Kary Mullis and colleagues in the early 1980s at the Cetus Corporation, and was formally presented at the American Society of Human Genetics Conference in October, 1985. The first clinical trial was done to analyze sickle cell anemia and the same work was published in 1985 and the U.S. Patent 4683202 was awarded to Kary for the PCR.

The PCR technique is simple in design with seemingly endless applications. The objectives of PCR can be achieved with great specificity, efficiency, and accuracy as PCR reactions show great efficacy. This can be achieved by repetitive bidirectional DNA synthesis via primer extension of a region of nucleic acid.

In general, specific DNA sequences are used as markers, grouped as hybridization-based markers and PCR-based markers. In hybridization-based methods, species-specific DNA profiles are discovered by hybridizing DNA that is digested by restriction enzymes and then by comparing it with labeled probes (DNA fragments of known origin or sequence). Whereas, in PCR-based methods, the amplification of target loci using specific or arbitrary primers and a DNA polymerase enzyme gives rise to DNA fragments which can further be separated electrophoretically and then the banding patterns are detected by autoradiography or any other staining method.

The RTi-PCR assay is considered as the most accurate quantification method for nucleic acids. It is a commonly used and the most reliable method for the detection of foreign DNA in genetically modified food (Baeumler et al. 2006, Leimanis et al. 2006). The detection and quantification of transgenes form genetically modified food can be specifically amplified with high accuracy and sensitivity by RTi-PCR (Nakamura et al. 2013). In case of Basmati adulteration, a long-grain rice of cheap quality is used as an adulterant. In case of Basmati rice, there is no Basmati-specific allele. But for aromatic rice, a specific 8 bp deletion in the betaine-aldehyde dehydrogenase-2 gene (*bad-2*) has been exploited to amplify an 80 bp product, exclusively in a non-Basmati variety, by capillary electrophoresis (Archak et al. 2007). In a similar way, TaqMan-based RTi-PCR has been used for the quantification of adulteration in Basmati by using *bad-2* gene-specific primers (Lopez 2008, Mathure et al. 2014). But this method was not found suitable for practical application because many unnotified Basmati carry the same genomic Basmati allele. Herńandez et al. (2005) detected quantitatively 20 non-transgenic rice varieties by RTi-PCR. This method was based on the *gos 9* sequence and it specifically detected and quantified DNA from rice plants with a detection limit of 3.3 genome copies and the limit of quantification was approximately 100 haploid genomes with a confidence level of 95%. It has been reported that the performance assay of this method revealed 100% identification for 12 commercial food products.

19.2.2.2 DNA Barcoding

Lack of standardization and universality in DNA-based techniques led to the development of a new system, DNA barcoding. In 2003, it was developed by researchers at the University of Guelph (Canada). This method is based on the analysis of variability within a standard region of the genome called "DNA barcode" (Hebert et al. 2003). This technique proved handy and useful for solving taxonomic problems in theoretical and practical applications (Rasmussen et al. 2009, Valentini et al. 2009, Hollingsworth et al. 2011). Three important innovative processes: molecularization of identification process, i.e., investigation of DNA variability; standardization of the procedure from sample preparation to analysis; and computerization give DNA barcoding an advantage over the already known molecular identification approaches (Casiraghi et al. 2010).

DNA barcoding refers to the identification of the product using an infrared scanner and the black stripes of the universal product code (UPC). Ideally, DNA barcode requires two fundamental characteristics: high taxonomic coverage, i.e., correct amplification of the genomic region chosen, and high resolution (Hebert et al. 2003). A high resolution in a DNA barcode ensures the identification of different taxa based on a high interspecific and low intraspecific variability in DNA barcode sequences. With a higher effectiveness/cost ratio, DNA barcoding can be a reliable alternative to DNA fingerprinting approaches in plant identification.

The DNA-based methods are able to differentiate rice at inter- and intra-varietal levels. Archak et al. (2007) differentiated three varieties of rice such as traditional Basmati, evolved Basmati, and non-Basmati varieties with eight microsatellite markers (RM1, RM55, RM44, RM72, RM348, RM241, RM202, and RM171) of cultivar-specific allele properties. Vemireddy et al. (2014) later developed DNA barcodes/fingerprints for 90 high yielding varieties of rice by using eight hvRM markers (RM11313, RM13584, RM15004, RM5844, RM22250, RM22565, RM24260, and RM8207).

19.2.2.3 Hybrid Seed Technology

Hybrid seed technology is successful because of the purity of seeds along their parental lines. Nowadays, hybrid seeds are gaining popularity over conventionally developed rice varieties because of the advantage of yield. Nandakumar et al. (2004) and Sundaram et al. (2008) have identified informative microsatellite markers that are capable of distinguishing the hybrid rice parental lines. These microsatellite markers were used for assessing the purity of hybrid seeds by using a two-dimensional bulked DNA sampling strategy (Sundaram et al. 2008). Maintaining the purity of hybrid seeds is very important because it has been estimated that a loss of approximately 100 kg/ha can occur if concentration is 1% (Mao et al. 1996).

19.2.2.4 Fragmentation Separation Methods

The fragment separation methods greatly influence the accuracy in detection and quantification of the adulterants in the samples (Tang et al. 2004, Spaniolas et al. 2006). It has been observed that an ambiguity exists in choosing the best fragment separation method. For example, two different methods have been adopted for fragment separation, such as capillary electrophoresis (Archak et al. 2007) and slab-gel (Bligh 2000). From the experimental observations it has been reported that capillary electrophoresis is more reproducible as compared to other methods for fragment separation. The capillary electrophoresis method produces less error in the determination of allele size as compared to the slab-gel and agarose gel electrophoresis (Vemireddy et al. 2007).

19.2.3 Determining the Age of Rice Samples

The storage of rice materials over a long period of time is also another technique of adulteration. The unscrupulous traders mix rice of different storage periods. It is a general trend that the rice stored for a longer period of time commands a higher price. This problem cannot be solved by DNA-based methods because there are no genotypic differences between the different age groups in these cases. But this variation can be observed at the metabolic level. It has been observed that longer storage of rice in air results in the increase in volatile compounds because of the accumulation of aldehydes and ketones by the oxidation of lipids (Widjaja et al. 1996). A simple, fast, and nondestructive method has been developed for the classification of rice grains based on storage periods with an accuracy of 97.5% by using near-infrared reflectance spectroscopy (NIR) (Dengsheng and Xiaoli 2007).

19.2.4 Determining the Geographical Areas of Rice Samples

Another important parameter for identifying the quality of rice is the region of origin. Many unscrupulous traders mislabel inferior/low quality rice grown outside their regular cultivation areas. At the beginning, grain proteins (Montalvan et al. 1998, Lan et al. 2002) and isozymes (Li and Rutger 2000) were used for the identification of geographical regions. This method is based on the principle that trace elements and compositions of isotopes in the rice grains predict the types of soils and the latitudes of the growing areas. The trace elements can easily be transported into plants and can act as indicators of the geographical regions, and also their concentrations vary with topography and soil

TABLE 19.2 Methods for the Determination of Geographical Areas

Method	Rice/Area of Origin	Target Elements	References
Inductively Coupled Plasma Atomic Emission Spectrometry	*Oryza sativa*	Amount of P, K, Mg, Ca, Mn, Zn, Fe, Cu	Yashui and Shindoh (2000)
Inductively Coupled Plasma High Resolution Mass Spectrometry (ICP-HRMS-MS)	*Oryza sativa*, Tohoku/Kanto, and Hokuriku	Amount of Rb, Mo, Ba, Sr, Ni, Cd, Cs, Pb, Al, Cr, Co	Yashui and Shindoh (2000)
Inductively Coupled Plasma Optical Emission Spectrometry	Basmati rice	Trace elements	Kelly et al. (2002)
Multiple Collector Inductively Coupled Plasma Mass Spectrometry (MC-ICP-MS)	Brown rice, Japanese, Chinese, and Vietnamese rice	Provenance based on Strontium isotope Sr87/Sr86	Kawasaki et al. (2002)
Atomic Fluorescence Spectroscopy (AFS)	Polished rice from China	Selenium concentration	Chen et al. (2002)
Isotope Ratio Mass Spectrometry and ICP MS	Rice from USA, Europe, Basmati regions of India/Pakistan	Carbon 13, Oxygen 18, and other elements	Kelly et al. (2005)

characteristics. Some of the important methods used for the determination of geographical area are listed in Table 19.2. The methods like multi-isotopic and multi-elemental analyses have been widely used for detecting the geographical locations of rice samples (Kelly et al. 2002, 2005, Suzuki et al. 2008). The techniques like inductively coupled plasma atomic emission spectrometry (ICP-AES) and high resolution mass spectrometry (HRMS) were used for the determination of different elements in rice samples for differentiating the areas of cultivation.

19.3 CONCLUSION

Authentication and traceability of rice cultivars are considered as important issues. A wide variety of scientific techniques have been introduced to identify and quantify the adulterants and to check the traceability of food products. Currently, the traditional as well as advanced methods of analysis are available for the detection and quantification of rice adulteration. A variety of sophisticated methods have been discussed that can be effective in determining and quantifying adulteration. The implementation of such techniques by the food control laboratories largely depends on factors such as accuracy of detection, sophistication of the techniques, and cost effectiveness. In response to consumer concerns, the regulations on food products have been increased, leading to an introduction of new advanced techniques for the same.

The DNA-based methods were found to be the most effective for food authentication and traceability, particularly the combination of capillary electrophoresis and microsatellite markers such as RAPD, AFLP, ISSR, and SSR. Further, the DNA-based methods can also be used to detect the adulterants along the parental lines of the hybrid seed production and processed food. However, they have a limitation in detecting the age of the samples and also the geographical areas of the cultivated rice. Protein-based methods and multi-element and multi-isotopic analyses techniques are found very effective for the detection of age. Another powerful technique in food authentication and traceability is isotope analysis. This technique has several advantages over other methods, such as high precision and the fact that it can be applied to almost any type of food production. Recent technological advances in IRMS have introduced the possibility of performing simultaneous rapid analyses of multiple isotope ratios. DNA barcoding can be used as a universal tool for food traceability because it is a sensitive, fast, cheap, and reliable method for authentication and traceability. Due to its universality, DNA barcoding can be used in different contexts and by different operators. DNA barcoding is the future of authentication and traceability of food products.

REFERENCES

Archak, S., Lakshminarayanareddy, V., and Nagaraju, J. 2007. High-throughput multiplex microsatellite marker assay for detection and quantification of adulteration in Basmati rice (Oryza sativa). *Electrophoresis* 28: 2396–2405.

Arvanitoyannis, I. S. 2008. Trends in *Food Authentication* in *Modern Techniques* for *Food Authentication*, Sun, D. (ed.). Academic (Elsevier Ltd), Amsterdam, the Netherlands, p. 643.

Asensio, L., González, I., García, T., and Martin, R. 2008. Determination of food authenticity by enzyme-linked immunosorbent assay (ELISA). *Food Cont.* 19: 1–8.

Ashfaq, M., and Khan, A. S. 2012. Genetic diversity in Basmati rice (Oryza sativa L.) germplasm as revealed by microsatellite (SSR) markers. *Genetika* 48: 62–71.
Baeumler, S., Wulff, D., Tagliani, L., and Song, P. 2006. A real-time quantitative PCR detection method specific to widestrike transgenic cotton (event 281-24-236/3006-210-23). *J. Agric. Food Chem.* 54: 6527–6534.
Beerh, O. P., and Srinivas, T. 1991. Some histological factors associated with Basmati rice. *Oryza* 28: 399–401.
Berrini, A., Tepedino, V., Borromeo, V., and Secchi, C. 2006. Identification of freshwater fish commercially labelled "perch" by isoelectric focusing and two-dimensional electrophoresis. *Food Chem.* 96: 163–168.
Bligh, H. F. J. 2000. Detection of adulteration of Basmati rice with nonpremium long-grain rice. *Int. J. Food Sci. Technol.* 35: 257–265.
Carter, R. M., Yan, Y., and Tomlins, K. 2006. Digital imaging based classification and authentication of granular food products. *Meas. Sci. Technol.* 17: 235–240.
Casiraghi, M., Labra, M., Ferri, E., Galimberti, A., and De Mattia, F. 2010. DNA barcoding: a six-question tour to improve users' awareness about the method. *Brief Bioinform.* 11: 440–453.
Castro-Puyana, M., and Herrero, M. 2013. Metabolomics approaches based on mass spectrometry for food safety, quality and traceability. *Trends Anal. Chem.* 52: 74–87.
Cavaliere, B., De Nino, A., Hayet, F., Lazez, A., Macchione, B., Moncef, C., Perri, E., Sindona, G., and Tagarelli, A. 2007. A metabolomic approach to the evaluation of the origin of extra virgin olive oil: A convenient statistical treatment of mass spectrometric analytical data. *J. Agric. Food Chem.* 55: 1454–1462.
Chen, L., Yang, F., Xu, J., Hu, Y., Hu, Q., Zhang, Y., and Pang, G. 2002. Determination of selenium concentration of rice in china and effect of fertilization of selenite and selenate on selenium content of rice. *J. Agric. Food Chem.* 50: 5128–5130.
Choudhury, G., Ranjitkumar, N., Malathi, S., Deborah, D. A. K., Abhilash, V., Anuradha, G., Siddiq, E. A., and Vemireddy, L. R. 2013. Molecular genetic diversity of major Indian rice cultivars over decadal periods. *PLoS One* 8: e66197.
Choudhury, P. R., Kohli, S., Srinivasan, K., Mohapatra, T., and Sharma, R. P. 2001. Identification and classification of aromatic rices based on DNA fingerprinting. *Euphytica* 118: 243–251.
Cuny, M., Vigneau, E., Le Gall, G., Colquhoun, I., Lees, M., and Rutledge, D. N. 2008. Fruit juice authentication by 1H NMR spectroscopy in combination with different chemometrics tools. *Anal. Bioanal. Chem.* 390: 419–427.
Dabbene, F., Gay, P., and Tortia, C. 2014. Traceability issues in food supply chain management: A review. *Biosyst. Eng.* 120: 65–80.
Dengsheng, Z., and Xiaoli, L. 2007. Determination of storage time of rice seed using ANN Based on NIRS lecture notes in computer. *Sci. Adv. Neural Networks* 4493: 1043–1048.
E. Parliament. 2002. European Council, Regulation (EC) No 178/2002 of the European Parliament and of the Council of 28 January 2002 laying down the general principles and requirements of food law, establishing the European Food Safety Authority and laying down procedures in matters of food safety. *Off. J. Eur. Communities* L31: 1–24.
E. Parliament. 2003. European Council, Regulation (EC) No 1830/2003 of 22 September 2003 concerning the traceability and labelling of genetically modified organisms and the traceability of food and feed products produced from genetically modified organisms and amending Directive 2001/18/EC. *Off. J. Eur. Union* L268: 24–28.

E. Parliament. 2012. European Council, Regulation (EU) No 1151/2012 of the European Parliament and of the Council of 21 November 2012 on quality schemes for agricultural products and foodstuffs. *Off. J. Eur. Union* L343: 1–29.

E. Parliament. 2013. European Council, Regulation (EU) No 1379/2013 of the European Parliament and of the Council of 11 December 2013 on the common organisation of the markets in fishery and aquaculture products, amending Council Regulations (EC) No 1184/2006 and (EC) No 1224/2009 and repealing. *Off. J. Eur. Union* L354: 1–21.

European Commission. 2012. Commission Implementing Regulation (EU) No 579/2012 of 29 June 2012 amending Regulation (EC) No 607/2009 laying down certain detailed rules for the implementation of Council Regulation (EC) No 479/2008 as regards protected designations of origin and geographical indications, traditional terms, labelling and presentation of certain wine sector products. *Off. J. Eur. Union* L171: 4–6.

European Parliament. European Parliament Resolution of 14 January 2014 on the Food Crisis, Fraud in the Food Chain and the Control thereof (2013/2091(INI)). P7_TA (2014)0011, 2014.

Findlay, I. T. A., Quirke, P., Frazier, R., and Urquhart, A. 1997. DNA fingerprinting from single cells. *Nature* 389: 555–556.

Glaszman, J. C. 1987. Isozymes and classification of Asian rice varieties. *Theor. Appl. Genet.* 74: 21–30.

Hebert, P. D. N., Ratnasingham, S., and deWaard, J. R. 2003. Barcoding animal life: Cytochrome c oxidase subunit 1 divergences among closely related species. *Proc. Roy. Soc. B: Biol. Sci.* 270: S96–S99.

Herńandez, M., Esteve, T., and Pla, M. 2005. Real-time polymerase chain reaction based assays for quantitative detection of barley, rice, sunflower, and wheat. *J. Agric. Food Chem.* 53: 7003–7009.

Hollingsworth, P. M., Graham, S. W., and Little, D. P. 2011. Choosing and using a plant DNA barcode. *PLoS One* 6: e19254.

Jain, S., Jain, R. K., and McCouch, S. R. 2004. Genetic analysis of Indian aromatic and quality rice (Oryza sativa L) germplasm using panels of fluorescently-labeled microsatellite markers. *Theor. Appl. Genet.* 109: 965–977.

Kawasaki, A., Oda, H., and Hirata, T. 2002. Determination of strontium isotope ratio of brown rice for estimating its provenance. *Soil Sci. Plant Nutr.* 486: 635–640.

Kelly, S., Baxter, M., Chapman, S., Rhodes, C., Dennis, J., and Brereton, P. 2002. The application of isotopic and elemental analysis to determine the geographical origin of premium long grain rice. *Eur. Food Res. Technol.* 214: 72–78.

Kelly, S., Heaton, K., and Hoogewerff, J. 2005. Tracing the geographical origin of food: The application of multi-element and multi-isotope analysis. *Trends Food Sci. Technol.* 16: 555–567.

Kim, S. S., Jo, J. S., Kim, Y. J., and Sung, N. K. 1997. Authentication of rice by three-sided image analysis of kernels using two mirrors. *Cereal Chem.* 74: 212–215.

Lai, V. M. F., Lu, S. N., He, W. H., and Chen, H. H. 2007. Non-starch polysaccharide compositions of rice grains with respect to rice variety and degree of milling. *Food Chem.* 101: 1205–1210.

Lan, Y., Fang, Q., Kocher, M. F., and Hanna, M. A. 2002. Detection of fissures in rice grains using image enhancement. *Int. J. Food Prop.* 5: 205–215.

Le Gall, G., Puaud, M., and Colquhoun, I. J. 2001. Discrimination between orange juice and pulp wash by 1H nuclear magnetic resonance spectroscopy: Identification of marker compounds. *J. Agric. Food Chem.* 49: 580–588.

Leimanis, S., Hernandez, M., Fernandez, S., Boyer, F., Burns, M., Bruderer, S., Glouden, T., Harris, N., Kaeppeli, O., Philipp, P., Pla, M., Puigdomenech, P., Vaitilingom, M., Bertheau, Y., and Remacle, J. 2006. Amicroarray-based detection system for genetically modified (GM) food ingredients. *Plant Mol. Biol.* 61: 123–139.

Li, X., Bai, H., Wang, X., Li, L., Cao, Y., Wei, J., Liu, Y., Liu, L., Gong, X., Wu, L., Liu, S., and Liu, G. 2011. Identification and validation of rice reference proteins for western blotting. *J. Exp. Bot.* 14: 4763–4772.

Li, Z., and Rutger, J. N. 2000. Geographic distribution and multilocus organization of isozyme variation of rice (Oryza sativa L). *Theor. Appl. Genet.* 101: 379–387.

Lopez, S. J. 2008. TaqMan based real time PCR method for quantitative detection of Basmati rice adulteration with non-Basmati rice. *Eur. Food Res. Technol.* 227: 619–622.

Mao, C. X., Virmani, S. S., and Kumar, I. 1996. Technological innovations to lower the costs of hybrid rice seed production. In: Marker for fragrance genotyping in rice. *Mol. Breed.* 16: 279–283.

Martinez, I., Aursand, M., Erikson, U., Singstad, T. E., Veliyulin, E., and van der Zwaag, C. 2003. Destructive and non-destructive analytical techniques for authentication and composition analyses of foodstuffs. *Trends Food Sci. Technol.* 14: 489–498.

Mathure, S. V., Jawali, N., Thengane, R. J., and Nadaf, A. B. 2014. Comparative quantitative analysis of headspace volatiles and their association with BADH2 marker in non-Basmati scented, Basmati and nonscented rice (Oryza sativa L.) cultivars of India. *Food Chem.* 142: 383–391.

McCouch, S. R., Teytelman, L., Xu, Y., Lobos, K. B., Clare, K., Walton, M., Fu, B., Maghirang, R., Li, Z., Xing, Y., Zhang, Q., Kono, I., Yano, M., Fjellstrom, R., DeClerck, G., Schneider, D., Cartinhour, S., Ware, D., and Stein, L. 2002. Development and mapping of 2240 new SSR markers for rice (Oryza sativa L). *DNA Res.* 9: 257–279.

McLeod, R. J., Garland, M., Hale, R. V., Steiman, S., and Frew. R. D. 2013. Determining the most effective combination of chemical parameters for differentiating the geographic origin of food products: an example using coffee beans. *J. Food Chem. Nutr.* 1: 49–61.

Moco, S., Forshed, J., De Vos, R. C., Bino, R. J., and Vervoort, J. 2006. Intra-and inter-metabolite correlation spectroscopy of tomato metabolomics data obtained by liquid chromatography-mass spectrometry and nuclear magnetic resonance. *Metabolomics* 4: 202–215.

Montalvan, R., Ando, A., and Echeverrigaray, S. 1998. Use of seed protein polymorphism for discrimination of improvement level and geographic origin of upland rice cultivars. *Genet. Mol. Biol.* 21: 531–535.

Nagaraju, J., Kathirvel, M., Kumar, R. R., Siddiq, E. A., and Hasnain, S. E. 2002. Genetic analysis of traditional and evolved Basmati and non- Basmati rice varieties by using fluorescence-based ISSR-PCR and SSR markers. *Proc. Natl. Acad. Sci. U. S.* 99: 5836–5841.

Nakamura, K., Akiyama, H., Kawano, N., Kobayashi, T., Yoshimatsu, K., Mano, J., Kitta, K., Ohmori, K., Noguchi, A., Kondo, K., and Teshima, R. 2013. Evaluation of real-time PCR detection methods for detecting rice products contaminated by rice genetically modified with a CpTI-KDEL-T-nos transgenic construct. *Food Chem.* 141: 2618–2624.

Nakamura, S., and Ohtsubo, K. 2010. PCR method for the detection and identification of cultivars of rice flours used in yeast leavened breads containing both wheat and rice flours. *J. Cereal Sci.* 52: 16–21.

Nandakumar, N., Singh, A. K., Sharma, R. K., Mohapatra, T., Prabhu, K. V., and Zaman, F. U. 2004. Molecular fingerprinting of hybrids and assessment of genetic purity of hybrid seeds in rice using microsatellite markers. *Euphytica* 136: 257–264.

Narshimulu, G., Jamaloddin, M., Vemireddy, L. R., Anuradha, G., and Siddiq, E. A. 2011. Potentiality of evenly distributed hypervariable microsatellite markers in marker-assisted breeding of rice. *Plant Breed.* 130: 314–320.

Ogrinc, N. K. I. J., Košir, I. J., Spangenberg, J. E., and Kidrič, J. 2003. The application of NMR and MS methods for detection of adulteration of wine, fruit juices, and olive oil. A review. Anal. *Bioanal. Chem.* 2003(376): 424–430.

Ohtsubo, K., and Nakamura, S. 2007. Cultivar identification of rice (Oryza sativa L.) by PCR method and its application to processed rice products. *J. Appl. Glicosci.* 54: 235–243.

Ohtsubo, K., Suzuki, K., Haraguchi, K., and Nakamura, S. 2008. Novel method for preparation of the template DNA and selection of primers to differentiate the material rice cultivars of rice wine by PCR. *J. Biochem. Biophys. Methods* 70: 1020–1028.

Olsen, P., and Borit, M. 2013. How to define traceability. *Trends Food Sci. Technol.* 29: 142–150.

Pal, S., Jain, S., Saini, N., Aarti, N., and Jain, R. K. 2004. Identification of microsatellite markers for differentiating some Basmati and non-Basmati rice varieties. *Indian J. Biotechnol.* 3: 519–526.

Petersen, A., and Green, D. 2005. Seafood Traceability: A Practical Guide for the U.S. Industry, North Carolina Sea Grant. UNC-SG-06-04, p. 28.

Primrose, S., Woolfe, M., and Rollinson, S. 2010. Food forensics: methods for determining the authenticity of foodstuffs. *Trends Food Sci. Technol.* 21: 582–590.

Rahman, S. N., Islam, M. S., and Nasiruddin, K. M. 2007. Genetic polymorphismin rice (Oryza sativa L) through RAPD analysis. *Indian J. Biotechnol.* 6: 224–229.

Rasmussen, R. S., Morrissey, M. T., and Hebert, P. D. N. 2009. DNA barcoding of commercially important salmon and trout species (Oncorhynchus and Salmo) from North America. *J. Agric. Food Chem.* 57: 8379–8385.

Shirasawa, K., Shiokai, S., Yamaguchi, M., Kishitani, S., and Nishio, T. 2006. Dot-blot-SNP analysis for practical plant breeding and cultivar identification in rice. *Theor. Appl. Genet.* 113:147–155.

Siddiq, E. A. 1982. Breeding for quality improvement in rice-present state and strategy for the future. In: *Golden Jubilee Symposium on Increase in the Productivity of Rice, Rice Research Stations*, Chinsurah and Ban Kura, West Bengal, India.

Siddiq, E. A., Nerkar, Y. S., and Mehta, S. L. 1972. Intra and inter subspecific variation in soluble proteins of Oryza sativa L. *Theor. Appl. Genet.* 42: 351–356.

Siddiq, E. A., Vemireddy, L. R., and Nagaraju, J. 2012. Basmati rices: Genetics, breeding and trade. *Agric. Res.* 1: 25–36.

Singh, R. K., Sharma, R. K., Singh, A. K., Singh, V. P., Singh, N. K., Tiwari, S. P., and Mohapatra, T. 2004. Suitability of mapped sequence tagged microsatellite site markers for establishing distinctness, uniformity and stability in aromatic rice. *Euphytica* 135: 135–143.

Siwach, P., Jain, S., Saini, N., Chowdhury, V. K., and Jain, R. K. 2004. Allelic diversity among Basmati and non-Basmati long-grain Indica rice varieties using microsatellite markers. *J. Plant Biochem. Biotechnol.* 13:25–32.

Spaniolas, S., May, S. T., Bennett, M. J., and Tucker, G. A. 2006. Authentication of coffee by means of PCR-RFLP analysis and lab-on-a-chip capillary electrophoresis. *J. Agric. Food Chem.* 54: 7466–7470.

Spink, J., Moyer, D. C., and Speier-Pero, C. 2016. Introducing the food fraud initial screening model (FFIS). *Food Control* 69: 306–314.
Spink, J., and Moyer, D. C. 2011. Defining the public health threat of food fraud. *J. Food Sci.* 76: R157–R163.
Steele, K. A., Ogden, R., McEwing, R., Briggs, H., and Gorham, J. 2008. InDel markers distinguish Basmatis from other fragrant rice varieties. *Field Crop Res.* 105: 81–87.
Storck, C. R., Picolli da Silva, L., and Fagundes, C. A. 2005. Categorizing rice cultivars based on differences in chemical composition. *J. Food Compos. Anal.* 18: 333–341.
Sundaram, R. M., Naveenkumar, B., Biradar, S. K., Balachandran, S. M., Mishra, B., IlyasAhmed, M., Viraktamath, B. C., Ramesha, M. S., and Sarma, N. P. 2008. Identification of informative SSR markers capable of distinguishing hybrid rice parental lines and their utilization in seed purity assessment. *Euphytica* 163: 215–224.
Suzuki, Y., Chikaraishi, Y., Ogawa, N. O., Ohkouchi, N., and Korenaga, T. 2008. Geographical origin of polished rice based on multiple element and stable isotope analyses. *Food Chem.* 109: 470–475.
Tang, T., Huang, J., Zhong, Y., and Shi, S. 2004. High-throughput S-SAP by fluorescent multiplex PCR and capillary electrophoresis in plants. *J. Biotechnol.* 114: 59–68.
Thind, G. K., and Sogi, D. S. 2005. Identification of coarse (IR-8), fine (PR-106) and superfine (Basmati-386) rice cultivars. *Food Chem.* 91: 227–233.
U.S. Food and Drug Administration. 2014. U.S. Federal Food, Drug, and Cosmetic Act. Chapter IV: Food.
Vaingankar, N. M., and Kulkarni, P. R. 1989. A cooking quality parameter as an indicator of adulteration of Basmati rice. *J. Sci. Food Agric.* 48: 381–384.
Valentini, A., Pompanon, F., and Taberlet, P. 2009. DNA barcoding for ecologists. *Trends Ecol. Evol.* 24: 110–117.
Vemireddy, L. R., Archak, S., and Nagaraju, J. 2007. Capillary electrophoresis is essential for microsatellite marker based detection and quantification of adulteration of Basmati rice (Oryza sativa). *J. Agric. Food Chem.* 55: 8112–8117.
Vemireddy, L. R., Ranjithkumar, N., Vipparla, A., Surapaneni, M., Choudhary, G., Sudhakarrao, K. V., and Siddiq, E. A. 2014. Molecular profiling of major Indian rice cultivars using a set of eight hypervariable microsatellite markers. *Cereal Res. Commun.* 43: 189–203.
Vlachos, A., and Arvanitoyannis, I. S. 2008. A review of rice authenticity/ adulteration methods and results. *Crit. Rev. Food Sci. Nutr.* 48: 553–598.
Voorhuijzen, M. M., van Dijk, J. P., Prins, T. W., Van Hoef, A. M., Seyfarth, R., and Kok, E. J. 2012. Development of a multiplex DNA-based traceability tool for crop plant materials. *Anal. Bioanal. Chem.* 402: 693–701.
Widjaja, R., Craske, J. D., and Wootton, M. 1996. Changes in volatile components of paddy, brown and white fragrant rice during storage. *J. Sci. Food Agric.* 71: 218–224.
Woolfe, M., and Primrose, S. 2004. Food forensics: Using DNA technology to combat misdescription and fraud. *Trends Biotechnol.* 22: 222–226.
Yashui, A., and Shindoh, K. 2000. Determination of the geographic origin of brown-rice with trace-element composition. *Bunseki Kagaku* 49: 405–410.
Zhang, Q., Maroof, M. A., Lu, T. Y., and Shen, B. Z. 1992. Genetic diversity and differentiation of indica and japonica rice detected by RFLP analysis. *Theor. Appl. Genet.* 83: 495–499.
Zhao, Y., Zhfang, B., Chen, G., Chen, A., Yang, S., and Ye, Z. 2014. Recent developments in application of stable isotope analysis on agro product authenticity and traceability. *Food Chem.* 145: 300–305.

CHAPTER 20

Fingerprinting Techniques in Food Authentication and Traceability of Marine Species

Abdul Moheman
M. J. P. Rohilkhand University

Showkat Ahmad Bhawani and Ahmad Husaini
UNIMAS

Mohammad Sarwar Alam
Jamia Hamdard

Abu Tariq
Universiti Sains Malaysia

CONTENTS

20.1 INTRODUCTION

The authentication of food is growing rapidly because of the increasing awareness about the quality and safety of food. Authentication of food deals with the process which verifies that the label descriptions on the food are in compliance with safety regulations. Because of the globalization of food markets, consumers have become keenly interested in the geographical origins and also in the assured quality of food and drinks. Since ancient times, authenticity of food has been a global concern for producers, regulators, and consumers (Posudin et al. 2015). The authentication process provides information about the food products regarding their origins, methods of production, and processing technologies. Many developed countries have imposed traceability systems that can trace the quality of food and can reduce the information asymmetry problems (Shogren 2002; Sykuta 2005; Peres et al. 2007; Zhao and Zhao; 2007). As a consequence of this, the traceability system has come to the forefront worldwide (Moises 2007; Popper 2007). Food traceability is actually based on the enterprise credibility. In other words, all the information stored about a product shows the enterprise willingness and credit. Thus, if there is no proper supervision, they can easily provide counterfeit information. In the seafood industry, the authentication of fish and seafood is a rising issue due to the increase in international trade. The increasing consumption of fish and seafood worldwide with different levels of supply and demand for certain species leads to economic frauds where one species is substituted illegally with another species (Civera 2003; Martinez et al. 2005; Reilly 2018). In order to control such economic fraud, the European Union has established laws regarding the labeling of fish and aquaculture products with traceability information (identification, origin, and method of production) (Moretti et al. 2003; Martinez et al. 2005; Reilly 2018). The substitution of seafood has also been prohibited in the United States and a product is considered as misbranded if it is available in the market under the name of another product (United States code, title 21, chapter 9, subchapter IV, section 343) (USFDA 2006). Seafood adulteration has detrimental effects on the commercial market and also puts consumers at risk since they might end up consuming mislabeled and harmful products. This also leads to the decreasing effectiveness of marine conservation programs in protecting endangered species (Moran and Garcia-Vazquez 2006). The enforcement of labeling regulations for frozen fillets and precooked seafood has become complicated due to the difficulty in identifying morphological characteristics (Moran and Garcia-Vazquez 2006). Therefore, to prevent product substitution and to impose labeling regulations, a reliable method is required that can analyze the species of a seafood product that has no detectable external features (Gil 2007; Mafra et al. 2007). In the same way, in order to impose laws against poaching and trade of overexploited species, most reliable methods are required for the diagnosis of species (Baker et al. 2000; Kyle and Wilson 2007).

Fingerprinting is a promising and emerging technology for the detection of adulteration in food. Fingerprinting technology provides an image and a spectrum is obtained from certain analytical tools. The fingerprinting technology has been in use for a long time for the detection of several parameters in a sample, such as source, component, additives, microorganisms, etc. (Valeria et al. 2005; Serge et al. 2007; Zarkadas et al. 2007; Santonico et al. 2008; Casale et al. 2010; Ahad and Nissar 2017).

20.2 CLASSIFICATION

The fingerprinting techniques can be broadly divided into four classes as shown in Figure 20.1. Electrophoretic fingerprinting is widely used for the identification of species and the analysis of purity of food materials. Electrophoretic fingerprinting can further be classified into biochemical fingerprinting and DNA fingerprinting. Biochemical fingerprinting is subdivided into two types: protein electrophoretic and isoenzyme electrophoretic. Protein electrophoresis distinguishes different species based on the characteristic protein markers, while the isoenzyme electrophoresis is based on the isoenzyme specificity. DNA fingerprinting enables identification based on the differences in the patterns of DNA

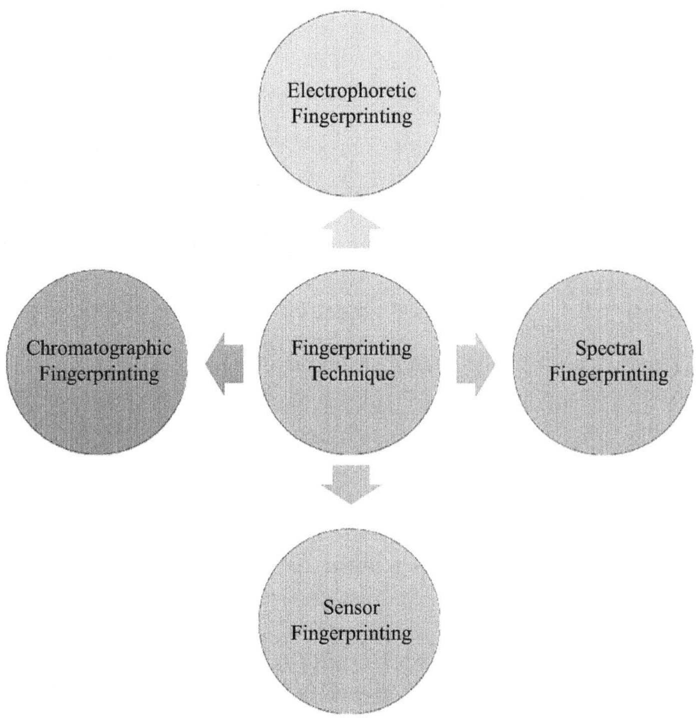

FIGURE 20.1 **(See color insert.)** Classification of fingerprinting techniques.

molecular markers obtained from the electrophoretic process. Spectrum fingerprinting can be divided into five groups, namely nuclear magnetic resonance spectroscopy (NMR), infrared spectroscopy (IR), Raman spectroscopy, fluorescence spectroscopy (FS), and mass spectrometry (MS). All the five techniques are used widely in the evaluation of structures and the compositions of samples. Chromatographic fingerprinting is a kind of separation technology that includes gas chromatographic (GC) and liquid chromatographic (LC) fingerprinting. The chromatographic fingerprinting is widely used for the analysis of trace contents in samples, such as pesticide residues, food additives, etc. Up to date, instrumental sensors like electronic nose (e-nose), electronic tongue (e-tongue), and electronic eye (e-eye) are being used as fingerprinting techniques to provide information about the color, taste, and odor of fish and other marine food items.

20.3 APPLICATION

20.3.1 Electrophoretic Fingerprinting

The interchange between fish and other marine food is a type of financial fraud where highly prized species are interchanged with those of lesser value. To prevent illegal species interchange, different electrophoretic fingerprinting techniques have been proposed to detect species substitution in marine food products. The procedure of PCR-RFLP (polymerase chain reaction-restriction fragment length polymorphism) is showed in Figure 20.2 to demonstrate electrophoretic fingerprinting (Zhang et al. 2011).

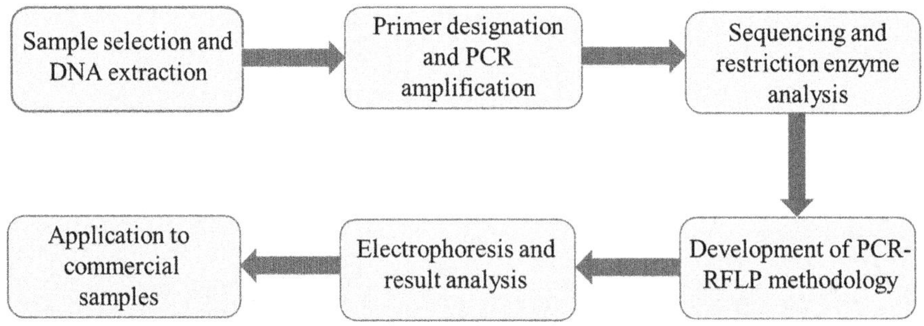

FIGURE 20.2 The procedure of PCR-RFLP fingerprinting method.

20.3.1.1 Biochemical Fingerprinting

20.3.1.1.1 Isoenzyme Electrophoretic Isoenzyme electrophoretic fingerprinting is used in the identification of species based on the specificity of an isoenzyme. Zhihua et al. (2009) identified the specific enzymes in six isoenzymes from two organizations for four populations of natural and hatched meretrix by using isoenzyme electrophoretic fingerprinting. In another study, Meng et al. (2009) investigated two isoenzymes from six tissues of *Channa argus* and *Channa asiatica*. Isoenzyme electrophoretic fingerprinting also gives valid identification for the loach groups from different regions (Wu and Chang 2009). However, several studies revealed that isoenzyme electrophoretic fingerprinting often fails in detecting adulteration (Zhang et al. 2011).

20.3.1.1.2 Protein Electrophoretic Protein electrophoretic techniques, such as SDS-PAGE (sodium dodecyl sulfate-polyacrylamide gel electrophoresis), IEF (isoelectric focusing), 2DE (two-dimensional electrophoresis), PMF (peptide mass fingerprint), and PFF (peptide fragmentation fingerprint) are extensively used for the detection and authentication of species present in marine food products (Mackie et al. 2000; Ortea et al. 2010; Zhang et al. 2011; Ortea et al. 2016; Jagadeesh et al. 2017). Due to great phenotypic similarity between allied species, different kinds of intentional or unintentional adulterations can happen, such as the substitution of a highly valuable species with a similar but low-priced one or the labeling of an aquaculture-obtained product as fished wild. Therefore, all over the world, legislation has become mandatory in order to assure authenticity and traceability in the interchange of these species. For example, the European legislation (EU 2013) mandates that marine food products should be labeled, showing the commercial name of the species, the production method (wild or farmed), and the geographical area (country) in which the fish and other marine species were caught or cultivated. Therefore, accurate and reliable analytical fingerprinting methods are needed in order to assess these parameters. Electrophoretic protein profiles can be used as markers for differentiating very closely related species (An et al. 1988; An et al. 1989). SDS-PAGE or IEF of the proteins from muscle samples were used for the differentiation of three shrimp species (Ortea et al. 2016). However, SDS-PAGE and IEF gels have also been used for the differentiation of shrimp from crab meat and fish, lobster, and cephalopod meat (Civera and Parisi 1991; Rehbein 1995). These studies revealed that no single extraction method and electrophoretic system was suitable for all applications. For example, SDS-PAGE and IEF of the sarcoplasmic, water-soluble proteins performed well while differentiating the raw samples, whereas

SDS extracts worked better for the cooked muscles. In a study, water-soluble muscular proteins were used for differentiating closely related shrimp and prawn species with native IEF, with consequent identification of species-specific proteins by LC/MS (Ortea et al. 2010). In another study, two-dimensional gel electrophoresis (2DE) has been used to investigate the quantitative differences between two marine mussel species (López et al. 2002). The differences noticed between the two species could be due to climate changes as well as genetic variability. Pineiro et al. (2010) has extensively reviewed the use of various protein electrophoretic fingerprinting techniques for evaluating the effects of climate change on marine food products. 2DE has also been used for finding the potential biomarkers which can be used for differentiating between two populations of the great scallop (Artigaud et al. 2014). Ortea et al. (2009) has used sarcoplasmic protein, arginine kinase (AK), as a biomarker for the identification of six different species belonging to the order Decapoda. However, the PMF spectra of AK have also been used as a marker for the authentication of six shrimp species of commercial relevance. In a study, SDS-PAGE and urea-IEF methods were used for the identification of ten commercial fish species after cooking (Etienne et al. 2000). The authors concluded that the two methods performed differently depending on the species. Renon et al. (2005) used IEF sarcoplasmic protein pattern for the discrimination between swordfish, blue marlin, and Mediterranean spearfish in fresh, frozen, and cold-smoked fillets. 2DE profiles of water-soluble muscular proteins were used for the differentiation between fish species, such as gadoid fish (Piñeiro et al. 1998), flat fish (Piñeiro et al. 1999), and hake species (Piñeiro et al. 2001). Nucleoside diphosphate kinase A (NDK A) has also been used as a marker protein for the identification of hake species. Proteins such as triose phosphate isomerase, pyruvate kinase, troponin T, and beta-enolase have been used as potential markers for discriminating tuna species, like *Thunnus thynnus*, *Thunnus alalunga*, *Thunnus albacares*, and *Thunnus obesus*, with 2DE (Pepe et al. 2010; Pepe et al. 2012). Chen et al. (2004) were able to differentiate five puffer fish species with 2DE using urea-soluble protein extracts from the fish muscles. In another study, 2DE was also used to distinguish the wild cod fish from the farmed variety (Martínez et al. 2007). It has been demonstrated that PMF and PFF provide some advantages over 2DE because the former allows the identification of the protein, which is useful for the direct comparison and classification of species (Carrera et al. 2006; Carrera et al. 2007; Carrera et al. 2011). Ortea et al. (2009) used AK as a marker for the discrimination of shrimp species by combining 2DE with PMF. In a study, Mazzeo et al. (2008) applied the "molecular profiling" approach based on matrix-assisted laser desorption/ionization-time of flight-mass spectrometry (MALDI)-TOF-MS for fish authentication.

20.3.1.2 DNA Fingerprinting

DNA fingerprinting is a technique that simultaneously detects large numbers of minisatellites in the genome to produce a pattern characteristic of an individual. It was first invented by Professor Sir Alec Jeffreys of Leicester University in 1984. DNA-based fingerprinting methods are generally more advantageous and are currently widespread in species authentication and traceability of marine food products (Rasmussen and Morrissey 2009; Espineira et al. 2010; Zhang et al. 2011; Scarano and Rao 2014; Danezis et al. 2016; Galal-Khallaf et al. 2016; Tagliavia et al. 2016; Ahad and Nissar 2017). Some examples of these methods are multiplex or species-specific PCR, RFLP, random amplified polymorphic DNA (RAPD), amplified fragment length polymorphism (AFLP), single-stranded conformational polymorphism (SSCP), simple sequence repeat (SSR), or short tandem repeat (STR), inter-simple sequence repeat (ISSR), sequence tagged site (STS), forensically informative nucleotide sequencing (FINS), and single nucleotide polymorphisms (SNP), etc. These methods largely depend on the highly specific amplification of DNA fragments. A large variety of gene targets from both nuclear and mitochondrial (mt) DNA have been employed in DNA fingerprinting methods for the identification of species in marine food products. RFLP was the first DNA-based marker used to identify DNA polymorphisms among different individuals. It is also one of the most widely used markers in the detection of food adulteration for close genetic relationship species. PCR-based fingerprinting methods are fast, extremely sensitive, and are widely used in marine food authentication. Several authors reported that RFLP combined with PCR (PCR-RFLP) is very useful for the identification and traceability of species in marine food products (Arahishi 2005; Hung-Sheng et al. 2007; Wen-Feng and Deng-Fwu 2007; Ferrito and Pappalardo 2017). Carrera et al. (1999a, b) reported that the PCR-RFLP fingerprinting method allowed the identification of raw and smoked meats of two fish species belonging to the Salmonidae family. However, the main drawbacks of RFLP are that it is an expensive and time-consuming process, not amenable to automation, and now it is considered outdated. RAPD is a simple, quick, and cost effective method, providing more information than RFLP fingerprinting. It can be used for taxonomic and phylogenetic studies in species differentiation and also for studying the phylogeographic patterns of genetic variations (Scarano and Rao 2014). RAPD is a generally dominant genetic marker and can be carried out for the whole genome. It does not require the steps of hybridization, isotope labeling, or cloning preparation. Zhou et al. (2006) were able to differentiate between yellow catfish (*Pelteobagrus fulvidraco*) and long yellow catfish (*Pelteobagrus eupogon*) with RAPD and cluster analysis. The AFLP marker is developed by integrating RAPD and RFLP. Compared to RFLP and RAPD, the AFLP method has high efficiency, high resolution of polymorphism, and requires less DNA template. It is mainly used to distinguish closely related species that are generally used in food adulteration. SSCP markers are used for the detection of mutations in genes as well as polymorphisms occurring in DNA. The PCR-SSCP fingerprint method was also employed to identify fish species (Hayashi 1992; Rehbein et al. 1999). SSR or STR (also known as microsatellite DNA marker) is commonly used for the study of genetic variations among and within species (Zietkiewicz et al. 1994). It has the same advantages as RFLP and has higher repeatability and stability than RAPD. Compared to AFLP, SSR is a codominant marker. ISSR analysis is considered beneficial in gene mapping studies. It can be applied in various studies, for example, genetic identity, origin, clone and strain identification, and taxonomic studies of closely related species. The main advantage of ISSR is that no sequence data for primer construction are needed since they are randomly dispersed throughout the genome. STS is used to detect a single, unique, sequence-defined point in the genome. It requires sequence information, but it does not require cloning. FINS is also considered as a suitable technique for species identification in the marine food industry (Espiñeira et al. 2010). It was first described by Bartlett and Davidson (1992) who proposed the genetic identification of species by the use of phylogenetic analysis. Over the past few years, the SNP method, also called Snip marker, has gained in popularity for genetic characterization studies as well as for translational genomic studies (Johnston et al. 2013).

Various studies have been carried out to apply electrophoretic fingerprinting methods for species authentication of fish and other marine food (Table 20.1).

TABLE 20.1 Studies Employing Electrophoretic Fingerprinting Techniques for Marine Species Authentication

Technique	Purpose of Analysis	Target	Reference
SDS-PAGE and urea-IEF	Identification of ten commercially important fish species (Atlantic salmon, sea trout, rainbow trout, turbot, Alaska pollock, pollack, pink salmon, Arctic char, chum salmon, and New Zealand hake) after cooking	Protein	Etienne et al. 2000
Native IEF	Differentiation of shrimp species	Sarcoplasmic calcium-binding proteins	Ortea et al. 2010
2DE	Discrimination of two great scallop (*Pecten maximus*) populations	Mantle proteins	Artigaud et al. 2014
2DE	Differentiation of tuna species	Muscular proteins	Pepe et al. 2010; Pepe et al. 2012
MS/MS DDA + PFF; 2DE and PMF	Differentiation of shrimp species	Arginine kinase	Ortea et al. 2009a, Ortea et al. 2009b
Multiplex PCR	Differentiation of Iceland scallop and American sea scallop	mt COI	Marshall et al. 2007
Species-specific monoplex and quadruplex PCR	Differentiation of six Gadoid fish species	Pantophysin I	Hubalkova et al. 2008
Species-specific PCR	Identification of scallop adductor muscles (Zhikong scallop, noble scallop, yesso scallop, and Atlantic bay scallop)	Microsatellites	Zhan et al. 2008
PCR	Authentication of six shellfish (*Pleoticus robustus*, *Metapenaeopsis barbata*, *Parapenaeus fissuroides*, *Hymenopenaeus debilis*, *Metapenaeus affinis*, and *Sepia aculeate*)	Cytochrome oxidase COI and 16S rDNA	Galal-Khallaf et al. 2016
RFLP	Differentiation of *Scomber* mackerel fish species	5S ribosomal DNA (5S rDNA)	Arahishi 2005
PCR-RFLP (4 enzymes)	Differentiation of nine cod fish species (Pacific cod, Alaska pollock, Saffron cod, Arctic cod, Southern blue whiting, Southern hake, South Pacific hake, longfin codling, and blue grenadier)	mt $tRNA^{Glu}$ cytochrome b	Akasaki et al. 2006
PCR and PCR-RFLP (1 to 2 enzymes)	Identification of four species of mud crab (*S. serrata*, *S. olivacea*, *S. paramamosain*, and *S. tranquebarica*)	mt 16S rRNA nuclear ITS1	Imai et al. 2004
PCR-RFLP (2 enzymes)	Differentiation of four species of penaeid shrimps (green tiger prawn, caramote prawn, deep-water rose shrimp, and *M. monoceros*)	mt cytochrome b	Hisar et al. 2008

(*Continued*)

TABLE 20.1 (*Continued*) Studies Employing Electrophoretic Fingerprinting Techniques for Marine Species Authentication

Technique	Purpose of Analysis	Target	Reference
PCR-RAPD	Identification of fish species (Nile perch, Grouper, wreck fish, and Blackfin goosefish)	Random amplification of genome to produce "DNA fingerprint"	Asensio et al. 2002; Ramella et al. 2005
PCR-SSCP	Authentication of closely related fish species (tuna, eel, salmon, trout, and sturgeon)	mt cytochrome b	Rehbein et al. 1997
PCR-SSCP	Genetic differentiation between two clam species (grooved carpet shell and pullet carpet shell)	Nuclear α-actin	Fernandez et al. 2002
PCR-SSCP	Identification of ten salmonids (Atlantic, chum, coho, pink, sockeye, chinook, rainbow trout, Arctic char, brook trout, and brown trout) fish species (raw or cold smoked)	mt cytochrome b nuclear parvalbumin and nuclear growth hormone	Rehbein 2005
FINS	Genetic identification of about thirty species belonging to the class cephalopods	cytochrome b	Espiñeira et al. 2010
SNP genotyping assay	Differentiation of the two species of crabs (Tanner crab, queen scallop) along with their hybrids	mt 16S rRNA nuclear ITS1	Smith et al. 2005

20.3.2 Spectral Fingerprinting

Over the past few years, fingerprinting techniques have gained more and more popularity due to the advancements in the analytical instruments and the application of chemometrics methods. A large number of processed marine food products have lost their morphological appearances. Therefore, accurate and reliable analytical methods are needed to determine the species, chemical compositions, geographical origins (country), production methods (i.e., wild/farmed), processing conditions (fresh/frozen thawed fish), and substitutions with fish and other marine food products. Spectral fingerprinting techniques such as NMR, IR, Raman, FS, and MS have been successfully employed to analyze these parameters. The advantages of these spectral techniques are that they are rapid, nondestructive, and non-offensive; there is a possibility of carrying out simultaneous determination of several parameters; and they have the ability to replace costly and time-consuming reference techniques (Karoui et al. 2010; Karoui and Blecker 2011; Cozzolino and Murray 2012; Liu et al. 2013; Hassoun and Karoui 2017). In addition, spectroscopic fingerprinting techniques often require little or no sample preparation and avoid sample destruction (Karoui et al. 2006, 2010; Cheng et al. 2013). NMR spectroscopy has been used as a tool for authenticity purposes in food science. The best analytical strategy for the authentication of geographical origins (country) of marine food products is the determination of elemental concentrations and isotopic variations (hydrogen-2/hydrogen-1, nitrogen-15/nitrogen-14, carbon-13/carbon-12, oxygen-18/oxygen-16) in different biological tissues. It is well known that freshwater ecosystems are generally deficient in carbon-13 and nitrogen-15 as compared to the marine ecosystems. So, the analysis of these stable isotope ratios can be used for differentiating between freshwater and marine species. Furthermore, stable isotope analysis can also be used to distinguish between different environments. NMR in conjugation with isotopic ratio mass spectrometry (NMR/IRMS) and site-specific natural isotope fractionation via nuclear magnetic resonance (SNIF-NMR) are commonly used for determining the ratios of stable isotopes. Thus, stable isotope ratio (SIR) analysis can provide effective methods for the identification and tracking of food products. Lipid is considered as a good candidate for identifying the geographical origin (country) of marine species with NMR spectroscopy (Aparicio et al. 1998; Aursand et al. 2000; Aursand and Axelson 2001). In a study, the "fingerprints" that result from the chemical shift positions and peak heights of ^{13}C NMR spectra of lipids has been used to identify the species and origins of purified marine oils (Aursand et al. 1995a, b). Wild salmon is considered superior than the farmed variety in terms of quality and nutritional properties. Capuano et al. (2012) utilized ^{1}H NMR to distinguish the wild from the farmed salmon. ^{1}H NMR has also been used to authenticate 114 salmon fish oils originating from eight different areas, such as Canada, Alaska, the Faroes, Ireland, Iceland, Norway, Scotland, and Tasmania (Masoum et al. 2007). NMR techniques have also been applied to estimate stress and freshness of the fish (Erikson et al. 1997; Sitter et al. 1999) and whether the fish (cod) had been frozen (Howell et al. 1996). During the past few decades, high resolution NMR spectroscopy has been utilized as a tool to evaluate the quality of fish and other seafood (Bekiroglu 2011). IRMS and high resolution ^{2}H SNIF-NMR spectroscopy have been utilized to classify different kinds of fish oils and lipids extracted from the muscles of wild and farmed salmons (Aursand et al. 2000).

During the past few years, IR spectroscopic fingerprinting technique has become one of the most attractive and commonly used techniques to be applied in diverse fields of food science. Several studies have been performed to evaluate the potential of IR spectroscopy in the authentication of fish and other marine food species. IR spectroscopy has been used to distinguish the fresh from the frozen fish (Nilsen et al. 2002) by determining the water content in the product and to detect fraud replacements (Cozzolino et al. 2002). The fatty acid profile obtained with IR spectroscopy has been used to differentiate the wild from the farm fish. For instance, a high level of arachidonic acid, unsaturated n-3, and saturated fatty acids in muscles are characteristic of the product fished, while a high amount of C18 fatty acids (particularly oleic, linoleic, and α-linolenic acids), due to the intake of vegetable oils from feedings, characterizes the farm fish. Chemometrics are multivariate data analyses where mathematical and statistical approaches are employed to reduce the dimensionality of the dataset. IR spectroscopy combined with chemometric tools such as PCA (principal component analysis), PLS (partial least squares), FDA (factorial discriminate analysis), PLSDA (partial least squares discriminant analysis), SVM (support vector machine), and so on could be considered as powerful tools to determine freshness, chemical, and microbiological parameters of fish and other marine food (Cozzolino and Murray 2012; Liu et al. 2013). Tristep IR spectroscopy (Fourier transform infrared spectroscopy (FTIR) integrated with second derivative infrared (SD-IR) spectroscopy and two-dimensional correlation infrared spectroscopy (2DCOS-IR) coupled to PCR has been employed for the characterization and discrimination of three marine fish *surimi* (white croaker *surimi* (WCS), hairtail *surimi* (HS), and red coat *surimi* (RCS)) (Liu et al. 2015). Freshness is essential for the quality assessment of marine food, since this quality parameter is affected by several factors such as rigor mortis, autolysis processes, and microbiological spoilage after death (Alasalvar et al. 2011; Cheng et al. 2014; Cheng et al. 2015). Madigan et al. (2013) reported that storage time and sample

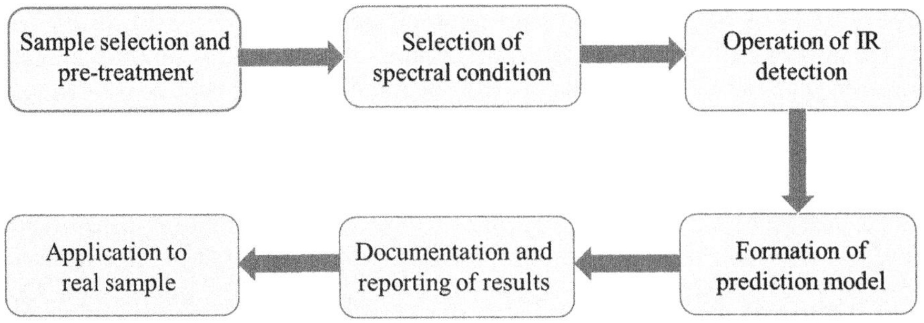

FIGURE 20.3 The procedure of IR.

odor are good indicators to distinguish the fresh from the frozen fish and other marine species with IR spectroscopy. Chemical parameters, for instance, moisture, protein, total fat, and free fat of fish and other marine food have been successfully and accurately analyzed with IR spectroscopy (Xiccato et al. 2004; Khodabux et al. 2007; ElMasry and Wold 2008). Karoui et al. (2010) enabled the characterization of fish and other seafood quality with IR spectroscopy. Uddin et al. (2005) has demonstrated that VIS (visible)/NIR (near-infrared) spectroscopy in reflectance mode could be used to differentiate the fresh fish samples from the frozen-thawed ones. Hernández-Martínez et al. (2013) has used FTIR in reflectance mode to predict total fat, fatty acids, content of omega-3, and fish lipid quality index of three fish species (Atlantic bluefin tuna, crevalle jack, and Atlantic Spanish mackerel). The procedure of IR is shown in Figure 20.3 to demonstrate the spectrum fingerprinting (Zhang et al. 2011).

Like IR, Raman spectroscopy can also be a promising tool for the rapid discrimination of fish and other marine food products depending on their species and freshness (Velioglu et al. 2015). In addition, Raman spectroscopy has been used to measure the fat content and monitor the lipid oxidation and protein structure in marine food products (Marquardt and Wold 2004). PCA is a powerful and versatile chemometric method, providing an overview of complex multivariate data. By applying PCA approach to Raman spectra, good data about fat, pigment, and collagen can be obtained. Raman spectroscopy has also been used to predict the unsaturated fatty acid levels expressed as iodine values (Afseth et al. 2006).

FS is a fast and nondestructive technique used for screening a large number of food products (Karoui and Blecker 2011). FS can be used for monitoring the molecular changes occurring during handling and processing or storing of fish and other marine species (Aubourg et al. 1998; Aubourg and Medina 1999; Airado-Rodriguez et al. 2010; Eaton et al. 2012). FS allows to discriminate fish samples according to their geographical origins (Eaton et al. 2012). This method was used to determine the pyrene concentrations in marine fish samples (Liu et al. 2012). It has been reported that "fingerprints" that result from tryptophan, aromatic amino acids and nucleic acids (AAA + NA), and nicotinamide adenine dinucleotide (NADH) fluorescence spectra can be used in the discrimination of fresh and aged fish fillets (Dufour et al. 2003; Hassoun and Karoui 2015). Hence, tryptophan, AAA + NA, and NADH fluorescence spectra along with chemometric tools may be considered as powerful tools for authentication issues.

In proteomic research, electrospray ionization (ESI) and MALDI are the two techniques most commonly used for MS (Yates et al. 2009). MALDI-TOF-MS, and ESI-MS/MS-based methods have been used for the authentication of shrimp species of commercial value (Ortea et al. 2009). Several researches have revealed that the distribution of mineral elements varies between and among marine species, depending on age, development, tissues, and several physiological and geographical factors (Khansari et al. 2005; Toppe et al. 2007; Sfar et al. 2012). Atomic absorption spectroscopy (AAS) and inductive couple plasma mass spectrometry (ICP-MS) have been used for the authentication of marine food species on the basis of fingerprint data of elements (Medeiros et al. 2012; Guo et al. 2013).

Some examples of spectroscopic fingerprinting techniques used for the food authentication and traceability of marine species are listed in Table 20.2.

20.3.3 Chromatographic Fingerprinting

Chromatography is a very popular analytical technique mainly used for separating and identifying different substances and for determining the constituents of a sample. It can be used for describing the chemical, physical, sensory characteristics, adulteration, and contamination of fish and other marine food, in order to ensure their quality and safety for the consumers, and also for establishing fair pricing policies (Antoine et al. 2001; Gallo and Ferranti 2016; Ahad and Nissar 2017). Chromatographic techniques are extensively used for analyzing the fingerprints of foodstuffs and identifying the fish species (Ashoor and Knox 1985; Knuutinen and Harjula 1998). Biogenic amines such as histamine, cadaverine, tyramine, and agmatine are produced from the decarboxylation of free amino acids such as histidine, lysine, tyrosine, and arginine, respectively, liberated from

TABLE 20.2 Spectral Fingerprinting Techniques Used for Authentication and Traceability of Fish and Other Marine Food

Species	Technique	Purpose of Analysis	Reference
Cod, mackerel, salmon, and whiting	FS/PCA	Determination of fish freshness	Dufour et al. 2003
Salmon paté	FS/PCA and PLSR	Detection of early stages of lipid oxidation	Olsen et al. 2006
Whiting	FS/PCA and FDA	Authentication issues (fresh/frozen thawed fish)	Karoui et al. 2006
Shrimp	FS/PARAFAC-SIMCA	Authentication of species/origin (country)	Eaton et al. 2012
Color carp	FS	Determination of pyrine in fish gills	Liu et al. 2012
Whiting	FS	Authentication issues (storage conditions, presence/absence of light, total/partial vacuum)	Hassoun and Karoui 2015
Cod and salmon	VIS/NIR coupled to PCA and PLS	Inspection of fish freshness (visible is applied for cod, while NIR is used for salmon)	Nilsen et al. 2002
European sea bass	NIR/PLSDA	Authentication issues (wild/farmed fish)	Ottavian et al. 2012
Atlantic salmon	NIR/PCA and PLS	Prediction of microbial spoilage	Tito et al. 2012

(Continued)

TABLE 20.2 (*Continued*) Spectral Fingerprinting Techniques Used for Authentication and Traceability of Fish and Other Marine Food

Species	Technique	Purpose of Analysis	Reference
Halibut	VIS/NIR coupled to PCA and SVM	Authentication issues (fresh/ frozen thawed fish)	Zhu et al. 2013
Tuna	NIR/PCA and PLS	Prediction of some chemical parameters	Khodabux et al. 2007
Pacific oysters	NIR/Linear mixed models	Freshness assessment	Madigan et al. 2013
Whiting	MIR/PCA and FDA	Authentication issues (fresh/ frozen thawed fish)	Karoui et al. 2007
Atlantic bluefin tuna, crevalle jack, and Atlantic Spanish mackerel	MIR/PCA and PLS	Prediction of some chemical quality parameters	Hernández-Martínez et al. 2014
European sea bass	^{1}H NMR	Authentication issues (wild/ farmed fish)	Vidal et al. 2012
Haddock, saithe, Pollack, north-east arctic cod, and Norwegian coastal cod	^{13}CNMR/PCA, LDA, and BBN	Authentication issues (Gadoid fish species)	Standal et al. 2010
Horse mackerel, European anchovy, red mullet, bluefish, Atlantic salmon, and flying gurnard	Raman/PCA	Authentication issues (fresh/ frozen thawed fishes) and species	Velioglu et al. 2015
Atlantic and horse mackerel	Raman/ANOVA	Changes in lipid and protein structure during freeze-drying	Sarkardei and Howell 2007
Rainbow trout	MS/MALDI-TOF	Presence and stability of low molecular weight peptides in post mortem rainbow trout muscle during ice storage and cooking	Bauchart et al. 2007
Different fish species	MALDI-TOF-MS	Molecular profiling methodology to fish authentication	Mazzeo et al. 2008
Shrimps	MALDI-TOF-MS, and ESI-MS/MS	Identification of seven commercial, closely related species of Decapoda shrimps	Ortea et al. 2009

Source: Hassoun and Karoui 2017.

FS = Fluorescence Spectroscopy; PCA = Principal Component Analysis; PLSR = Partial Least Squares Regression; FDA = Factorial Discriminate Analysis; PARAFAC= Parallel Factor Analysis; SIMCA = Soft Independent Modeling of Class Analogy; VIS = Visible; NIR = Near-infrared; PLSDA = Partial Least Squares Discriminant Analysis; SVM = Support Vector Machine; PLS = Partial Least Squares; MIR = Mid-infrared; 1H NMR = Proton Nuclear Magnetic Resonance; 13C NMR = Carbon Nuclear Magnetic Resonance; LDA = Linear Discriminant Analysis; BBN = Bayesian Belief Networks; ANOVA = Analysis of Variance; MS = Mass Spectrometry; MALDI-TOF = Mass Matrix-assisted Laser Desorption Ionization–Time-of-Flight; ESI = Electrospray Ionization.

proteins and peptides. They are found in very low concentrations in the fresh muscles of fish and other marine food products, but their concentrations increase due to microbial decomposition of muscles. Therefore, the identification and quantification of biogenic amines in the muscles of fish and other seafood are important due to their toxicological threats and also because of their use as indicators of spoilage. Chromatographic methods have been successfully utilized for determining the levels of biogenic amines in the muscles of fish and other marine food products (Antoine et al. 2001; Onal et al. 2013). GC and high performance liquid chromatography (HPLC) provide high resolution compound separations and can be used in conjugation with different detectors such as diode array detector (DAD) or flame ionization detector (FAD). It has been reported that GC-MS methods have been used widely in metabolomics because of their high separation efficiencies and the easy interfacing of GC with MS. It can also be used in the analysis of a wide range of volatile and chemically derivatized semi-volatile compounds. However, LC-MS methods have been widely used instead of GC-MS because derivatization of polar/nonvolatile metabolites for GC-MS analysis is time consuming and requires more material as compared to LC-MS (Koal and Deigner 2010). Figure 20.4 shows the steps of HPLC-MS method as an example (Zhang et al. 2011). A recent development in the field of metabolomics is ultra performance liquid chromatography (UPLC) system coupled to MS/MS instruments. It has been seen that compared to HPLC-MS, UPLC-MS has higher sensitivity, resolution, and signal-to-noise (S/N) ratio (Lu et al. 2008). Manthey-Karl et al. (2016) reported that chromatography is useful for analyzing the approximate composition of fish, its nutritional quality, and the presence of contaminants while identifying variations from the natural composition. Several authors have used GC, HPLC, Reversed phase HPLC (RP-HPLC), silver ion HPLC with mass spectrometry (MS), and HPLC with fast atom bombardment-MS (FAB-MS) to elucidate the structure of triacylglycerols (TAG) in dietary omega-3 (ω-3) rich oils (Mcgill and Moffat 1992; Hori et al. 1994; Perona and Ruiz-Gutierrez 1999). Liquid chromatography electrospray tandem mass spectrometry (LC–ESI–MS/MS) has been extensively used in the structural characterization of naturally occurring TAG in cod liver oil. This study revealed that it is possible to discriminate the genuine from the adulterated cod liver oil (Zeng et al. 2010). Chromatographic fingerprinting technique coupled to chemometric tools, such as PCR, can be considered as a powerful tool to authenticate fish and other marine food (Solaesa et al. 2014). Fatty acid composition of fish muscle fat is largely affected by the feeding and geographical origin. It is possible to discriminate the wild and farmed Atlantic salmon (*Salmo salar*) from the wild turbot (*Psetta maxima*) by fatty acid analysis combined with chemometrics (Busetto et al. 2008; Axelson et al. 2009).

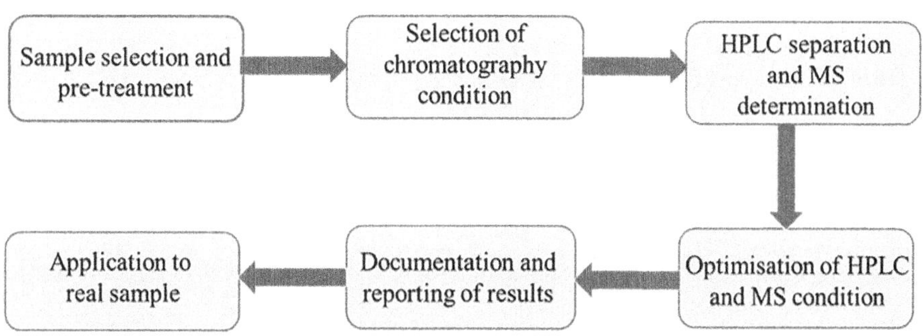

FIGURE 20.4 Steps of HPLC-MS method.

20.3.4 Sensor Fingerprinting

Sensory analysis has been considered as the most reliable and versatile method to provide information regarding smell, taste, and color of fish and other marine food. Instrumental sensors like e-nose, e-tongue, and e-eye have been successfully employed to analyze these parameters. E-nose has been considered as a simple, fast, reliable, and versatile technique for evaluating the freshness of fish and other marine species. For instance, two e-noses, namely FreshSense and LibraNose, were tested for the freshness assessment of cod fish fillets (Di Natale et al. 2001). The author applied a chemometric tool such as partial least squares dimension reduction (PLSDR) to assess the potentiality of e-noses in predicting fish freshness. Haugen et al. (2006) has examined the smoked salmon process in industrial surroundings by using a prototype solid state-based gas sensor array system called fish nose. In another study, Tian et al. (2012) used metal oxide sensors (e-nose) to monitor the freshness of hairtail fish at different storage temperatures and times. In a similar approach, an e-nose based on four integrated metal oxide microsensors were applied to examine the freshness of sardine (El Barbri et al. 2009). The shelf life of European sea bass (Limbo et al. 2009) was examined by using a commercial, portable e-nose composed of ten metal oxide semiconductors. E-tongue is a device made of sensors responding to the taste (soluble) of some food through the transduction of a signal or a pattern of signals. Several electrochemical sensors comprising amperometric, potentiometric, and voltammetric sensor arrays have been proposed for this purpose (Cosio et al. 2012). The versatility of e-tongue depends largely on the characteristics of the sensors. Gil et al. (2008) has developed an e-tongue, comprising 16 potentiometric electrodes of the metal, metal oxide, and insoluble metal salt type for assessing the freshness of sea bream stored at 4°C for up to 15 days. Similarly, Barat et al. (2008), employed an e-tongue composed of gold and silver wires to determine the degree of freshness of sea bream fish. It has been reported that volumetric e-tongues can be used to monitor freshness, shelf life, and spoilage of fish and other seafood (Rodríguez-Méndez et al. 2009; Ruiz-Rico et al. 2013). By applying chemometric tools such as PCA, PLSDA, and PLSR to e-tongue data, it was possible to classify the day of fish degradation and to predict pH parameter with more accuracy. More recently, Han et al. (2014) checked the efficiency of e-nose and e-tongue for 150 large yellow croaker samples. The authors demonstrated that e-nose and e-tongue can be used as potential tools to detect fish freshness rapidly and nondestructively. Color is one of the most important parameters of fish and other marine food products because of its relation with product freshness and it has a direct impact on consumer perceptions. Colorimeters and spectrophotometers have been used widely for

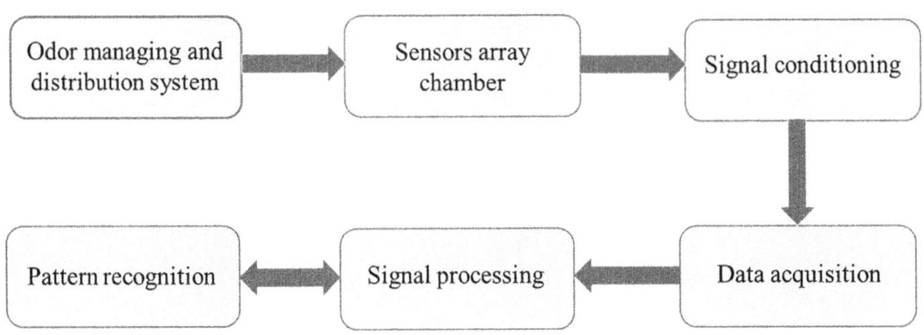

FIGURE 20.5 The process of electronic nose (e-nose) fingerprinting method.

color measurement. However, over the past few years, many instrumental sensors, such as computer vision system (CVS) and machine vision system (MVS), have been developed for the color evaluation of fish and other seafood products. In a recent study, MVS was used for the investigation of color parameters including lightness, redness, yellowness, total color change, and chroma of farmed and wild gilthead sea bream (Dowlati et al. 2013). In another study, CVS was employed to assess the color of salmon fillet (Quevedo et al. 2010). Xiong et al. (2010) has reported that an e-eye can also be used to determine the quality of fish and other sea food products by using shape and size parameters. On the basis of the above facts it can be concluded that instrumental sensors for fingerprinting can be alternatives to time-consuming, laborious, and expensive sensorial panels for the evaluation of odor, taste, and color. Figure 20.5 shows the steps of an e-nose method as an example to demonstrate sensor fingerprinting.

20.4 CONCLUSION

Food authentication and traceability of marine species are considered as important issues in preventing commercial fraud as well as health problems. This chapter provides a comprehensive and critical overview of the application of the different fingerprinting techniques such as electrophoretic fingerprinting, spectral fingerprinting, chromatographic fingerprinting and sensor fingerprinting to determine freshness, authenticity, quality, and safety parameters of fish and other marine species. The literature compiled in this chapter revealed that PCR-based fingerprinting method is reliable, effective and therefore, could be employed for routine marine species analysis. It has been noticed that fingerprinting including spectroscopic and instrumental sensory techniques combined with chemometric tools provides new possibilities of analysis data without any prior information. It is hoped that the contents of this chapter will contribute to build a current picture of the fingerprinting techniques and comprehensive knowledge used for marine food species authentication and traceability, consequently facilitating to support future developments in the marine food industry.

REFERENCES

Afseth, N. K., Wold, J. P., and Segtnan, V. H. 2006. The potential of Raman spectroscopy for characterisation of the fatty acid unsaturation of salmon. *Anal. Chim. Acta.* 572: 85–92.

Ahad, T., and Nissar, J. 2017. Fingerprinting in determining the adultration of food. *J. Pharmacogn. Phytochem.* 6: 1543–1553

Airado-Rodriguez, D., Galeano-Díaz, T., Durán-Merás, I., and Wold, J. P. 2010. Assessment of the quality attributes of cod caviar paste by means of front-face fluorescence spectroscopy. *J. Agric. Food Chem.* 58: 5276–5285.

Akasaki, T., Yanagimoto, T., Yamakami, K., Tomonaga, H., Sato, S. 2006. Species identification and PCR-RFLP analysis of cytochrome *b* gene in cod fish (order *Gadiformes*) products. *J. Food Sci.* 71: C190–C195.

Alasalvar, C., Grigor, J. M., and Zulfiqur, A. 2011. Practical evaluation of fish quality by objective, subjective, and statistical testing. *In Handbook of Seafood Quality, Safety and Health Applications*, Alasalvar, C., Miyashita, K., Shahidi, F., Wanasundara, U. (Eds.). Chichester: John Wiley and Sons Ltd., pp. 13–28.

An, H., Marshall, M. R., Otwell, W. S., and Wei, C. L. 1988. Electrophoretic identification of raw and cooked shrimp using various protein extraction systems. *J. Food Sci.* 53: 313–318.

An, H., Marshall, M. R., Otwell, W. S., and Wei, C. L. 1989. Species identification of raw and boiled shrimp by a urea gel isoelectric focusing technique. *J. Food Sci.* 54: 233–236.

Antoine, F. R., Wei, C. I., Littell, R. C., Quinn, B. P., Hogle, A. D., and Marshall, M. R. 2001. Free amino acids in dark- and white-muscle fish as determined by O-phthaldialdehyde precolumn derivatization. *J. Food Sci.* 66: 72–77.

Aparicio, R., McIntyre, P., Aursand, M., Eveleigh, L., Marighetto, N., Rossell, B., Sacchi, R., Wilson, R., and Woolfe, M. 1998. Fish oil products. In: Lees, M. (ed.) Food Authenticity Issues and Methodologies. Final report of the concerted action AIR3-CT94-2452, Eurofins Scientific, Nantes, France, pp. 213–218.

Arahishi, F. 2005. PCR-RFLP analysis of nuclear nontranscribed spacer for mackerel species identification. *J. Agric. Food Chem.* 53: 508–511.

Artigaud, S., Lavaud, R., Thébault, J., Jean, F., Strand, Ø., Strohmeier, T., Milan, M., and Pichereau, V. 2014. Proteomic based comparison between populations of the great scallop, Pectenmaximus, *J. Proteome* 105: 164–173.

Asensio, L., Gonzalez, I., Fernandez, A., Rodriguez, M. A., Lobo, E., Hernandez, P. E., Garcia, T., Martin, R. 2002. Application of random amplified polymorphic DNA (RAPD) analysis for identification of grouper (*Epinephelus guaza*), wreck fish (*Polyprion americanus*), and Nile perch (*Lates niloticus*) fillets. *J. Food Prot.* 65: 432–435.

Ashoor, S. H., and Knox, M. J. 1985. Identification of fish species by high-performance liquid chromatography. *J. Chromatogr. A* 324: 199–202.

Aubourg, S. P., and Medina, I. 1999. Influence of storage time and temperature on lipid deterioration during cod Gadus morhua and haddock (*Melanogrammus aeglefinus*) frozen storage. *J. Sci. Food Agric.* 79: 1943–1948.

Aubourg, S. P., Sotelo, C. G., and Pérez-Martín, R. 1998. Assessment of quality changes in frozen sardine (*sardina pilchardus*) by fluorescence detection. *J. Am. Oil Chem. Soc.* 75: 575–580.

Aursand, M., and Axelson, D. 2001. Origin recognition of wild and farmed salmon (Norway and Scotland) using ^{13}C NMR spectroscopy in combination with pattern recognition techniques. In: Webb, G. A., Belton, P. S., Gil, A. M., and Delgadillo, I. (eds.) *Magnetic Resonance in Food Science: A View to the Future*, RSC books, London, pp. 227–231.

Aursand, M., Jørgensen, L., and Grasdalen, H. 1995a. The use of high resolution^{1}H and^{13}C nuclear magnetic resonance NMR spectra to obtain information on the composition of fats and oils. *Lipid Forum* 49: 14–17.

Aursand, M., Jørgensen, L., and Grasdalen, H. 1995b. Positional distribution of ω3 fatty acids in marine lipid triacylglycerols by high-resolution ^{13}C nuclear magnetic resonance spectroscopy. *J. Am. Oil Chem. Soc.* 72: 293–297.

Aursand, M., Mabon, F., and Martin, G. J. 2000. Characterization of farmed and wild salmon Salmo salar by a combined use of compositional and isotopic analyses. *J. Am. Oil Chem. Soc.* 77: 659–666.

Axelson, D. E., Standal, I. B., Martinez, I., and Aursand M. 2009. Classification of wild and farmed Salmon using Bayesian Belief Networks and Gas Chromatography-Derived fatty acid distributions. *J. Agric. Food Chem.* 57: 7634–7639.

Baker, C. S., Lento, G. M., Cipriano, F., Palumbi, S. R. 2000. Predicted decline of protected whales based on molecular genetic monitoring of Japanese and Korean markets. *Proc. R. Soc. Lond. B* 267: 1191–1199.

Barat, J. M., Gil, L., Garcia-Brejio, E., Aristory, M., Toldra, F., Martinez-manez, R., and Soto, J. 2008. Freshness monitoring of sea bream (*Sparus aurata*) with a potentiometric sensor. *Food Chem.* 108: 681–688.

Bartlett, S. E. and Davidson, W. S. 1992. FINS forensically informative nucleotide sequencing - a procedure for identifying the animal origin of biological specimens. *Biotechniques* 12: 408–411.

Bauchart, C., Chambon, C., Mirand, P. P., Savary-Auzeloux, I., Remond, D., and Morzel, M. 2007. Peptides in rainbow trout (Oncorhynchus mykiss) muscle subjected to ice storage and cooking. *Food Chem.* 100: 1566–1572.

Busetto, M. L., Moretti, V. M., Moreno-Rojas, J. M., Caprino, F., Giani, I., Malandra, R., Bellagamba, F., and Guillou, C. 2008. Authentication of farmed and wild Turbot (Psetta maxima) by fatty acid and isotopic analyses combined with chemometrics, *J. Agri. Food Chem.* 56: 2742–2750.

Capuano, E., Lommen, A., Heenan, S., de la Dura, A., Rozijn, M., and Van-Ruth, S. 2012. Wild salmon authenticity can be predicted by 1H-NMR spectroscopy. *Lipid Technol.* 24: 251–253.

Carrera, E., García, T., Céspedes, A., González, I., Fernández, A., Hernández, P. E., and Martín, R. 1999a. PCR-RFLP of the mitochondraial cytochrome oxidase gene: A simple method for discrimina-tion between Atlantic salmon (*Salmo salar*) and rainbow trout (*Oncorhynchus mykiss*). *J. Sci. Food Agr.* 79: 1654–1658.

Carrera, E., García, T., Céspedes, A., González, I., Fernández, A., Hernández, P. E., and Martín, R. 1999b. Salmon and trout analysis by PCR-RFLP for identity authentication. *J. Food Sci.* 64: 410–413.

Carrera, M., Cañas, B., López-Ferrer, D., Piñeiro, C., Vázquez, J., Gallardo, J. M. 2011. Fast monitoring of species-specific peptide biomarkers using high-intensity-focused-ultrasound-assisted tryptic digestion and selected MS/MS ion monitoring. *Anal. Chem.* 83: 5688–5695.

Carrera, M., Cañas, B., Piñeiro, P., Vázquez, J., and Gallardo, J. M. 2006. Identification of commercial hake and grenadier species by proteomic analysis of the parvalbumins fraction. *Proteomics* 6: 5278–5287.

Carrera, M., Cañas, B., Piñeiro, C., Vázquez, J., and Gallardo, J. M. 2007. De novo mass spectrometry sequencing and characterization of species-specific peptides from nucleoside diphosphate kinase B for the classification of commercial fish species belonging to the family Merlucciidae. *J. Proteome Res.* 6: 3070–3080.

Casale, M., Oliveri, P., and Armanino, C. 2010. NIR and UV-vis spectroscopy, artificial nose and tongue: comparison of four fingerprinting techniques for the characterisation of Italian red wines. *Anal. Chim. Acta* 668: 143–148.

Chen, T.-Y., Shiau, C.-Y., Wei, C.-I., and Hwang, D.-F. 2004. Preliminary study on puffer fish proteome-species identification of puffer fish by two-dimensional electrophoresis. *J. Agric. Food Chem.* 52: 2236–2241.

Cheng, J.-H., Dai, Q., Sun, D.-W., Zeng, X.-A., Liu, D., and Pu, H.-B. 2013. Applications of non-destructive spectroscopic techniques for fish quality and safety evaluation and inspection. *Trends Food Sci. Tech.* 34: 18–31.

Cheng, J.-H., Qu, J.-H., Sun, D.-W., and Zeng, X.-A. 2014. Texture and structure measurements and analyses for evaluation of fish and fillet freshness quality: A review. Compr. Rev. *Food Sci. Food Safety.* 9: 52–61.

Cheng, J.-H., Sun, D.-W., Zeng, X.-A., and Liu, D. 2015. Recent advances in methods and techniques for determination and evaluation of fish and fish fillets freshness quality: A review. *Crit. Rev. Food Sci. Nutr.* 55: 1012–1225.

Civera, T. 2003. Species identification and safety of fish products. *Vet. Res. Commun.* 27: 481–489.

Civera, T., and Parisi, E. 1991. The use of sodium dodecyl sulphate polyacrylamide gel electrophoresis for the identification of rawand cooked shrimps and crabs. *Ital. J. Food Sci.* 3: 149–157.

Cosio, M. S., Scampicchioy, M., and Benedetti, S. 2012. Electronic noses and tongues. In: Pico, Y. (ed.) *Chemical Analysis of Food: Techniques and Applications*, Academic Press, Boston, pp. 219–243.

Cozzolino, D., and Murray, I. 2012. A review on the application of infrared technologies to determine and monitor composition and other quality characteristics in raw fish, fish products, and seafood. *Appl. Spectrosc. Rev.* 47: 207–218.

Cozzolino, D., Murray, I., and Scaife, J. R. 2002. Near infrared reflectance spectroscopy in the prediction of chemical characteristics of minced raw fish. *Aquac. Nutr.* 8: 1–6.

Danezis, G. P., Tsagkaris, A. S., Camin, F., Brusic, V., Georgiou, C. A. 2016. Food authentication: Techniques, trends & emerging approaches. *TrAC Trends Anal. Chem.* 85: 123–132.

Di Natale, C., Olafsdotti, G., Einarsson, S., Martinelli, E., Paolesse, R., and D'Amico, A. (2001). Comparison and integration of different electronic noses for freshness evaluation of cod-fish fillets. *Sens. Actuators B.* 77: 572–578.

Dowlati, M., Mohtasebi, S. S., Omid, M., Razavi, S. H., Jamzad, M., and De La Guardia, M. 2013. Freshness assessment of gilthead sea bream (*Sparus aurata*) by machine vision based on gill and eye color changes. *J. Food Eng.* 119: 277–287.

Dufour, É., Frencia, J. P., and Kane, E. 2003. Development of a rapid method based on front-face fluorescence spectroscopy for the monitoring of fish freshness. *Food Res. Int.* 36: 415–423.

Eaton, J. K., Alcivar-Warren, A., and Kenny, J. E. 2012. Multidimensional fluorescence fingerprinting for classification of shrimp by location and species. *Environ. Sci. Technol.* 46: 2276–2282.

El Barbri, N. E., Mirhisse, J., Ionescu, R., Bari, N. E., Correig, X., Bouchikhi, B., and Llobet, E. 2009. An electronic nose system based on a micro-machined gas sensor array to assess the freshness of sardines. *Sens. Actuators B: Chem.* 141: 538–543.

ElMasry, G., and Wold, J. P. 2008. High-speed assessment of fat and water content distribution in fish fillets using online imaging spectroscopy. *J Agric. Food Chem.* 56: 7672–7677.

Erikson, U., Beyer, A. R., and Sigholt, T. 1997. Initial distribution of white muscle high-energy phosphates related to handling stress affects K values during ice storage of Atlantic salmon *Salmo salar. J. Food Sci.* 62: 43–47.

Espiñeira, M., Vieites, J. M., and Santaclara, F. J. 2010. Species authentication of octopus, cuttlefish, bobtail and bottle squids (families Octopodidae, Sepiidae and Sepiolidae) by FINS methodology in seafoods. *Food Chem.* 121: 527–532.

Etienne, M., Jérome, M., Fleurence, J., Rehbein, H., Kündiger, R., Mendes, R., Costa, H., Pérez-Martin, R., Pineiro-Gonzalez, C., Craig, A., Mackie, I., Malmhedem-Yman, I., Ferm, M., Martinez, I., Jessen, F., Smelt, A., and Luten, J. 2000. Identification of fish species after cooking by SDS-PAGE and urea IEF: A collaborative study. *J. Agric. Food Chem.* 48: 2653–2658.

EU. 2013. Regulation No 1379/2013 of the European Parliament and of the Council of 11 December 2013 on the common organisation of the markets in fishery and aquaculture products, amending Council Regulations (EC) No 1184/2006 and (EC) No 1224/2009 and repealin. *Off. J. Eur. Union* L354: 1–21.

Fernandez, A., Garcia, T., Asensio, L., Rodriguez, M. A., Gonzalez, I., Lobo, E., Hernandez, P. E., Martin, R. 2002. Genetic differentiation between the clam species *Ruditapes decussatus* (grooved carpet shell) and *Venerupis pullastra* (pullet carpet shell) by PCR-SSCP analysis. *J. Sci. Food. Agric.* 82: 881–885.

Ferrito, V., and Pappalardo, A. M. 2017. Seafood species identification by DNA barcoding, a molecular tool for food traceability. *Biodivers. J.* 8: 65–72.

Galal-Khallaf, A., Ardura, A., Borrell, Y. J., Garcia-Vazquez, E. 2016. PCR-based assessment of shellfish traceability and sustainability in international Mediterranean seafood markets. *Food Chem.* 202: 302–308.

Gallo, M., and Ferranti, P. 2016. The evolution of analytical chemistry methods in foodomics. *J. Chromatogr. A* 1428: 3–15.

Gil, L. A. 2007. PCR-based methods for fish and fishery products authentication. *Trends Food Sci. Technol.* 18: 558–566.

Gil, L., Barat, J. M., Escriche, I., Garcia-Beijo, E., Maınez-Manez, R., and Soto, J. 2008. An electronic tongue for fish freshness analysis using a thick-film array of electrodes. *Microchim. Acta* 163: 121–129.

Guo, L., Gong, L., Yu, Y., and Zhang, H. 2013. Multi-element fingerprinting as a tool in origin authentication of four east China marine species. *J. Food Sci.* 78: C1852–1857.

Han, F., Huang, X., Teye, E., Gu, F., and Gu, H. 2014. Nondestructive detection of fish freshness during its preservation by combining electronic nose and electronic tongue techniques in conjunction with chemometric analysis. *Anal. Methods* 6: 529–536.

Hassoun, A., and Karoui, R. 2017. Quality evaluation of fish and other seafood by traditional and nondestructive instrumental methods: Advantages and limitations. *Crit. Rev. Food Sci. Nutr.* 57: 1976–1998.

Hassoun, A., and Karoui, R. 2015. Front-face fluorescence spectroscopy coupled with chemometric tools for monitoring fish freshness stored under different refrigerated conditions. *Food Control* 54: 240–249.

Haugen, J. E., Chanie, E., Westad, F., Jonsdottir, R., Bazzo, S., Labreche, S., Marcq, P., Lundbyc, F., and Olafsdottir, G. 2006. Rapid control of smoked Atlantic salmon (*Salmo salar*) quality by electronic nose: Correlation with classical evaluation methods. *Sens. Actuators, B* 116: 72–77.

Hayashi, K., 1992. PCR-SSCP: A method for detection of mutations. *Genet. Anal.: Tech. Appl.* 9: 73–79.

Hernández-Martínez, M., Gallardo-Velázquez, T., Osorio-Revilla, G., Almaraz-Abarca, N., Ponce-Mendoza, A., and Vásquez-Murrieta, M. S. 2013. Prediction of total fat, fatty acid composition and nutritional parameters in fish fillets using MID-FTIR spectroscopy and chemometrics. *LWT - Food Sci. Technol.* 52: 12–20.

Hernández-Martínez, M., Gallardo-Velázquez, T., Osorio-Revilla, G., Almaraz-Abarca, N., and Castañeda-Pérez, E. 2014. Application of MIR-FTIR spectroscopy and chemometrics to the rapid prediction of fish fillet quality. *CYTA – J. Food* 12: 369–377.

Hisar, S. A., Aksakal, E., Hisar, O., Yanik, T., and Mol, S. 2008. Discrimination of penaeid shrimps with PCR-RFLP analysis. *J. Shellfish Res.* 27: 917–920.

Hori, M., Sahashi, Y., Koike, S., Yamaoka, R., and Sato, M. 1994. Molecular species analysis of polyunsaturated fish triacylglycerol by high-performance liquid chromatography/fast atom bombardment mass spectrometry. *Anal. Sci.* 10: 719–724.

Howell, N., Shavila, Y., Grootveld, M., and Williams, S. 1996. High-resolution NMR and magnetic resonance imaging MRI studies on fresh and frozen cod (*Gadus morhua)* and haddock (*Melanogrammus aeglefinus). J. Sci. Food Agric.* 72: 49–56.

Hubalkova, Z., Kralik, P., Kasalova, J., and Rencova, E. 2008. Identification of gadoid species in fish meat by polymerase chain reaction (PCR) on genomic DNA. *J. Agric. Food Chem.* 56: 3454–3459.

Hung-Sheng, H., Tuu-Jyi, C., and Deng-Fwu, H. 2007. Using the PCR-RFLP method to identify the species of different processed products of billfish meats. *Food Control* 18: 369–374.

Imai, H., Cheng, J. H., Hamasaki, K., Numachi, K. I. 2004. Identification of 4 mud crab species (genus Scylla) using ITS-1 and 16S rDNA markers. *Aquat. Living Resour.* 17: 31–4.

Jagadeesh, D. S., Kannegundla, U., and Reddy, R. K. 2017. Application of proteomic tools in food quality and safety. *Adv. Anim. Vet. Sci.* 5: 213–225.

Johnston, S. E., Lindqvist, M., Niemelä, E., Orell, P., Erkinaro, J., Kent, M. P., Sigbjørn, L., Vähä, J., Vasemägi, A., and Primmer, C. R. 2013. Fish scales and SNP chips: SNP genotyping and allele frequency estimation in individual and pooled DNA from historical samples of Atlantic salmon (*Salmo salar*). *BMC Genomics* 14, 439–452.

Karoui, R., and Blecker, C. 2011. Fluorescence spectroscopy measurement for quality assessment of food systems—A Review. *Food Bioprocess Technol.* 4: 364–386.

Karoui, R., Downey, G., and Blecker, C. 2010. Mid-Infrared spectroscopy coupled with chemometrics: a tool for the analysis of intact food systems and the exploration of their molecular structure-quality relationships - a Review. *Chem Rev.* 110: 6144–6168.

Karoui, R., Lefur, B., Grondin, C., Thomas, E., Demeulemester, C., De Baerdemaeker, J., and Guillard, A. S. 2007. Mid-infrared spectroscopy as a new tool for the evaluation of fish freshness. *Int. J. Food Sci. Technol.* 42: 57–64.

Karoui, R., Mouazen, A. M., Dufour, E., Pillonel, L., Picque, D., De Baerdemaeker, J., and Bosset, J.-O. 2006. Application of the MIR for the determination of some chemical parameters in European Emmental cheeses produced during summer. *Eur. Food Res. Technol.* 222: 165–170.

Karoui, R., Thomas, E., and Dufour, E. 2006. Utilisation of a rapid technique based on front-face fluorescence spectroscopy for differentiating between fresh and frozen–thawed fish fillets. *Food Res. Int.* 39: 349–355.

Khansari, F. E., Ghazi-Khansari, M., Abdollahi, M. 2005. Heavy metals content of canned tuna fish. *Food Chem.* 93: 293–296.

Khodabux, K., L'Omelette, M. S. S., Jhaumee-Laulloo, S., Ramasami, P., and Rondeau, P. 2007. Chemical and near-infrared determination of moisture, fat and protein in tuna fishes. *Food Chem.* 102: 669–675.

Knuutinen, J., and Harjula, P. 1998. Identification of fish species by reversed-phase high-performance liquid chromatography with photodiode-array detection. *J. Chromatogr. B Biomed. Sci. App.* 705: 11–21.

Koal, T., and Deigner, H. P. 2010. Challenges in mass spectrometry based targeted metabolomics. *Curr. Mol. Med.* 10: 216–226.

Kyle, C. J., Wilson, C. C. 2007. Mitochondrial DNA identification of game and harvested freshwater fish species. *Forensic Sci. Int.* 166: 68–76.

Limbo, S., Sinelli, N., Torri, L., and Riva, M. 2009. Freshness decay and shelf life predictive modelling of European sea bass (*Dicentrarchus labrax*) applying chemical methods and electronic nose. *LWT-Food Sci. Technol.* 42: 977–984.

Liu, D., Zeng, X. A., and Sun, D. W. 2013. NIR spectroscopy and imaging techniques for evaluation of fish quality—A Review. *Appl. Spectrosc. Rev.* 48: 609–628.

Liu, X., Jing, J., Li, S., Zhang, G., Zou, T., Xia, X., and Huang, W. 2012. Measurement of pyrene in the gills of exposed fish using synchronous fluorescence spectroscopy. *Chemosphere* 86: 198–201.

Liu, Y., Hu, W., Guo, X. X., Wang, X. C., Sun, S. Q., Xu. C. H. 2015. Rapid discrimination of three marine fish surimi by Tri-step infrared spectroscopy combined with Principle Component Regression. *Spectrochim. Acta Part A: Mol. Biomol. Spectrosc.* 149: 516–522.

López, J. L., Marina, A., Vázquez, J., and Álvarez, G. 2002. A proteomic approach to the study of the marine mussels Mytilus edulis and M. galloprovincialis, *Mar. Biol.* 141: 217–223.

Lu, X., Zhao, X., Bai, C., Zhao, C., Lu, G., Xu, G. 2008. LC-MS--based metabonomics analysis. *J. Chromatogr. B* 866: 64–76.

Mackie, I., Craig, A., Etienne, M., Jérôme, M., Fleurence, J., Jessen, F., Pérez-Martin, R., Piñeiro, C., Rehbein, H., and Kündiger, R. 2000. Species identification of smoked and gravad fish products by sodiumdodecylsulphate polyacrylamide gel electrophoresis, urea isoelectric focusing and native isoelectric focusing: A collaborative study. *Food Chem.* 71: 1–7.

Madigan, T., Kiermeier, A., Carragher, J., De Barros Lopes, M., and Cozzolino, D. 2013. The use of rapid instrumental methods to assess freshness of half shell Pacific oysters, *Crassostrea gigas*: A feasibility study. *Innov. Food Sci. Emerg. Technol.* 19: 204–209.

Mafra, I., Ferreira, I. M., Oliveira, M. B. 2007. Food authentication by PCR-based methods. *Eur. Food Res. Technol.* 227: 649–665.

Manthey-Karl, M., Lehmann, I., Ostermeyer, U., and Schröder, U. 2016. Natural chemical composition of commercial fish species: Characterisation of pangasius, wild and farmed Turbot and Barramundi. *Foods* 5: 58.

Marquardt, B. J., and Wold, J. P. 2004. Raman analysis of fish: A potential method for rapid quality screening. *Lebensm. Wiss. Technol.* 37: 1–8.

Marshall, H. D., Johnstone, K. A., and Carr, S. M. 2007. Species-specific oligonucleotides and multiplex PCR for forensic discrimination of two species of scallops, *Placopecten magellanicus* and *Chlamys islandica*. *For. Sci. Int.* 167: 1–7.

Martinez, I., James, D., Loreal, H. 2005. Application of modern analytical techniques to ensure seafood safety and authenticity. FAO fisheries technical paper nr 455. Rome, Italy: FAO. 73 p.

Martínez, I., Slizyte, R., and Dauksas, E. 2007. High resolution two-dimensional electrophoresis as a tool to differentiate wild from farmed cod (Gadus morhua) and to assess the protein composition of klipfish. *Food Chem.* 102: 504–510.

Masoum, S., Malabat, C., Jalali-Heravi, M., Guillou, C., Rezzi, S., and Rutledge, D. N. 2007. Application of support vector machines to 1H NMR data of fish oils: Methodology for the confirmation of wild and farmed salmon and their origins. *Anal. Bioanal. Chem.* 387: 1499–1510.

Mazzeo, M. F., Giulio, B. d., Guerriero, G., Ciarcia, G., Malorni, A., Russo, G. L., and Siciliano, R. A. 2008. Fish authentication by MALDI-TOF mass spectrometry. *J. Agric. Food Chem.* 56: 11071–11076.

McGill, A. S., and Moffat, C. F. 1992. A study of the composition of fish liver and body oil triglycerides. *Lipids* 27: 360–370.

Medeiros, R. J., Santos, L. M., Freire, A. S., Santelli, R. E., Braga, A. M., Krauss, T. M., Jacob, S. C. 2012. Determination of inorganic trace elements in edible marine fish from Rio de Janeiro State, Brazil. *Food Control* 23: 535–541.

Meng, Y., Zhang, Y., and Xu, Y.-F. 2009. Comparative studies on the isoenzymes of Channa asiatica and Channa argus. *J. Anhui Agric. Sci.* 37: 12951–12964.

Moises, R. F. 2007. A principal-agent model for investigating traceability systems incentives on food safety. 105th EAAE (European Association of Agricultural Economists) Seminar 'International Marketing and International Trade of Quality Food Products', Bologna, Italy, 8–10.

Moran, P., Garcia-Vazquez, E. 2006. Identification of highly prized commercial fish using a PCR-based methodology. *Biochem. Mol. Biol. Educat.* 34: 121–124.

Moretti, V. M., Turchini, G. M., Bellagamba, F., Caprino, F. 2003. Traceability issues in fishery and aquaculture products. Vet. *Res. Commun.* 27(Suppl 1): 497–505.

Nilsen, H., Esaiassen, M., Heia, K., Sigernes, F. 2002. Visible/near infrared spectroscopy. A new tool for the evaluation of fish freshness? *J. Food Sci.* 67: 1821–1826.

Olsen, E., Veberg, A., Vogt, G., Tomic, O., Kirkhus, B., Ekeberg, D., and Nilsson, A. 2006. Analysis of early lipid oxidation in salmon pâté with cod liver oil and antioxidants. *J. Food Sci.* 71: S284–S292.

Onal, A., Tekkeli, S. E. K., and Onal, C. 2013. A review of the liquid chromatographic methods for the determination of biogenic amines in foods. *Food Chem.* 138: 509–515.

Ortea, I., Canas, B., and Gallardo, J. M. 2009a. Mass spectrometry characterization of species-specific peptides from arginine kinase for the identification of commercially relevant shrimp species. *J. Proteome Res.* 8: 5356–5362.

Ortea, I., Cañas, B., Calo-Mata, P., Barros-Velázquez, J., and Gallardo, J. M. 2009b. Arginine kinase peptide mass fingerprinting as a proteomic approach for species identification and taxonomic analysis of commercially relevant shrimp species. *J. Agric. Food Chem.* 57: 5665–5672.

Ortea, I., Cañas, B., Calo-Mata, P., Barros-Velázquez, J. and Gallardo, J. M. 2010. Identification of commercial prawn and shrimp species of food interest by native isoelectric focusing. *Food Chem.* 121: 569–574.

Ortea, I., O'Connor, G., and Maquet, A. 2016. Review on proteomics for food authentication. *J. Proteom.* 147: 212–225.

Ottavian, M., Facco, P., Fasolato, L., Novelli, E., Mirisola, M., Perini, M., and Massimiliano, B. 2012. Use of near-infrared spectroscopy for fast fraud detection in seafood: application to the authentication of wild European sea bass (*Dicentrarchus labrax*). *J. Agric. Food Chem.* 60: 639–648.

Pepe, T., Ceruso, M., Carpentieri, A., Ventrone, I., Amoresano, A., and Anastasio, A. 2010. Proteomics analysis for the identification of three species of Thunnus, *Vet. Res. Commun.* 34: S153–S155.

Pepe, T., Ceruso, M., Carpentieri, A., Ventrone, I., Amoresano, A., Anastasio, A., and Cortesi, M. L. 2012. Differentiation of four tuna species by two-dimensional electrophoresis and mass spectrometric analysis. In: Heazlewood, J. L., and Petzold, C. J. (eds.) Proteomic Application in Biology, In Tech.

Peres, B., Barlet, N., Loiseau, G., and Montet, D. 2007. Review of the current methods of analytical traceability allowing determination of the origin of foodstuffs. *Food Control* 18: 228–235.

Perona, J. S., and Ruiz-Gutierrez, V. 1999. Characterization of the triacylglycerol molecular species of fish oil by reversed-phase high performance liquid chromatography. *J. Liq. Chromatogr. Relat. Technol.* 22: 1699–1714.

Piñeiro, C., Barros-Velázquez, J., Sotelo, C. G., and Gallardo, J. M. 1999. The use of two-dimensional electrophoresis in the characterization of the water-soluble protein fraction of commercial flat fish species. *Zeitschrift Für Leb. Und-Forsch. A* 208: 342–348.

Piñeiro, C., Barros-Velázquez, J., Sotelo, C. G., Pérez-Martín, R. I., and Gallardo, J. M. 1998. Two-dimensional electrophoretic study of the water-soluble protein fraction in white muscle of gadoid fish species. *J. Agric. Food Chem.* 46: 3991–3997.

Pineiro, C., Canas, B., Carrera, M. 2010. The role of proteomics in the study of the influence of climate change on seafood products. *Food Res. Int.* 43: 1791–1802.

Piñeiro, C., Vázquez, J., Marina, A., Barros-Velázquez, J., and Gallardo, J. M. 2001. Characterization and partial sequencing of species-specific sarcoplasmic polypeptides from commercial hake species by mass spectrometry following two-dimensional electrophoresis. *Electrophoresis* 22: 1545–1552.

Popper, D. E. 2007. Traceability: Tracking and privacy in the food system. *Geogr. Rev.* 97: 365–388.

Posudin, Y. I., Peiris, K. S., and Kays, S. J. 2015. Non-destructive detection of food adulteration to guarantee human health and safety. *Ukrainian Food J.* 4: 207–260.

Quevedo, R. A., Aguilera, J. M., and Pedreschi, F. 2010. Color of salmon fillets by computer vision and sensory panel. *Food Bioprocess Technol.* 3: 637–643.

Ramella, M. S., Kroth, M. A., Tagliari, C., Arisi, A. C. M. 2005. Optimization of random amplified polymorphic DNA protocol for molecular identification of *Lophius gastrophysus. Cienc. Tecnol. Aliment Campinas* 25: 733–5.

Rasmussen, R. S., and Morrissey, M. T. 2009. Application of DNA-based methods to identify fish and seafood substitution on the commercial market. *Compr. Rev. Food Sci. Food Saf.* 8: 118–154.

Rehbein H. 2005. Identification of the fish species of raw or cold-smoked salmon and salmon caviar by single-stranded conformation polymorphism (SSCP) analysis. *Eur. Food Res. Technol.* 220: 625–32.

Rehbein, H. 1995. Food control by applied biochemistry of marine organisms: comparison of proteins and metabolites from fish and invertebrate muscle. *Helgoländer Meeresun* 49: 747–757.

Rehbein, H., Kress, G., and Schmidt, T. 1997. Application of PCR-SSCP to species identification of fishery products. *J. Sci. Food Agric.* 74: 35–41.

Rehbein, H., Mackie, I., Pryde, S., González-Sotelo, C., Medina, I., Pérez-Martín, R., Quinteiro, J., and Rey Mendez, M. 1999. Fish species identification in canned tuna by PCR-SSCP: validation by a collaborative study and investigation of intra-species variability of the DNA patterns. *Food Chem.* 64: 263–268.

Reilly, A. 2018. Food and Agriculture Organization of the United Nations (FAO).

Renon, P., Bernardi, C., Malandra, R., and Biondi, P. A. 2005. Isoelectric focusing of sarcoplasmic proteins to distinguish swordfish, blue marlin and Mediterranean spearfish. *Food Control* 16: 473–477.

Rodríguez-Méndez, M. L., Gay, M., Apetrei, C., and De Saja, J. A. 2009. Biogenic amines and fish freshness assessment using a multisensor system based on voltammetric electrodes. Comparison between CPE and screen-printed electrodes. *Electrochim. Acta* 54: 7033–7041.

Ruiz-Rico, M., Fuentes, A., Masot, R., Alcaniz, M., Fernandez-Segovia, I., and Barat, J. M. 2013. Use of the voltammetric tongue in fresh cod Gadus morhua quality assessment. *Innov. Food Sci. Emerg. Technol.* 18: 256–263.

Santonico, M., Pittia, P., and Pennazza, G. 2008. Study of the aroma of artificially flavoured custards by chemical sensor array fingerprinting. *Sens. Actuaors B Chem.* 133: 345–351.

Sarkardei, S., and Howell, N. K. 2007. The effects of freeze-drying and storage on the FTRaman spectra of Atlantic mackerel (Scomber scombrus) and horse mackerel (Trachurus trachurus). *Food Chem.* 103: 62–70.

Scarano, D., and Rao, R. R. 2014. DNA markers for food products authentication. *Diversity* 6: 579–596.

Schröder, U. 2008. Challenges in the traceability of seafood. *J. Verbr. Lebensm.* 3: 45–48.

Serge, R., Ivan, G., and Karoly, H. 2007. Classification of gilthead sea bream (Sparus aurata) from^{1}H NMR lipid profiling combined with principal component and linear discriminant analysis. *J. Agric. Food Chem.* 55: 9963–9968.

Sfar, S., Heni, J. E., Laporte, F., Braham, H., Jawed, A., Amor, S., Sfar, M. T., and Kerkeni, A. 2012. Trace element status and fatty acids metabolism during healthy ageing: An example of a population from the Tunisian eastern coast. *Exp. Gerontol.* 47: 243–249.

Shogren, J. F. 2002. Food-safety economics: consumer health and welfare. Proceedings of the Frontis workshop on new approaches to food-safety economics, Wageningen, the Netherlands 14–17.

Sitter, B., Krane, J., Gribbestad, I., Jørgensen, L., and Aursand, M. 1999. Quality evaluation of Atlantic halibut *Hippoglossus*, L. during ice storage using ^{1}H NMR spectroscopy. In: Belton, P. S., Hills, B. P., and Webb, G. A. (eds.) *Advances in Magnetic Resonance in Food Science*, Royal Society of Chemistry, MPG Books, Ltd, Bodmin, Cornwall, UK, pp. 226–237.

Smith, C. T., Grant, W. S., and Seeb, L. W. 2005. A rapid, high-throughput technique for detecting tanner crabs (*Chionoecetes bairdi*) illegally taken in Alaska's snow crab fishery. *Trans. Am. Fish Soc.* 134: 620–623.

Solaesa, A. G., Bucio, S. L., Sanz, M. T., Beltran, S., and Rebolleda, S. 2014. Characterization of triacylglycerol composition of fish oils by using chromatographic techniques. *J. Oleo Sci.* 63: 449–460.

Standal, I., Axelson, D., and Aursand, M. 2010. ^{13}C NMR as a tool for authentication of different gadoid fish species with emphasis on phospholipid profiles. *Food Chem.* 121: 608–615.

Sykuta, M. 2005. Agricultural organization in an era of traceability. *J. Agric. Appl. Econ.* 37: 365–377.

Tagliavia, M., Nicosia, A., Salamone, M., Biondo, G., Bennici, C. D., Mazzola, S., and Cuttitta, A. 2016. Development of a fast DNA extraction method for sea food and marine species identification. *Food Chem.* 203: 375–378.

The Millennium Prize Laureate 2008 Professor Sir Alec Jeffreys Department of Genetics University of Leicester, United Kingdom.

Tian, X.-Y., Cai, Q., and Zhang Y.-M. 2012. Rapid classification of hairtail fish and pork freshness using an electronic nose based on the PCA method. *Sensors* 12: 260–277.

Tito, N. B., Rodemann, T., and Powell, S. M. 2012. Use of near infrared spectroscopy to predict microbial numbers on Atlantic salmon. *Food Microbiol.* 32: 431–436.

Toppe, J., Albrektsen, S., Hope, B., and Aksnes, A. 2007. Chemical composition, mineral content and amino acid and lipid profiles in bones from various fish species. *Comp. Biochem. Physiol. B Biochem. Mol. Biol.* 395–401.

Uddin, M., Okazaki, E., Turza, S., Yumiko, Y., Tanaka, M., and Fukuda, Y. 2005. Nondestructive Visible/NIR spectroscopy for differentiation of fresh and frozen-thawed fish. *J. Food Sci.* 70: 506–510.

USFDA. 2006. Seafood substitution. College Park, Md.: U.S. Food and Drug Administration Center for Food Safety and Applied Nutrition. Available from: www.cfsan.fda.gov/~frf/econ.html.

Valeria, T., Caterina, M., and Antonio, G. 2005. DNA-based methods for identification and quantification of small grain cereal mixtures and fingerprinting of varieties. *J. Cereal Sci.* 41: 213–220.

Velioglu, H. M., Temiz, H. T., and Boyaci, I. H. 2015. Differentiation of fresh and frozen thawed fish samples using Raman spectroscopy coupled with chemometric analysis. *Food Chem.* 172: 283–290.

Vidal, N. P., Manzanos, M. J., Goicoechea, E., and Guillén, M. D. 2012. Quality of farmed and wild sea bass lipids studied by^{1}H NMR: Usefulness of this technique for differentiation on a qualitative and a quantitative basis. *Food Chem.* 135: 1583–1591.

Wen-Feng, L., and Deng-Fwu, H. 2007. Application of PCR-RFLP analysis on species identifcation of canned tuna. *Food Control* 18: 1050–1057.

Wu, Y., and Chang, C. 2009. Biochemical genetic analysis of isozymes in three populations of Misgurnus anguillicaudatus. *J. Northwest A & F Univ. (Natural Science Edition)*, 37: 37–42.

Xiccato, G., Trocino, A., Tulli, F., and Tibaldi, E. 2004. Prediction of chemical composition and origin identification of European sea bass (*Dicentrarchus labrax* L.) by near infrared reflectance spectroscopy NIRS. *Food Chem.* 86: 275–281.

Xiong, G., Lee, D. J., Moon, K. R., and Lane, R. M. 2010. Shape similarity measure using turn angle cross-correlation for oyster quality evaluation. *J. Food Eng.* 100: 178–186.

Yates, J. R., Ruse, C. I., Nakorchevsky, A. 2009. Proteomics by mass spectrometry: Approaches, advances, and applications. *Annu. Rev. Biomed. Eng.* 11: 49–79.

Zarkadas, C. G., Gagnon, C., and Poysa, V. 2007. Protein quality and identification of the storage protein subunits of tofu and null soybean genotypes, using amino acid analysis, one and two-dimensional gel electrophoresis, and tandem mass spectrometry. *Food Res. Int.* 40: 111–128.

Zeng, Y. X., Araujo, P., Du, Z. Y., Nguyen, T. T., Frøyland, L., and Grung, B. 2010. Elucidation of triacylglycerols in cod liver oil by liquid chromatography electrospray tandem ion-trap mass spectrometry. *Talanta* 82: 1261–1270.

Zhan, A., Hu, J., Hu, X., Lu, W., Wang, M., Peng, W., Hui, M., Bao, Z. 2008. Fast identification of scallop adductor muscles using species-specific microsatellite markers. *Eur. Food Res. Technol.* 227: 353–359.

Zhang, J., Zhang, X., Dediu, L., and Victor, C. 2011. Review of the current application of fingerprinting allowing detection of food adulteration and fraud in China. *Food Control* 22: 1126–1135.

Zhao, Y., and Zhao, G. 2007. Study on the traceability system in the food supply chain. *Food Ferm. Indus.* 33: 146–149.

Zhihua, L., Shouju, J., Yinghui, D., Xueliang, C., Rong-mao, L., and Yongpu, Z. (2009). The isozymes in different populations of clam Meretrix. *J. Dalian Fisheries Univ.* 6: 525–530.

Zhu, F., Zhang, D., He, Y., Liu, F., and Sun, D. W. 2013. Application of visible and near infrared hyperspectral imaging to differentiate between fresh and frozen–thawed fish fillets. *Food Bioprocess Tech.* 6: 2931–2937.

Zietkiewicz, E., Rafalski, A., and Labuda, D. 1994. Genome fingerprinting by simple sequence repeat (SSR) anchored polymerase chain reaction amplification. *Genomics* 20: 176–183.

Index